The Use of Clay in the Upper Palaeolithic of Europe

Symbolic Applications of a Material

Estelle J. Bougard

BAR International Series 2069
2010

Published in 2016 by
BAR Publishing, Oxford

BAR International Series 2069

The Use of Clay in the Upper Palaeolithic of Europe

ISBN 978 1 4073 0476 2

BAR Publishing is the trading name of British Archaeological Reports (Oxford) Ltd. British Archaeological Reports was first incorporated in 1974 to publish the BAR Series, International and British. In 1992 Hadrian Books Ltd became part of the BAR group. This volume was originally published by John and Erica Hedges Ltd. in conjunction with British Archaeological Reports (Oxford) Ltd / Hadrian Books Ltd, the Series principal publisher, in 2010. This present volume is published by BAR Publishing, 2016.

Printed in England

BAR
PUBLISHING

BAR titles are available from:

BAR Publishing
122 Banbury Rd, Oxford, OX2 7BP, UK
EMAIL info@barpublishing.com
PHONE +44 (0)1865 310431
FAX +44 (0)1865 316916
www.barpublishing.com

FOREWORD

The present publication presents the results from a doctoral research achieved in 2007 in the Department of Archaeology of the University Of Liverpool, UK. The thesis itself is published here in its integral version with some added updates.

For the purpose of this research, a synthetic review of the uses of one specific material, clay, in the Upper Palaeolithic period (ca 30,000 to 10,000 BP) in Europe was undertaken. Because of the relative rareness of clay artefacts for this time period or of the difficulty to read and analyse the material excavated, the study of clay has been fairly neglected or ignored in archaeological research in comparison to the more 'classic' archaeological materials such as stone, bone, antler and ivory. Some recent research, however, (Soffer and Vandiver 1989, 1990; Boureux 2004) has started to emphasise the potential born by clay artefacts in some specific contexts, especially the quantity and quality of material actually available.

The present research intends to go a step further by arguing that clay itself as a material has an important potential as a medium of symbolic expression and was used as such in the Upper Palaeolithic contexts where its use is demonstrated. If we accept this argument, it implies that the study of the use made of clay in the Upper Palaeolithic in Europe has a strong potential in contributing to our understanding of issues of behavioural complexity, as symbolic expression is seen as a very important marker of this complexity. Conkey (1985:300) reminds us that 'Palaeolithic art is often seen as the strongest of all archaeological evidence of increasing ritual behaviour' for example. As clay use has been little studied for this period of time, its potential in contributing new insights to Upper Palaeolithic studies can be seen as very important.

Technology has been chosen here as the framework for the research as the *chaînes opératoires* and technical schemes of the working of clay are particularly revealing: clay as a material possesses unique qualities, such as plasticity and its capacity to harden permanently under the action of heat; but is has sometimes also been worked in the same manner as other more commonly studied materials used in the Upper Palaeolithic. The choices made in its working can provide a basis for the analysis of the material. Gamble (1999:82) reminds us that the concept of operational sequence used in the study of technology fuses social and technical activities, and this concept has been much developed by a number of researchers (Leroi-Gourhan 1943; Lemonnier 1990; Schlanger 1994...).

Together with technology, issues of style have also been considered as we are dealing here with a material applied exclusively to symbolic expression for the time period considered. Therefore, shapes and forms need to be looked at together with the techniques seen.

The technological and stylistic study of the use of clay has been used here in particular to consider the context of knowledge visible through the working of the material. This has allowed addressing the problem of the transmission of knowledge and know-how and its possible social contexts as seen in the two regions studied.

Among the inventoried data, it appeared that two specific cultural contexts with important traditions of using clay did exist in the Upper Palaeolithic. One is found in Central Europe in the Czech region of Moravia during the Gravettian period; the second in the French Pyrénées during the Magdalenian. In spite of the major differences in time period and locations of these two traditions, it was decided to study them both comparatively in order to determine what information can be gained from clay studies. These regional contexts provided the framework for detailed case studies, as well as opportunities for temporal and regional comparisons, especially regarding the capacity of clay as a material to witness behavioural complexity in the archaeological record.

The study of clay uses in these two contexts has allowed underlying two very different situations with regard to the technological knowledge applied and possibly also to its transmission, with what appears a standardised technology of clay in Central Europe opposed to a situated individual application of knowledge related to clay technologies in the French Pyrénées. The global technological context for each material culture studied then becomes revealing for our wider understanding of these cultures. It is certainly a topic that deserves more emphasis in future research than has been possible to do here.

The scope of the archaeological material considered for this study being extremely large, it has not been possible to carry out a first hand study of all the material mentioned, especially in the Pyrenean context. In that case, a thorough bibliographical research was compiled. Objects from Enlène, Mas d'Azil and Bédeilhac were however studied. I wish to thank especially Robert Bégouen, Yanick Leguillou, Pascal Allard, the Mas d'Azil museum, Georges Sauvet, René Gailli, Pierre-Élie Moullé and the Musée de Préhistoire de Menton for making access to these materials possible.

As for the Moravian record (Czech Republic), it was possible to study first hand the excavated fashioned objects from Dolni Vestonice I and II and Pavlov I, a total of 316 pieces that form my database. A large number of these pieces are still unpublished or very little known, and I am extremely grateful to Martin Oliva from the Anthropos Institute in Brno, Czech Republic and to Jiri Svoboda from the Archaeological Institute, Academy of Sciences, Dolní Vestonice, for allowing me access to this material and for permitting their publication. I wish to emphasise the importance of their respective work in the study and conservation of the collections from these sites.

I also would like to thank my supervisors Dr A. Sinclair and Professor J. Gowlett for their help with carrying out this research project, and Professor Larry Barham for his precious support in shaping the final reflection.

This research has benefited from a John Lennon Scholarship grant from the University of Liverpool.

CHAPTER 1: INTRODUCTION: DEFINITIONS AND STATE OF THE RESEARCH

The present research intends to be a technological synthesis of the use of one material, clay, in the Upper Palaeolithic of Europe, in a similar manner that others have been done in the past for stone (Bordes 1961; Tixier *et al.* 1990) or osseous materials (Ramseyer 2004). A synthetic study has not been carried out before, to the best of my knowledge, for clay material in the Upper Palaeolithic. A possible reason for this state of things is that clay is not traditionally associated with material studies for this time period, one of the dogmas of traditional archaeology being that ceramic-making - the major technique to which clay is applied – was first invented and used in Neolithic contexts for the making of pottery (utilitarian containers) (Hodges 1964).

Another possible reason could be that the use of clay in the Upper Palaeolithic appears deeply connected to symbolic expression, a field that traditional technological studies have avoided until recently (White 1997; Fritz 1999). This is a striking point and a main difference with the use of clay made in later periods (Neolithic onwards) in which pottery became a major technology with a primarily utilitarian function. Because the properties of clay as a material (see below) make it a very useful utilitarian material (especially for increasing the variety of food preparation techniques), the question of why it was mostly used in symbolic expression, although its utilitarian qualities were clearly known and noticed by the Palaeolithic people (Bahn and Vertut 1997: 98), is one underlying the present research. This remark finds an echo in recent trends in theories of prehistoric invention (Jeunesse 2008) which are showing that simple utilitarian motivations are not the main reason behind technical innovations. The meaningful technological choices visible in the Upper Palaeolithic uses of clay can thus be analysed with the hope to help our understanding of Upper Palaeolithic art.

The apparent scarcity of the cases of clay uses in comparison to other materials such as stone and osseous materials might have also contributed to the relative neglect surrounding the study of this topic. As will be seen below, however, this scarcity is regional rather than global; the material available for the study is abundant enough (table 1.1 and 1.2) to justify undertaking this research; although this fact certainly needs publicising and remains little known.

An initial inventory of all the cases of use of clay in the Upper Palaeolithic of Europe was achieved through a thorough literature search and through contacting the persons in charge of the various relevant collections in the institutions in which some artefacts are kept. This allowed gathering the data base for the present study which is presented in the form of a catalogue of inventoried cases of use of clay in the Appendixes at the end of this volume. Any reference number of use of clay mentioned in the present research corresponds to the reference numbers given in the catalogue and is our own.

Through this process, it became clear that most of the cases of use of clay during the Upper Palaeolithic were found in two separate regions at distinct periods in time: in the French Pyrénées in the Magdalenian (material presented in Appendix 1) and in Central Europe in the Gravettian period (material presented in Appendix 2). Table 1.1 summarises the inventory of clay uses found in these two regions with a minimum of 11,632 recorded cases in central Europe and 289 in the French Pyrénées. This record is a minimal number as, because of the current state of the research on this topic, it is certainly not exhaustive, especially in the French Pyrénées[1]. It is, to the best of my knowledge, the most complete existing up to date.

The recorded data base for these two regions, Central Europe and the French Pyrénées, was therefore chosen as the background on which to carry out a general study of the use of clay in the Upper Palaeolithic. Focusing on two regions with a coherent cultural context seemed most likely to contribute to increasing our knowledge of the material complexity for the Upper Palaeolithic. It must be emphasised, however, that the record of use of clay in the Upper Palaeolithic is not restricted to these two regions. The other cases encountered are inventoried in Table 1.2.

The map in Figure 1.1 indicates the locations of all the sites mentioned in Tables 1.1 and 1.2, and underlines clearly the two regional concentrations of sites mentioned above.

1 Clay: introducing a material

Definitions

'Clay'

There are many definitions of the term clay. Some insist on the strict geological definition of the mineral, some rather more on its properties as a material.

- **Chemical and geological definitions**

Clay is in general considered to be hydrated aluminium silicate with a sheet-like particle structure. Its purest form, kaolinite, can be represented by the general formula $Al_2O_32SiO_22H_2O$. But most types of clay contain many forms of impurities.

Another meaning of the word clay is based on a criteria of particle size, as it qualifies particles smaller than two microns, whatever their nature (Hodges 1964; Gibson and Woods 1990; Rapp and Hill 1998; Bravard *et al.* 1999).

Clays are formed by the weathering of certain rocks like granites, diorites and basalts. Two main clay types are generally recognised: residual or primary clay that has not been transported from its site of formation, and sedimentary, or secondary clay, which has been re-deposited by the action of water (river or lake deposition).

[1]Most of the related excavations are ancient and the resulting assemblages are extremely dispersed in various collections: for the site of Bédeilhac only for example the exact inventory of clay objects is not known nor published but is certainly in excess of 200 (Georges Sauvet personal communication August 2004; Marie-Sylvie Larguèze personal communication March 2007).

Table 1.1: Inventory of the number of clay uses for the two studied regions: the French Pyrénées and Central Europe. A clay use is considered as a recognisable graphic unit or as a portable object made in clay. FT indicates the presence of non-quantifiable macaroni Finger-Tracings in the caves of the French Pyrénées.

Region	Site Name	Number of clay uses	Region	Site Name	Number of clay uses
French Pyrénées	Tuc d'Audoubert	15+FT	**Moravia**	Dolni Vestonice I	Over 5,751
	Enlène	3		Dolni Vestonice II	13
	Labouiche	1		Dolni Vestonice III	At least 1
	Niaux	65		Pavlov I	Over 5,716
	Mas d'Azil	2		Pavlov II	135
	Fontanet	11		Pavlov V	At least 1
	Bédeilhac	At least 138		Predmost	At least 1
	Massat	13 + FT		Spytihněv	At least 1
	Montespan	22 + FT		Jarošov	At least 1
	Labastide	1		Boršice	At least 1
	Bois du Cantet	1		Petrkovice I	At least 2
	Erbérua	12 + FT	**Slovakia**	Moravany-Lopata	1
	Oxocelhaya-Hariztoya	3 + FT		Kasov	3
	Etcheberri	2		Cejkov	1
Total		289 + FT	**Austria**	Krems-Watchberg	At least 3
			Hungary	Sagvar	1
			Total		**Over 11,632**

Figure 1.1: Distribution map of all the sites with Upper Palaeolithic clay uses in Europe[2]

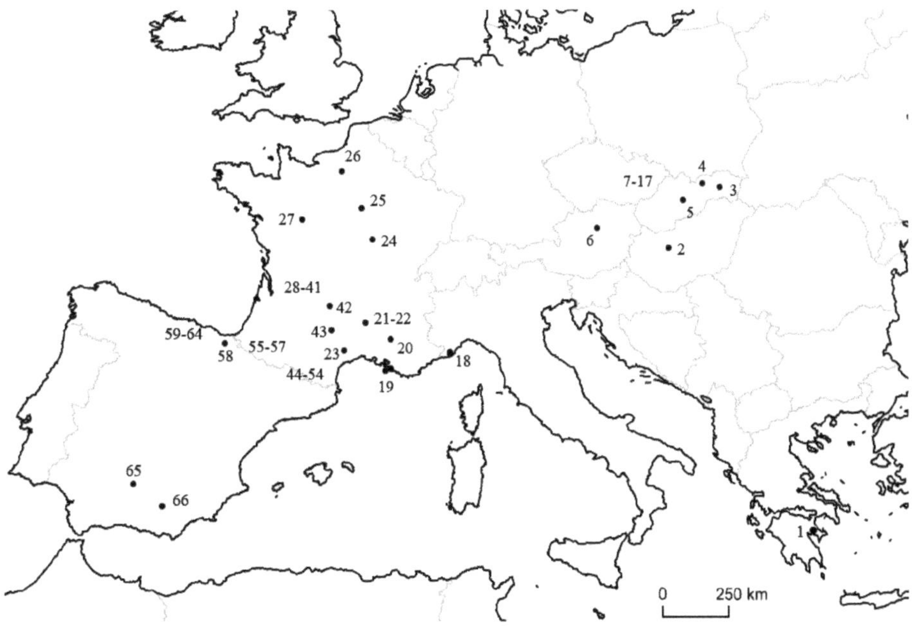

Greece: 1 Klissoura; **Hungary**: 2 Sagvar; **Slovakia**: 3 Cejkov, 4 Moravany-Lopata, 5 Kasov I; **Austria**: 6 Krems-Watchberg ; **Moravia**: 7 Predmosti, 8 Dolni Vestonice I, 9 Dolni Vestonice II, 10 Dolni Vestonice III, 11 Pavlov I, 12 Pavlov II, 13 Pavlov V, 14 Spytihněv, 15 Jarošov, 16 Boršice; 17 Petrkovice I ; **Italy**: 18 Barma Grande ; **Bouches-du-Rhône :** 19 Cosquer ; **Gard :** 20 Baume-Latrone ; **Ardèche :** 21 Chauvet, 22 Vacheresse ; **Hérault :** 23 Aldène ; **Yonne :** 24 Grotte du cheval ; **Seine et Marne :** 25 Croc-Marin ; **Haute-Normandie :** 26 Le Renard ou Orival ; **Mayenne :** 27 Mayenne-Sciences ; **Dordogne :** 28 La Gravette, 29 Bara Bahau, 30 Bernifal, 31 Font de Gaume, 32 La Calévie, 33 La Croze à Gontran, 34 La Forêt, 35 La Martine, 36 Lascaux, 37 Paulin-Cournazac, 38 Rouffignac, 39 Saint-Cirq, 40 Sous-Grand-Lac, 41 Cussac ; **Lot :** 42 Pech Merle ; **Tarn-et-Garonne** : 43 Bruniquel ; **Ariège :** 44 Tuc d'Audoubert, 45 Enlène, 46 Labouiche, 47 Bédeilhac, 48 Mas d'Azil, 49 Niaux, 50 Fontanet, 51 Massat ; **Haute-Garonne** : 52 Montespan ; **Hautes-Pyrénées** : 53 Labastide, 54 Bois du Cantet ; **Pyrénées-Atlantiques** : 55 Erbérua, 56 Oxocelhaya-Hariztoya, 57 Etcheberri ; **Guipuzcoa** : 58 Ekain ; **Cantabria :** 59 Altamira, 60 Las Chimineas, 61 Hornos de la Peña, 62 La Clotilde de Santa Isabel ; **Asturias** : 63 El Pindal , 64 El Quintanal ; **Andalucia** : 65 La Pileta, 66 Ardales.

[2] Except for the three Russian sites of Kostenki 1, Kapovaya and Ignatievskaya (see references in Table 1.2 below).

Table 1.2: Inventory of sites with clay uses in the Upper Palaeolithic of Europe outside Central Europe and the French Pyrénées.

Type of clay use found	Region	Sites names	References
Portable objects	Bouches-du Rhone	Cosquer	Clottes *et al.* 2005
	Dordogne	La Gravette	Lacorre 1960, Perlès 1977
	Tarn-et-Garonne	Bruniquel	Piette 1875; Breuil 1903
	Italy	Barma-Grande	Palès 1972
	Russia	Kostenki Ignatievskaya Kapova	Absolon 1949; Soffer *et al.* 1993; B. Bradley personal communication January 2003; Vandiver *et al.* 1990; Valoch 1996 Abramova 1995 Shchelinsky 1989; Shchelinsky and Sirokov 1999
Non Portable use (modelage, impression, possible non-portable container)	Ardèche	Chauvet	Clottes 1996b, Clottes *et al.* 2003
	Lot	Pech Merle	Méroc 1959; Lemozi 1929; Lorblanchet 1984
	Dordogne	Font-de-Gaume La Calévie, Paulin-Cournazac La Foret	G. and B. Delluc personal communication 2001 Aujoulat 1984a Roussot 1984 Aujoulat 1984b; G. and B. Delluc personal communication 2001
	Cantabria	Altamira	Bernaldo de Quiros 1991; Beltran *et al.* 1999
	Basque Country	Ekain	Boureux 2005
	Greece	Klissoura	Pawlikowski *et al.* 2000
Finger-tracing/engraving on clay	Yonne	Le Cheval	Leroi-Gourhan 1984; Baffier and Girard 1998
	Ardèche	Vacheresse Chauvet	Combier and Roudil 1984 Clottes 1996b, Clottes *et al.* 2003
	Bouches-Du-Rhone	Cosquer	Clottes 1996a; Baffier and Girard 1998
	Hérault	Aldène	Palès and Vialou 1984
	Haute-Normandie	Le Renard (Orival)	Fosse 1984; A. Tomat January 2007 personal communication
	Seine-et-Marne	Croc-Marin	Poignant 1984
	Lot	Pech Merle	Lemozi 1929; Lorblanchet 1984
	Dordogne	Bara-Bahau Croze a Gontran La Martine Paulin-Cournazac Rouffignac Saint-Cirq Sous-Grand-Lac Cussac	Aujoulat and Dauric 1984; Delluc and Delluc 1997- Delluc and Delluc 1984a Delluc and Delluc 1984b Roussot 1984 Barrière 1984, Plassard and Plassard 2000 G. and B. Delluc personal communication. 2001 G. and B. Delluc personal communication 2001 Boureux 2005
	Cantabria/Asturias	Altamira Hornos de la Pena, Las Chimineas La Clotilde de Santa Isabel El Pindal El Quintanal	Bernaldo de Quiros 1991; Beltran *et al.* 1999 Breuil 1952; Baffier and Girard 1998 Baffier and Girard 1998 Breuil *et al.* 1911; Breuil 1952 Breuil 1952 Breuil 1952; Baffier and Girard 1998
	Andalusia	La Pileta Ardales	Breuil 1952; Baffier and Girard 1998 Breuil 1952
Use as Pigment	Dordogne	Bernifal, Lascaux, Rouffignac	G. and B. Delluc personal communication 2001 Leroi-Gourhan 1979; G. and B. Delluc personal communication 2001 Barrière 1984, Plassard and Plassard 2000
	Gard	La Baume-Latrone	Drouot 1984; Roudil 1984
	Mayenne	Mayenne-Sciences	Pigeaud 2002

Residual clay, like kaolinite, is very pure but is not very plastic and requires a high firing temperature. It was not used for firing in Prehistory, as this would require a complex technology that was not mastered then. However, kaolinite has sometimes been used as a white pigment in cave art in the Upper Palaeolithic such as at Lascaux (Couraud and Laming-Emperaire 1979). Sedimentary or secondary clay is the type of clay mostly used in Prehistory as it is more plastic and its varying content of impurities gives it qualities that can be used in its working (Hodges 1964; Gibson and Woods 1990).

- **Definitions linked to material properties/ the potter's definition**

Definitions of clay are also found that put an emphasis on its material properties; they can be helpful in identifying the material, bearing in mind the very wide range of different possible chemical compositions. Gibson and Wood (1990: 117) for example say that the American Ceramic Society has defined clay as 'a fine-grained rock which when suitably crushed and pulverized, becomes plastic when wet, leather hard when dried and on firing is converted to a permanent rock-like mass'. This definition emphasises the idea that a certain preparation of the material is necessary to eventually produce fired clay. However, as will be seen further in this study, this is far from always being the case as naturally suitable material can be found.

The following definition is possibly more suitable, as it is much more human-orientated and may approach how Palaeolithic people saw clay; it was used in a ceramic exhibition held in the Tate Gallery in Liverpool in the summer 2004. It states that clay is a form of mud, of moist sticky dirt, a fine-grained, firm, earthy material that is plastic when wet, brittle when dry and very hard when heated.

Other soft materials also called 'clay'

The present research focuses on the way the studied material is used. In a number of cases, materials which are not strictly clay by their chemical definition are included in the present study as they possess similar properties to clay (plasticity, possibility of alteration of their texture by adjusting their water content, resulting in a capacity to harden through drying and firing) and have been used in the same manner as clay by Upper Palaeolithic people. These materials are silt, mondmilch and loess.

Silt is fine sand, clay, or other material carried by running water and deposited as sediment (Oxford English Dictionary 1989). It is found as naturally forming the floor of some caves in the French Pyrénées and was sometimes used to make art, notably the plaquettes found in the cave of Bédeilhac (Mons 1974; Mons and Delporte 1973; Sauvet 2006).

In a cave context, the surface of the limestone walls sometimes precipitate into a white soft layer of calcium carbonate which is called mondmilch (Bahn and Vertut 1997: 106). This material is suitable for finger-tracings in the same manner as a clay surface. As the exact nature of the wall surface on which the finger-tracing are found is not always described in the relevant literature, it is must be noted that this surface was soft and used for its qualities but that it can be clay or mondmilch.

Loess is the material that was used in the sites of Central Europe for the making of fired clay figurines (Soffer and Vandiver 1994, 1997). Loess is a loosely compacted yellowish-grey deposit of wind-blown sediment (Oxford English Dictionary 1989). When wetted and kneaded, it acquires similar properties to clay and has been used just like it.

All these materials will be called clay throughout the present research as they can be assimilated to clay by the qualities they possess and by the use that was made of them in the Upper Palaeolithic of Europe. However, it is understood that they are not clay by their chemical definition.

Clay after firing

Firing changes the nature of clay and makes it permanently hard, transforming it into fired clay. Few words are used to describe fired clay, the most common being the term '**ceramic**', which is applicable to any form of fired clay. It can be pointed out that the definition of ceramics by scientists who study ancient ceramic technologies differs greatly from the general meaning given to the word. These specialists define ceramics as inorganic oxides subjected to a pyrotechnology that include refractories, plasters, cements, some pigments, glazes, enamels, glasses, metallurgic slags and clay-based products like pottery, porcelain, bricks and tiles (Vandiver 1988; Sayre *et al.* 1988). In the present research, the term ceramic will only be applied to clay-based products.

It must be noted that in American English a distinction between pottery and ceramic material is not made, and both simply qualify as any type of fired clay (D. Gehlsen personal communication October 2002). I shall use the term **ceramic** in this research to describe any type of clay, or assimilated material, fired over the point of ceramic change of around 500°C, above which the internal structure of the material changes in a non-reversible manner (Gibson and Woods 1990).

The word **pottery** is used to describe fired clay pots or vessels, utilitarian containers by their nature (Hodges 1964), and is therefore not suitable for the types of clay use found in the Upper Palaeolithic. It will not be used in this research.

The Properties of clay

Clay as a material offers properties that are unique such as its plasticity and its capacity to harden through drying or firing. Through these properties, clay changes nature and then acquires properties similar to other materials widely used in the Upper Palaeolithic such as soft and hard stone. These properties are summarised in Table 1.3 below.

Recognising the extent to which the properties unique to clay were understood and used in the Upper Palaeolithic, and to what extent the material was treated just like any other soft or hard stone is important in terms of indicating if a new technology is recognised or if there is a transfer of knowledge from the working of other materials. These elements can contribute to our knowledge of the technological complexity of Upper Palaeolithic people. This question is one that will underline the entire research presented here.

Table 1.3: The properties of clay as visible in the archaeological record

<table>
<tr><td rowspan="7">Properties of clay</td><td rowspan="4">Unique to clay</td><td>Plasticity (capacity to retain a shape, possibility to change any shape; reversible)</td></tr>
<tr><td>Hardening by drying (reversible)
Hardening by firing (non-reversible)</td></tr>
<tr><td>Capacity to burst in fire</td></tr>
<tr><td>Reversibility</td></tr>
<tr><td rowspan="3">Shared with other materials</td><td>Soft stone/osseous material hardness</td></tr>
<tr><td>Hard stone hardness</td></tr>
<tr><td>Colouring properties</td></tr>
</table>

Properties unique to clay

Plasticity is the property of clay that permits it to be deformed by pressure and to retain a shape given to it. Plasticity arises from the presence of water between the clay particles that allows them to glide over each other. The smaller the clay mineral involved the more water that can be absorbed between the particles in any given volume of clay. This is why sedimentary clay that has been broken down by transport is more plastic than residual clay (Gibson and Woods 1990; Rapp and Hill 1998).

The plasticity of clay can be controlled by adjusting the water content in the material as long as the material has not been fired: any work done on unfired clay is thus fully reversible. Water can be added to render the texture softer; or the material can be left to dry slightly, with various means of controlling the speed of this process. The plasticity of the clay can also be kept constant by insulating the material, for example by wrapping it into a waterproofed animal skin.

The quality of plasticity is unique to clay (and associated materials here). It is easy to shape by direct contact with the hands without any need for tools, although it can of course also be worked with tools. It is extremely flexible and allows modelling into any shape. Clay also allows an additive working process, as well as a reductive one, but an equal quantity of material can also be kept, simply by transforming its shape.

Another important property of clay is that it can harden by simple air-drying. The water contained in the clay, which is not chemically combined with the clay particles in any way, evaporates with drying (Hodges 1964).

This process can harden clay artefacts in a significant manner, to the extent that even some air-dried pieces have survived from Upper Palaeolithic periods, such as some of the artefacts from Mas d'Azil in the French Pyrénées (Péquart 1960). However, the material remains susceptible to being re-hydrated at any time if it is placed in a humid environment or in water, returning to soft and plastic clay. Only firing will permanently change the material.

Sedimentary clay becomes permanently hard under the action of heat when fired over 500°C to 550°C. This process completely changes the nature of clay as a material by removing the chemically combined water in the clay molecules. It turns the clay into fired clay, or ceramic (Hodges 1964; Gibson and Woods 1990).

Different firing temperatures have different effects on the finished ceramic objects. The table below (Table 1.4) describes the terms that will be used in correlation to these different temperatures.

Table 1.4: Terms used to describe clay fired at different temperatures

Firing temperature	0- 550°C	550 - 700°C	700-800°C
Term used	Unfired or lightly fired	Low fired	Well fired

This table is based on the fact that the ceramic change in clay occurs when it is heated to or beyond 500°C to 550°C. That different sources (Gibson and Woods 1990; Vandiver *et al.* 1990) seem to indicate variations in this temperature of ceramic change can be explained by the fact that the exact chemical composition of each type of clay is slightly different, and therefore each reacts differently to firing. It must also be pointed out that the complex reaction occurring in the clay structure at the time of firing depends both on the firing temperature and on the length of the firing process (Hodges 1964), a combination that is hard to determine on archaeological specimens.

Unfired or lightly fired clay has not undergone a permanent change in structure and can still return to a plastic stage if re-hydrated. Clay fired below the temperature of 550°C, the temperature for ceramic change, can be included in this category as it has not permanently changed its structure. The separation between low-fired and well-fired clay is determined by the fact that around 700°C (again with variations depending on the exact nature of the material), sintering starts happening between the clay particles (Vandiver *et al.* 1990). Sintering is the initial stage in the vitrification process of the material in which the edges of the clay particles begin to melt and stick together. This results in visible changes in the texture of the ceramic material; low-fired clay remains porous and less hard than well-fired clay, which is both harder, and denser (Gibson and Woods 1990).

Table 1.4 does not include ceramics fired over 800°C as this has been established as the upper limit for the range of temperatures applied to the firing of the Moravian ceramics, as determined by Differential Thermal Analysis (DTA), Scanning Electron Microscope (SEM) and x-ray diffraction analyses (Soffer *et al.* 1993: 265; Vandiver 1988: 94; Vandiver *et al.* 1989: 1006). 800°C is also the maximal firing temperature normally obtained in a bonfire (Gibson and Woods 1990: 28).

Another property of clay associated with the firing process is its capacity to burst under the action of heat. This capacity is normally seen as a nuisance that potters try to control and avoid as it destroys the final pottery product; the lengthy preparation of the clay material, notably by the addition of temper, is done primarily with this purpose in mind. This property of clay does not carry any practical application. It appears, however, that it was sought out and used durably in some Palaeolithic contexts, especially in Moravia during the Gravettian period, in relation with the making/destruction of figurines (Soffer *et al.* 1993; Vandiver *et al.* 1997).

Finally, another property unique to the clay material is the complete reversibility of its working before any firing process. Any form fashioned can be erased or altered by playing on the degree of hydration of the material, and any previous traces of working can be made to disappear. All this makes clay an extremely versatile material, especially suitable to symbolic associations.

Properties of clay shared with other materials

When clay undergoes changes in its nature such as hardening by drying or firing it acquires a certain hardness which allows its working exactly in the same manner as other soft or hard materials known to have been worked in the Upper Palaeolithic. This property allowed Palaeolithic people to transfer techniques they already knew such as the carving, fashioning, engraving, polishing, and scraping of soft and hard stone or osseous materials (bone, ivory, and antler) to the working of clay. The technique of knapping, although technically possible for some ceramics, does not seem to have been applied to clay in the Upper Palaeolithic.

There are a number of instances, however, both in Central Europe in the site of Pavlov I for example (see Chapter 3) and in the French Pyrénées (see Chapter 2) in which fired clay objects have clearly been reworked after firing to fashion details on a figure of portable art.

Clay also has colouring properties which vary according to its exact mineral content. For example clay is often found to be rich in iron oxide that gives it a red to brown colour (Clottes *et al.* 1990: 171). This quality of the material has sometimes clearly been used in the making of cave art in a number of the Pyrenean sites and is mentioned as such in the relevant literature (Clottes 1989: 77; Laplace and Larribau 1984: 285; Vialou 1986). However, when looking at this property of clay, it begins to overlap with another material used in the Upper Palaeolithic: ochre. And in the same manner that the definition of clay can vary from source to source, according to the context it is connected to (for example for a geologist, a chemist or a potter), the definition of ochre is fairly variable: in Clottes *et al.* (1990: 174) they define it as 'a clay rich in iron oxide', naturally coloured from yellow to red, from which a more intense red can be obtained by heating up the pigment.

Except for the few cases mentioned in the French Pyrénées, the use of clay as a pigment will not be considered in the present research.

Some other properties of clay were possibly used but are not visible from the archaeological record. For example, clay is waterproof; this property could have been used for lining basketry containers to make them watertight. This hypothesis has been much used, to the point of becoming a cliché, to explain the invention of fired clay containers after accidental firings of such baskets (Leroi-Gourhan 1943: 224). More recently, some cases of impressions of weaved fibre material found on fired clay fragments in Central and Eastern Europe (Adovasio *et al.* 1996, 1997) have also been explained by such an application. No other evidence of the use of this property is found in the archaeological record however.

The various existing types of clay also have numerous medicinal properties (antiseptic, anti-inflammatory, cleansing, absorbent to toxins...) which might have been used for healing wounds in the Upper Palaeolithic. However, this type of use does not leave evidence in the archaeological record. There is evidence for medicinal uses of clay from the very beginning of historical times, in Sumerian tablets as well as in the writings of Ancient Egypt, Ancient Greece or Rome; and many people today still swear by it as a near miraculous remedy, as any internet search can confirm (Estrade 1994; Dextreit 1997). A use as a healing substance might therefore be considered as a possible 'invisible' application of clay which might have had importance in the symbolism associated to the material: it should be remembered here that the use of clay visible in the Upper Palaeolithic is applied to symbolic expression only.

How to work clay

These properties of clay as a material allow its working according to certain patterns, which are flexible to an extent, but in which certain steps must be followed regardless of the context considered. The presence of constant elements in the process of working clay makes it very suitable to the reconstruction of *chaînes opératoires,* that is to say operational sequences describing the suite of actions necessary in order to achieve a wanted result (see theoretical developments in part 4 in this chapter). Karlin and Julien (1994: 154) consider that the observation of a number of *chaînes opératoires* allow the reconstruction of a technical scheme that defines all the different stages of a type of production, including all the possible methods like missing one stage or going back to a previous one. Such a technical scheme has been applied in this section to the production of ceramics which is the most complex technique for the working of clay found in the Upper Palaeolithic.

Since clay is not a material the Palaeolithic archaeologist is usually familiar with, it was thought useful to present here the techniques most commonly found in the working of clay: the making of pottery.

The 'flow chart' in Figure 1.2 considers the process of ceramic-making by prehistoric potters as seen in some regions in the early Neolithic period and indeed even much later, for example in Great Britain (Gibson and Woods 1990). This process indicates the simplest way possible of making ceramics, with the building of the pots by hand alone and with very rudimentary firing structures or even just bonfires.

This 'flow chart' is intended as a tool for comparison with the various ways clay was worked in Central Europe and in the sites of the French Pyrénées, but no pottery (as utilitarian container) is known in these regions for the Upper Palaeolithic. The possible ways to work clay, however, are similar even if only part of the technical scheme is followed.

The various phases of the technical scheme described by this flow chart are the phases of procurement, preparation, fashioning, drying, decorating, firing, use and discarding of the produced objects. They are detailed as follows:

- The phase of **procurement** of the raw material consists of the three sequences of finding a source of raw material; testing it for suitability, by touching or by working a small amount; then extracting the material, usually by digging it out at the source with a suitable tool or with the hands. Testing the material for suitability depends mostly on the potter's experience and on the result he or she aims to achieve or the exact use the clay is intended for.

Figure 1.2: Organization of the general technical scheme defining the possible stages for the production of ceramics in a Neolithic context.
(key to reading the flow chart: the numbers indicate the possible steps of working of the material in the order usually followed; numbers in a full circle are compulsory steps that cannot be missed in the process; numbers in dotted circle are steps that can be skipped; the arrows indicate the possibilities of skipping steps in the process or of reverting to a previous step in a reversible process)

Phase	Sequence	
Procurement	Finding source	1
	Testing material	2
	Extraction	3
Preparation	Adjust texture	4
	Adjust plasticity	5
Fashioning	Shape object	6
Drying	Partial drying	7
	Full drying	8
Decorating	Mechanical work on surface	9
	Addition of elements (slip, paint, handles)	10
Drying	Full drying	11
Firing	Preparation of firing structure	12
	Procurement of fuel	13
	Stacking of pieces to be fired	14
	Ignite fire	15
	Firing	16
	Retrieve objects	17
Use object	Use object	18
Discard	Crush and reuse as temper	19
	Discard	20

- The phase of **preparation** consists in adjusting the texture and plasticity of the material to make it fully suitable for the intended use. Adjusting the texture of the clay can be done by cleaning out inclusions (by hand; by drying, crushing and sieving the clay dust; by levigation in water tanks), by kneading (folding and pressing the clay to get rid of air bubbles and homogenise the texture), by mixing temper into the clay.

A temper, or filler, is any organic or inorganic opening material added to clay to help it stand the firing process. It also influences the texture of the clay and its plasticity. Adjusting the plasticity can therefore be done by the addition of a temper but also, and mostly, by the addition of water to the clay (Hodges 1964; Bravard *et al.* 1999; Henderson 2000).

The preparation of the clay can be a lengthy process, especially in recent times when it is intended to be as fine and pure as possible. However, the finer the clay is, the more complex and delicate the firing process. Some clays are naturally tempered enough to stand a basic firing process without any problem, and in this case the preparation of the material can be reduced to a simple addition of water, and the final product can be described as crude pottery (Gibson and Woods 1990; Eygun 2001; Rapp and Hill 1998).

- The phase of **fashioning** consists of shaping the object. This can be done by a wide variety of techniques that can be roughly classified as follows:

1) Additive technique that works by putting smaller quantities of the material together in, for example, coil-building (rolled coils of clay built spirally onto prepared base, then smoothed together), slab-building (flat slabs of clay are stuck together to form a vessel) or the shaping of the different parts of a figurine separately (Hodges 1964; Gibson and Woods 1990; Henderson 2000).

2) Reductive technique that takes off material from an original quantity (not normally used for vessel-making).

3) Techniques that do not change the original quantity of material but only its shape, for example, vessel-making by thumb-pinching (shaped by fingers from a lump of clay), moulding (pressing wet clay into a fired ceramic or plaster mould in single or more parts) or wheel-throwing (shaping of vessels on a rotating wheel) techniques (Hodges 1964; Gibson and Woods 1990; Henderson 2000).

These techniques can be combined and applied to different parts of an object. It is interesting to note that clay has this rare quality of being able to be worked and fashioned without altering the original quantity of material. However, loss of water in the drying process implies some later shrinkage.

- The phase of **drying** consists of leaving the fashioned clay object to air-dry for a certain period of time in order to obtain the desired hardness of the surfaces of the object. The drying can be partial, and in this case it aims at preparing the surface for a certain type of surface treatment that is more easily done in this state, mostly decorating. It can also be a full drying, in which the object becomes very hard. Some surface treatment can still be done at this stage, but full drying consists mostly in the last step before the firing process, as it will help the clay to stand the firing well. The length of time necessary for drying clay depends on many variables such as the nature of the clay, the thickness of the object and weather conditions and cannot be generalised (Henderson 2000).

- The phase of **decorating** can take the form of two main sequences: a mechanical work of the surface(s) of the object and the addition of elements on it. The mechanical work on the surfaces is done with the help of tools. It can remove material in the form of incisions, engraving, scraping, combing, carving, and perforation. It can simply impress, stamp, roulette the material to produce printed decorations, or smooth it by the paddle-and-anvil technique (beating the outside of the vessel with a stick while a stone or clay 'anvil' is held on the other side). The whole surface can be smoothed with a wet cloth, or burnished by polishing with a hard and smooth object to obtain a shiny finish. The addition of elements on the object can, for example, take the form of handles, feet, bases, or *appliqué* decorations added to a vessel. It can also be the application of liquid suspensions of clay particles to the surfaces that will form designs if they are thick, smooth and/or colour the whole surface when thinner (Hodges 1964; Henderson 2000).

- The phase of **firing** of the clay objects to obtain ceramic is the most complex phase, as it requires the combination of a range of material factors with the clay objects.

The first sequence is the preparation of a firing structure to contain and control the firing process. This ranges from a simple pit dug into the ground in which fuel and objects are stacked, to a pit with walls around it called a pit kiln, in which the fuel is directly in contact with the objects, to a variety of kiln structures in which the fuel and the objects are separated and that allows a higher temperature as well as a better control of temperature, firing conditions and length of firing for example. A firing structure can be complex to build, but it must be noted that a structure is not necessary to fire ceramics. A simple bonfire in which the objects and fuel are stacked together on the ground will produce rough low-fired ceramics easily if the clay is suitable, that is to say if it contains enough temper or opening material. An open bonfire reaches on average a temperature between 600°C and 900°C, which is well above the temperature of 500°C necessary to produce a non-reversible change in the structure of the clay and change it into ceramic (Henderson 2000).

A second sequence is the procurement of fuel for the firing. It can be wood, but dried animal dung, dry grasses, wood chip or coal can be used successfully (Hodges 1964; Gibson and Woods 1990). In the Upper Palaeolithic in some regions bone was the fuel used in fires as wood was scarce (Clottes 1989: 17; Bruce Bradley, personal communication January 2003), and it is likely that in some sites, such as Kostenki in Russia, clay was fired with this fuel.

Then, the fuel can then be stacked with the objects to be fired in a way that will provide a unique short and intense firing in the case of a bonfire, for example, or in a kiln that will need to be refuelled before the end of the firing.

The next step is to light the fire and wait for the required time while the firing takes place. This can be a lengthy process lasting for many hours, with refuelling if necessary.

The objects then need to be retrieved from the ashes or from the kiln when the firing is finished (Hodges 1964; Gibson and Woods 1990).

- The last phase of the technical scheme for ceramic manufacture is of course the **use of the objects** produced, until they are discarded.

It can be noted that the whole process of ceramic manufacture as seen in the technical scheme in figure 1.3 is reversible until the firing happens. After this, the only sequence that can fit back earlier in the process is the re-use of discarded pottery with sherds crushed and reintegrated into the process as a tempering material to adjust the texture or the plasticity of the clay.

2 Other definitions

A number of terms that will be used throughout this research are defined in this section.

The word **cupule** has been retained to describe marks frequently found on clay especially in the cave sites of the French Pyrénées. A cupule is a small cup-shaped depression on a surface (Oxford English Dictionary, second edition 1989). This has been decided in order to distinguish the cupules found in the Upper Palaeolithic from the cup-marks frequently found in the rock art of the Holocene across Europe (Bednarik 2001).

The French word ***modelage*** will be used to describe the result of the action of modelling clay that is to say of producing or fashioning a figure or the imitation of anything in clay (translation of the definition in Trésor de la langue française 1985). This has been decided as there is no satisfying equivalent to the term in English for the applications in this research; the noun 'model' implies the representation of a structure or a notion of imitation; the word 'modelling' used as a noun implies the action of making models, and thus also the implications of the word model; these terms are not usually found applied to clay (all English definitions from the Oxford English Dictionary, second edition 1989).

High relief will be used to describe sculpture or carved worked in which the figure project more than one half of their true proportions from the wall or surface on which they are carved (Oxford English Dictionary, second edition 1989).

The term **low relief** describes a sculpture or carved work in which the figures project less than one half of their true proportions from the surface on which they are carved (Oxford English Dictionary, second edition 1989).

The **graphic unit** has been chosen as the unit upon which the catalogue inventory in the Appendixes is based. An individual graphic unit can be a figure, or part of one, in the case of an animal representation for example; it can also be a sign. Quantifying signs in Palaeolithic art is a notoriously difficult exercise as the problem arises of deciding what a full sign is and what the components of a sign are (Bahn and Vertut 1997). It has been decided for this research to count as one graphic unit groups of markings that appear made in relation with each other; for example a series of short parallel lines is counted as one graphic unit. Each graphic unit was given a reference number in the catalogue in the appendixes. It must be noted that for the purpose of the present research all the recorded graphic units have been re-referenced when they previously had a reference number. The old reference number, however, is of course mentioned in the appendixes next to the new one.

3 State of current research

The topic of the use of clay in the Upper Palaeolithic has received relatively little attention to date in comparison to the other surviving materials (flint, ivory, antler, bone) used during the same period.

In the late 19th century and at the beginning of the 20th century, before the great periods of Prehistory and their characteristics became firmly established, some heated debate occurred about the question of the existence of pottery in the Upper Palaeolithic. A consequent amount of energy was spent on the topic with the means available at the time. P. Bahn (1978) and K. Absolon (1949) have researched this topic in details. Eventually, the artefacts that gave rise to these debates, these 'Palaeolithic potsherds' were rather identified as resulting from lighting hearths on clay floors, which accidentally baked the soils underneath into fragments similar to potsherds, or as intrusions from more recent archaeological layers. These findings were interpreted as proving that pottery, and therefore ceramic, did not exist before the Neolithic, an assumption that became dogma for 20th century archaeology.

This assumption is the probable cause of the scarcity of ceramic discovered in Upper Palaeolithic contexts (outside of the sites of Moravia). Fired clay was not looked for, and thus was not found except for a small number of reported pieces, soon forgotten, found among some published assemblages (see Table 1.2 above in the introduction of this chapter for a referenced list); any find would have been dismissed as intrusive or accidental, and therefore would not have been kept. By extension, any use of clay in a non-fired shape was also dismissed for the Upper Palaeolithic period. Recent discoveries in East Asia and Siberia have shown that pottery containers were made and used in these regions as early as 16,000 BP in Late Palaeolithic contexts (Kuzmin and Keally 2001, 2003; Keally *et al.* 2003; Kenrick 1995; Imamura 1996; Hyland *et al.* 2001; Derev'anko 1998; Kaner, S., 2003) and have seriously shaken the idea that pottery only appeared with the Neolithic. This context is questioning previous assumptions and encouraging new discoveries; it is likely future excavations will bring more information to the subject of clay uses in the Upper Palaeolithic as is already visible in the context of the last major discoveries of cave art in France, Chauvet cave (Clottes (ed.) 2003) and Cosquer cave (Clottes *et al.* 2005). The most recent publications for both of these sites have mentioned examples of the use of clay, perhaps minor in comparison with the rest of the art at the site, but which would almost certainly have been overlooked in publication 20 years ago.

The fact that clay was used in the Upper Palaeolithic in the making of art, however, was acknowledged as early as 1912 with the discovery of the two clay sculptures of bison in Tuc D'Audoubert cave in the Ariège, French Pyrénées (Bégouen 1912). This knowledge was confirmed a few years later, in 1924 (Absolon 1949; Valoch 1996), by the discovery of the first ceramics dated to the Gravettian in the site of Dolni Vestonice I in Moravia. Further punctual discoveries of use of clay were made in the caves of the French Pyrénées and in the region of the Pavlovské Hills in Moravia throughout the rest of the 20th century (see the detail of Chapter 2 and 3 for description and references of all the sites discovered) and each was reported individually, often in the wider context of their site.

For the French Pyrénées, for example, Clottes (1989) does mention some characteristics of the use of clay in the region in the wider context of the parietal and portable art of that region, but he does not embark on a synthetic study of this aspect of the art.

The only work that offers an attempt at synthesising the use of clay in the art of the Magdalenian in the French Pyrénées was published recently (Boureux 2005) in the context of a Masters research.

Therefore, the sources for the present research had to be found/extracted among the various existing publications and monographs dedicated to the study of single sites (i.e. Bégouen and Breuil 1958, Trombe and Dubuc 1947…).

The recent work by Boureux (2005) is very indicative of a certain renewal of the archaeological research in the French Pyrénées, a region in which most of the important sites were discovered anciently (late 19th-early 20th century): the studies of some of the major sites have been undertaken again. Some of these had never been published as a monograph, such as Tuc d'Audoubert (Robert Bégouen, personal communication March 2002) in the Ariège. In others, the progresses in method of study have allowed new discoveries, such as in the caves of Montespan in Haute-Garonne (Michel Garcia, personal communication March 2004) and Bédeilhac in the Ariège (Georges Sauvet, personal communication August 2004). As all these sites are of major importance for the use of clay in the region, it is most likely their publication will bring to light new discoveries in the field of the use of clay.

As for central Europe, the study of the ceramic found in quantity in the sites of Dolni Vestonice and Pavlov (Moravia) during the previous decades was undertaken in the early 1980s by O. Soffer, P. Vandiver, J. Svoboda and B. Klima (Soffer *et al.* 1990, 1993 and also: Vandiver 1988; Vandiver *et al.* 1989; Vandiver *et al.* 1990; Soffer and Vandiver 1994, 1997). Their work is remarkable for the thorough study of the archaeological objects carried out through the application of a wide range of relevant scientific methods of analysis. This very large corpus of fired clay pieces, over 10,000 fragments, had scarcely been studied before, and as late as 1977, some researchers were still questioning the existence of intentional clay firing in the Upper Palaeolithic (Perlès 1977:122). These publications concern the objects from Dolni Vestonice I (Vandiver *et al.* 1989; 1990), Dolni Vestonice II (Svoboda Dir. 1991), Pavlov I 1952-1953 excavations (Svoboda Dir. 1994), Pavlov I - Northwest excavations (Svoboda Dir. 1997a), and Pavlov I southeast (Svoboda Dir. 2005). Ceramic objects from other excavated areas of the Pavlov- Dolni Vestonice locale are still being studied and remain unpublished.

Soffer and Vandiver's work has been priceless in demonstrating the possibilities offered by an early ceramic assemblage consisting of figurines and fragments of figurines, something that had not been done before. However, their research work concentrates on the nature of the ceramic objects from Dolni Vestonice and Pavlov, and very little on the analysis of their forms or on their wider context.

Stylistic analysis of the Moravian ceramics undertaken in the present research underlines the actual importance of this corpus of objects in quantity as well as in quality. The inventory of themes usually found for this period (female figurines, dangerous animals) is much increased. Rare types of representation, notably what is called here 'body extremities', are also being found in quantity, with consequent repercussions on the symbolic background associated with the Moravian sites.

On the basis of further information about other Gravettian sites in the region, found in various articles (Gonysevova 1999; Neugebauer-Maresch 1999; Banesz 1961, 1996; Svoboda *et al.* 1996; Jarošova *et al.* 1996; Valoch 1996) and given by personal communications (Svoboda 2007), it is now possible to assert that the phenomenon of firing clay figurines in Central Europe in the Pavlovian/Gravettian period was regional rather than limited to the Dolni Vestonice-Pavlov locale, with a total of 16 sites presently recognised. These data are entirely new and indicate the importance of starting to analyse the Moravian ceramics in their wider context.

Beside the existing regional-centred publications, Bahn (1978, 1984, 1986, Bahn and Vertut 1997; Bahn and Otte 1985) appears to be the only researcher who has attempted some synthesis on the topic of the use of clay in the Upper Palaeolithic. He presents the most complete list of Upper Palaeolithic sites with ceramic material that has been found in the literature (Bahn 1986; Bahn and Vertut 1997) and I wish to acknowledge that his work on Palaeolithic ceramic has been an inspiration for me in starting the present research. However, his synthesis is mostly aimed at demonstrating a widespread high level of complexity among Palaeolithic hunter-gatherers and thus does not go deep into analysis.

The fairly limited amount of research encountered shows that the topic of the use of clay in the Upper Palaeolithic of Europe has been altogether little studied. If one ventures into the field of ceramic and pottery studies from ethnographic and anthropological point of view, or for more recent periods, on the other hand, the literature is abundant (see for example Van Der Leeuw and Pritchard 1984; Arnold 1989; Barnett and Hoopes 1995; Eygun 2001). Some useful elements can be gained, especially for technical comparisons, but the major difference of context and production soon becomes an obstacle.

For these reasons, the present research hopes to underline the potential for Upper Palaeolithic clay studies, as well as possibly to indicate some routes for continuing the research. The present work has been undertaken as a synthesis, and as such it merely underlines the main areas of research that were recognised. The richness of the topic, however, did not allow doing more in the time span dedicated to a single PhD. This leaves plenty of scope for future research.

4 Research Problematic/Methodology

The approach chosen to study the topic of the use of clay in the Upper Palaeolithic of Europe is technological as this is a factor that can link places and periods in a synthesis on the topic. Furthermore, the technological analysis has been combined with a formal study of the considered artefacts; as we are considering forms of symbolic expression, the shapes and styles given are necessarily extremely meaningful And as we are considering techniques apparently specifically dedicated to symbolic expression, it can be argued that technological and formal studies must be closely associated to gain the best insights possible on our topic. In each of the Chapters describing the database, each aspect will be presented in turn. The technological analysis however, will base the framework on which the research develops.

Technology here will be defined following Ingold (1990:7) as 'a body of generalised, objective knowledge insofar as it is capable of practical application', encompassing tools, technique, knowledge and know-how.

Knowledge and know-how related to the technical processes form two different levels of knowledge. Knowledge, or *connaissance*, is the conceptual, abstract knowledge applied in technology (Schlanger 1994: 148); it is explicit, objective, discursive and can be encoded in words (Ingold 1990: 8); it regroups the concepts of ideal forms and raw materials and the gestures associated to the action (Pélégrin 1991). Know-how, or *savoir-faire*, is the practical or procedural knowledge involved in the technological process which implies instrumental dexterity, evaluations and judgements (Schlanger 1994: 148); it is tacit, subjective and context-dependent (Ingold 1990: 8).

The material record for the use of clay in this research has been presented and analysed through a *chaîne opératoire* study with the purpose of determining the different contexts of knowledge visible in each of the specific cultural context considered. This approach allows looking at the contexts for the transmission of knowledge and know-how necessary for the working of clay and at their wider social implications. The two very different Upper Palaeolithic backgrounds considered, the Magdalenian in the French Pyrénées and the Gravettian period in Moravia, will thus be compared throughout the research. The elements of the theoretical approach chosen for the analysis are presented below.

The chaîne opératoire concept

The *chaîne opératoire* is a concept developed by A. Leroi-Gourhan in the 1950s; it considers a technical process as a meaningful sequence of operations and actions that can be very informative for identifying choices at every stage of the manufacturing process and to approach cognitive problems, including intentionality (Audouze 2002: 287). The following definition has been given by Karlin and Julien (1994: 164): 'in a technical activity, a *chaîne opératoire* takes the form of an ordered train of actions, gestures, instruments or even agents leading the transformation of a given material toward the manufacture of a product, through major steps that are more or less predictable'.

The *chaîne opératoire* was firstly developed for the study of lithics for which it allowed the reconstitution and organisation of all the actions concerning a block of raw material from the moment it was chosen through the refitting of the obtained pieces. Thus Karlin and Julien (1994: 158) describe the investigation of a *chaîne opératoire* as 'an assemblage of refitted products'.

It is only recently that the application of the *chaîne opératoire* has been broadened to elements other than lithics, and in particular to the making of art, which was previously studied through a primarily formal and stylistic analysis. It was not recognised that art could demonstrate a technological tradition in the same manner as stone tools, for example, because of a frequent lack of context (for some cave art or objects from ancient excavations) and of a given status as work of art in the modern sense that would tend to remove the object from any social or economical context (Fritz 1999, Moro Abadia 2006). This view has steadily been changing, and Conkey *et al.* (1997: 7) urge us to 'examine prehistoric images as we would any category of archaeological artefacts'. Gamble (1999: 29) emphasises that *chaîne opératoire* analysis is indeed a tool that permits doing away with the distinction between style and function that has plagued the study of archaeological artefacts for so long.

White (1993, 1997) as applied the *chaîne opératoire* concept to Upper Palaeolithic portable art, stating that detailed analysis of the materials will provide us with more contextualised, nuanced and local understandings of the imagery. He argues that it is in the analysis of the production of the images themselves that we may find important factors to start to understand what made the imagery meaningful for prehistoric people. White addresses the *chaîne opératoire* concept by including in it immaterial and cultural factors as sequences in the making. Fritz (1999) is also a researcher who has applied with success in recent years *chaînes opératoires* and technical schemes to the study of portable art in the form of Magdalenian engraved bone objects.

The *chaîne opératoire* concept has also been applied to materials other than flint as it was realised that even without the direct evidence provided by the refitting, the exact steps followed to make an object could be deduced from the analysis of the final artefact as well as from experimental replication and possibly some ethnographic comparisons (Karlin and Julien 1994: 153; Pélégrin 1991).

In the case of the uses made of clay in the Upper Palaeolithic, it is not usually possible to find much physical traces of the working process of the material in archaeological assemblages. However, thanks to the capacity of clay to preserve traces of its working, some steps can be deduced from observing the final result and interpreting the traces visible in the light of one's experience of the material, completed by relevant literature search. The steps of the working of the material are mostly deduced from these various elements, although a few cases are known in which direct evidence of the steps of the working of clay has been found; this is the case for example in the Tuc d'Audoubert cave in the French Pyrénées in which the area from which clay was extracted to make a large bison sculpture is still visible together with the tool (a broken stalactite) probably used for the extraction (Bégouen and Breuil 1958: 97) or in the site of Dolni Vestonice I in Moravia where ceramic firing structures were found in the excavation (Soffer et al. 1993: 270).

Specific *chaînes opératoires* have been reconstructed for the various graphic units in clay recognised in the present research; their combination in technical schemes allowed some generalisation about the contexts of knowledge considered.

Technical schemes

The *chaîne opératoire* itself develops in a linear way, one step after the other; in a technical scheme, the combination of all the possible *chaînes opératoires* for a single technical process is presented, including the possible choices which can be made, such as missing a stage or going back to a previous one. The constant evaluation and choices made by the person manufacturing the objects are thus visible and can be interpreted as they are all meaningful; they provide us with information about technical production and

Figure 1.3: Karlin and Julien's 'flow chart' showing the organisation of the technical scheme of Magdalenian blade production (From Karlin and Julien 1994, Fig. 15.1).
(key to reading the scheme: the numbers indicate the possible steps of working of the material in the order usually followed; numbers in a full circle are compulsory steps that cannot be missed in the process; numbers in dotted circle are steps that can be skipped; the arrows indicate the possibilities of skipping steps in the process or of reverting to a previous step in a reversible process. Here the sequences of the working are further divided into operations and gestures in order to detail it more).

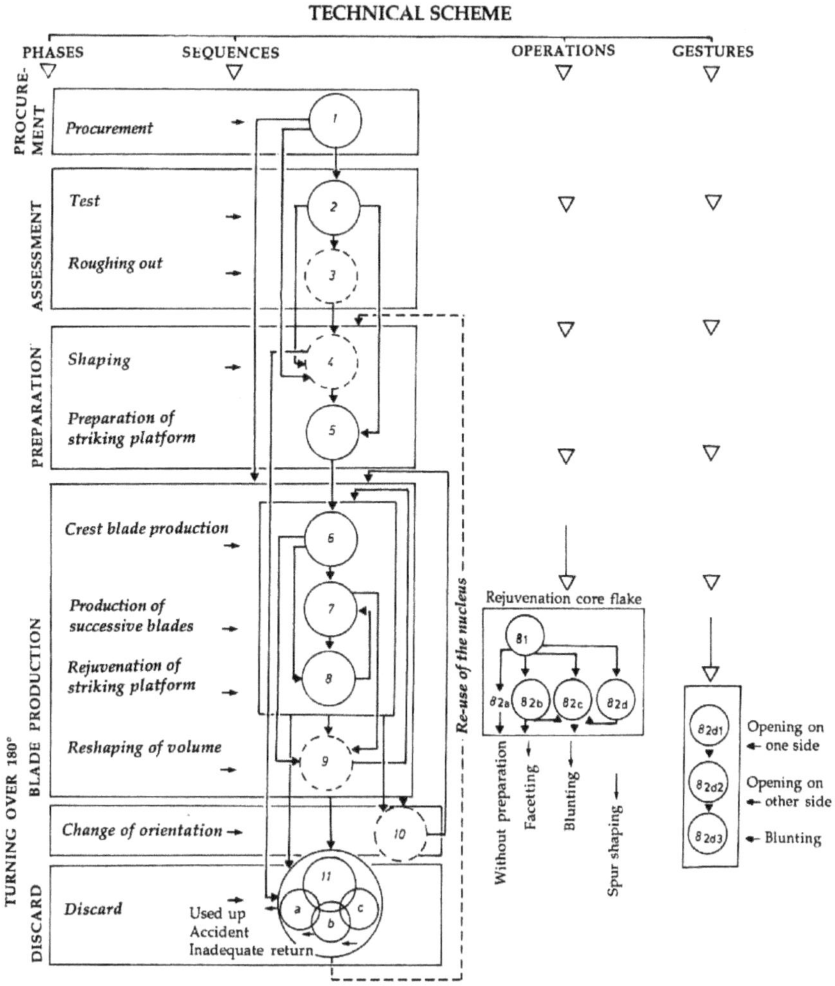

material culture, and also about all the cultural, spatial and economic implications upon which the production depends as well as individual, social and technical behaviours (Knecht 1997; Karlin and Julien 1994; Garanger 1992).

The model of technical scheme used in this study of clay follows the one developed by Karlin and Julien (1994) for the production of flint blades in the Magdalenian in the Paris Basin with a presentation as a 'flow chart' (figure 1.3 below).

In this 'flow chart', the main phases of the production and the various sequences constituting them are all indicated. The sequences are numbered and the potential linearity of the chaîne opératoire indicated by the arrows joining each of them. Some sequences, however, which are not essential and that could be skipped are indicated by a broken circle to differentiate them from the necessary sequences indicated by a full circle. Arrows between the various sequences indicate all possibilities to skip sequences or to go back to a previous one (Karlin and Julien 1994: figure 15.1).
Reconstituting technical schemes for the use of clay in the upper Palaeolithic has allowed determining some patterns in this use, from which elements concerning the corresponding contexts of knowledge can be deduced. The emerging data from this analysis then becomes interpretable, notably in the light of social context and in term of transmission of knowledge.

Constellations of technical knowledge

The idea that constellations of knowledge surround any technological process was used to synthesise the information obtained from the technological study of the database. The constellation of knowledge is a conjunction of the different elements involved in tool use or artefact manufacture (Dougherty and Keller 1982; Sinclair 2000: 200). These elements form four major groups concerning the raw material, the implements used in the manufacturing process, the particular techniques or processes involved and the desired end-points for the whole process. All the possible links between these elements are also considered, notably any monitoring criteria (Figure 1.4).

Figure 1.4: diagram of the elements forming a constellation of knowledge (after Sinclair 2000, figure 13.3)

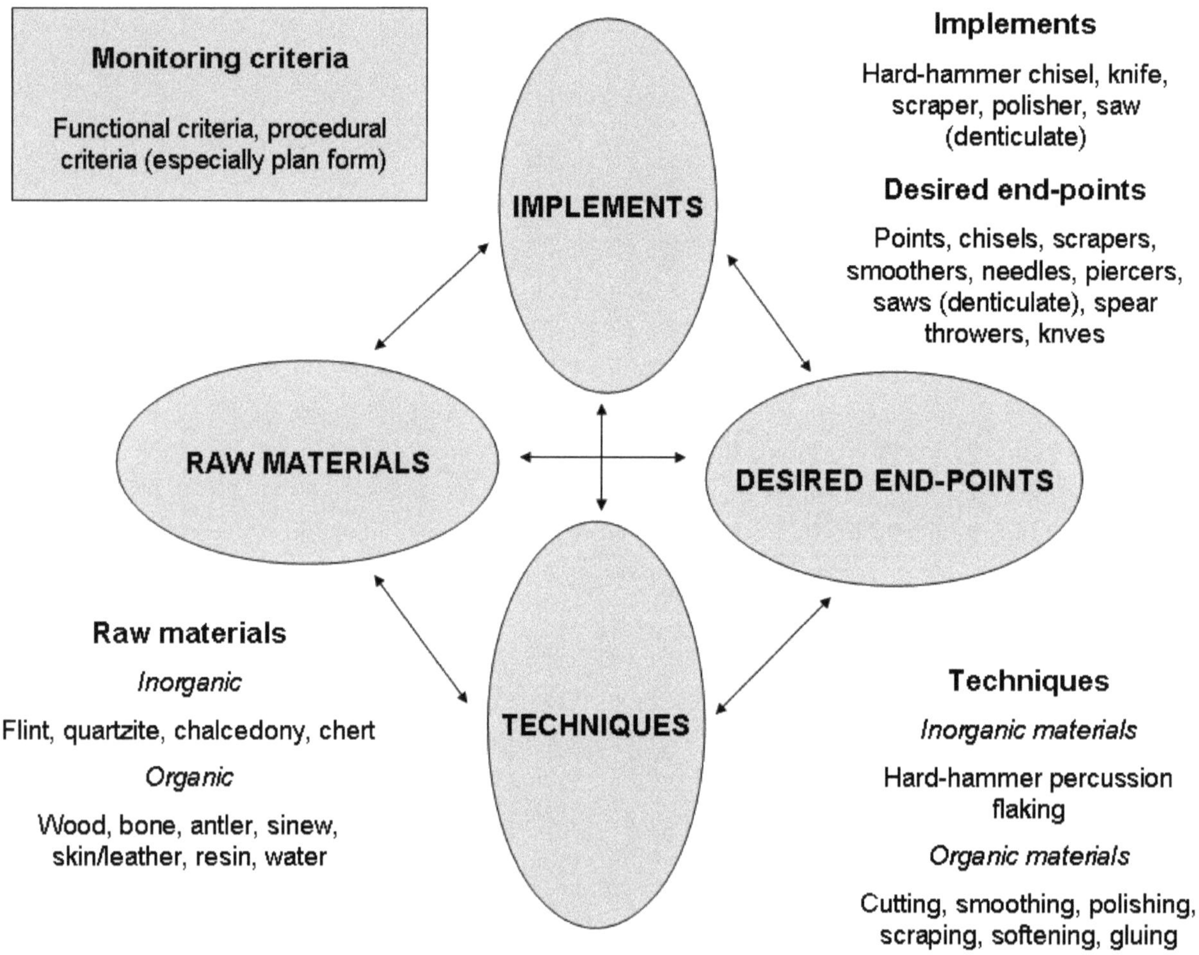

All the components of the constellation of knowledge depend upon the knowledge that an individual has acquired of them and each constellation is specific to a particular form of production (Sinclair 1995; 2000: 200). The use of constellations of knowledge allows understanding the manufacture and use of technology in their wider social context and notably seeing the individual behind the production. It is also a way of engendering the technical process and thus of reflecting on the social agency surrounding it (Dobres and Hoffman 1994; Dobres 2000; Sinclair 1995, 2000).

The constellation of knowledge is a useful approach as it permits addressing every aspect of the technical process, including those which are not visible any more from the archaeological record.

It is true this approach has previously mostly been applied to tool-making. In the context of the present research, it is applied to the technical process behind symbolic expression, for which it is also highly suitable.

The transmission of 'knowledge' and' know-how': the context for learning clay-working techniques

Through the technological analysis of the use of clay chosen, the elements related to the knowledge ('*connaissance*') and know-how (*savoir-faire*) involved in the making become visible at various levels as well as any possible patterns they might present. These elements allow approaching the context of learning surrounding clay techniques, notably with regard to any possible transmission of knowledge and know-how. Some implications on the social context considered can thus be underlined.

Lithic technological studies have been the first to emphasise the potential of the *chaîne opératoire* analysis for helping to recognise different levels of skills in the making of objects, thus demonstrating the fact that an apprenticeship could be seen from the archaeological record (Pigeot 1987; 1990).

The technological analysis of the lithic material in the large open-air sites of Etiolles and Pincevent in the Paris Basin region in France has shown spectacular results with regard to the variation of the level of skills visible from the archaeological record. In Etiolles, it was possible to recognise three levels of skill; the master knappers were working the best quality flint, producing the best quality tools and were seated closest to the hearths; the average knapper used imperfect knapping material and produced average tools; the bad knappers were, perhaps, apprentices or children: they were only using the material rejected by the others and only working at the periphery of the flint-working area. Elements of the social organisation of the Magdalenian people from Etiolles thus became visible (Pigeot 1987, 1990: 138; Karlin and Julien 1994: 156).

The situation at Etiolles and Pincevent is exceptional by the meticulousness of the excavations which allowed extracting as much information as possible from the archaeological record. This fact, combined with the potential for refitting of knapped flint, is the probable reason why such a successful analysis was made possible in these two sites, to the point of being able to consider apprenticeship. Apprenticeship there is described as a gradual acquisition of knowledge and know-how, as a transmission of knowledge and skills by Pigeot (1990: 132). In modern times, apprenticeship has the meaning of being a system of training a new generation of skilled craft practitioners in which a master craftsman employs young people as inexpensive workforce in exchange for formal training in the craft (Oxford English Dictionary 1989). This normally resulted in extremely codified social relationships between the actors of the apprenticeship. In the Upper Palaeolithic contexts considered here, it must be pointed out that the type of apprenticeship considered is different. It rather consists of a transmission of knowledge which can be formal or informal, passive by simple observation of the technical action, or more active by explanation, discussion and demonstration of the action (Pigeot 1990: 137).

The working of clay appears much more flexible than the working of flint, and accordingly would accept a lower standard of working without such a drastic impact on the final result; even if it is worked by people with rather low skill, the material permits reversibility and thus easily correcting mistakes before any firing is applied. The firing process on the other hand is elaborate and implies the mastery of a number of elements such as firing structure, fuel, temperature… The nature of the material, as well as the frequent lack of data regarding the corresponding archaeological contexts imply that it has not been possible in the study of the use of clay in the Upper Palaeolithic to determine degrees of apprenticeship similar to those in Etiolles or Pincevent. However, some elements underline a possible transmission of knowledge and skills and will be acknowledged as such when encountered in the research.

The data gathered on the topic of the use of clay is presented and analysed in Chapter 2 for the material from the French Pyrénées and in Chapter 3 for the material from Central Europe. Through the technological analysis, it has been possible to underline two very different contexts for the use of clay in these two regions. In the French Pyrénées, clay was commonly used in cave art during the Magdalenian and this art respects the conventions in style and themes visible in other art techniques in the region. When one looks at the detail of the techniques used for working clay, however, this uniformity disappears as these are extremely varied, even within the same site. In Central Europe, in the sites of Dolni Vestonice I and II and Pavlov I and II, on the opposite, the thousands of fired clay fragments found appear to result from the same complex technical process. The corresponding contexts of knowledge of the clay techniques are accordingly different; they are the topic of Chapter 4.

It must also be noted that the dates used in this research are uncalibrated dates. Recent developments in radiocarbon calibration are making it possible to calibrate dates up to circa 26,000 cal BP (Ramsey *et al.* 2006). However, most of the Gravettian record considered here in Central Europe still falls out of this range or is close to its limit (27,000 to 25,000 BP for the occupations at Dolni Vestonice and Pavlov) (Svoboda *et al.* 1996) making attempts at calibration difficult for the whole period. As for the Magdalenian in the French Pyrénées, although the period is more recent and dates for it should be easier to calibrate, the ancient age of much of the corresponding excavations and the lack of radiocarbon dates for part of the cave art context (Clottes 1989) make attempts at calibration rather useless in the context of the present attempt of synthesis. Any radiocarbon date mentioned in the research is therefore repeated as found in the corresponding source.

Also, unless stated otherwise, the photographs used are my own and their scale is in centimetres.

CHAPTER 2: THE USE OF CLAY IN THE MAGDALENIAN OF THE FRENCH PYRÉNÉES

In a famous telegram dated to 10th October 1912, Count Henri Bégouen announced to Emile Cartailhac the discovery of Magdalenian clay *modelages* (Absolon 1949: 28)[3]. They had just found the magnificent bison sculpted in clay in Tuc d'Audoubert cave in the Ariège (figure 2.2), and for the first time it was recognised that this material and technique had been used in the art of the Upper Palaeolithic.

Since then, many more discoveries have been made in the cave sites of the French Pyrénées, underlining a phenomenon that appears strictly regional and related to the Middle (circa 15,000 BP- 13,500 BP in radiocarbon dates) and Upper Magdalenian culture (13,500 BP- 12,000 BP in radiocarbon dates)[4] (*Djindjian et al.* 1999: 262): the working of clay (Bahn 1984; Boureux 2005).

The French Pyrénées have strong regional cultural specificities in the Magdalenian period. The habitation sites encountered are mostly caves in this region, when inhabiting caves is actually quite rare in the rest of the Upper Palaeolithic. They are often connected to cave art sites (Garanger Dir. 1992). The French Pyrénées appear mostly empty of human occupation before the Middle Magdalenian (Vialou 1986), although recent evidences from Enlène indicate an occupation layer there in the Gravettian (Clottes 1989: 37).

For the Middle Magdalenian in the Pyrénées, the lithic toolkit is quite homogeneous, with an important quantity of backed bladelets, truncated and denticulated, often triangular in shape. Tools are small in size, and typical tools like beaked burins, Teyjat points and Laugerie-Basse points are not found. The toolkit in osseous material is very abundant and diversified. Spear-points have simple bevelled bases and are often grooved. Short spear-points with a long bevelled base and a groove on the back (Lussac-Angles type) are still sometimes found in the Pyrénées in the Middle Magdalenian, when elsewhere they belong to the Magdalenian III (Garanger Dir. 1992; Clottes 1989: 36).

The use of clay in the region appears associated to the Magdalenian culture. Given that clay uses are not directly datable as they are mostly unfired, the association with the Magdalenian has to be inferred or assumed to be contemporary with directly dated imagery and other artefacts found in these sites. The chronological attribution of the various sites forming the present record can be found in Table 2.1.

The French Pyrénées are the only known region in which clay has been worked to such an extent at the end of the Upper Palaeolithic, the material being used principally for symbolic expression, mostly in unfired form (Clottes 1989 (1999): 80).

The record of clay uses comprises a total of 14 cave sites in the French Pyrénées, mostly in the Central Pyrénées (Ariège, Hautes-Pyrénées and Haute-Garonne) and also in the Pyrénées-Atlantiques to the west. These sites are Tuc d'Audoubert, Enlène, Labouiche, Niaux, Mas d'Azil, Fontanet, Bédeilhac and Massat in the Ariège, Montespan in Haute-Garonne, Labastide and Bois du Cantet in Hautes-Pyrénées, and Erbérua, Oxocelhaya-Hariztoya and Etcheberri in Pyrénées-Atlantiques (figure 2.1).

In Trois-Frères, which is a major site for cave art with a few hundreds of engravings on its walls (Breuil 1952), works on clay are also found in the part of the cave called the 'Tréfonds' in the shape of finger tracings (figurative and macaronis) (Bégouen and Clottes 1984a: 406) and at least one printed sign on a wall (Vialou 1986; Bégouen and Breuil 1958: 20). However, as a number of researchers (Breuil 1952; Bégouen and Clottes 1987: 186; Vialou 1986) agree that the art on soft surfaces in the 'Tréfonds' is likely to be much older[5] than the rest of the parietal art in Trois-Frères, it has been decided not to include this site in the inventory of clay uses in the Magdalenian in the French Pyrénées. Similarly, clay uses were also found in Gargas (Haute-Pyrénées) but because of the fairly certain attribution of the art of this site to the Gravettian period[6] (Clottes *et al.* 1992: 180), after which the mouth of the cave collapsed, sealing the entrance (Barrière 1984), there appears to be no direct connection with the later phenomenon of Magdalenian use of clay in the region. This site has therefore not been included in the present inventory.

For these 14 sites, a total of 289 individual graphic units[7] have been recognised that can be described as follow by the type of technique used: at least 128 portable clay objects, 35 modelages in three-dimensions, high and low relief, 10 engravings with tool on wall surfaces, 78 floor engravings, 19 finger engravings on walls and 3 on floors, 5 printed signs on walls and 12 on clay floors (table 2.2).

[3] Henri Bégouen wrote 'Les Magdaléniens modelaient aussi l'argile!' To which Emile Cartailhac immediately answered 'J'arrive' (I am coming) (Absolon 1949: 28).

[4] The Middle Magdalenian corresponds to Leroi-Gourhan's Magdalenian IV. The Magdalenian V (rare in the French Pyrénées) and VI are associated with the Upper Magdalenian (Garanger Dir. 1992).

[5] The problem of the existence of various phases in the making of the art in Trois-Frères has been raised anciently (Breuil 1952) on stylistic ground and because of some topographical separation visible in the cave. The stylistic comparison with an engraved plaquettes dated to the Perigordian V and found in stratigraphy in the cave of Gargas has usually been taken as evidence (Clottes 1989 (1999: 83). Recent discoveries of Perigordian V (Gravettian) layers in excavations at the Entrance of Enlène cave, which is connected to Trois-Frères, have brought for the first time some concrete elements to support this claim (Bégouen and Clottes 1987: 186) as they confirm that the sites were occupied at this time.

[6] A radiocarbon date from a bone splinter found in a crack in the wall near some hands stencils (the 'Grand Panneau') yielded 26,860±460 BP (Gif-A-92.369), corresponding to an Ancient Gravettian. (Clottes *et al.* 1992: 180).

[7] An individual graphic unit can be a figure or part of one, for an animal representation for example, or a sign: this can regroup a few lines or dots that appear to have been made in relation with each other: thus a series of short parallel lines will be counted as one graphic unit.

Figure 2.1: Location map of the sites with clay uses in the French Pyrénées
1. Erbérua, 2 Oxocelhaya-Hariztoya, 3 Etcheberri, 4 Bois du Cantet, 5 Labastide, 6 Montespan, 7 Tuc d'Audoubert, 8 Enlène, 9 Mas d'Azil, 10 Massat, 11 Niaux, 12 Fontanet, 13 Bédeilhac, 14 Labouiche.
(Main map after Clottes 1995: figure 13; map of France from website www.mapsofworld.com, accessed 16th May 2007)

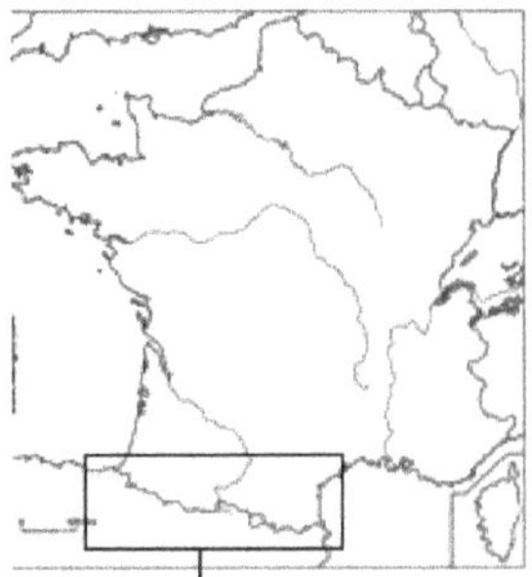

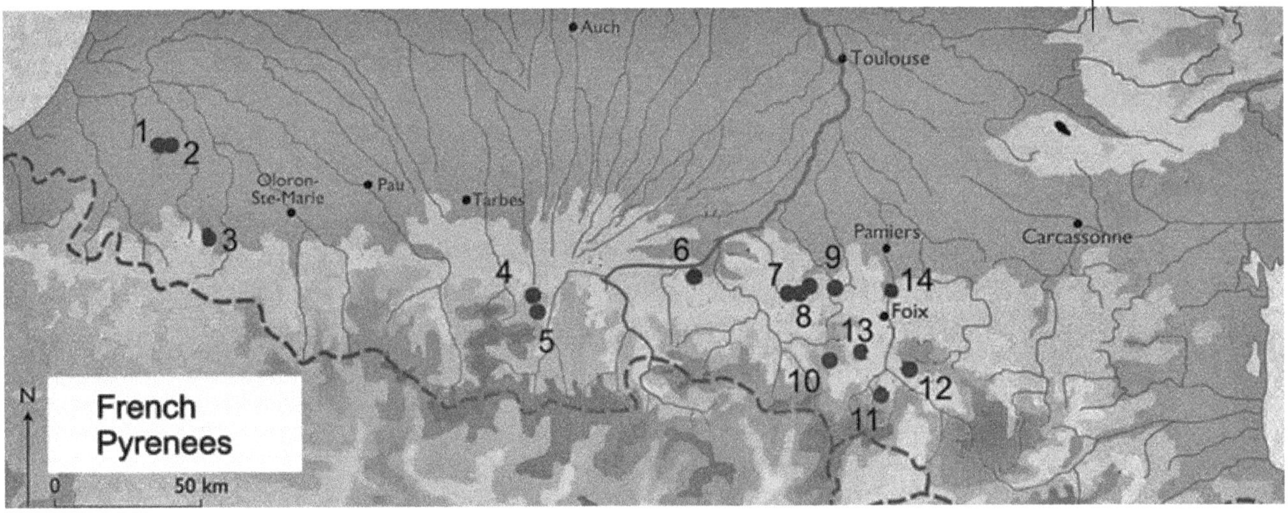

Table 2.1 Chronological attribution of sites with clay uses in the French Pyrénées

Location	Site name	Chronological attribution		Reference
		Radiocarbon	Stylistic	
Ariège	Tuc d'Audoubert	Middle/ Early middle Magdalenian		Lorblanchet 1995; Boureux 2005; Bégouen and Clottes 1983
	Enlène	Gravettian, Middle Magdalenian		Bégouen and Clottes 1981, Bégouen and Clottes 1987
	Labouiche		Middle Magdalenian	Boureux 2005
	Niaux	Middle Magdalenian / Upper Magdalenian		Lorblanchet 1995; Clottes et al. 1992; Clottes 195
	Mas d'Azil	Middle and Upper Magdalenian		Alteirac and Bahn 1982; Alteirac and Vialou 1984
	Fontanet	Middle Magdalenian/ Upper Magdalenian		Lorblanchet 1995; Clottes *et al.* 1984; Vialou 1986 ; Clottes 1993
	Bédeilhac		Middle and Upper Magdalenian	Gailli *et al.* 1984; Vialou 1986 ; Boureux 2005
	Massat		Middle Magdalenian /Upper Magdalenian	Lorblanchet 1995; Clottes and Gailli 1984
Haute-Garonne	Montespan		Middle and Upper Magdalenian	Rivenq 1984; Boureux 2005
Hautes-Pyrénées	Labastide	Middle Magdalenian		Lorblanchet 1995; Boureux 2005; Simmonet *et al* 1984
	Bois du Cantet	Upper Magdalenian		Clot 1984
Pyrénées-Atlantiques	Erbérua		Magdalenian	Larribau and Prudhomme 1989; Boureux 2005
	Oxocelhaya-Hariztoya		Magdalenian	Laplace and Larribau 1984
	Etcheberri		Middle and Upper Magdalenian	Laplace *et al.* 1984; Boureux 2005

This forms the record of clay uses analysed in this chapter. This database was collected from an in-depth literature search, especially of old French archaeological literature, as well as from personal fieldtrips and is, to the best of my knowledge, the most exhaustive list of clay uses presently existing for the French Pyrénées. Destroyed or disappeared cases of clay uses have been included in the inventory when mentioned and described by a reliable source such as Breuil (1952) for the site of Bédeilhac or Bégouen and Casteret (1923) for Montespan.

The inventory of clay uses is organised according to the techniques applied to the material (table 2.2). It will be seen that often a combination of techniques are used to make a figure or an object: in this case the dominant technique used will be retained for the inventory.

The first of these techniques considered is *modelage*, which is specifically suitable to clay as a soft material, as it implies the working of plastic materials by hand to build up form (Encyclopaedia Britannica 2007). *Modelage* has been separated into three-dimensional objects, high-relief and low-relief. In the three-dimensional objects grouping have been included a number of small portable sculptures in the round all made of dried or still wet clay that were found in a certain position in the caves, some stuck on the wall like in Montespan (Trombe and Dubuc 1947) and Labouiche (Méroc 1959), some part of a larger composition such as the small bison in Tuc d'Audoubert (Bégouen and Clottes 1977). They have been dissociated from the portable objects category as they do not appear to have been made with the intention of carrying them around. Haut-relief modelage consists of *modelage* that are attached to a supporting surface but that raise from it. Bas-relief modelage consists of figures that are attached to a surface and that only raise slightly from it. It is true that some cases of clay uses are really at the border between two techniques such as the Tuc d'Audoubert large bison *modelage* that are between three-dimensional sculptures and high-reliefs.

Engraving is a technique that consists of marking a shape onto a surface by removing material by incision with a tool. This is often found in cave art for rocky wall surfaces but is also applied to clay surfaces, which are soft enough to be worked simply with a finger. In these cases a technological choice was made by the artist which obviously bears some meaning.

Finger-tracings allow marking a clay covered surface simply with the fingers in order to draw figures or simply to intensively cover a surface with meandering lines, the 'macaronis' tracings[8]. Figurative and non-figurative finger tracings have been separated here with the idea that figurative finger-tracings have similarities with engravings that allow some comparison and a reflection on the choice of the technique used. Also they can be quantified, which is not the case of the macaroni tracings.

Impressing is another technique found among the use of clay in the French Pyrenees, and it appears restricted to the drawing of patterns with cupules that can be interpreted as signs. These prints can be done with tools or with fingertips.

In the case of engravings, figurative finger-tracings and printing on clay, the inventory takes into account whether they were made on walls or on the floors, as figures drawn on clay floors appear, up to the present, a specificity of the Magdalenian French Pyrénées (Clottes 1989), just as the phenomenon of using clay in symbolic expression altogether. The relation of the clay art on floors with the clay art on the walls is also a topic that will be looked at in this research.

Another technique looked at is sculpture in the form of portable objects. As some objects in this category are made of fired clay or of well dried material, they were distinguished from the three-dimensional modelages which are unfired and rather soft.

Finally, in some cases, the clay has been directly used from a soft lump as a pigment to draw a figure on the cave wall (for example in Bédeilhac, Gailli *et al.* 1984). This is indicated in the inventory when it was clearly mentioned as such in the relevant literature. However, this can be seen as a tricky category as it is known that clay has often been used in the Upper Palaeolithic as an extender in pigment recipes used in cave painting (Couraud 1983; Couraud and Laming-Emperaire 1979) in what is perhaps the most common and rather invisible use of this material for the whole of the period. It might also be useful here to remind that the barrier between what is defined as clay and what is ochre can be rather blurred as ochre is often considered as a clay extremely rich in iron oxide. A strict geological definition clearly does not apply most of the time when archaeological researchers are considering the materials Palaeolithic paintings are made of.

To this list can be added numerous traces, found on various surfaces, of what has been called here 'human activity' in the caves. This category gathers traces that have been left unintentionally such as prints of feet, hands, fingers, others that could result from some utilitarian purpose of the clay such as the lining of hearth structures, other structures or possibly body ornaments as beads; and finally traces that could be connected to symbolic expression on clay but which are too damaged or imprecise to be sure such as clay removals, worked lumps, possible 'melted' modelage, perforated clay masses and clay balls. These traces are important to emphasise the fact that the present-day record of the use of clay in the French Pyrénées is probably only a small portion of what existed originally because of the fragility of the unfired clay works themselves, as well as of the easily inflicted damage which this form of art has often suffered in modern times: the clay is still as plastic as when it was worked.

The first part of this chapter will present the analysis of the clay record for each of the sites studied, including any finger-tracings and traces of human activity on clay when present. The second part will consider the technology of these clay uses with the purpose of addressing the wider Magdalenian cultural context regarding issues of behavioural complexity, intra and inter regional contacts and transmission of technical knowledge, in particular in relation to the art context, issues that will be presented in the third part of this chapter. All the sites with clay uses known in the French Pyrénées are cave sites, and in all of them techniques other than the working of clay were also applied to make Magdalenian art. It is important to point out that the clay uses recorded in the French Pyrénées are rarely isolated, but usually combined with other techniques in the process of symbolic expression.

[8] After Henry Breuil's words (1952).

Table 2.2: Quantitative inventory of the clay uses found in the sites of the French Pyrénées according to their techniques. X indicate the presence of use as pigment; FT indicates the presence of macaroni finger-tracings that cannot be quantified

Site name				Tuc d'Audoubert	Enlène	Labouiche	Niaux	Mas d'Azil	Fontanet	Bédeilhac	Massat	Montespan	Labastide	Bois du Cantet	Erbérua	Oxocelhaya-Hariztoya	Etcheberri	Total number of clay uses	% of total (289)
Type of clay uses	*Modelage*	Three-dimensional		6		1						2						9	3.1
	Modelage	High-relief		2						2		6			4			14	4.8
	Modelage	Low relief		1					3	5		2	1					12	4.1
	Engraving	On Wall							1		8	1						10	3.4
	Engraving	On Floor		2			65			9							2	78	27
	Figurative finger-tracing	On wall									4	7			8			19	6.6
	Figurative finger-tracing	On floor														3		3	1
	Impressing	With tool	On wall									4						4	1.4
	Impressing	With tool	On floor						4					1				5	1.7
	Impressing	With fingers	On wall								1							1	0.3
	Impressing	With fingers	On floor	4					3									7	2.4
	Portable object				3			2		121								127	43.9
	Use as pigment									X			X			X	X		
	'macaroni' finger-tracings			FT							FT	FT			FT	FT			
Total for each site				15+ FT	3	1	65	2	11	138	13+ FT	22+ FT	1	1	12+ FT	3+ FT	2	289	
% of total quantity of clay uses				5.2	1	0.3	22.5	0.7	3.8	47.7	4.5	7.6	0.3	0.3	4.1	1	0.7		100

1 Description of the material record in clay in the French Pyrénées

The full inventory of clay uses for each of the site mentioned can be found in the Appendix 1 of the present research. Significant elements have merely been underlined in this part. All the reference numbers used in the table correspond to the reference numbers given in the catalogue in Appendix 1.

Tuc d'Audoubert, Ariège

Tuc d'Audoubert is one of the decorated caves of the Ariège region and was discovered in 1912. Together with the caves of Trois-Frères and Enlène, it is part of the karstic network created by the Volp River and this group of three caves is often referred to as 'the Volp Caves'[9]. Tuc d'Audoubert has been owned and managed, together with the other Volp caves, for four generations by the Bégouen family in an exemplary manner. This has led to excellent conservation of the sites, and especially the fragile works on clay which are found in both Tuc d'Audoubert and Enlène (Bégouen and Clottes 1984).

The entrance to the cave of Tuc d'Audoubert is through the lower level of the karst in which the Volp River still flows today: a boat is needed to access it. The cave has a lower, middle and an upper level. The traces of prehistoric activities are concentrated in its middle and upper parts. A

[9] Les cavernes du Volp (Bégouen and Breuil 1958)

Magdalenian habitat site was excavated in the middle part, with typical portable objects for the period of the Middle Magdalenian, among which were works of art (Bégouen and Clottes 1984). It has been dated by radiocarbon to 14,350 ± 160 BP on bone collagen, which makes it an early example of the Middle Magdalenian, the earliest found at the Volp cave sites (Bégouen and Clottes 1983: 121).

The art of Tuc d'Audoubert is found for the majority of it in side galleries in the middle part of the cave (off the 'Salle Nuptiale') and for some in the upper part of the cave where there are no traces of habitation. This upper gallery is accessed by climbing and passing through a narrow 'cat flap' and thus topographically rather separated from the rest of the cave (Bégouen and Clottes 1984: 410). The art is dominated by engravings (at least 24 figurative representations), and some red or black paintings but the site is famous for the two remarkable bison *modelages* in clay found at the very end of the upper gallery of the cave. A number of works and traces on clay are also found in the vicinity of the bison in the terminal chamber and in the nearby 'Chamber of the Heels' (Bégouen and Clottes 1984: 413).

Tuc d'Audoubert is indeed a major site for the use of clay in symbolic expression in the Magdalenian period with nine instances of *modelage* recorded, among which six are three-dimensional, two are in high relief and one is in low-relief; at least two floor engravings; four printed signs done with finger tips; and some panels of macaroni finger-tracings. The descriptive inventory of the individual graphic units at the site is found in table 2.3. This inventory is likely to increase in the near future as a monograph of the site is in preparation and much in Tuc d'Audoubert has never yet been published[10].

Figure 2.2: The two bison sculptures in clay from Tuc d'Audoubert, Ariège (from Vialou 1998: 96).They measure 61 and 63 cm long.

The two figures represent a female bison, which is clearly sexed, followed by a male, which is implied from the larger body of the animal (Clottes *et al.* 1994: 126). They are magnificently staged at the very end of the upper gallery of Tuc d'Audoubert: they seem to be running away from the viewer arriving in the terminal chamber. The two bison are in fact part of a larger composition (figure 2.3), as the outline of a partly modelled bison is also visible on the clay floor behind the rock on which the large bison are resting, and as a small portable clay sculpture of a bison was also found between them and the engraving (Bégouen *et al.* 1977). All this forms a complex three-dimensional scene.

The two clay bison (figure 2.2) are among the largest known sculptures from the Upper Palaeolithic period, a form of monumental art on clay possibly comparable to what is found on stone in other regions[11] at the same period. Because they are attached to the prepared rock surface supporting them, they are considered in the present research as high-relief *modelage*s.

Figure 2.3: sketch of the position of the four bison representation in the final gallery of Tuc d'Audoubert (from Bégouen *et al.* 1977, figure 1)

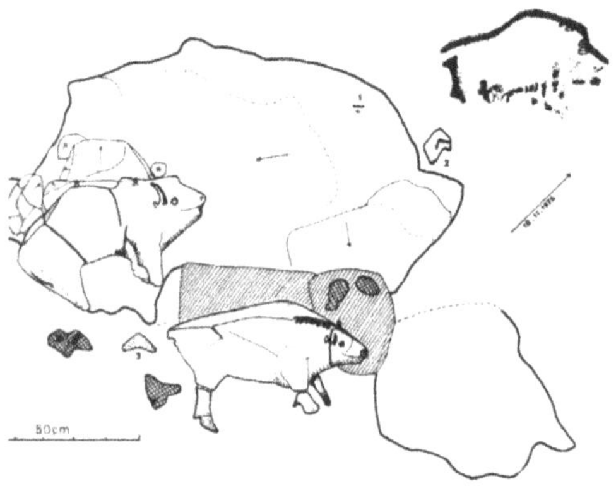

A small side chamber called the 'Chamber of the Heels' is found about 25 m away from the group of four bison, before reaching the end of the upper gallery, in a lower position (Bégouen and Clottes 1984b: 413). This is where the rest of the identified graphic units on clay in Tuc d'Audoubert are gathered, as well as many traces resulting from human activity.

On the floor of the chamber, five clay 'sausages' are found grouped by two and three. They form coil-like fragments that were originally interpreted as phalluses (Bégouen and Breuil 1958: 98), but that are now thought to be possible remains left while testing the plasticity of the clay (Beasley 1986; Bahn and Vertut 1997). Because they are unique in the record of clay uses in the French Pyrenees they have been considered as possible three-dimensional *modelages*. The other graphic units found on the clay floor of this chamber are signs. An oval shape and the 'pseudo-barbed' sign (a line with large commas on each side) (Bégouen and Breuil 1958) are tool-engraved. The four other signs, a circular shape and three linear designs, are made of juxtaposed small holes made by prints of fingertips (Bégouen and Breuil 1958: 101).

As for the finger-tracings or 'macaronis' in Tuc D'Audoubert, they are found in two different locations, one panel being in a side gallery in the vicinity of the 'Salle Nuptiale' in the middle level of the cave; the second being found in the middle of the upper gallery, in a side gallery, and seems to cover a whole ceiling (figure 2.4). It is to note for this second panel that it is the only form of art found in this part of the upper gallery before reaching the clay bison (Bégouen and Clottes 1984b: 413; Bégouen and Clottes 1981b: 235).

[10] Personal communication Robert Bégouen July 2002

[11] For example, the sculpted rock shelter of Abri Bourdois in Angles-sur-l'Anglin (Vienne, France), a decorated Middle and Late Magdalenian living site which was adorned with dozens of large bas-relief sculptures (White 2003: 103-104).

Figure 2.4: Tuc d'Audoubert cave map with the location of the various art techniques found at the site. (After Bégouen and Clottes 1984b: figure 1)

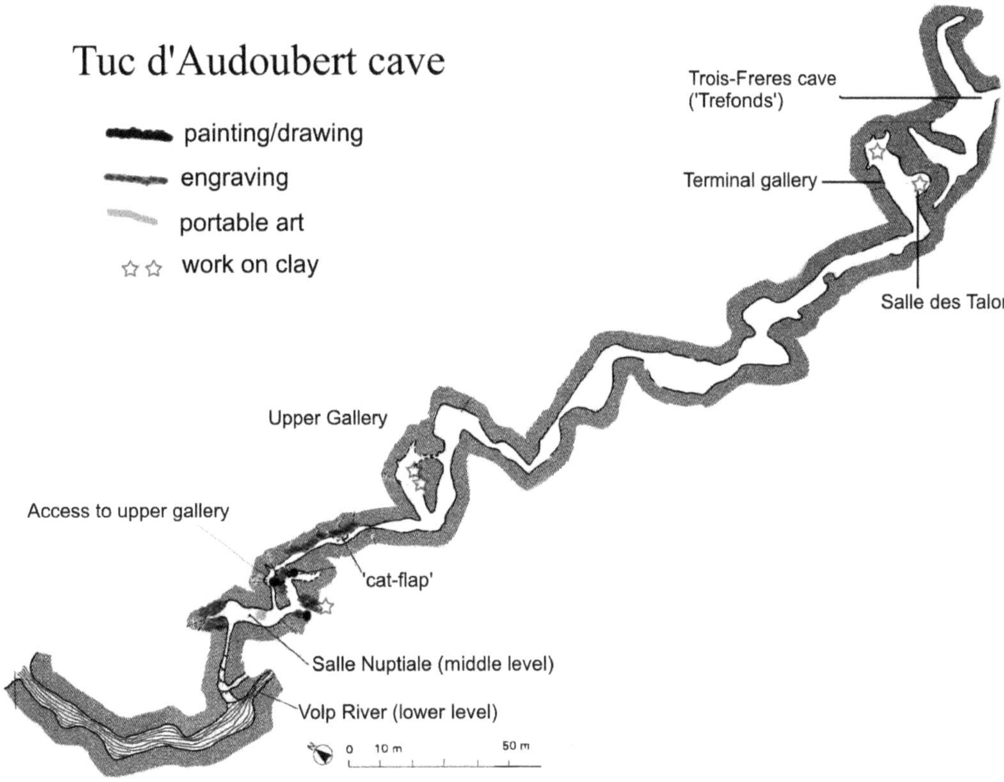

Table 2.3: Inventory of the individual recognised graphic units on clay in Tuc d'Audoubert, Ariège, with their position and the technique used. For the techniques used, a large X indicates the main technique used to make the graphic unit, * indicate that another technique is also found often to make details

Graphic unit reference	Description		Position			Technique used							
						Modelage					Impressing		
	Figurative	Non-figurative	Wall	Floor	other	Three-dimension	High relief	Low relief	Engraving	Finger-tracing	With tool	With fingers	Portable Sculpture
TA 1	Female bison				X rock	*	X		*	*			
TA 2	Male bison				X rock	*	X		*	*			
TA 3	Small bison sculpture			X		X			*				*
TA 4	Bison engraving			X				X		*			
TA 5 a		Clay 'sausage' 1		X		X							
TA 5 b		Clay 'sausage' 2		X		X							
TA 5 c		Clay 'sausage' 3		X		X							
TA 5 d		Clay 'sausage' 4		X		X							
TA 5 e		Clay 'sausage' 5		X		X							
TA 6		Oval sign		X					X				
TA 7		'Pseudo-barbed' sign		X					X				
TA 8		Printed circular sign		X								X	
TA 9		Printed linear sign 1		X								X	
TA 10		Printed linear sign 2		X								X	
TA 11		Printed linear sign 3		X								X	

Table 2.4: Position of the art of clay in Tuc d'Audoubert in relation with the cave and the rest of the art in the cave. A large X indicates the presence of the relevant category.

Graphic unit reference	Position in cave		Relation with other art techniques				Relation with other clay works	
	Passage area	Terminal part	Apart	Near paintings	Near engravings	Near portable art	Close	Isolated
TA 1		X	X				X group of 4 bison, terminal gallery	
TA 2		X	X				X group of 4 bison, terminal gallery	
TA 3		X	X				X group of 4 bison, terminal gallery	
TA 4		X	X				X group of 4 bison, terminal gallery	
TA 5 a		X	X				X in Chamber of the Heels	
TA 5 b		X	X				X in Chamber of the Heels	
TA 5 c		X	X				X in Chamber of the Heels	
TA 5 d		X	X				X in Chamber of the Heels	
TA 5 e		X	X				X in Chamber of the Heels	
TA 6		X	X				X in Chamber of the Heels	
TA 7		X	X				X in Chamber of the Heels	
TA 8		X	X				X in Chamber of the Heels	
TA 9		X	X				X in Chamber of the Heels	
TA 10		X	X				X in Chamber of the Heels	
TA 11		X	X				X in Chamber of the Heels	

Thanks to the very good conditions of preservation of the Tuc d'Audoubert cave, many traces of what can be described as human activity are visible. Some simply are the unintentional traces left by the passages of people, such as footprints and hand and fingerprints. A few hundreds of these are visible in the whole Tuc d'Audoubert cave (Bégouen and Breuil 1958: 96 and 101). Something perhaps more enigmatic is the presence of about 50 heel prints in the small Chamber of the Heels, apparently belonging to children or teenagers. These marks have initiated much discussion about the potential activities that led to their creation. Some interpreted this as resulting from some possible initiation ritual that concerned only young people (Bégouen and Breuil 1958: 101), but one cannot fully rule out a simple game carried out by children.

Other traces of human activity include marks that could have been associated with the making of art: removal of various quantities of clay (Bégouen and Breuil 1958: 97); or that could be remains of damaged works on clay: four shapeless worked lumps near the large bison *modelage*, (Bégouen *et al* 1977:116) which are possible remains of 'melted modelling', sometimes perforated (Bégouen and Clottes 1984b: 414).

The inventoried graphic units on clay in Tuc d'Audoubert are all found on the floor, except for the two large bison *modelages* that are lying against a rocky outcrop in the middle of the terminal chamber (see table 2.3). The finger-tracings are on wall and ceilings.

As for the techniques used, *modelage* appears the main technique used at the site with nine examples, then impressing with fingertips is the second most numerous with four examples and finally two examples of engraved signs. It must be noted that for the four graphic units that are figurative (all bison) more than one technique was used, especially to indicate details (see table 2.3).

The location of the clay art in Tuc d'Audoubert is also well delimited inside the cave: all the graphic units from the inventory are found in the terminal part where there is no other art technique used. They seem in relation with each other in two main groups: the four bison composition in the terminal chamber and the works in the 'Chamber of the Heels' (see table 2.4). Only finger tracings are found closer to the rest of the art in this site.

Enlène, Ariège

Enlène is the third of the Volp caves and has been excavated since the second half of the 19th century. In contrast with the other two caves of the group, it is not a site with parietal art[12] but it was a long-term habitation site with extremely rich Middle Magdalenian levels that include portable art and most notably over a thousand engraved sandstone plaquettes. Enlène communicates with Trois-Frères by a narrow corridor that marks the limit between the living site and the 'sanctuary' (figure 2.5) (Bégouen and Clottes 1984: 404).

Enlène has been dated by radiocarbon from excavated material. One date from a hearth gave 13,940 ± 250 BP (Bégouen and Clottes 1981) and confirms the Middle Magdalenian occupation especially in the deeper parts of the cave where no other culture was found (Clottes 1989: 31). Trois-Frères is also attributed to this period by association with the material found in Enlène and by stylistic attribution.

[12] Although wall surfaces covered with red ochre have recently been identified in the deeper part of the cave ('Salle du fond') (Clottes 1981b)

Figure 2.5: Enlène and Trois-Frères cave map with the location of the various art techniques found at the site. Some work on clay is indicated in Trois-Frères, but it was not studied in the present research as it is doubtful it was done in the Magdalenian period (after Bégouen and Clottes 1984a: figure 1).

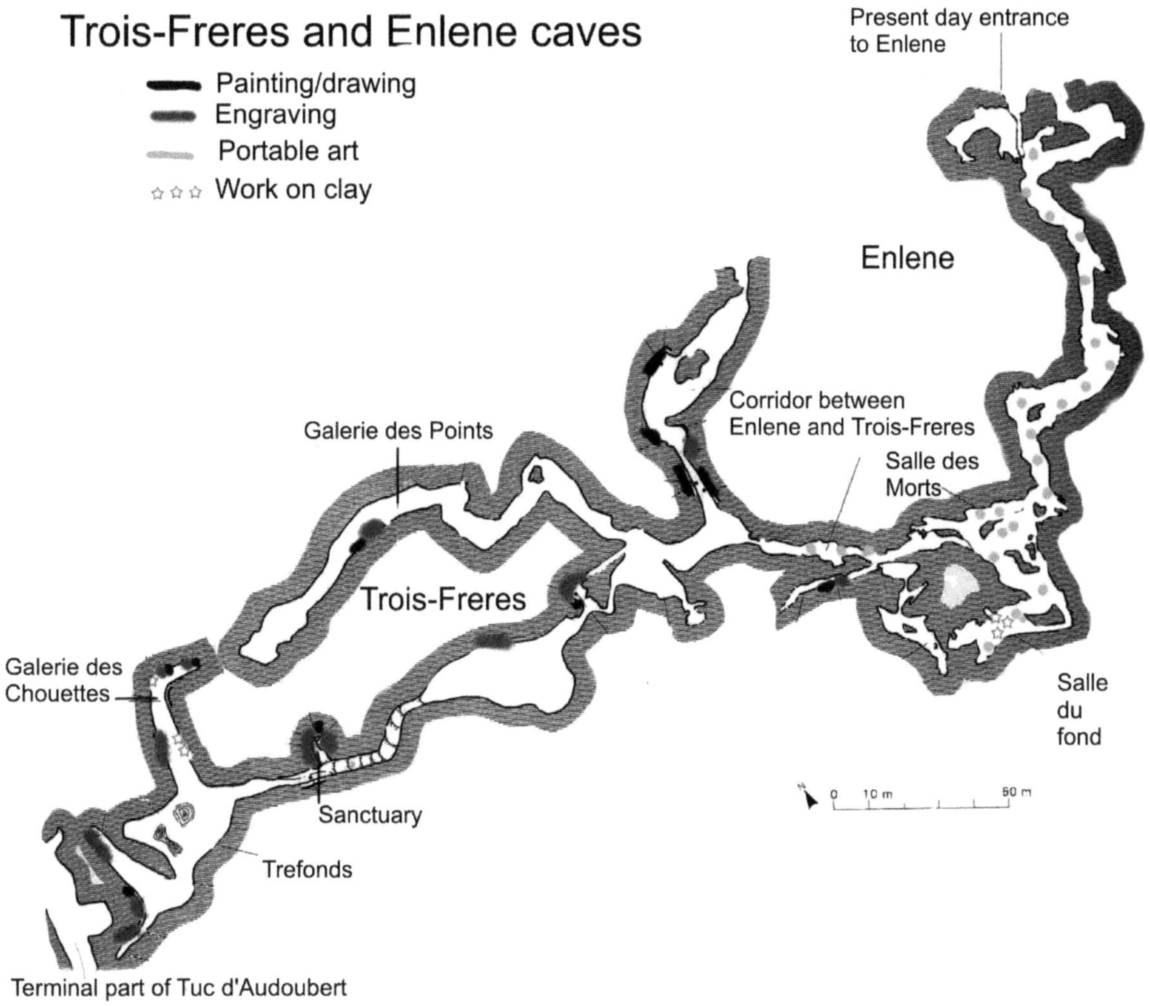

Table 2.5: Inventory of the graphic units on clay from Enlène with their position and the technique used. For the techniques used, a large X indicates the main technique used to make the graphic unit, * indicates that another technique is also found, often to make details

Object reference	Description		Position		Technique used							
					Modelage					Impressing		Sculpture
	Figurative	Non-figurative	Wall	Floor	Three-dimensions	High relief	Low relief	Engraving	Finger-tracing	With tool	With fingers	
E1	Animal body			X								X fired clay
E2	Possible animal head			X				?				X fired clay
E3		Facetted engraved bloc		X				*				X

Table 2.6: Position of the art of clay in Enlène in relation with the cave and the rest of the art in the cave. A large X indicates the presence of the relevant category. For the 'other art techniques', a large X indicates the dominant form of art, * indicates the presence of other art forms as well.

Graphic unit reference	Position in cave		Other art techniques at site			Relation with other art techniques				Relation with other clay works	
	Passage area	Terminal part	painting	Engraving	Portable art	Apart	Near paintings	Near engravings	Near portable art	Close	Isolated
E1		X	*		X				X		X
E2		X	*		X				X	?	
E3		X	*		X				X	?	

Among the Magdalenian occupation layers in the deeper part of Enlène ('Salle du Fond'), three objects made of clay have been discovered.

One of them is the headless body of a small animal (E1), measuring 2.7 cm long and 1.9cm high. The two back legs are represented by a single fashioned stump, which is quite massive. A rounded protuberance marks the tail. The line of the belly is rather straight; it seems fashioned by scraping. The two front legs are also represented as a single stump, this time shorter than the back one, quite rounded at the front and shaped as a sort of 'hook' where it meets the line of the belly. The area of the head does not seem broken off as the surfaces are rounded: whether the head was neatly cut off intentionally from the body, whether it was never fashioned.

Figure 2.6: Headless animal figurine E1 from Enlène: drawing (private collection Bégouen Family).

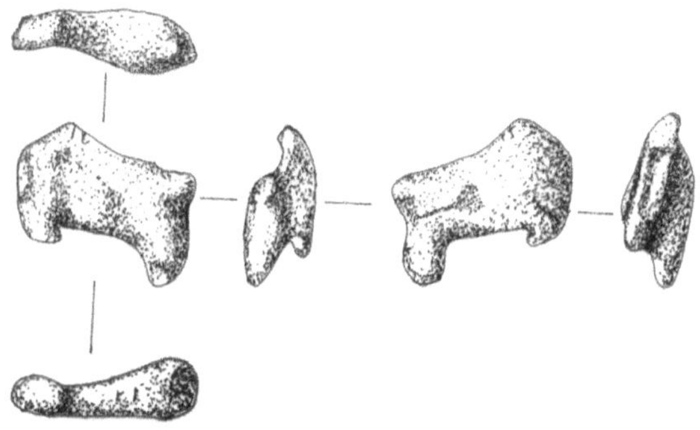

A view of the figurine from above shows that the spine is realistically curved toward one side as in the depiction of an animal in movement (figure 2.6). This impression of movement is reinforced by the 'hook' under the front legs which seems to represent the folded legs of a jumping animal. Two striations with a flint tool are visible at the neck. The general volume of the different parts of the body combined with the wide neck area, the movement of the animal and the short rounded tail allow us to propose an interpretation as a male ibex in a jumping position (figure 2.7). This figurine was published shortly after its discovery in 1936 by Bégouen and Bégouen. The material of the figurine is well fired clay that shows traces of fashioning both before and after being fired.

Figure 2.7: The fired clay figurine E1 from Enlène: a possible headless ibex figure (private collection Bégouen Family).

The two other objects have not been published. They were found by Robert Bégouen during the excavations in Enlène in the 1980s (personal communication March 2002).

The object E 2 (figure 2.8) is a fashioned piece that also appears made of fired clay, although the surfaces have suffered much from alteration. It measures 5.5 cm long, 4.7 cm wide and 3.5 cm maximal thickness and it was found in the 'Salle des Morts' area of the cave. One side of the object is better preserved than the other: this is where traces of fashioning allow interpreting it as a likely animal head. These traces are indicated on figure 2.8: they consist of a flattened forehead, a dent for the mouth, a curved area for the chin. A side forming a generally oval surface, about 5 cm by 3.5 cm, is also very roughly flattened. If we admit this is an animal head, it would be the surface of a cut-off neck.

Figure 2.8: Possible animal head E2 from Enlène, in fired clay (private collection Bégouen Family).

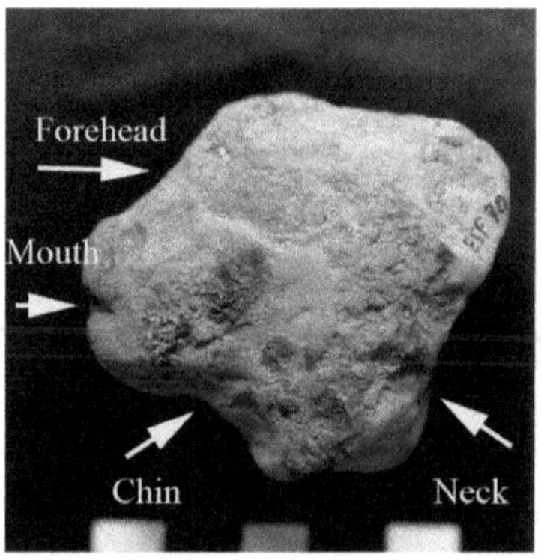

E3 is a small bloc that shows flattened facets all around, some of them being engraved (figure 2.9). It measures 4 cm by 3.8 cm by 3.6 cm. It has a base, a top face not exactly parallel to the base, with four flat faces and two small irregular ones (probably breakages) connecting them. Three of the faces bear engravings of lines that sometimes combine to form a chevron motif. The whole object is not very regular and the surfaces are very worn and fragile. It is a yellow-brown in colour, with darker patches in places. The broken base shows that the material is much harder inside the object that on the surfaces. This object does not appear to have been fired. It is possible, however, that it was exposed to fire enough to harden it slightly and to produce the breakages by bursting under the effect of the heat.

Figure 2.9: Engraved facetted bloc E3 in clay from Enlène: view of the most decorated side with drawing of the engraved chevron pattern on it (private collection Bégouen Family).

The three clay objects found in Enlène are figurative for two of them (E1 and E2), and geometrical for the other (E3). The fact that E1 and E2 are made of fired clay is remarkable as examples of fired clay are very rare in the French Pyrenees (see the site of Mas d'Azil also). However, the rarity of these objects as well as their study does not seem enough to support the idea that they were fired intentionally and that the elaborate process necessary to make fired clay objects was indeed consciously followed here. This is why they are simply considered as sculptures in the inventory for the site (Table 2.5). It is also noteworthy that they were all found in the deep part of the cave among other forms of portable art (table 2.6 and figure 2.5).

In Enlène, some other traces of clay uses have also been found. They are notably some brick-like fragments of fired clay (Bégouen 1921), and what looks like the levelling of the floor of a living area with clay (Clottes 1995).

Labouiche, Ariège

Labouiche is a cave site in the Ariège that follows an underground river with numerous side galleries branching off from the main one. Méroc (1959) excavated a part of the cave called the Dunac gallery, located above the present-day river level and found traces of a Magdalenian occupation with lithics and faunal remains, two ornaments and sandstone plaquettes, as well as some remains directly on the floor surface. The entrance to this area is situated about 200 m from the resurgence of the river, out of reach of daylight (Méroc 1959: 2). This is where, at one extremity of Dunac gallery (figure 2.9), the clay sculpture of a bison (figure 2.10) was found lying on the surface of a clay bank together with traces of clay removal and fingers.

In Méroc's opinion, the old age of these marks on clay is certified by what he calls a 'black dirt', accumulated over centuries[13], that covers everything (Méroc 1959: 19). He then associates them to the Magdalenian he found in the excavations. However it is true the cave has never been sealed and they were found on surfaces, unburied.

The discovery of a small headless bison statuette, made of dried clay, implies a more definite chronological attribution to the Upper Palaeolithic as this animal is traditionally much represented in the art of this period. It was found in 1952. It is easily recognisable by the marked hump on the back of the animal. The object measures about 12 cm long, 8 cm wide and 3 cm maximum thickness. The statuette modelled in clay was meant to be put flat on the floor or lying against a wall. Méroc says it was found at eye level of someone kneeling at the end of the terminal gallery, on the slope of a clay bank. He mentions that the 'dirt of centuries' also covered it (Méroc 1959: 20). The sculpture is very flat and could be considered as a *contour découpé.* One side of the object is more smoothed and worked than the other: it might have been made to be seen from one side only. The head has possibly been removed from the figure but the front leg simply appears not to have been fashioned as the surface where it should be on the animal is smoothed (my remark). Méroc indicates that the breakages are old as they were as dirty and blackened as the rest of the figure. The tail is represented as raised and rounded (1959: 20). Méroc indicates that one side of the bison is more 'finished' than the other as if it was meant to be looked at from this side. Also the reverse side is very dirty with various dirt incrustations in the clay as if it has been moved, dragged and rested on dirty surfaces various times in Méroc' s opinion. There are also finger prints on it: three on the abdomen, two on the hump (Méroc 1959: 22).

Four engraved parallel lines on the front of the hump are interpreted by Méroc as arrows directed towards the heart of the animal, in a form of hunting magic (Méroc 1959: 22) in the light of which he also interprets the mutilation of the head, in an interpretation commonly found around the time of his publication (Bégouen 1936; Breuil 1952; Breuil and Bégouen 1958).

[13] What Henri Breuil has been calling 'poussière des siècles' (Breuil 1952)

Figure 2.9: Labouiche cave map with the location of the various art techniques found at the site (after Méroc 1959: figure 4).

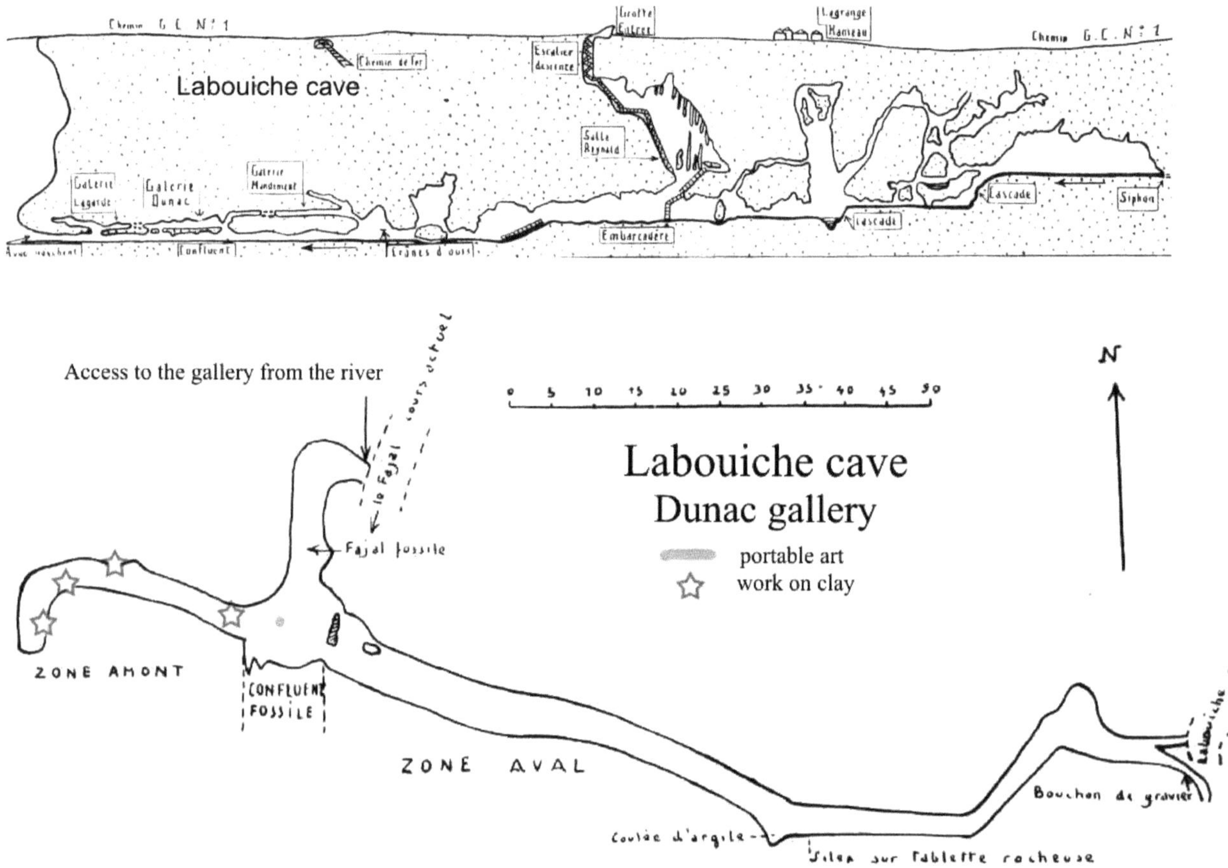

Table 2.7: Inventory of the graphic units on clay in the site of Labouiche, Ariège, with their position and the technique used. For the techniques used, a large X indicates the main technique used to make the graphic unit, a * indicate that another technique is also found often to make details

Graphic unit reference	Description		Position			Technique used							
						Modelage					Printing		
	Figurative	Non-figurative	Wall	Floor	Other	Three-dimensions	High relief	Low relief	Engraving	Finger-tracing	With tool	With fingers	Sculpture
Labouiche bison	Bison				Clay bank	X			*				

Table 2.8: Position of the art of clay in Labouiche, Ariège, in relation with the cave and the rest of the art in the cave. A large X indicates the presence of the relevant category. For the 'other art techniques', a large X indicates the dominant form of art, * indicates the presence of other art forms as well.

Graphic unit reference	Position in cave		Other art techniques at site			Relation with other art techniques				Relation with other clay works	
	Passage area	Terminal part	painting	Engraving	Portable art	Apart	Near paintings	Near engravings	Near portable art	Close	Isolated
Labouiche bison		X			*				In same end of side gallery		X

Other traces of clay uses, traces of human activity, are mentioned by Méroc, which are all situated in the terminal part of the Dunac Gallery, in the vicinity of the place where the clay bison was found. They are: a basin-like structure in clay, 1.1 m wide; a series of irregular holes dug out in the floor with the excess material accumulated nearby; clay removals; clay balls; worked lumps of clay; and finger-traces (Méroc 1959).

Labouiche is the only site together with Enlène (but see footnote 9) among the Pyrenean caves studied in the present research that is not also a parietal art site. Portable art, however, has also been found in Labouiche in the shape of engraved sandstone plaquettes (Méroc 1959).

The bison model was found at the very end of the Dunac gallery, in its terminal part, isolated from any other graphic unit on clay. Some human traces on clay were however visible in its close vicinity (see tables 2.7 and 2.8).

Figure 2.10: the headless bison sculpture in clay from Labouiche: view from each side (from Méroc 1959: figure 11)

Niaux, Ariège

The cave of Niaux is located near Tarascon-sur-Ariège, in a high valley that was not accessible to humans before about 14,000 years ago when glaciers started to retreat (Clottes 1984: 417). Niaux is a very large cave with over two kilometres of galleries in which traces of the passage of prehistoric people are found (Clottes 1984: 416). The art of the cave was officially discovered in 1906. There are no traces of habitation in Niaux, with only a few rare tools and hearths originally found alongside the galleries (Clottes 1984: 418).

Most of the art in Niaux is found concentrated in the part called the 'Salon Noir' which is well-known for the numerous black drawings on its walls. It is much less known that at the foot of the decorated walls of the 'Salon Noir' a series of engravings of signs and animals were made on the clay floor. Art is also found scattered in other parts of the cave in much smaller numbers, on the walls as well as on the clay floors for two more engravings of animals (Clottes 1984: 418). Clottes (1984: 421) notices that the animal representations are elaborate and detailed in the 'Salon Noir' but that in the rest of the cave the art is done more rapidly and many more signs are found, as if less time was invested in making the art.

The art of Niaux was attributed by style to the Middle Magdalenian by Leroi-Gourhan (his Magdalenian IV) (Clottes 1984: 421). Recent progresses in AMS dating partly confirmed this attribution, but also confirmed that the art on the walls of the 'Salon Noir' was made over a thousand year period, between about 14,000 and 13,000 BP, bringing it to the transition with the Upper Magdalenian (Clottes *et al.* 1992: 176; Clottes 1993: 173). This originally came as a surprise as the great stylistic uniformity of the Salon Noir's panels seemed to point to a short period of painting (Clottes 1995).

In Niaux, only a handful of engravings are found on the walls of the cave but a number of them were made on the clay floors. These are the only form of clay uses in this cave. 22 animals are represented engraved on the clay floor at the foot of the decorated walls of the 'Salon Noir', plus two more found in other parts of the cave. 41 abstract representations have also been counted (Vialou 1986; Clottes 1995) (see Table 2.9). The animals represented engraved on the clay floor are bison (9) (Figure 2.11), horses (4), fish (4), ibex (3), aurochs (2), one bear footprint, and one indeterminate.

There appears to be much uniformity among the works on clay found in the cave of Niaux; they are all found on floor surfaces and all are engraved with tools (Table 2.9). Nearly all except for two are located in the Salon Noir area, where the majority of the figurative parietal art is also located (figure 2.11). There seems to be a topographical separation of the use of the art techniques in this part of Niaux; painting seems reserved for the walls (where only one engraving exists), and engraving for the floors, even if the two are not necessarily found together: for example some engravings on clay are found at the foot of undecorated walls. As far as style is concerned, however, there is much uniformity between the figures made with the various techniques (Clottes 1995: 124).

Figure 2.11: Engravings on clay floor from Niaux: bison head 119 and horse 122, Salon Noir (from Clottes 1995: fig. 166)

The two floor engravings found outside the Salon Noir, a representation of an aurochs with three circles drawn on it and the well-known bison with cupules, seem to relate differently to the rest of the art of the cave as they are found isolated (Table 2.10).

It is likely many more floor engravings did exist in Niaux and were destroyed by early visitors to the cave, unaware of their presence, simply by walking on them. The remaining engravings today are all located in corners and under overhangs where it is not possible to stand (Clottes 1984: 420)

Traces of human activities left on clay were found in the form of numerous footprints and finger prints (Clottes 1984, 1995).

Table 2.9: Inventory of the recognised graphic units on clay in Niaux cave, with their position and the technique used. The reference numbers are the ones used in the catalogue in the Appendix 1.

Graphic unit reference	Description		Position:	Technique used
	Figurative	Non-figurative	on Floor	Engraving
N1	Trout		X	X
N2 (hardly visible)	Trout		X	X
N3		a dozen short parallel lines (destroyed)	X	X
N4	Fish		X	X
N5	Fish		X	X
N6		13 lines, 1 T-shaped, 1 barbed	X	X
N7 Group of signs		1 barbed line, 1 curved	X	X
N8	Horse		X	X
N9	Indeterminate animal		X	X
N10	Bison		X	X
N11	Bison		X	X
N12	Bison		X	X
N13 (destroyed)	Ibex		X	X
N14		6 series of parallel lines	X	X
N15	Bison		X	X
N16	Bison		X	X
N17	Bison		X	X
N18	Ibex		X	X
N19	Horse		X	X
N20 (destroyed)	Horse		X	X
N21	Ibex		X	X
N22	Bear print		X	X
N23	Aurochs		X	X
N24	Horse		X	X
N25	Bison		X	X
N26 aurochs with circles	Aurochs		X	X
N27 the bison with cupules	Bison		X	X

Figure 2.12: Niaux cave map with the location of the various art techniques found at the site (after Clottes 1984: Figure 2).

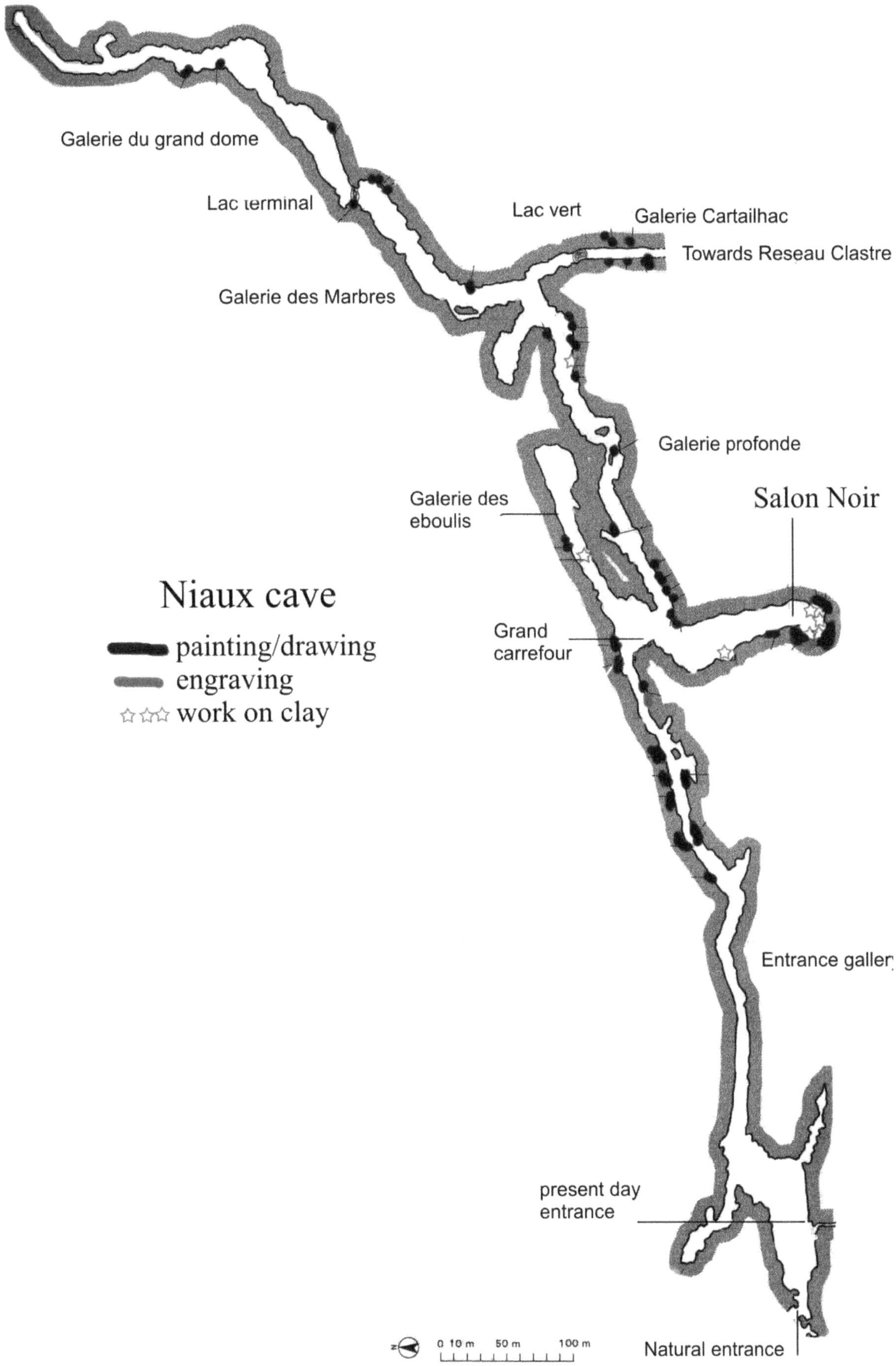

Table 2.10: Position of the art of clay in Niaux in relation with the cave and the rest of the art in the cave.

Graphic unit reference	Position in cave		Relation with other art techniques				Relation with other clay works	
	Passage areas	Terminal part	Apart	Near paintings	Near engravings	Near portable art	Close	Isolated
N1		X		X			X	
N2		X		X			X	
N3		X		X			X	
N4		X		X			X	
N5		X		X			X	
N6		X		X			X	
N7		X		X			X	
N8		X		X			X	
N9		X		X			X	
N10		X		X			X	
N11		X		X			X	
N12		X		X			X	
N13 (destroyed)		X		X			X	
N14		X		X			X	
N15		X		X			X	
N16		X		X			X	
N17		X		X			X	
N18		X		X			X	
N19		X		X			X	
N20 (destroyed)		X		X			X	
N21		X		X			X	
N22		X		X			X	
N23		X		X			X	
N24		X		X			X	
N25		X		X			X	
N26 aurochs with circles	X		X					X
N27 bison with cupules	X			X				X

Mas d'Azil, Ariège

Mas d'Azil (Ariège) is an enormous cave-tunnel, 420 m long, its southern entrance being large enough (70 m high by 80 m wide) to accommodate the Arize river and a secondary road (Alteirac and Vialou 1984: 389). Many galleries branch off from the main tunnel to form a fossil cave network. Both this network and the main tunnel have been extensively excavated over time from as early as the 1880s, uncovering a wealth of material from the Upper Palaeolithic, and notably becoming the eponymous site for the Mesolithic Azilian Culture that was first recognised there by E. Piette (Djindjian *et al.* 1999). The site has been occupied at all periods throughout Prehistory and history.

The Magdalenian levels of the cave were very rich with a large quantity of exceptional portable art objects: *contours découpés* in bone, discs, engravings, sculptures, pierced batons, many ornaments and decorated spear-throwers (Lumley 1984). A few radiocarbon dates have been obtained for the Middle Magdalenian levels, all approximately between 13,300 BP and 13,500 BP[14]. These dates are very close to the ones at Enlène, which is only a few kilometres away and shows many similarities with Mas d'Azil (Bégouen and Clottes 1983: 122).

Parietal art (paintings and engravings) was also found in some of the Galleries by H. Breuil and H. Bégouen as early as 1901-1902. It is dated to the Middle Magdalenian by comparison with the other known sites in the region and by the organisation of the themes of representation (Alteirac and Vialou 1984: 394).

As for work on clay, a small series of ten portable objects were found by Marthe and Saint-Just Péquart in their excavation of the 'Galerie des Silex' between 1937 and 1942 (Péquart 1960), among which two can be described as

[14] Radiocarbon dates for the Middle Magdalenian level of Mas d'Azil: 13,200 ±110 (Gif. 5680) 13,400 ±1000 (Gif. 5679); 13, 640± 110 (Gif. 5522) (Alteirac and Bahn 1982; Bégouen and Clottes 1983: 120)

graphic units, and the rest as resulting from some form of human activity, possibly utilitarian. One graphic unit, MA 1 (figure 2.13) has been interpreted in the past as the head of a feline by Péquart (1960) although it is rather shapeless. It appears made of fired clay but it is too worn to see any details that could inform us about the stages in its manufacture or confirm that the firing was intentional and post-fashioning. The other graphic unit was a small headless bison figure (MA 9) which unfortunately did not survive post-excavation treatment: the material bison was too soft and fragile (Péquart 1960).

All the clay objects are described as being made of dried clay, possibly mixed with fat and red ochre for some, by the Péquart (1960). However, it will be suggested in part 3 on the technology of the use of clay that a number of them are probably made of fired clay, which is another common point with the cave of Enlène. The clay objects were all found in the 'Galerie des Silex' by Péquart (1960) (figure 2.14). Half of this corridor is occupied by a rich habitation, that stops in the second half where only a very large hearth is found in a pit next to a side gallery with engravings on the walls: a possible separation between a living area and a 'sanctuary' is found there (Clottes 1989: 64).

Some of the series of clay objects found by the Péquart have been designated here as possible traces of human activity. Among them are found three beads, two large and one very small, one clay ball, a fashioned lump, a 'tattooing plaque' (Péquart 1960), and two shapeless worked lumps. They could attest of the use of clay as a material in the making of ornaments and of body decoration, if the interpretation as tattooing equipment is correct.

From table 2.12, it is possible to see that the two graphic units in clay were found among the habitation site in the first part of the 'Galerie des Silex', as well as the rest of the clay objects. These objects were also among other forms of portable art. Unfortunately it is difficult to know more about their precise provenience because of the excavation methods used at the time of recovery.

Figure 2.13: The possible feline head in fired clay from Mas d'Azil. The legend indicates the elements recognised by Péquart (1960) in their interpretation as a feline head (Péquart collection, Mas d'Azil Museum).

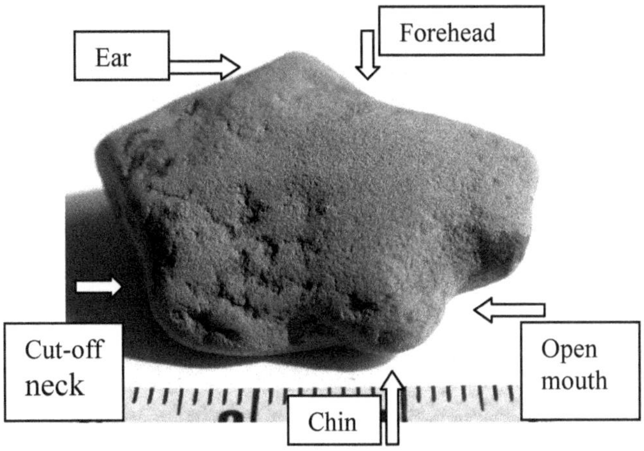

Table 2.11: Inventory of the recognised graphic units on clay in Mas d'Azil, with their position and the technique used. For the techniques used, a large X indicates the main technique used to make the graphic unit.

Graphic unit reference	Description		Position		Technique used							
					Modelage			Engraving	Finger-tracing	Impressing		Sculpture
	Figurative	Non-figurative	Wall	Floor	Three-dimensions	High relief	Low relief			With tool	With fingers	
MA 1	Possible feline head			X								X fired clay
MA 2 (destroyed)	Bison			X								X

Table 2.12: Position of the art of clay in Mas d'Azil in relation with the cave and the rest of the art in the cave. A large X indicates the presence of the relevant category. For the 'other art techniques', a large X indicates the dominant form of art, a *the presence of other art forms as well.

Graphic unit reference	Position in cave: in galerie des Silex		Relation with other art techniques				Relation with other clay works	
	Passage area	Terminal part	Apart	Near paintings	Near engravings	Near portable art	Close	Isolated
MA 1	X				*	X	X	
MA 2 (destroyed)	X				*	X	X	

Figure 2.14: Mas d'Azil cave map with the location of the various art techniques found at the site (after Alteirac and Vialou 1984: figure 8).

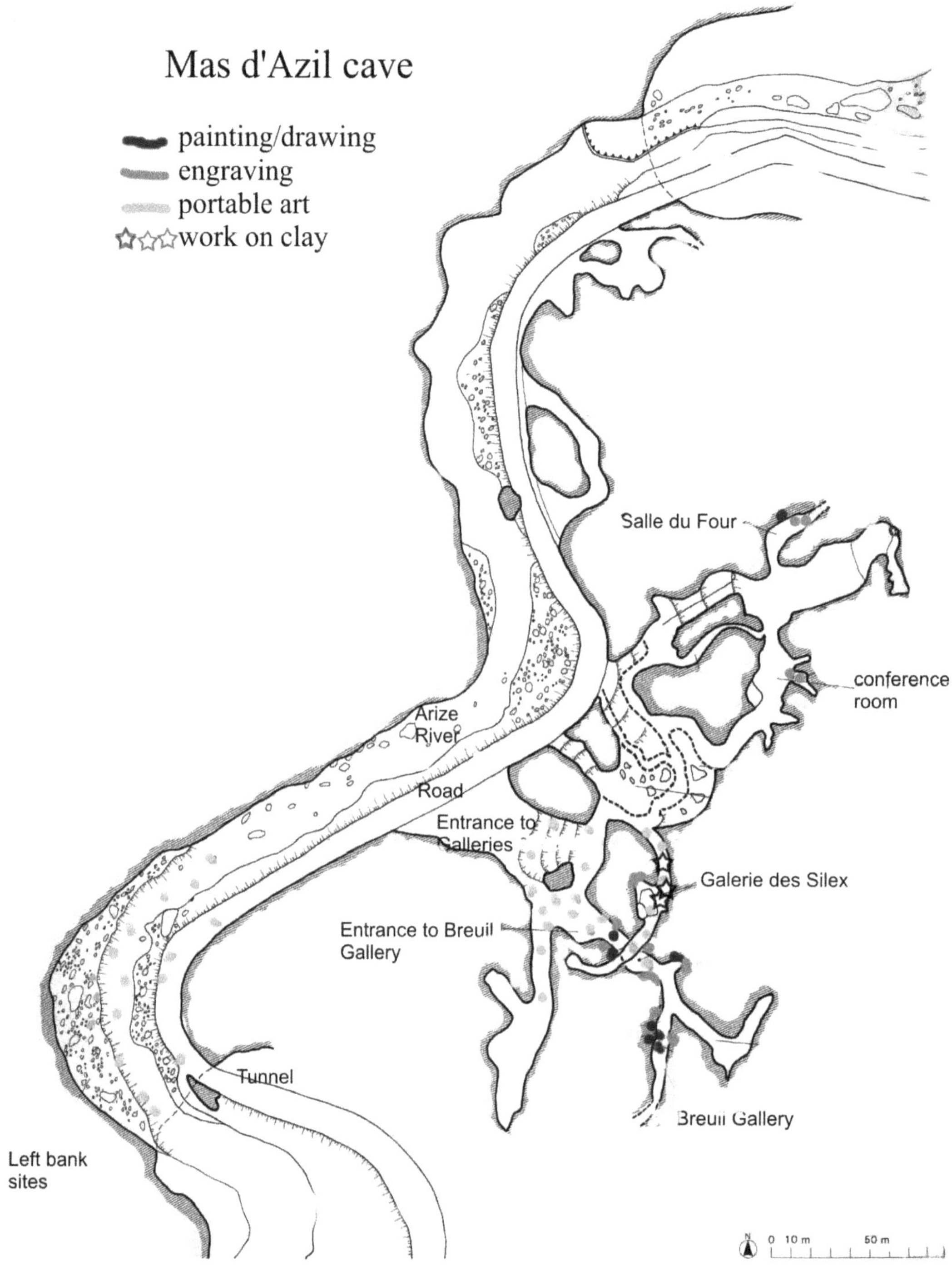

Fontanet, Ariège

The decorated cave of Fontanet is found in the vicinity of the sites of Niaux and Bédeilhac in the Ariège. It is part of a much larger (3 km long) karst network, out of which only the part called the Wahl gallery yielded prehistoric traces. It is thought that this part of the cave had a separate entrance, now collapsed, and that Magdalenian people never went further than the Wahl Gallery as there are no more traces of human presence and as access to the lower galleries is very difficult (Clottes *et al.* 1984: 434).

The Magdalenian part of the cave was discovered in 1972 and much care was taken to preserve every possible remain, especially the floors. Thus the cave presents itself as an intact chronocultural space (Vialou 1986). Traces of a Magdalenian habitat are concentrated in the first 60 m of the cave (from the original Magdalenian entrance) with three hearths. Flint was knapped at the site but hardly any tools are found there (Clottes *et al.* 1984: 434).

A radiocarbon date for one of the hearths was determined as 12,770 ± 42 BP (Ly.2184) (Clottes *et al.* 1984: 434), which indicates an Upper Magdalenian use that makes Fontanet contemporaneous with the second phase of occupation in Niaux. This seems confirmed by pigment analyses that indicate that pigment 'recipes' known elsewhere in the Upper Magdalenian were used here (Clottes 1995). Before these analyses, Fontanet was attributed to the Middle Magdalenian on stylistic criteria such as the presence of claviforms (Clottes *et al.* 1984: 437).

Parietal art was also found in Fontanet with paintings and engravings. As for the art on clay, 10 signs made of various patterns of punctuations (alignments, circle, snake-like shape, triangular) or cupules varying in sizes, printed or modelled, are found on the clay floors (Vialou 1986: 71-72). Another sign made of a group of ten vertical parallel lines is found closer of the inhabited area, this time on a wall surface (Clottes 1989: 87) (see table 2.14).

There seems to be a marked difference between the paintings and engravings constituting most of the art in Fontanet and the patterns found on the floors: those are located deeper in the cave after the main art panels. No engravings or painting were made in the parts were those are found. It is also true that the nature of the floor was not suitable for this type of patterns in the first part of the cave (Clottes *et al.* 1984: 437).

The Wahl Gallery itself is blocked off about 200 m from the collapsed entrance by a deep well presenting first a 5 m drop leading onto a platform and then a 20 m vertical drop (Clottes *et al.* 1984: 433). This is a dangerous obstacle but it did not stop the Magdalenian people as they left their marks beyond: patterns printed on the clay floors and footprints (Clottes *et al.* 1984: 434).

In Fontanet, it appears that the presence of printed signs on the floors is connected to changes in the morphology of the gallery or to narrow or dangerous passages. Also these signs are only found in the second part of the cave where there is no parietal art (figure 2.16). Decorating cave floors is a speciality of the Pyrenean caves (Vialou 1986).

Figure 2.15: printed sign 173 on the floor in Fontanet, near a shoed footprint (from Clottes 1995: figure 25).

Various traces of human activity have been found in Fontanet. In a crack on the wall further in the gallery a horizontal oval hole (3 cm by 2 cm) with a rounded bottom was dug into the clay filling, 1.8 m above ground level. Traces of clay removals can be seen, as well as small clay balls in at least two different places. Also many footprints, some showing that people slid, hand and finger prints are visible, most of them attributable to small individuals, teenagers or children. One footprint seems to be of a foot wearing a soft shoe and this is unique up to date for the Upper Palaeolithic (figure 2.15). Animal prints are also visible closer to the original entrance of the cave, now collapsed (Clottes *et al.* 1984).

Figure 2.16: Fontanet cave map with the location of the various art techniques found at the site (after Clottes *et al.* 1984: figure 1).

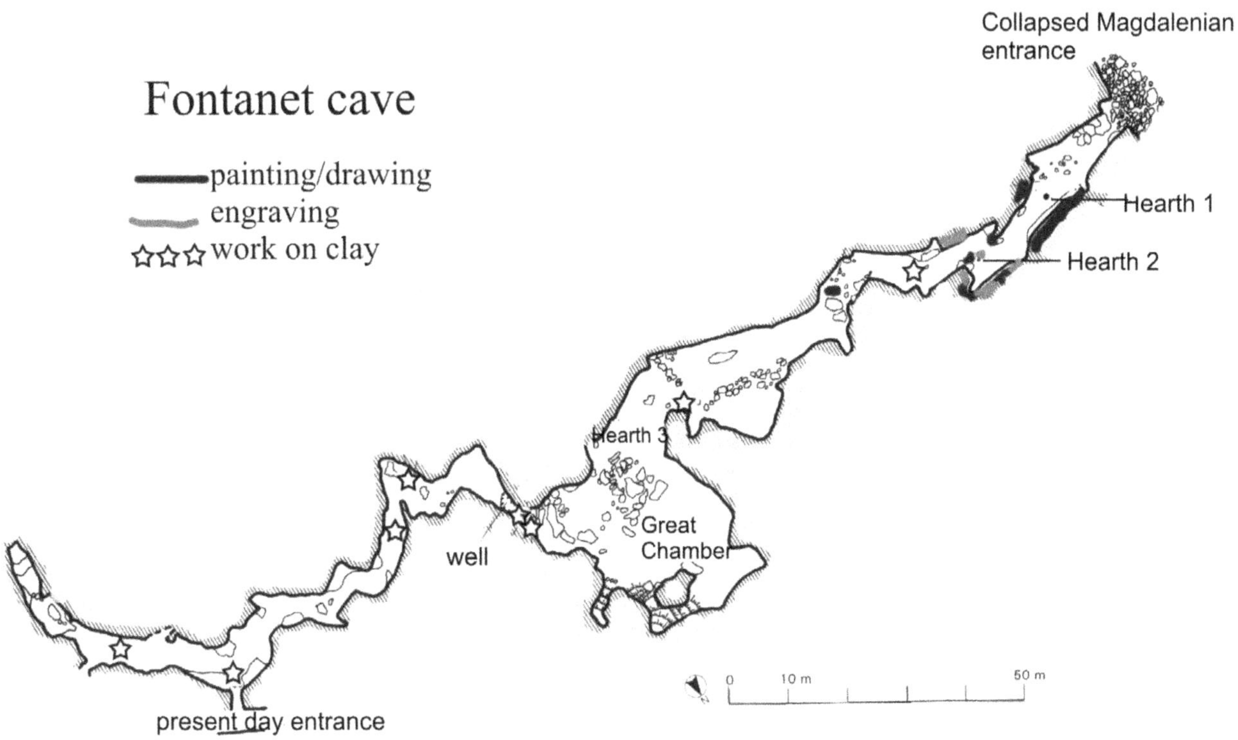

Table 2.13: Inventory of the recognised graphic units on clay in Fontanet, with their position and the technique used. For the techniques used, a large X indicates the main technique used to make the graphic unit.

Graphic unit reference	Description	Position		Techniques used							
	Non-figurative			Modelage			Engraving	Finger-tracing	Impressing		Sculpture
		Wall	Floor	Three-dimensions	High relief	Low relief			With tool	With fingers	
F1	Group of punctuations		X							X	
F2	Alignment of 12 cupules		X			X			?		
F3	Alignment of 4 cupules		X			X			?		
F4	Circular pattern		X						X	?	
F5	Alignment of 5 finger prints		X							X	
F6	pattern with 7 cupules		X			X					
F7	Rectangular group		X						X	?	
F8	Triangular pattern		X			?			X	?	
F9	Alignment of 5 cupules		X							X	
F10	Long snake-like composition		X						X	?	
F11	Vertical parallel lines	X					X				

Table 2.14: Position of the art of clay in Fontanet in relation with the cave and the rest of the art in the cave. A large X indicates the presence of the relevant category. For the 'other art techniques', a large X indicates the dominant form of art.

Graphic unit reference	Position in cave		Relation with other art techniques				Relation with other clay works	
	Passage area	Terminal part	Apart	Near paintings	Near engravings	Near portable art	Close	Isolated
F1	X				X			X
F2	X		X				Near 168	
F3	X		X				Near 167	
F4	X		X				Near 170	
F5	X		X				Near 171	
F6	X		X					X
F7	X		X					X
F8	X		X					X
F9	X		X					X
F10		X	X					X
F11	X			?	?			X

Bédeilhac, Ariège

Bédeilhac is a monumental cave found in the vicinity of Niaux and Fontanet in the Ariège. The enormous cave mouth is a marker in the landscape and has been at all times, which helps explain the very poor state of conservation of the art at the site (Vialou 1986). The cave was excavated as early as the 1860s, and many times afterwards, by different archaeologists, mostly without method or adequate recording of the excavations, and with very few publications. The resulting collections have been scattered around various institutions and private collections. All this means that this major Magdalenian site in the region is rather badly known (Vialou 1986). A monograph in preparation by Georges Sauvet (personal communication May 2004) will hopefully help remedy this.

In spite of these problems, it is known that in at least 11 locations in various places inside the cave Middle and Upper Magdalenian occupations layers were found, which were recognised by their typological characteristics. Beside a very rich assemblage of habitation debris in the first part of the cave, a characteristic of Bédeilhac is the quantity and variety of portable art objects it yielded: Georges Sauvet mentions over 900 portable art objects found (Sauvet 2006), kept in at least four different museum collections[15] and mostly unpublished. They include sculpted and engraved objects, among which a large series of worked plaquettes (the exact number is not known as they are not all studied and published, but at least 77 are described in Sauvet 2006) made of a variety of materials: soft and hard sandstone in a variety of shades (nearly white, yellow, orange, brown and red), limestone and the silt that makes the floor of the last part of the cave. This silt has been assimilated to clay here because of its similar properties (notably plasticity and capacity to harden to various stages). The use of this material to turn into decorated plaquettes is absolutely unique up to date in the art of the Upper Palaeolithic (Gailli *et al.* 1984: 370; Vialou 1986).

A key feature of the art of Bédeilhac is that nearly all the techniques used in the Franco-Cantabrian cave art, except for sculpture, are found at the site. This includes engraving on walls, monochrome and polychrome painting, often with the integration of natural reliefs, as well as a number of clay techniques: modelage, engraving on clay, drawing with clay on the walls (Gailli *et al.* 1984: 375). This statement finds an echo in some of the points mentioned by Conkey (1980: 612) as indicators of aggregation sites, notably the greater total range of activities than in other sites, and the fact that some activities are different from those that took place at other sites in the region (i.e. the making of clay plaquettes). This would indicate a special social context for the use of clay at the site of Bédeilhac, a point that will be looked at again in Chapter 4.

The works on clay found in Bédeilhac are located in two distinct parts of the cave (figure 2.18). First in the Jauze-Mandement gallery, also called the 'modelling gallery', are found a group of *modelages* in low relief on a clay bank, representing four bison, among which only one is still well preserved today (figure 2.17), and a vulva. Very close to them is an engraved sign in the shape of a double bracket (Gailli *et al.* 1984: 373; Octobon 1939; Clottes *et al.* 1994: 120). A diamond-shape sign engraved on the clay floor and an engraved animal represented with arrows on its body were published in 1939 (Octobon 1939) but have disappeared since. A small horse head described as being shaped in clay and stuck on the wall is also found (Gailli *et al.* 1984: 373). Clay also seems to have been used as a pigment for a horse figure on the walls of the Jauze-Mandement gallery (Clottes 1989: 77).

The second location where the works of clay are found in Bédeilhac is in the end gallery of the main part of the cave. There a bison, a horse head, the leg of a cervid, a possible horse and a sign have been found engraved on the surface of the clay floor in areas with low ceilings or under rocky overhangs that have prevented them being walked on. The engravings of another horse head and a possible bison head have also been published (Octobon 1939) but they cannot be found any more today.

Figure 2.17: The best preserved modelled bison *modelage* from the Jauze-Mandement Gallery in Bédeilhac (From Gailli *et al* 1984: figure 3).

It is also on the floor surfaces of the end chamber that the series of engraved and fashioned clay plaquettes (more precisely made of the silt forming the floor of the site) was found in Bédeilhac. The silt of the floor naturally fractures into plaquettes (thin flat objects with two parallel faces) (Sauvet 2006) and some of them have been worked and transformed with various techniques according to their state of dryness: modelled with fingers or engraved and fashioned with flint tools (see table 2.15) (Mons 1974; Mons and Delporte 1973). Seven of these clay plaquettes are published and have been studied (Mons 1974; Mons and Delporte 1973; Vialou 1986). At least 116 other exist in the collections of the Musée de Préhistoire Régionale de Menton which remain unpublished (Georges Sauvet, personal communication June 2004). Out of these it was possibly to carry out a short study of 51 of the pieces, the results of which is summarised in the relevant sections in Parts 2 and 3 further in this Chapter. All the studied pieces are also described in the catalogue in the Appendix 1 on the French Pyrenean clay material.

[15] In Musée des Antiquités Nationales in Saint-Germain en L'Haye, Musée de Préhistoire Régionale de Menton, Musée de Foix and Field Museum in Chicago, U.S.A.

Table 2.15: Inventory of the graphic units on clay in Bédeilhac, Ariège, with their position and the techniques used. For the techniques used, a large X indicates the main technique used to make the graphic unit, * indicates that another technique is also found often to make details. The clay plaquettes have not been included here.

Graphic unit reference	Description		Position		Technique used					
					Modelage					
	Figurative	Non-figurative	Wall	Floor	Three-dimensions	High relief	Low relief	Engraving	Finger-tracing	Sculpture
B1	Bison 1			X			X			
B2	Bison 2 (damaged)			X			X			
B3	Bison 3 (damaged)			X		X (contour découpé?)	?			
B4	Bison 4 (damaged)			X			X			
B5	Vulva			X			X	*		
B6		Double bracket sign		X				X		
B7	Horse head 1		X			X (contour découpé?)				
B8		Diamond-shape sign (destroyed)		X				X		
B9	Animal with arrows (destroyed)			X				X		
B10	Engraved bison			X				X		
B11	Horse head 2			X				X		
B12	Horse head 3 (destroyed)			X				X		
B13		Sign		X				X		
B14	Leg of cervid			X				X		
B15	Possible horse			X			?	X		
B16	Possible bison head (destroyed)			X			X			

Table 2.16: Position of the art of clay in Bédeilhac in relation with the cave and the rest of the art in the cave. A large X indicates the presence of the relevant category. For the 'other art techniques', a large X indicates the dominant form of art, * indicates the presence of other art forms as well. The clay plaquettes have not been included here.

Graphic unit reference	Position in cave		Relation with other art techniques				Relation with other clay works	
	Terminal part of Side gallery	Terminal part	Apart	Near paintings	Near engravings	Near portable art	Close	Isolated
B1	X			*	*		X	
B2	X			*	*		X	
B3	X			*	*		X	
B4	X			*	*		X	
B5	X			*	*		X	
B6	X			*	*		X	
B7	X			*	*		X	
B8	X			*	*		X	
B9	X			*	*		X	
B10		X			*	X	X	
B11		X			*	X	X	
B12		X			*	X	X	
B13		X			*	X	X	
B14		X			*	X	X	
B15		X			*	X	X	
B16		X			*	X	X	

Figure 2.18: Bédeilhac cave map with the location of the various art techniques found at the site (after Gailli *et a*l. 1984: figure 5).

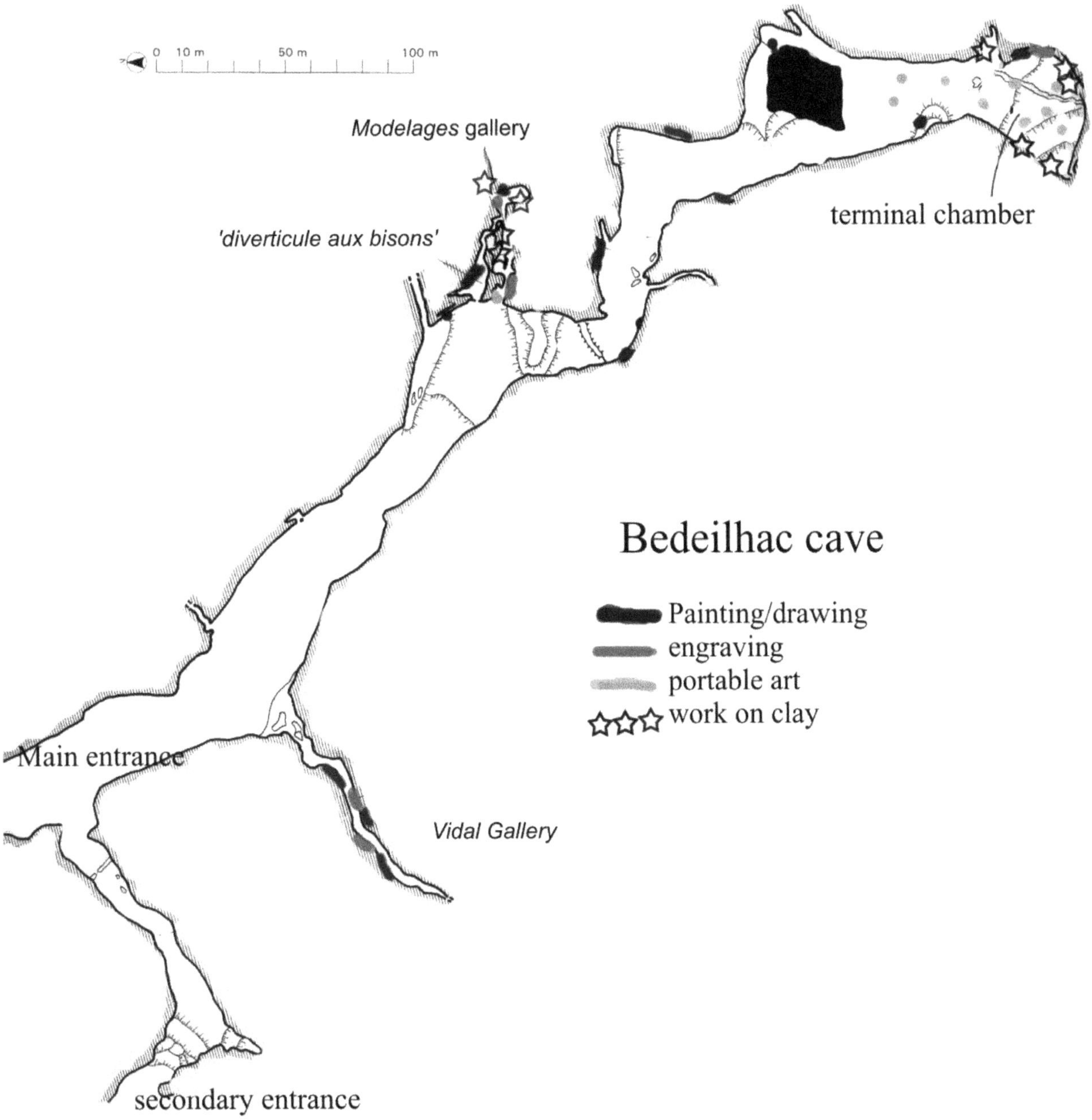

Out of the studied plaquettes, B19, B20, B21 and B25 are clearly representing bison, B17 shows a horse, B22 and aurochs and B68 a feline. Ten of the plaquettes also show marks that could be interpreted as possible abstract patterns (B24, B26, B30, B33, B35, B36, B38, B53, B58 and B60). The rest of the studied plaquettes is more difficult to interpret; caution has been exercised by keeping them as indeterminate, although if one follows Sauvet (2006) in his interpretation of the themes of the Bédeilhac sandstone plaquettes, a number of unrealistic or grotesque animal heads could be recognised as well in the clay objects, often by the simple addition of an eye or nose to their natural shape. As these interpretations remain highly subjective, it has been decided not to apply them here to the clay plaquette inventory.

The clay objects in Bédeilhac are part of a wider context of decorating plaquettes made in other materials that seem to have been brought from outside the cave. The clay objects were mostly found on the surface in the Terminal part of the cave where the floor has little changed since the Magdalenian period (figure 2.19). Floor engravings are also known in this area (Gailli *et al*. 1984: 370).

From tables 2.15 and 2.16 it can be seen that in Bédeilhac all the art on clay, with one exception is found on the floors. In the Jauze-Mandement gallery, however, the clay floor is not even, and the series of modelage is found on floor surface at roughly 45° angle with the horizontal. It can also be seen that the art on clay is not isolated from the other techniques of art found in the cave but is rather integrant part of the wider art context at the site.

Massat, Ariège

Massat cave, also called 'Ker de Massat', is a complex network of galleries and chambers situated in the isolated Massat Valley in the Ariège, 615 m in altitude. It is geographically apart from other decorated cave art sites in the region. An important habitat site was excavated at the mouth of the cave as early as the 1860s by E. Lartet but the site was never excavated with modern methods. Middle and upper Magdalenian layers seem present in the excavation, which also yielded portable art (Clottes and Gailli 1984: 397; Clottes 1989: 67).

All the cave art is found concentrated in a side gallery with low ceiling separated into two small chambers by a very narrow crawling passage (figure 2.19). This gallery was only discovered in 1953 and the art noticed in 1957. It seems the level of the floor has not changed since the Upper Palaeolithic (Clottes and Gailli 1984: 395). Engraving is the technique mostly used in the art at Massat, which shows bison, horses, mountain ibex, cervid, some possible anthropomorphs and many signs (Clottes 1989: 79).

The art on clay was only found in the second small chamber, probably as the walls are covered in clay in this part of the cave. It shows some difference in the themes represented with the art of the first chamber: there are many more signs for example, and the figures are only heads without body (Clottes and Gailli 1984: 399). The technique of engraving with tools is continued on the clay surface for a number of figures (three ibex head, two bovine heads and a fish) and signs (grid pattern, a group of 9 parallel lines and a group of 14 vertical lines), but techniques more adapted to clay are also used: finger tracing for a fish shape, a snake-like sign, a group of signs and a claviform shape; and printing with fingertips for a sign on a wall (all works described in Gailli 2004).

The figurative art has for the most part already been studied in Massat, but there remain countless scrapings, lines and traces to study, both on the rock and on the clay surfaces (Clottes and Gailli 1984: 397). Similarly, Clottes (1995) mentions some engravings on floors in Massat. It is likely much more remains to be found and published in Massat.

Other than the slight variation in the techniques used seen in table 2.17, there is much uniformity in the context of the art on clay found in Massat (table 2.18): it is found on the walls only, in the same part of the cave, in the vicinity of some engravings.

Figure 2.19: Massat cave map with the location of the various art techniques found at the site (after Clottes and Gailli 1984: figure 2)

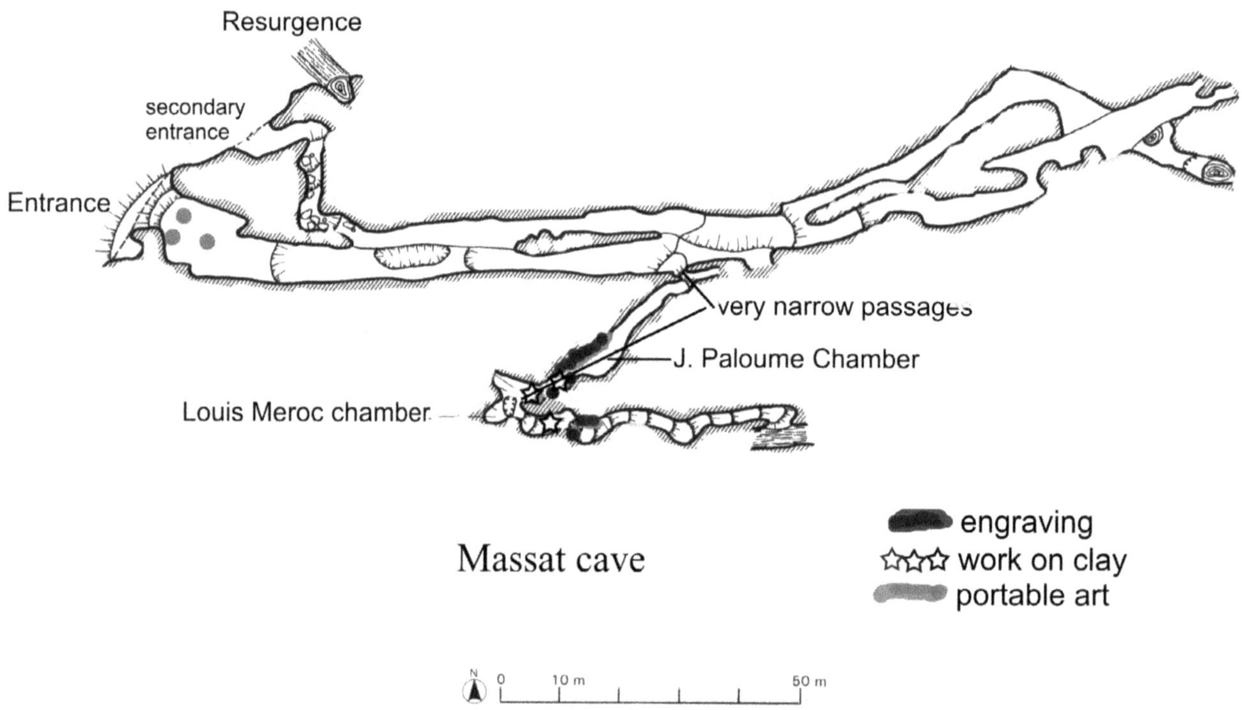

Table 2.17: Inventory of the graphic units on clay in Massat, with their position and the techniques used. For the techniques used, a large X indicates the main technique used to make the graphic unit.

Graphic unit reference	Description		Position		Technique used							
	Figurative	Non-figurative			Modelage			Engraving	Finger-tracing	Impressing		Sculpture
			Wall	Floor	Three-Dimensions	High relief	Low relief			With tool	With fingers	
M 1		Large grid pattern	X					X	X			
M 2		Snake-like sign	X						X			
M 3		Group of 9 parallel lines	X					X				
M 4	Ibex head 1		X					X				
M 5	Ibex head 2		X					X				
M 6	Ibex head 3		X					X				
M 7	Bovine head 1		X					X				
M 8	Bovine head 2		X					X				
M 9		Printed pattern	X								X	
M 10	Fish		X						X			
M 11		Group of signs	X						X			
M 12		Claviform sign	X						X			
M 13		Group of 14 vertical lines	X					X				

Table 2.18: Position of the art of clay in Massat in relation with the cave and the rest of the art in the cave. A large X indicates the presence of the relevant category.

Graphic unit reference	Position in cave		Relation with other art techniques				Relation with other clay works	
	Passage area in side gallery	Terminal part	Apart	Near paintings	Near engravings	Near portable art	Close	Isolated
M1	X				X		X	
M2	X				X		X	
M3	X				X		X	
M4	X				X		X	
M5	X				X		X	
M6	X				X		X	
M7	X				X		X	
M8	X				X		X	
M9	X				X		X	
M10	X				X		X	
M11	X				X		X	
M12	X				X		X	
M13	X				X		X	

Montespan, Haute-Garonne

The cave of Montespan (Haute-Garonne), also called Gantiès-Montespan, follows an underground river which forms siphons restricting the entrance to the galleries. The cave itself was discovered in August 1923 (Rivenq 1984: 440), and the parietal art very shortly after; the art is mostly concentrated in the Casteret-Godin side gallery (85 % of the art is found there) (Rivenq 1984: 438).

Some Palaeolithic objects were found in the Casteret-Godin Gallery (flint blades, points and scrapers) and some animal bones (horse, reindeer, and deer). Most of them were found in the clay of the walls and the floors, some intentionally stuck in cracks, as in other Pyrenean caves (Enlène or Erbérua, Larribau and Prudhomme 1989: 486). The site has been attributed to the Pyrenean Middle Magdalenian period, relying mainly on the style of the art, and also on the typology of some of the tools found (Trombe and Dubuc 1947; Rivenq 1984).

The art of Montespan is principally made of engravings and of works on clay, with painting reduced to some red marks. The themes represented are the following: 37 horses, 25 bison, 6 cervid, 5 bears, 1 bird, 3 anthropomorphs, 13 signs, 13 lines of mane or backs, 11 hind sides of animals, 5 eye-shaped marks (Rivenq 1984: 444)

Montespan is one of the main sites for the working of clay in the French Pyrénées with 22 graphic units identified as well as countless finger-tracings and traces of human activity visible in the clay (table 2.19). The art on clay is situated in two places in the cave: the Casteret-Godin gallery and the 'Hunting Frieze' gallery (Rivenq 1984: 440) (figure 2.21). In the Casteret-Godin Gallery, one finds 16 separate graphic units on clay which represent a series of arches, 3 alignments of cupules, a diamond-shape sign, two horse heads, three horses (among them two are headless), three horses or felines models, a vulva, a mammoth and the large sculpture of a headless bear. 6 of these representations are made in high relief modelling, two in three dimensional *modelage*, two in low relief modelling, three by finger-tracing and three by printing with tools. It can be pointed out that in the Casteret-Godin gallery, all the techniques used for the working of clay are fully adapted to the material: engraving is not found there on clay for example, although it is much used on the rocky surface nearby. Also in this part of the cave, works on clay are the only form of art that can be found in all the middle part of the gallery, then they mix again with engravings around the bear sculptures, and finally only engravings are found, with a few red signs, in the very last part of the gallery (table 2.22 and figure 2.21).

In the 'Hunting Frieze' gallery, only works on clay have been found. They make up a scene on the right hand side that has been interpreted at the time of its discovery as a hunting scene representing horses being driven towards fences with projectiles thrown at them. A group of vertical lines, finger-engraved, are first found, and then a detailed horse head (figure 2.20) is engraved vertically on the wall at the beginning of the scene. Two more sketchy horse silhouettes, headless, are then represented with finger-tracing, and partly obliterated by finger-traced vertical lines and many holes printed some with a pointed tool, some with fingertips (Trombe and Dubuc 1947: 51).

Figure 2.20: The vertical horse head engraved on clay at the beginning of the Hunting Frieze in Montespan (from Lorblanchet 1995: 101).

The art on clay in Montespan really shows a monumental dimension by the scale of the work and effort involved in its making as well as by the size of some of the figures. Large quantities of clay have been moved, worked, prepared to allow the making of the figures. The hunting scene, for example is drawn on a 3-metres long clay surface by about 0.3 m high which was smoothed to prepare the making of the figures (Trombe and Dubuc 1947: 51). The three horses or felines *modelages* in the Casteret-Godin gallery all measure about 1.5 m long by 0.6 m to 0.7 m high and were made in a 20 cm high relief (Trombe and Dubuc 1947: 72-74). As for the bear *modelage*, it is 1.1 m long, about 0.6 m high, and it has been estimated that about 700 kilograms of clay were transported to make it (Bahn and Vertut 1997). And in a number of other instances, clay surfaces were worked and prepared to make art (Trombe and Dubuc 1947: 95).

It must be noted that this effort in preparing surfaces is also visible on the rock surfaces in the Casteret-Godin gallery: many of the engravings can in fact be described as low-reliefs as they are very deeply engraved and shaped by scraping deeply around their contour, in a much time-consuming technique (Trombe and Dubuc 1947: 95, 119). As Rivenq (1984: 444) points out 'the support, rock or clay, did not matter much and the modelling need to be closely associated to the engravings'. This underlines the necessity to look at the use of clay in the art of the Magdalenian in the French Pyrénées in its wider art context.

There are also many non-figurative finger-tracings in Montespan, also in the Casteret-Godin Gallery in at least two locations: fairly close to the entrance of the gallery on the right wall (Trombe and Dubuc 1947: 96), and in a panel around the mammoth figure in the middle of the gallery (Trombe and Dubuc 1947: 98).

Figure 2.21: Montespan cave map with the location o f the various art techniques found at the site (after Rivenq 1984: figure 2).

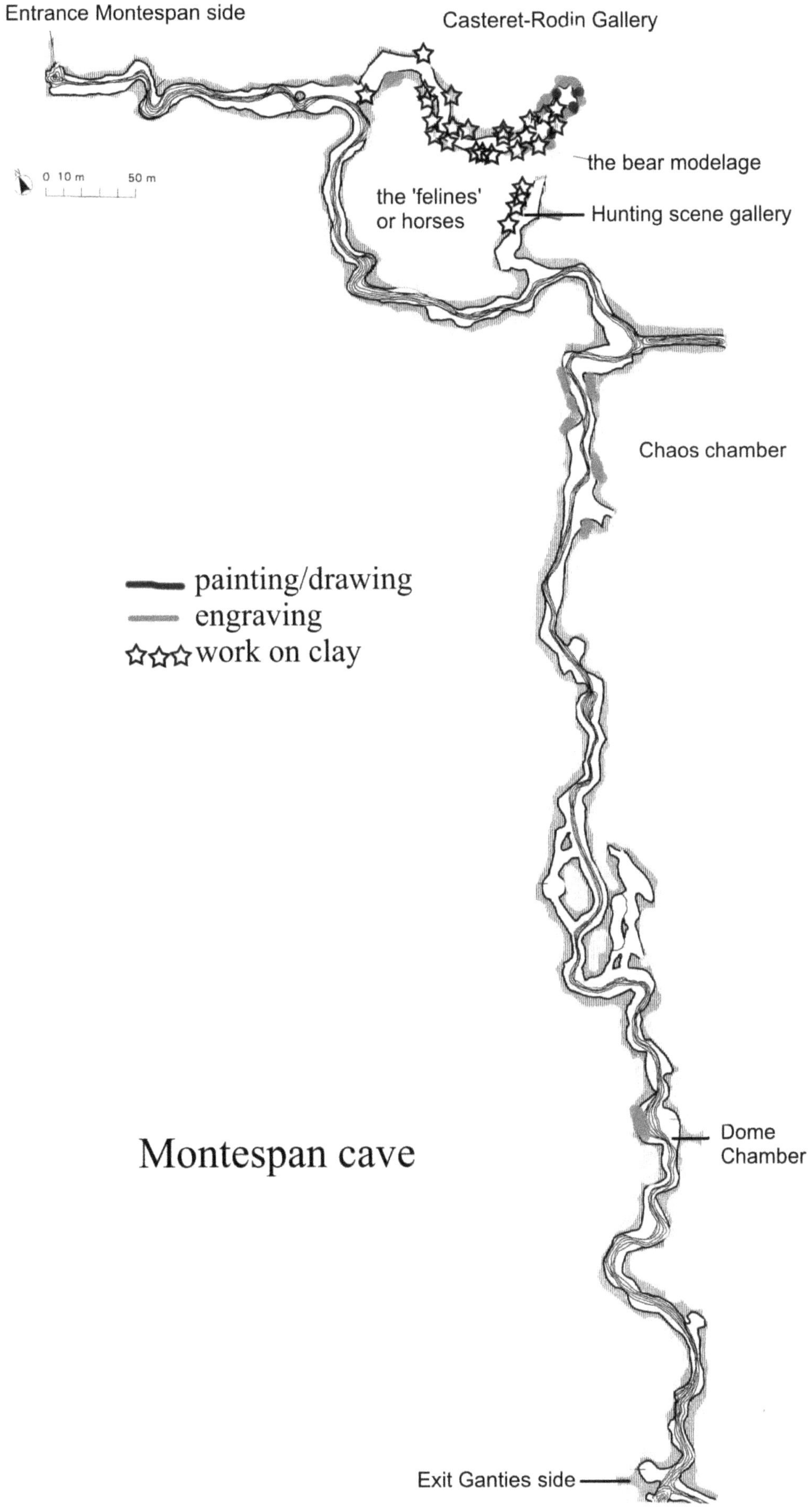

Table 2.19: Inventory of the graphic units on clay in Montespan, with their position and the techniques used. For the techniques used, a large X indicates the main technique used to make the graphic unit, * indicates that another technique is also found often to make details

Description		Position			Technique used							
Figurative	Non-figurative				Modelage					Impressing		
		Wall	Floor	Clay bank	Three-Dimensions	High elief	Low relief	Engraving	Finger-tracing	With tool	With fingers	Sculpture
	Series of arches	X							X			
	Alignment of cupules 1	X								X		
Horse head 1		X			X							
	Alignment of cupules 2	X								X		
Vulva				X		X		*				
Mammoth		X							X			
	Diamond-shape sign	X							X			
Horse/feline *modelage* 1				X		X						
Horse/feline *modelage* 2				X		X						
Horse/feline *modelage* 3				X		X						
Horse head 2				X			X	*				
	Alignment of cupules 3	X								X		
The bear			X		X	*						
Modelled horse			X				X	*				
Headless horse			X			X						
The pregnant mare			X			X		*				
	Group of vertical lines	X							X			
Vertical horse head		X						X				
	Vertical lines	X							X			
	holes	X								X	X	
Horse 2 hunting frieze		X							X			
Horse 3 hunting frieze		X							X			

Table 2.20: Position of the art of clay in Montespan in relation with the cave and the rest of the art in the cave. A large X indicates the presence of the relevant category.

Graphic unit reference	Position in cave		Relation with other art techniques				Relation with other clay works	
	Passage area	Terminal part	Apart	Near paintings	Near engravings	Near portable art	Close	Isolated
MO1	X				X			X
MO2	X				X			X
MO3	X		X				X	
MO4	X		X				X	
MO5	X		X				X	
MO6	X		X				X	
MO7	X		X				X	
MO8	X		X				X	
MO9	X		X				X	
MO10	X		X				X	
MO11	X		X				X	
MO12	X		X				X	
MO13		X		X	X		X	
MO14		X			X		X	
MO15		X			X		X	
MO16		X			X		X	
MO17		X	X				X	
MO18		X	X				X	
MO19		X	X				X	
MO20		X	X				X	
MO21		X	X				X	
MO22		X	X				X	

Traces of various human activities are extremely numerous in Montespan. In the Casteret-Godin gallery only at least 9 clay balls are mentioned (most of them destroyed today) as well as numerous traces of clay removals, holes, short lines, hand, feet and finger prints (Trombe and Dubuc 1947; Rivenq 1984). Two 'structures' are also found: a sort of large shaped basin with a rim, attached and carved into a wall surface (Bégouen and Clottes 1988: 29) and a small modelled cup attached to the wall towards the end of the gallery (Trombe and Dubuc 1947: 86) . Finally a group of 20 'molehill-like' accumulations of clay are found on the floor in the vicinity of the bear sculpture (Rivenq 1984: 442) and a second smaller group in the middle of the gallery (Trombe and Dubuc 1947: 84). They are thought to be possible remains of 'melted' modelling that deteriorated naturally in the damp conditions of the cave. Feet, hands and finger prints are also found in other part of the cave of Montespan (Trombe and Dubuc 1947: 44).

Labastide, Hautes-Pyrénées

The decorated site of Labastide is a very large cave found in the Hautes-Pyrénées region in the vicinity of Bois du Cantet, Tibiran and Gargas (Simmonet *et al* 1984: 527). An important Magdalenian habitation site was found there with much portable art. Parietal art was found deeper in the galleries in 1932. In 1945 a small side gallery was excavated in the part with cave art. It yielded important pieces of Magdalenian portable art including engraved plaquettes, an ensemble of *contours découpés* on hyoid horse bones found together, a broken foal similar to the one from the spear thrower with foal and birds from Mas d'Azil as well as ochre pigment (Simmonet *et al* 1984: 529).

The art was attributed to the Middle Magdalenian by the characteristic style of the bison legs and the sinuous M-shaped line on the bellies of horses. This appears confirmed by a radiocarbon date of 14,260 ± 440 BP on the archaeological material (Simmonet *et al* 1984: 533). This area of the cave was never fully excavated though. There are also more recent traces of occupation in the cave (Simonnet *et al.*1984).

Figure 2.22: The line of the back and the rear leg of a bison on clay from Labastide, partly engraved, partly modelled on c lay (from Clottes 1977: 12).

The parietal art is mostly constituted of engravings done with very fine lines, with some scraping for wider areas; this technique allows modelling the volume of the figures represented by playing on the colours of the wall. There are also some rare paintings, the most famous being a life-size horse in black and red. Horses are a dominant theme in the animal representations in Labastide with 23 figures represented. 11 bovines, 3 cervid, 3 caprine, 1 feline, 1 bird, some unidentifiable animals, some composite figures and 2 anthropomorphs are also found, as well as signs (Simonnet *et al.* 1984: 530).

Only one graphic unit is found on clay in Labastide: the figure of a bison (figure 2.22) partly modelled partly engraved on a boulder covered with clay (table 2.21). Only the back and the rear of the animal are clearly visible, and are sufficient to recognise the typical profile of the bison. It is found in the very last part of the cave (table 2.22 and figure 2.23), the work of art the furthest from the entrance. (Simonnet *et al.* 1984: 532).

Clottes (1989: 77) also mentions that clay is used as pigment to draw a horse figure in the cave of Labastide.

As for possible traces of human activity in this cave, they are limited to some groups of cupules marks found on the clay floor (Clottes 1989: 87) and some finger traces on a clay bank (Simmonet *et al.* 1984: 529).

Figure 2.23: Labastide cave map with the location of the various art techniques found at the site (after Simmonet *et al.* 1984: figure 1)

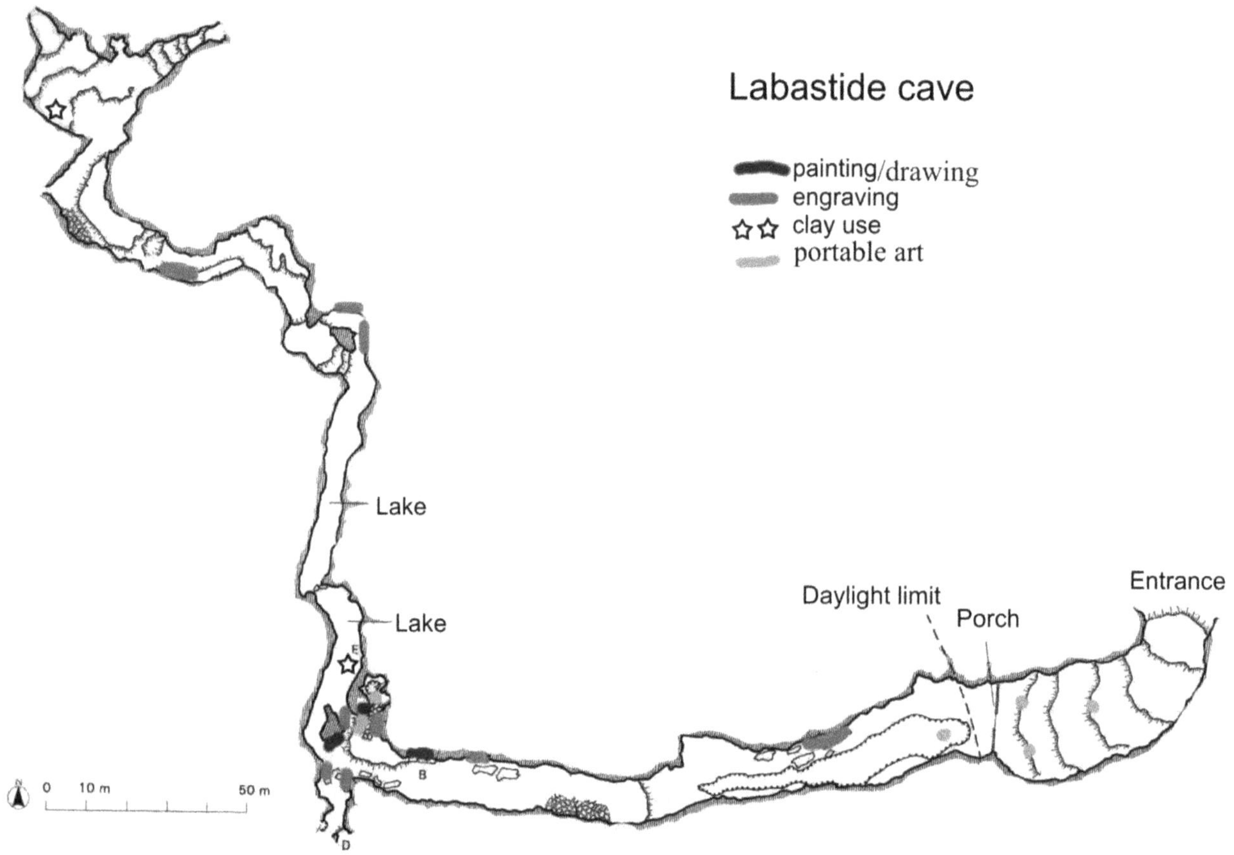

Table 2.21: Inventory of the recognised graphic units on clay in Labastide with their position and the techniques used. For the techniques used, a large X indicates the main technique used to make the graphic unit.

Graphic unit reference	Description		Position			Technique used							
						Modelage			Engraving	Finger-tracing	Impressing		Sculpture
	Figurative	Non-figurative	Wall	Floor	On rock	Three-dimensions	High relief	Low relief			With tool	With fingers	
L1	bison				X			X	X				

Table 2.22: Position of the art of clay in Labastide in relation with the cave and the rest of the art in the cave. A large X indicates the presence of the relevant category. For the 'other art techniques', a large X indicates the dominant form of art.

Graphic unit reference	Position in cave		Relationwith other art techniques				Relation with other clay works	
	Side gallery	Terminal part	Apart	Near paintings	Near engravings	Near portable art	Close	Isolated
L1		X	X					X

Bois Du Cantet, Hautes-Pyrénées

Bois du Cantet, also called Espèche, is a small decorated cave located in the Hautes-Pyrénées region, about 5 km from Labastide. A Magdalenian habitation site was found in the entrance of the cave with many faunal and lithic remains. Dates obtained on bone from the excavation correspond to the beginnings of the Upper Magdalenian, around 13,000 BP. Dates on bone found closer to the art on the floor surface are older, around 13,600 BP, still in the Middle Magdalenian period (Clot 1984: 524).

The parietal art was discovered in 1962. Bois du Cantet is a minor cave art site with only three engraved animal profiles as well as many signs and various undetermined engraved traces. Some red and black stains are visible on the walls. They indicate that paintings probably existed but have disappeared. The calcification on the walls is very thick and widespread, and could be hiding much of the original Magdalenian art (Clot 1984:526).

Two masses of clay are described as having been used by Palaeolithic people in the cave. One bears a printed pattern with 16 holes made with a conical object, possibly a spear head (table 2.23). The other shows traces of fingers (Clot 1984: 525). They are both found in the terminal part of the cave, well away from the rest of the art (table 2.24 and figure 2.24). Clottes (1989: 87) also mentions that in Bois du Cantet (also called Espèche) groups of cupules are found on the floor, but he does not give more detail.

Table 2.23: Inventory of the recognised graphic units on clay in Bois du Cantet, with their position and the techniques used. For the techniques used, X indicates the main technique used to make the graphic unit.

Graphic unit reference	Description		Position			Technique used							
						Modelage			Engraving	Finger-tracing	Impressing		Sculpture
	Figurative	Non-figurative	Wall	Floor	On clay bloc	Three-dimensions	High relief	Low relief			With tool	With fingers	
BO1		sign			X						X		

Table 2.24: Position of the art of clay in Bois du Cantet in relation with the cave and the rest of the art in the cave. X indicates the presence of the relevant category.

Graphic unit reference	Position in cave		Relation with other art techniques				Relation with other clay works	
	Side gallery	Terminal part	Apart	Near paintings	Near engravings	Near portable art	Close	Isolated
BO1		X	X					X

Figure 2.24: Bois du Cantet cave map with the location of the various art techniques found at the site (after Clot 1984: figure 1).

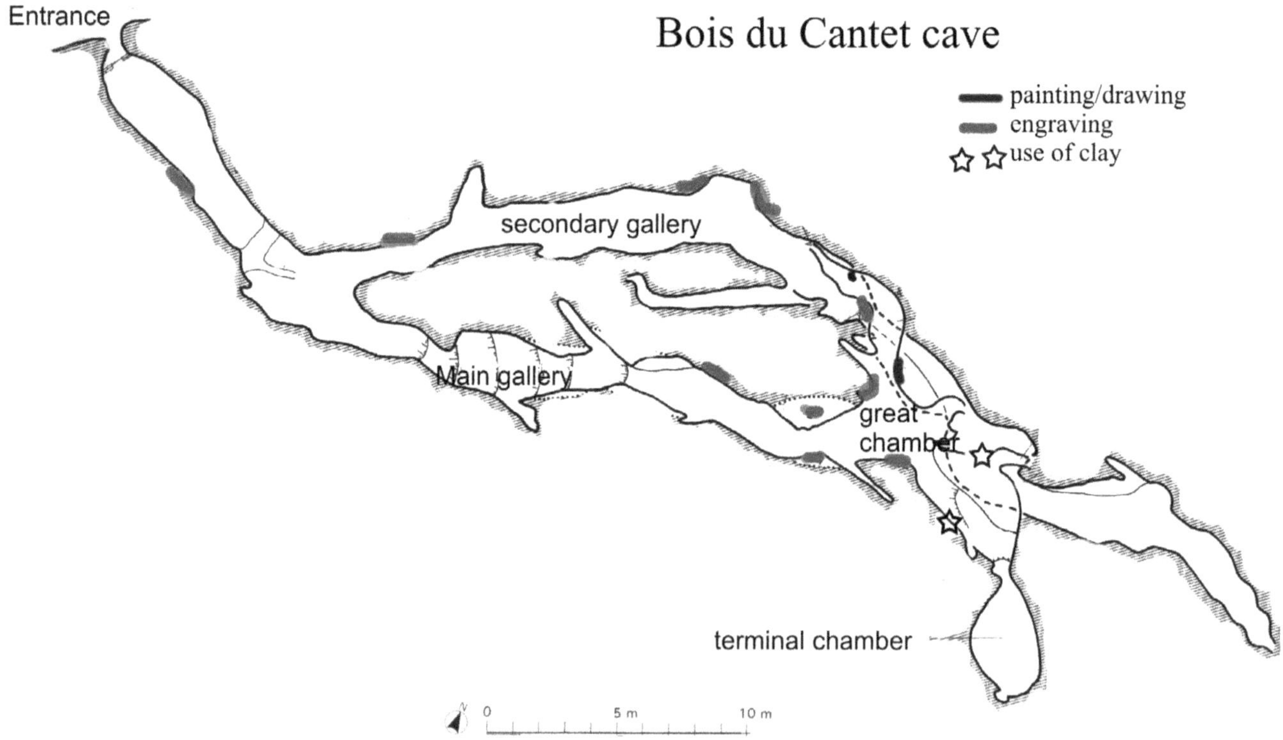

Erbérua, Pyrénées-Atlantiques

The cave of Erbérua is part of the Isturitz karst network on the western side of the French Pyrénées, the cave of Isturitz forming the upper level, the cave of Oxocelhaya-Hariztoya the middle level and Erbérua the lower and still active level. Isturitz yielded a major stratigraphy for the whole of the Upper Palaeolithic as well as some parietal art and is considered as a possible aggregation site. Oxocelhaya-Hariztoya is also a cave art site with traces of clay uses (see below). Erbérua was only discovered in 1973 as the access to the cave is extremely difficult and requires diving. Is also appears as a major site for cave art (Larribau and Prudhomme 1984:276).

The original entrance of the cave has probably collapsed and would correspond to an area where concentrations of artefacts are found on the floor indicating a habitat of which many remains (stone structures, bones, knapped flint, and hearths) are still visible. Because of the difficulty of access, there has been no excavation in this cave up to date (Larribau and Prudhomme 1989: 486). By style the art is attributed to the Magdalenian period but there is no absolute date available (Larribau and Prudhomme 1989: 492).

Erbérua is a major cave art site with at least 124 graphic units found up to now (Larribau and Prudhomme 1989: 490). The art is very rich and varied by the themes of representation as well as by the techniques used to make them. Animals are found (mostly horses, but also rare animals such as bear and feline), numerous signs, hand stencils and possible anthropomorphs. They are done by the techniques of red or black painting, drawing, engravings, scraping, finger-tracing and *modelage* of clay, techniques often combined with each other except for the art on clay. Natural rock shapes are often integrated into the figures (Larribau and Prudhomme 1984:277; Larribau and Prudhomme 1989: 486).

The art on clay in Erbérua consists of a group of 4 modelled horses and of 8 graphic units finger-traced on walls (Table 2.25). The modelled horses are much damaged from natural phenomena but their silhouettes can still be seen on the wall. They were made by accumulating clay on the wall and then shaping it to make the figures stand out in relief (Larribau and Prudhomme 1984:277 and 278). They are found in a small side gallery where two signs are also traced with fingers on clay and described as a cross-shaped sign and a series of short lines. Engravings on rock surfaces are also seen in this part of the cave. (Larribau and Prudhomme 1989: 481). The six other tracings on clay are seen in another chamber on the ceiling surface next to the passage connecting with the main gallery of the cave. They represent the head of a doe, two radiating signs, one sign with parallel lines, one sign described as a rectangle with handles and a horse figure (Larribau and Prudhomme 1989: 485).

A panel of macaroni tracing is also found in the cave in the part where the clay modelage is located. Other finger tracings are visible in another gallery (Larribau and Prudhomme 1989: 480).

The distribution of the art on clay in Erbérua seems well delimited: it is found only in one part of the cave (figure 2.25) and is associated with engraved figures only. Paintings are very numerous in Erbérua but they are all found in other parts of the cave (table 2.26 and figure 2.25). The art on clay is found in two separate concentrations. Larribau and Prudhomme (1989: 480) indicate this could be due to natural conservation conditions as this part of the cave is higher up and possibly drier than the rest of the network and is protected by very narrow access passages.

Table 2.25: Inventory of the graphic units on clay in Erbérua, with their position and the technique used. For the techniques used, a large X indicates the main technique used to make the graphic unit.

Graphic unit reference	Description		Position		Technique used							
					Modelage					Impressing		
	Figurative	Non-figurative	Wall	Ceiling	Three-dimensions	High relief	Low relief	Engraving	Finger-tracing	With tool	With fingers	Sculpture
ER1	Horse 1		X			X						
ER2	Horse 2		X			X						
ER3	Horse 3		X			X						
ER4	Horse 4		X			X						
ER5		Cross-shaped sign	X						X			
ER6		Series of short lines	X						X			
ER7	Head of a doe			X					X			
ER8		Radiating sign 1		X					X			
ER9		Radiating sign 2		X					X			
ER10		Sign with parallel lines		X					X			
ER11		Rectangle with handles sign		X					X			
ER12	Horse 5			X					X			

Table 2.26: Position of the art of clay in Erbérua in relation with the cave and the rest of the art in the cave. A large X indicates the presence of the relevant category. For the 'other art techniques', a large X indicates the dominant form of art.

Graphic unit reference	Position in cave		Relation with other art techniques				Relation with other clay works	
	Entrance Of side gallery	Terminal part of side gallery	Apart	Near paintings	Near engravings	Near portable art	Close	Isolated
ER1		X			X		X	
ER2		X			X		X	
ER3		X			X		X	
ER4		X			X		X	
ER5		X			X		X	
ER6		X			X		X	
ER7	X				X		X	
ER8	X				X		X	
ER9	X				X		X	
ER10	X				X		X	
ER11	X				X		X	
ER12	X				X		X	

Figure 2.25: Erbérua cave map with the location of the various art techniques found at the site (after Larribau and Prudhomme 1984: figure 3)

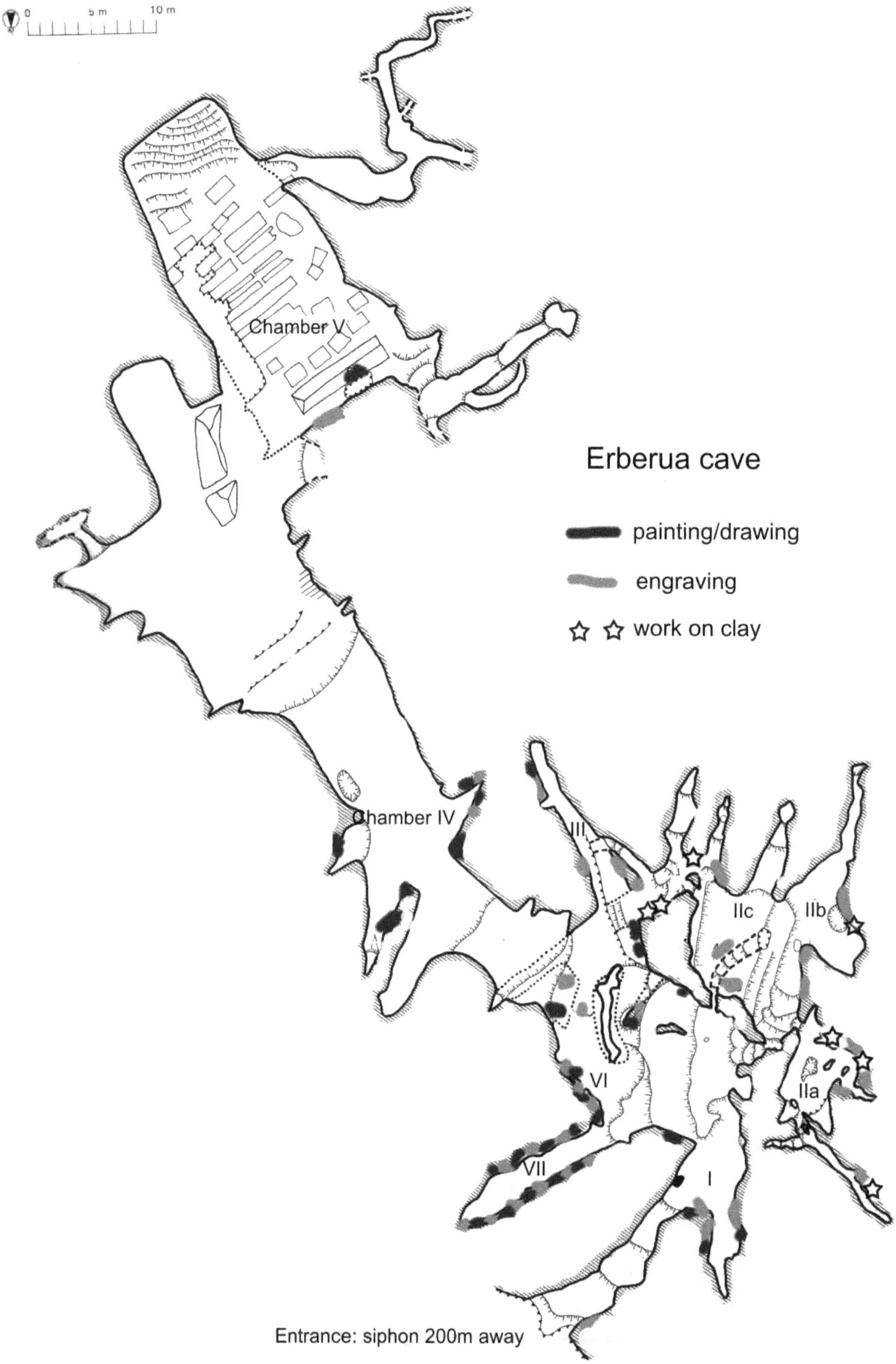

Oxocelhaya-Hariztoya, Pyrénées-Atlantiques

The cave of Oxocelhaya-Hariztoya in the western Pyrénées forms the middle part of the Isturitz karst network. The cave art was discovered in 1955, after which excavations were undertaken that yielded important archaeological layers spanning all periods between the Mousterian and the Iron Age. The art of the cave has been attributed to the Magdalenian by stylistic comparisons with other decorated caves and because some of the excavated material belongs to this culture (Laplace and Larribau 1984: 286). Another decorated gallery was discovered in 1982 with paintings and engravings (Laplace and Larribau 1984: 283).

Altogether, 38 graphic units are recognised in the cave: they are mostly horses, and also two bison, one doe, two bears or felines, two vulvas and signs (Laplace and Larribau 1984: 284). The techniques used are mostly painting and engraving.

Figure 2.26: Oxocelhaya-Hariztoya cave map with the location of the various art techniques found at the site (after Laplace and Larribau 1984: figure 1)

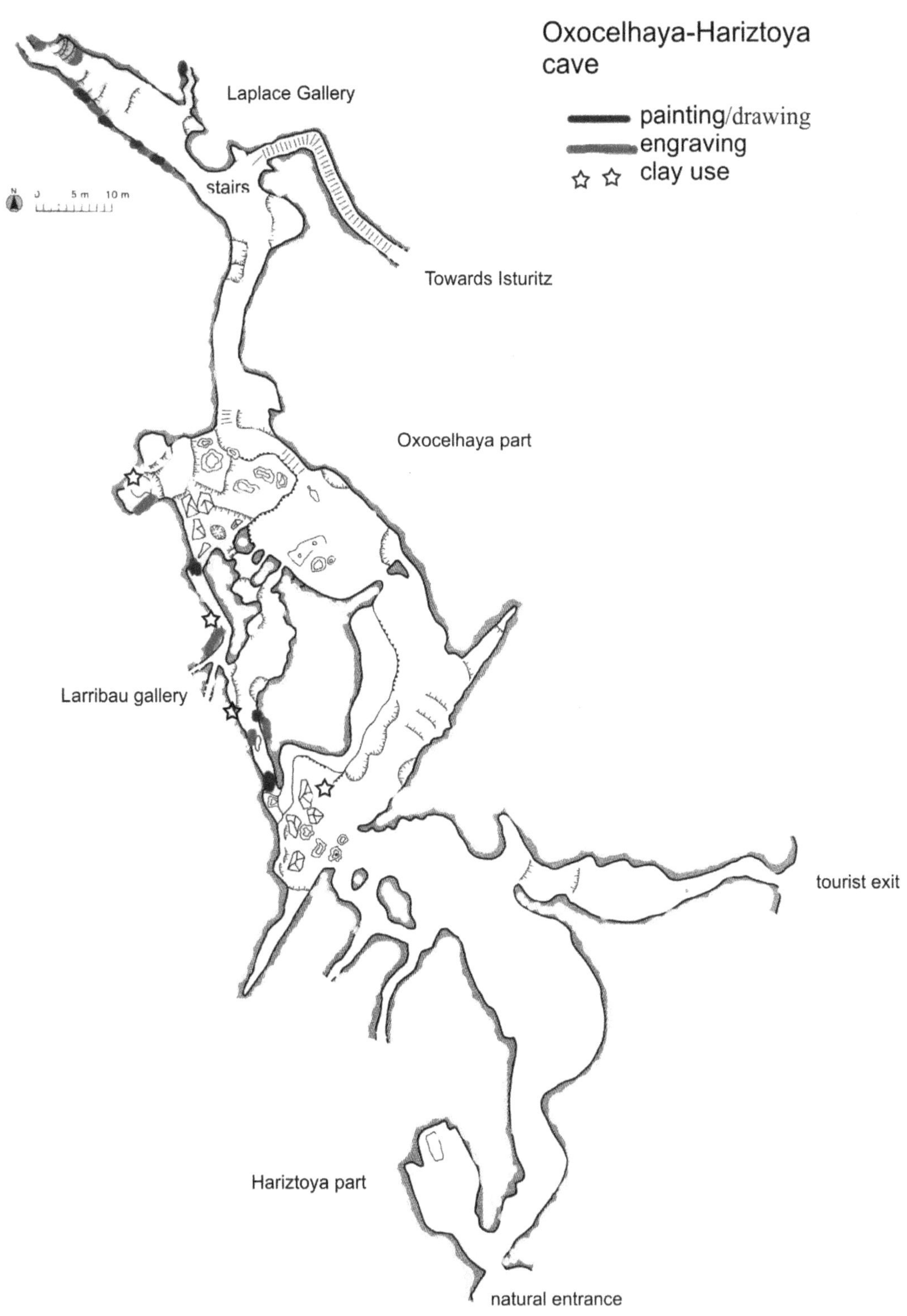

Table 2.27: Inventory of the recognised graphic units on clay in Oxocelhaya- Hariztoya, with their position and the techniques used. For the techniques used, a large X indicates the main technique used to make the graphic unit, * indicates that another technique is also found often to make details

Graphic unit reference	Description		Position			Technique used							
						Modelage					Impressing		
	Figurative	Non-figurative	Wall	Floor	On bloc	Three-dimensions	High relief	Low relief	Engraving	Finger-tracing	With tool	With fingers	Sculpture
O1		Triple line sign			X					X			
O2	Horse 1			X						X			
O3	Horse 2			X					*	X			

Table 2.28: Position of the art of clay in Oxocelhaya- Hariztoya in relation with the cave and the rest of the art in the cave. A large X indicates the presence of the relevant category. For the 'other art techniques', a large X indicates the dominant form of art.

Graphic unit reference	Position in cave		Relation with other art techniques				Relation with other clay works	
	Passage area	Terminal part	Apart	Near paintings	Near engravings	Near portable art	Close	Isolated
O1	X		X					X
O2		X			X		X	
O3		X			X		X	

Some finger tracings on clay are also found. There is a triple digital line done on the clay-covered surface of a large boulder (30 cm by 60 cm) figuring a large sign. Two horses are also represented on a layer of clay by finger tracing and incision (for the mane of the second one) (table 2.27). They seem to follow each other on a surface 1.2 m long. There are also remains of tracings in clay, with fingerprints. A large panel (1.5 m by 1.3 m) of triple and quadruple finger tracings forms meanders on a clay-covered surface. All these works are located in the Larribau Gallery of the cave (Laplace and Larribau 1984: 283) (figure 2.26).

Etcheberri, Pyrénées-Atlantiques

Etcheberri cave is situated in the western Pyrénées, in the French Basque Country, at an altitude of 440 m. The art in this cave was discovered in 1950, and studied in 1951 and 1952. More discoveries were made in 1978 and 1979, among which were horses incised on clay (Paillet 1989: 495).

The only Palaeolithic materials found in the cave are three burins recovered on the surface and a large ochre crayon. They are probably contemporaneous to the parietal art but it has not been possible to find more archaeological traces of any Upper Palaeolithic occupation. The art is thought to be Magdalenian purely on stylistic comparisons (Laplace *et al.* 1984: 271; Paillet 1989).

All the parietal art is located in the final part of the cave, accessible after crossing three underground lakes and crawling through a very narrow passage (Paillet 1989: 495). At least 68 graphic units have been found painted in red, brown or black (Paillet 1989: 501). Engravings are much rarer (Laplace *et al.* 1984: 270).

Work on clay is found in the shape of two superimposed horses both engraved with a tool and finger-traced (figure 2.27 and table 2.29). The clay surface on which they are made was previously prepared by smoothing and levelling. Their heads are turned to the right. They are very schematic. There are also three holes done on and near the animals (Laplace *et al.* 1984: 271).

The horses are found in dangerous and very difficult to access location in the cave: they are on a cornice just above the mouth of a 7 m deep abyss which explains why they were discovered much later than the rest of the art of the cave (Laplace *et al.* 1984: 269) (figure 2.27).

Clay was used as a pigment in the site, producing a brown colour. This was first recognised from the contour of a 1.6 m long brown horse figure but this colour is also found in 24 of other figures (Paillet 1989: 502). The difference in pigment colour, red, black and brown being found, has been related to different periods in the making of the art (Laplace *et al.* 1984: 271)

Figure 2.27: two superimposed horses engraved and finger-traced on clay in Etcheberri, 20 cm and 30 cm long. Left: photograph (from Boureux 2005: figure. 4); right: tracing of the figures (from Laplace *et al.* 1984: figure 6)

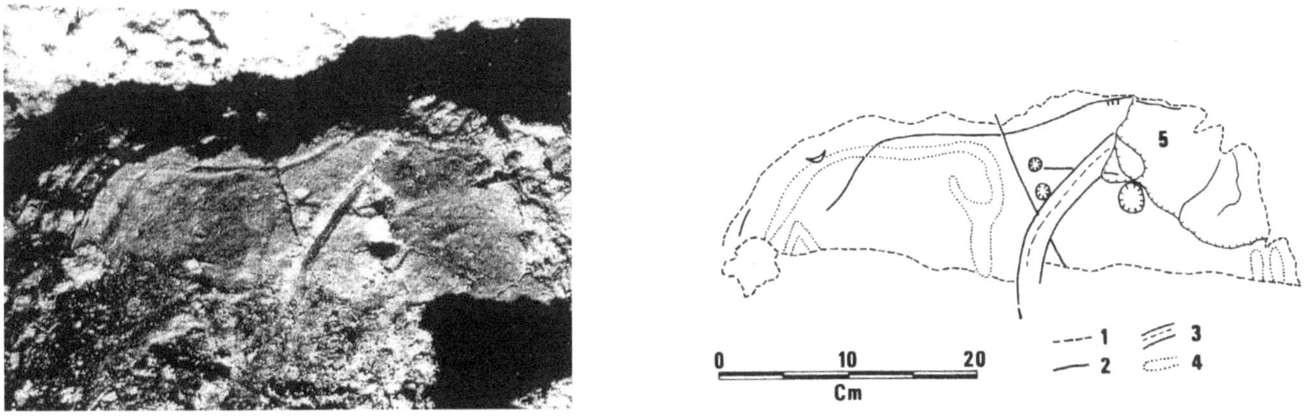

Figure 2.28: Etcheberri cave map with the location of the various art techniques found at the site (after Laplace *et al.* 1984: figure 1).

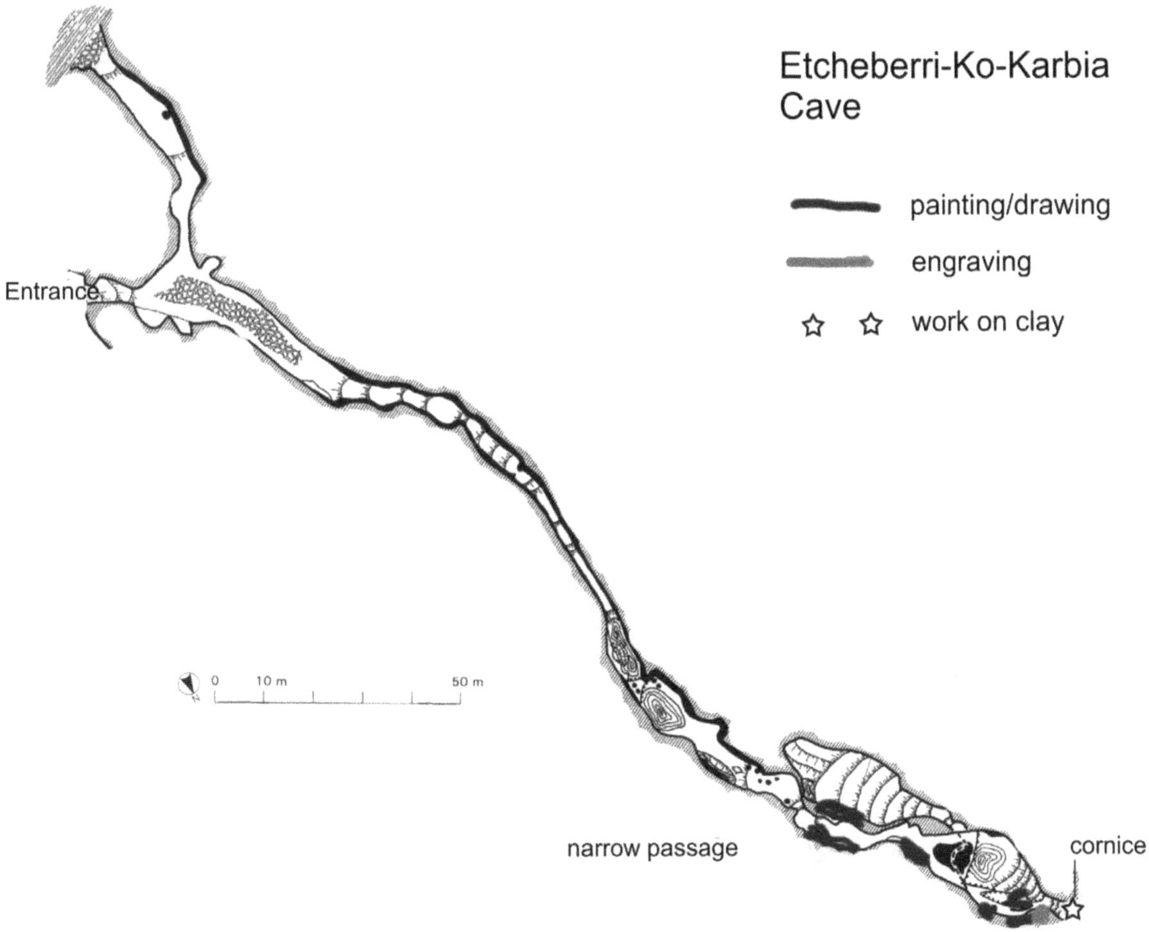

Table 2.29: Inventory of the recognised graphic units on clay in Etcheberri, with their position and the techniques used. For the techniques used, a large X indicates the main technique used to make the graphic unit.

Graphic unit reference	Description		Position	Technique used							
				Modelage			Engraving	Finger-tracing	Impressing		Sculpture
	Figurative	Non-figurative	On prepared clay mass	Three-dimensions	High relief	Low relief			With tool	With fingers	
ET1	Horse		X				X	X			
ET2	Horse		X				X	X			

Table 2.30: Position of the art of clay in Etcheberri in relation with the cave and the rest of the art in the cave. A large X indicates the presence of the relevant category. For the 'other art techniques', a large X indicates the dominant form of art.

Graphic unit reference	Position in cave		Relation with other art techniques				Relation with other clay works	
	Side gallery	Terminal part	Apart	Near paintings	Near engravings	Near portable art	Close	Isolated
ET1		X	X				X	
ET2		X	X				X	

2 The themes of representation of the art on clay in the French Pyrénées

The art found on clay in the French Pyrénées has been inventoried according to the recognisable graphic units. From this inventory, it results that 53.8 % of these graphic units are non-figurative representations or signs, 45.1 % are animal representations and 1.1 % are human (table 2.31). These themes of representation are the ones normally found in the art of the Upper Palaeolithic. The overall dominance of signs, the importance of the animal representations and the rarity of the human figures also follows the trends usually seen in the art of the period (Lorblanchet 1995) but it will be seen that a detailed analysis underlines some specific regional characteristics in the French Pyrénées.

For example, the animal theme is dominated by horse, bison and ibex representations above all. This together with the large number of signs found is perfectly representative of the general Middle Magdalenian art context to which the art on clay in the French Pyrénées belongs. There are also some marked differences within the region: in the eastern part the bison representations are the most commonly found; as one goes west, the horse become the most represented theme (Clottes 1989). The art on clay does follow these regional patterns.

The animal themes

Animal representations make up 45.1 % of the all the works on clay found in the French Pyrénées (Table 2.31). The detail of these representations is as follows in Table 2.32. On average, horse representations dominate slightly with 35.4 % of all animal representations. Bison are close behind with 31.7 % of the representations. However, these statistics hide important regional variations, which seem to follow the geography of the Pyrénées: the bison representations on clay neatly dominate in all the sites with clay uses found in the Ariège, on the East (Tuc d'Audoubert, Labouiche, Niaux, Mas d'Azil and Bédeilhac, which are listed first on Table 2.32). This is clearly in good accord with the work of Vialou on the cave art of the Ariège (1986) that identified 41.4 % of the representations as bison, 20.2 % as horse and 5.5 % as ibex representations. Horse representations on clay become dominant when looking at sites more to the West of the Pyrénées (see map on figure 2.1) such as Montespan (figure 2.29), Erbérua, Oxocelhaya-Hariztoya and Etcheberri. In all these sites, the bison is actually absent from representations made on clay and the animal theme is restricted with only horses identified in Etcheberri (2) and Oxocelhaya-Hariztoya (2), only five horses and one cervid in Erbérua and eleven horses, one bear and one mammoth in Montespan.

Table 2.31: Inventory of the themes represented in the art on clay in the French Pyrénées. For Montespan the large damaged modelage described as felines or horses in different literatures (Rivenq 1984; Trombe and Dubuc 1947) have been counted as horses here as this animal is the most commonly represented at the site. Some of the clay plaquettes from Bédeilhac which display recognisable animals or patterns interpretable as signs have been added to the inventory even if many more exist which are difficult to interprete.

Site name			**Tuc d'Audoubert**	**Enlène**	**Labouiche**	**Niaux**	**Mas d'Azil**	**Fontanet**	**Bédeilhac**	**Massat**	**Montespan**	**Labastide**	**Bois du Cantet**	**Erbérua**	**Oxocelhaya- Hariztoya**	**Etcheberri**	Total	% of total
Figurative representations	Animal	Bison	4		1	9	1		10			1					26	14.3
		Horse				4			5		11			5	2	2	29	15.9
		Aurochs				2			1	2							5	2.8
		Ibex				3				3							6	3.3
		Fish				4				1							5	2.8
		Cervid							1					1			2	1.1
		Bear				1					1						2	1.1
		Feline					1		1								2	1.1
		Mammoth									1						1	0.5
		Indetermined		2		1			1								4	2.2
	Human	Vulva							1		1						2	1.1
Non-Figurative	Engraved sign		2	1		41		1	13	1							59	32.4
	Finger-traced sign									3	4			6	1		14	7.7
	Modelled sign		5					3									8	4.4
	Printed signs	Circular	1					1									2	1.1
		Linear	3					2			3						8	4.4
		Other						4		1	1		1				7	3.8
Total			15	3	1	65	2	11	33	11	22	1	1	12	3	2	182	
% of total			8.2	1.7	0.5	35.7	1.1	6.1	18.1	6.1	12.1	0.5	0.5	6.6	1.7	1.1		100

Table 2.32: inventory of the animal themes of representation found on clay in the French Pyrénées

Site Name	Animal representations										
	Bison	Horse	Aurochs	Ibex	Fish	Cervid	Bear	Feline	Mammoth	Indeterminate	Total
Tuc d'Audoubert	4										4
Enlène										2	2
Labouiche	1										1
Niaux	9	4	2	3	4		1			1	24
Mas d'Azil	1							1			2
Fontanet											
Bédeilhac	10	5	1			1		1		1	19
Massat			2	3	1						6
Montespan		11					1		1		13
Labastide	1										1
Bois du Cantet											
Erbérua		5				1					6
Oxocelhaya- Hariztoya		2									2
Etcheberri		2									2
Total	26	29	5	6	5	2	2	2	1	4	82
% of total of animals	31.7	35.4	6.1	7.3	6.1	2.4	2.4	2.4	1.2	4.9	100

Figure 2.29: Montespan horse MO14 in bas-relief and engraving on the clay floor (Drawing Henri Breuil, published in Trombe and Dubuc 1947, figure 64)

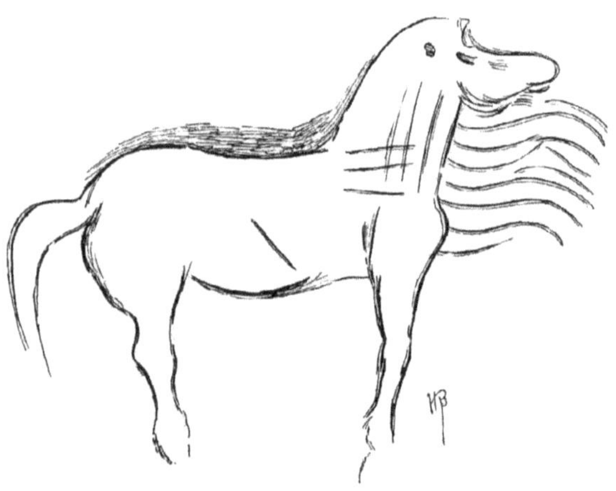

Thus it can be pointed out that the usual association of both bison and horse as a dominant theme of representation in the art of the French Pyrénées (Clottes 1989 (1999: 81); Vialou 1986; Lorblanchet 1995) does not work when only the representations on clay are considered. These two species are only found associated in two sites (Niaux and Bédeilhac) out of the 14 with clay uses in the Pyrénées. However, the art on clay is not isolated but usually found in sites with other forms of art as well (engravings, paintings, portable art) that may allow completing the association of animals. Also, Clottes 1989 (1999: 81) warns against too much generalisation as some individual sites do upset the general patterns seen.

When looking at the general statistics, the ibex theme is third most common with 7.3 % of the animal total, which again follows the general pattern seen in Magdalenian art representations (Clottes 1989 (1999: 81). This animal is found represented on clay, however, only in two sites, in Niaux and Massat with three figures each.

Other animals represented in clay are also aurochs (6.1%), fish (6.1%), cervid (2.4%), bear (2.4%), felines (2.4%), mammoth (1.2%) and undetermined animals (4.9%). Fish, bear, felines and mammoths are fairly rare themes of representation in the art of the Magdalenian. Aurochs and cervid are part of the large herbivore grouping which like the ibex can complement the main horse-bison association (Lorblanchet 1995). It should also be noted that in the sites of Fontanet and Bois du Cantet, art on clay is strictly non-figurative: these are the only sites of the inventory in which there is no figurative art on clay at all.

As for the variety of the representations, it appears to be restricted except for the two major sites of Niaux (7 animal themes) and Bédeilhac (6 animal themes). Massat and Montespan show three different animal themes each (aurochs, ibex and fish in Massat and horse, bear and mammoth in Montespan). In Mas d'Azil and Erbérua, two themes are found, and then only one in each of the other sites studied.

The themes of the animal representations found in the art on clay in the Magdalenian of the French Pyrénées allow seeing that this art is not isolated but clearly part of a much wider art context obtained with other techniques.

Looking at the relation between the animal themes found on clay and the type of clay-working technique used to make them is also very informative (Table 2.33).

Table 2.33: The correspondence between animal themes and techniques at Pyrenean sites.

Site name	Animal theme	*Modelage*	Engraving	Finger tracing	Impressing	Sculpture
Tuc d'Audoubert	Bison	4				
Enlène	Indeterminate					2
Labouiche	Bison	1				
Niaux	Bison		9			
	Horse		4			
	Aurochs		2			
	Ibex		3			
	Fish		4			
	Bear		1			
	Indeterminate		1			
Mas d'Azil	Bison	1				
	Feline					1
Bédeilhac	Bison	5	5			
	Horse	2	3			
	Aurochs		1			
	Cervid		1			
	Feline		1			
	Indeterminate		1			
Massat	Aurochs		2			
	Ibex		3			
	Fish			1		
Montespan	Horse	8	1	2		
	Bear	1				
	Mammoth			1		
Labastide	Bison	1				
Erbérua	Horse	4		1		
	Cervid			1		
Oxocelhaya	Horse			2		
Etcheberri	Horse		2			
Total		27	44	8	0	3

From this table it is possible to see that out of 29 horse representations, 14 are modelled, 10 are engraved and 5 are finger-traced. Out of 26 bison representations, 12 are modelled, 14 are engraved, none are found in finger-tracing. No other animal representation, with the exception of the large bear modelage in Montespan, is found done with the technique of *modelage*. That is to say this technique has been used only for representing the most commonly found animal species in the cave art of the French Pyrénées with the result of giving them a three-dimensional volume. The other animal figures are drawn in two dimensions with lines made with a tool or a finger. This might be seen as a way to emphasise the importance of these two animals in the belief system of the Palaeolithic people of the region (Clottes 1989(1999): 80). It is also noteworthy that the technique of impressing was only used for non-figurative representations, never for figurative.

Human representations

There are no complete representations of humans known on clay in the French Pyrénées. What is found in two cases is the sexual representation of a female by the symbol of the vulva which counts for 1.1% of the figurative representations on clay. One case is known in Bédeilhac (B5) where the vulva is modelled in a realistic manner (Figure 2.30) (Gailli *et al.* 1984: 373). The other vulva, MO5, is found in Montespan: it is more stylised and is also modelled (Trombe and Dubuc 1947: 96).

Figure 2.30: Modelled vulva from Bédeilhac (B5 in the catalogue) (from Sacco and Sauvet 1998: figure 32)

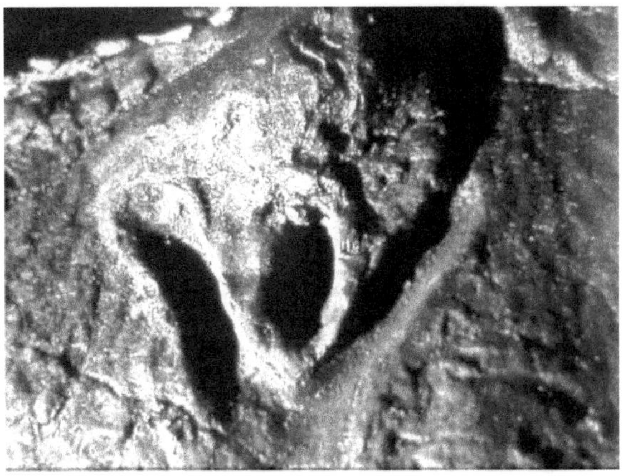

Human representations in the Magdalenian art of the French Pyrénées are varied, relatively numerous, and made with a variety of techniques. They consist of isolated heads, human silhouettes, composite part human/ part animal beings, vulvas, possible phalluses, hand stencils and hand prints (Clottes 1989 (1999: 82). The human representations are not realistic and are often reduced to a part of the body (hand or sexual organs) (Lorblanchet 1995). The two vulvae on clay are representative of this wider art context by their theme as well as by the proportions of the art on clay they represent as human representations are found consistently but in small numbers (Lorblanchet 1995: 53).

Signs

Non-figurative representations, or signs, on clay represent slightly more than half of the inventory of graphic units with 98 examples making 53.8% of the total. This proportion of signs is very similar to the one found by Vialou in his study of the decorated caves of the Ariège (Vialou 1986).

23 different types of non-figurative graphic units are found (Table 2.34) for the 61 cases for which it has been possible to find a detailed description. The remaining 37 non-figurative graphic units are mentioned as existing in Niaux by Clottes (1995), but are not described in detail in any publication that we know of. Of these 23 types, 14 are found just once, and 4 twice. Outside of simple motives like lines and parallel lines, which are the most commonly found with 19 examples, the only types of sign made on clay that seem to reoccur in a number of sites are the ones made with an association of cupules, especially linear ones like in Tuc d'Audoubert, Fontanet and Montespan and also groups of cupules.

The simple motives made with lines and dots are the most abundant types of signs usually found in cave art (Clottes 19898 (1999: 82). This appears true also of the art on clay as the cupules used for making some of the signs can be likened to the dots used in painting (as for example in a circular sign made of dots in Niaux, similar to the one in figure 2.31) for the visual effect they produce.

Table 2.34: inventory of the types of non-figurative graphic units found in the sites of the French Pyrénées

Figure description		Tuc d'Audoubert	Niaux	Fontanet	Bédeilhac	Massat	Montespan	Bois du Cantet	Erbérua	Oxocelhaya-Hariztoya	Total
'sausage'		5									5
Oval		1									1
'Pseudo-barbed'		1									1
Made with cupules	Circular	1		1							2
	Linear	3		4			3				10
	Group of punctuations			2		1	1	1			5
	Rectangular			1							1
	Triangular			1							1
	Snake-like			1							1
Group of parallel lines			8	1		2	1		2	1	15
T-shape			1								1
Barbed sign			2								2
Curved line			1								1
Double-bracket					1						1
Diamond-shape					1		1				2
Lines					2	1	1				4
Grid pattern						1					1
Snake-like						1					1
Claviforms						1					1
Arches							1				1
Cross-shaped									1		1
Radiating lines									2		2
Rectangular									1		1

Table 2.35: inventory of the non-figurative representations in relation with the clay techniques used to make them in the French Pyrénées

Site Name	Non-Figurative signs					
	Engraved	Finger-traced	Modelled	Impressed		
				Circular	Linear	Other
Tuc d'Audoubert	2		5	1	3	
Enlène	1					
Labouiche						
Niaux	41					
Mas d'Azil						
Fontanet	1		3	1	2	4
Bédeilhac	13					
Massat	1	3				1
Montespan		4			3	1
Labastide						
Bois du Cantet						1
Erbérua		6				
Oxocelhaya- Hariztoya		1				
Etcheberri						
Total	59	14	8	2	8	7
% of total of signs	60.2	14.3	8.2	2	8.2	7.1

Figure 2.31: Circular sign 169 made of a double circle of printed cupules on clay from Fontanet (from Vialou 1986: plate 13).

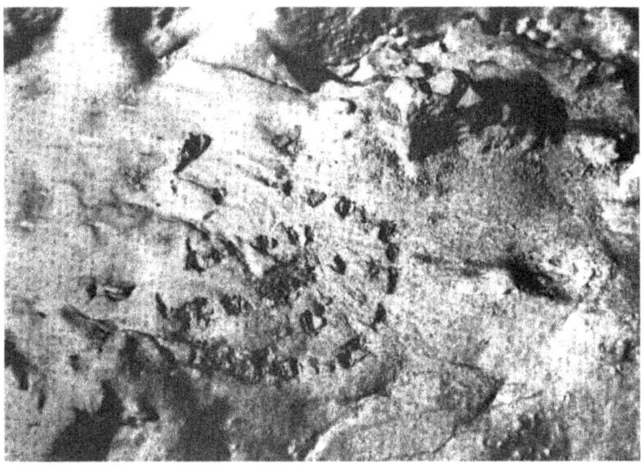

Characteristic signs of the Pyrenean Palaeolithic art are arrow-like and claviforms. Angular signs are also called barbed signs by researchers in order to disassociate them from the concept of hunting magic they have long been connected to. Claviforms signs are a regional marker in the Ariège caves (Clottes 19898 (1999: 82). Two barbed signs are found engraved on the clay floor in Niaux where they are associated with animals like in the well-known bison with cupules (N27 in the catalogue). These are the only examples inventoried on clay. As for the claviforms, only one possible example has been found in the cave of Massat (figure 2.32).

Thus it can be seen that whilst the non-figurative art on clay can be connected to the wider regional art context in some cases, for the large majority it shows original and isolated types of representations.

As for the types of clay techniques used for the working of non-figurative graphic units, they are shown in Table 2.35. Most of the signs are made of engraved representations (60.2% of the entire non-figurative inventory). Then impressed signs (circular, linear and other) follow with 17.3%, finger-traced signs with 14.3%, and modelled non-figurative graphic units with 8.2 %. It must be pointed out that the technique of modelage is used only in two sites, Tuc d'Audoubert and Fontanet for the non-figurative graphic units, each for a specific type of motif: 'sausages' in Tuc d'Audoubert and patterns of cupules in Fontanet.

Figure 2.32: A claviform-like sign (M12), drawn with a finger on clay from Massat, Ariège (from Gailli 2004, figure 41).

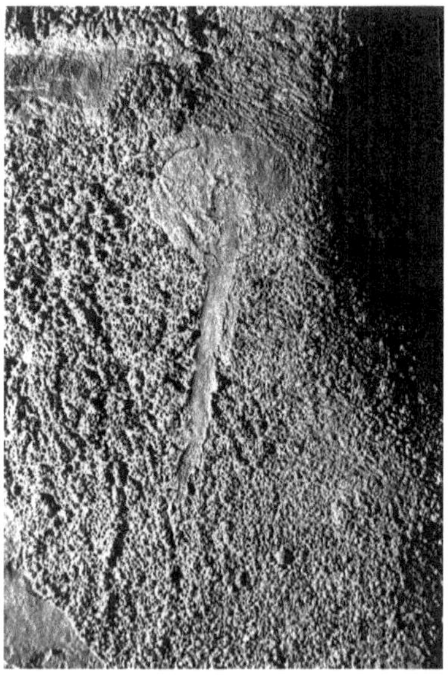

The use of the techniques in relation with the themes represented

It can be noted that the technique of clay *modelage* has been used primarily to represent figures (animals and vulva): this probably results from the need to give volume to the figurative representations, and thus a more realistic and life-like appearance. Signs, on the other hand, mostly remain two-dimensional which could confirm their abstract and highly symbolic nature.

A case of abstract clay *modelage* is found in Fontanet where 10 signs made of various patterns of cupules are present on the clay floors of the cave. The majority of the cupules are made by impressing the clay surfaces with tools or fingertips, exept for three of the signs which were modelled (Vialou 1986). This results into wider cupules, 4 to 5 cm in diameter, than the ones made by impressing.

It can also be noted than the technique of impressing was only used for non-figurative representations which seem to echo the signs made by patterns of dots found in the French Pyrénées.

The representations, realistic or abstract, done on clay by *modelage*, engraving, finger-tracings or impressing show similar patterns and conventions to those done with other techniques. They are often closely associated with parietal art and indeed sometimes appear to complement it, such as in the case of the floor engravings found at the foot of the painting on the walls of the Salon Noir in Niaux (Clottes 1989 (1999): 80). The art on clay needs to be looked at in the wider Pyrenean Magdalenian context and cannot be disassociated from it.

3 The techniques of the use of clay in the French Pyrénées

The techniques for the working of clay in the French Pyrenees will be presented here with a summary of their characteristics and a presentation of the corresponding technical scheme seen. For each technique, the actual use of the properties specific to clay will be looked at for the purposes of determining to what extent these properties were known and mastered, or to what extent techniques used on other materials were simply transferred to the use of clay.

These elements have been considered with the intention of determining possible traces of a transmission of knowledge behind the techniques applied. In the present chapter the factual elements necessary to carry out this analysis will be presented and underlined. The corresponding analysis, however, is presented further in Chapter 4.

Modelage: three-dimensional, low-relief and high-relief sculpture

The technique of clay *modelage* used in the cave sites of the French Pyrénées is a characteristic of the Magdalenian in this region. It can be seen in Table 2.36 that the technique of *modelage* is found in seven out of the fourteen recorded sites with clay use: Tuc d'Audoubert, Labouiche, Fontanet, Bédeilhac, Montespan, Labastide and Erbérua, with a total of 35 graphic units. These sites represent 12% of the total of the inventoried graphic units on clay in the French Pyrénées (Table 2.2).

Accumulations of clay gathered artificially in a number of sites (Tuc d'Audoubert, Labouiche and Montespan at least) have been indicated here as possible remains of 'melted *modelage*'. These are thought to be what is left of *modelage* that did not survive the damp conditions in cave sites and that consequently disintegrated. Some of the examples from Montespan still had some details visible on them at the time of the cave discovery which supports this interpretation (Trombe and Dubuc 1947: 84; Rivenq 1984: 442). Their interest here is to emphasise the fact that the present record of clay *modelage* for the French Pyrenees is made of very fragile works and that many more must have existed in the past.

It can be recalled here that *modelage* implies the working of plastic materials by hand to build up form (Encyclopaedia Britannica 2007). The various types of *modelage* have been separated into three-dimensional objects, high-relief and low-relief according to the degree of relief of the final representation.

The three-dimensional objects category includes *modelage* which are free-standing or nearly free-standing but which have not been made to be portable. This comprises a number of small sculptures in the round all made of dried or still soft clay. They were found in precise positions in the caves, some stuck on the wall like the horse (MO3) in Montespan (Trombe and Dubuc 1947) and the bison in Labouiche (Méroc 1959). Some are part of a larger composition such as the small bison (TA3) in Tuc d'Audoubert (Bégouen and Clottes 1977).

High-relief *modelage* consist of sculptured works in which the carving stands out very strongly from the background. Examples of this form of *modelages* are the most commonly found with 14 cases noted (Table 2.36). They occur in the caves of Tuc d'Audoubert, Bédeilhac, Montespan and Erbérua.

Low-relief *modelage* consist of sculptures in which the figure projects only slightly from the background. This technique is found in Tuc d'Audoubert, Bédeilhac, Montespan and Labastide.

Table 2.36: inventory of the various types of *modelage* techniques used for figurative and non-figurative representations in the French Pyrénées, plus the existing possible 'melted *modelage*'

Site name	Types of *modelage*						Possible Melted *modelage*	Total
	Figurative			Non-figurative				
	Three-dimensions	High-relief	Low-relief	Three-dimensions	High-relief	Low-relief		
Tuc d'Audoubert	1	2	1	5			At least 1	9
Labouiche	1						Some	1
Fontanet						3		3
Bédeilhac		2	5					7
Montespan	2	6	2				Over 20	10
Labastide			1					1
Erbérua		4						4
Total	4	14	9	5	0	3		35

Technical scheme for modelage in the French Pyrénées

The combination of the individual *chaînes opératoires* identifiable for each case of clay *modelage* results into a technical scheme that summarises the possible ways this material was worked in the Magdalenian in the French Pyrénées (Karlin and Julien 2004) (Table 2. 37). All the known examples fit into this scheme but with much variation in the exact number of steps applied as it is not necessary to follow all these steps in order to obtain a final modelage with clay. The essential steps to go through are few and have been underlined in the table. Most of the other steps are flexible and can be skipped or interchanged.

It can be emphasised that often the technique of modelage is complemented by engraving, finger-tracing or printing, mostly in order to indicate details on the figures represented.

Modelage and the properties of clay

The plasticity of clay is the property upon which the technique of *modelage* is reliant as it implies working a soft material by hand in order to shape it. Among all the 35 examples of *modelage* recorded (Table 2.36), the small bison sculpture (TA3) from Tuc d'Audoubert is the only one that shows some working of clay as a partially or fully dried material.

In a site like Montespan, the techniques applied to clay are fully adapted to the material used as they are mainly *modelage* and finger-tracings, with only one example of engraving on clay (horse head, MO18): the properties of the material are mastered and used (Table 2.21).

Also in Table 2.38, the 35 examples of clay *modelages* inventoried are categorised according to the presence or absence of transport of the clay material in the cave. In some cases, *modelages* on clay have been made in an opportunistic manner, simply where the material was available as clay is indeed a very common material naturally occurring in caves. But it is clear that in the majority of cases (22) clay has been transported. This underlines the intentionality of using this material in the making of art in the Pyrenean sites.

In Montespan, in several places in the cave, clay was even extracted and applied to the uneven walls to smooth surfaces prior to making the art. This is the case for example for the whole 'Hunting scene': the 3 metre-long vertical surface it is drawn on was carefully prepared (Trombe and Dubuc 1947: 95).

Table 2.37: Technical scheme for the making of clay *modelage* in the French Pyrénées

Phase	Sequence
Procurement of clay	Find source Extraction of the clay Testing clay for plasticity Choice of suitable final surface or support Transport to final position
Preparation of the material	Kneading/shaping Pre-forming/assembling suitable quantity of material
Preparation of final surface	Adding material Removing material Smoothing with hands or tool
Fashioning of the bulk of the object	Additive fashioning Reductive Fashioning Changing the shape of the same quantity of material
Fashioning of details	Modelage: additive (add-on ears, tails…) or reductive Engraving with tool Finger-tracing Impressing (eye)
Abandoning	Attaching to final surface by pressure, wetting… Left in chosen final position in cave

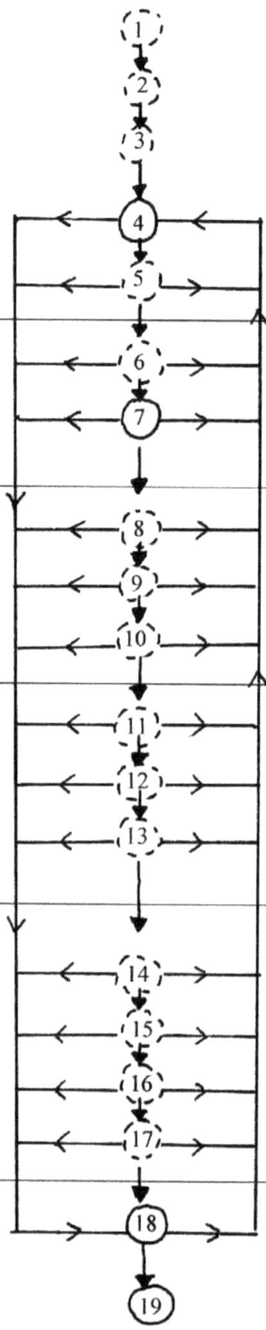

Table 2.38: presence or absence of transport of clay to make *modelage*

	Absence of transport	Presence of transport
References of clay *modelages* in the catalogue in Appendix 1	TA4 F2, F3, F6 B2, B3, B4, B16 MO11, MO14, MO15, MO16 L1	TA1, TA2, TA3, TA5a, TA5b, TA5c, TA5d, TA5e Labouiche bison B1, B5, B7 MO3, MO5, MO8, MO9, MO10, MO13 ER1, ER2, ER3, ER4
Total	13	22

Engraving on clay

Engraving consists in removing material from the worked surface by incising it with a sharp tool in order to mark it. Flint tools appear to have been used for engraving, especially burins as they are very suitable for working on hard surfaces. But any sharp tool can be used in engraving, especially if the surface to engrave is soft, as is the case for clay.

Engravings on clay surfaces are found in 7 sites out of the inventory of 14 (Table 2.39). They form a total of 88 recognised graphic units representing 30.3% of all the graphic units on clay (Table 2.2). It must be underlined however, that a single site, Niaux, completely dominates this record with 65 identified graphic units.

Another point to underline is the fact that the large majority of the known engravings on clay surfaces, 78 out of a total of 88, are found on clay floors rather than on walls. Again, the works found in the site of Niaux are responsible for this as all 65 graphic units engraved on clay at the site are found on floors.

Engravings on clay floors are regionally specific to the French Pyrénées as they are not found in any other region up to date (Vialou 1986). It is possible the clay floors used are slightly harder than some wall surface and therefore more adapted to engraving than finger-tracing, even if the resulting works remain extremely fragile.

Table 2.39: Inventory of the number and types of engravings on clay surfaces found in the sites of the French Pyrénées

Site name	Type of engraving				Total
	Figurative		Non-figurative		
	On wall	On floor	On wall	On floor	
Tuc d'Audoubert				2	2
Niaux		24		41	65
Fontanet			1		1
Bédeilhac		6		3	9
Massat	5		3		8
Montespan	1				1
Etcheberri		2			2
Total	6	32	4	46	88

Technical scheme for engraving on clay

This technical scheme shows the steps that may be followed in the most complex examples of engraving on clay such as those found in the cave of Etcheberri. However, in most of the other cases, preparation of the surface is not visible, and many of the sequences of the technical scheme can be overlooked.

Table 2.40: Technical scheme for the making of engravings on clay surfaces in the Magdalenian in the French Pyrénées.

Phase	Sequence
Procurement of material	Extraction →
	Transport to final position →
	Obtain suitable surface (artificially or naturally) →
Preparation of surface	Fashioning/shaping →
	→ Smoothing
Marking of surface	Removing material by Incision with hard tool →
	'Cleaning' of the edges of the incision free of clay smudges →
Abandoning	Left in chosen final position in cave →

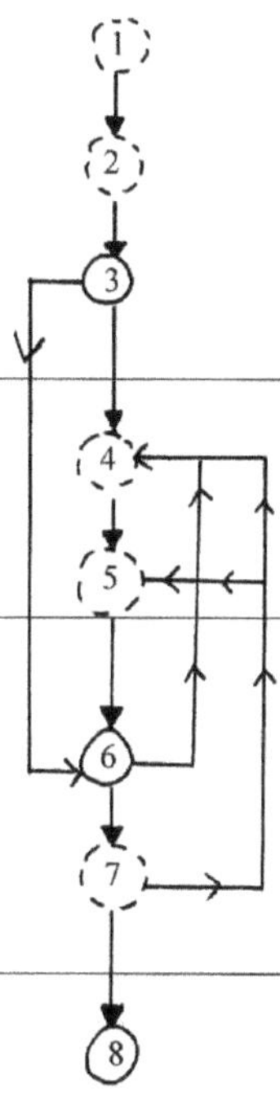

Engraving on clay and the properties of clay

The technique of engraving does not specifically use the properties of the clay material. It rather treats it like a soft stone or an organic bone or antler surface. As a technical choice it is meaningful as a soft clay surface can simply be worked with the finger but choice was made to use a tool, just like for the other hard surfaces possibly engraved at the same site. One advantage of engraving is the neatness of the obtained line which can be thin, deep and well marked.

In the site of Etcheberri, however, a mass of clay was specially gathered in order to bear the engraved figures of two horses (ET1 and ET2). In this case the final result needs to be considered as a combination of techniques as modelage was used to prepare the mass of clay before engraving it.

Finger-tracings

Finger-tracing is present in 25 sites, and is the only widespread clay technique used during the Upper Palaeolithic (see Table 1.2 in Chapter 1). It is found throughout the period in sites with cave art from eastern France to southern Spain (Baffier and Girard 1998). In the French Pyrénées, this technique is also used in the wider context of art on clay studied in the present research.

The technique of finger-tracing allows the marking of a soft clay covered surface simply by pressing on it with the fingers, without removing material in the majority of cases. Whilst figurative representations are done with this technique, most of the known examples are usually called 'macaronis' as they form non-representative unorganised intricate tracings that cover whole surfaces in an invasive manner (Lorblanchet 1999: 219).

Table 2.41: Inventory of the graphic units and 'macaronis' made by the technique of finger-tracing in the Magdalenian of the French Pyrénées. For the 'macaronis' tracings, their presence at the site was simply indicated as they are not usually possible to quantify.

Site name	Type of finger-tracing				
	Figurative		Non-figurative		Macaroni
	On wall	On floor	On wall	On floor	
Tuc d'Audoubert					Present
Massat	1		3		Present
Montespan	3		4		Present
Erbérua	2		6		Present
Oxocelhaya-Hariztoya		2		1	Present
Total	6	2	13	1	

The figurative finger-tracings in the French Pyrénées represent 7.5% of the total of the inventoried graphic units on clay in the region (Table 2.2), with 22 examples. This record is presented in Table 2.41 above and shows that only four sites in the French Pyrénées (Massat, Montespan, Erbérua, and Oxocelhaya-Hariztoya) reveal them for the Magdalenian period.

Figure 2.33: Snake-like sign M2, finger-traced, from Massat (from Gailli 2004, figure 32)

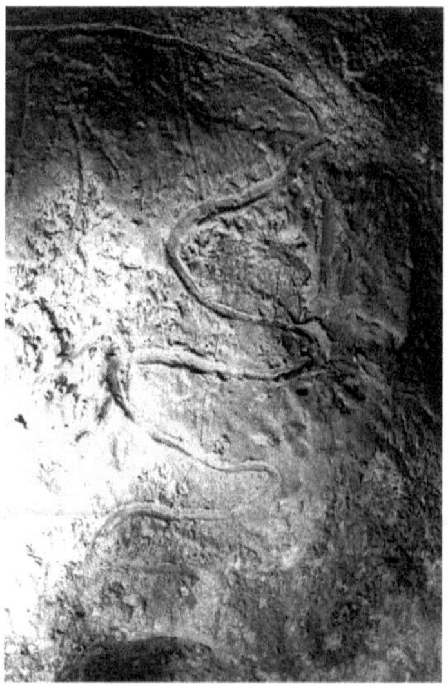

It can be noted that finger tracings are mostly found on wall surfaces (19 cases out of 22) and are mostly non-figurative (14 cases out of 22). These signs can be described as snake-like (Figure 2.33), claviform, arches, diamond-shaped, cross-shaped, radiating, rectangular or as simple lines, often parallel and in group. As for the figurative representations they show horses (in Montespan, Erbérua and Oxocelhaya-Hariztoya) (Trombe and Dubuc 1947: 96; Larribau and Prudhomme 1989: 485; Laplace and Larribau 1984: 285), a mammoth (in Montespan) (Trombe and Dubuc 1947: 96), a fish (in Massat) (Gailli 2004: 72), and a cervid (in Erbérua) (Larribau and Prudhomme 1989: 485). It can be noted that, outside of Tuc d'Audoubert, all the sites have both figurative and non-figurative finger-tracings.

Besides the figurative finger-tracings, some cave sites also have whole surfaces covered in macaroni finger-tracings. These are Tuc d'Audoubert (Bégouen and Clottes 1984b), Massat (Gailli 2004: 60), Montespan (Trombe and Dubuc 1947: 96), Erbérua (Larribau and Prudhomme 1984:277; 1989: 480) and Oxocelhaya-Hariztoya (Laplace and Larribau 1984: 285). It can be noted that, except for Tuc d'Audoubert, these are all site where figurative finger-tracings are found. As they cannot be exactly quantified, they have been recorded simply by their presence or absence in the sites. These macaronis can be made by one to four fingers held together in order to mark the wall surface.

Technical scheme for finger-tracing

Table 2.42 describes the phases and sequences necessary in the working of the clay material to make figurative finger-tracing in the most complex cases encountered among all the studied examples: in Montespan (Figure 2.34) (Trombe and Dubuc 1947) and Oxocelhaya-Hariztoya (Laplace and Larribau 1984). In these two sites, some clay surfaces have been artificially prepared before being finger-traced and the steps of the surface preparation need to be included in the making. In the majority of cases, however, like in three other cases of finger-tracing from Montespan, in Massat and Erbérua a naturally available surface was simply used and the sequence for obtaining finger-tracing becomes reduced to its simplest steps (1, 4 and 7 underlined in Table 2.42).

The making of 'macaronis' finger-tracing also follows a very simple technical process (Table 2.43); the only difference with the one for the figurative tracings is that more than one finger is used to draw parallel lines at the same time. 'Macaronis' tracings are also never found on artificially prepared surfaces: they take advantage of a natural phenomenon that sometimes allows cave walls to be covered by a thin layer of clay or to become soft themselves as the limestone transforms into mondmilch. There is no need for tools as one's hand can be used directly used on the soft surface. Some variation is visible in the number of fingers of the same hand used at the same time to mark the surface, from one to four fingers leaving four parallel lines.

Figure 2.34: Second horse MO 22, traced with fingers, facing right, from the 'Hunting Frieze' in Montespan: also traced with finger are the vertical lines in front of the horse; the holes are impressed; the clay surface of the wall has been artificially prepared and smoothed previously (from Lorblanchet 1995: 111).

Table 2.42: Technical scheme for the making of figurative finger-tracings on clay in the French Pyrenees

Phase	Sequence	
Procurement of material	Extraction →	1
	Transport to final position →	2
	Obtain suitable surface (artificially or naturally) →	3
Preparation of surface	Fashioning/shaping →	4
	Smoothing →	5
Marking of surface	Drawing line with finger(s) →	6
Abandoning	Left in chosen final position in cave →	7

Table 2.43: Technical scheme for the making of 'macaronis' finger-tracings

Phase	Sequence	
Procurement of material	Find naturally suitable surface →	1
Marking of surface	Drawing line with one to four finger(s) by pressing wall surface →	2
Abandoning	Left in chosen final position in cave →	3

Finger-tracing and the properties of clay

The technique of finger-tracing makes full use of the plasticity of clay material and is indeed possible only thanks to this property. No other material, with perhaps the exception of pigments once they have been prepared for application, can be worked without tool. Finger-tracing allows the Magdalenian artist the quickest form of symbolic expression of all the art techniques known for this period; one can go directly from the intention of marking a surface to doing it by the sole intermediary of one's hand. It is true however that the presence of a suitable surface is a prerequisite.

Impressing on clay

Impressing is another technique found among the uses of clay in the French Pyrénées. This technique has only been used, in the recorded inventory, for representing abstract patterns made of juxtaposed printed cupules. These impressions can be done with a tool (pointed or rounded) or with fingertips.

17 impressed graphic units are found in 5 sites, Tuc d'Audoubert, Fontanet, Massat, Montespan and Bois du Cantet, representing 5.8 % of the total of the inventoried graphic units on clay (table 2.2).

The majority (12 out of 17) of the impressed signs are found on floors in the sites of Tuc d'Audoubert, Fontanet and Bois du Cantet. Four more are found on the walls in Montespan and one in Massat (figure 2.35).

Figure 2.35: Impressed sign (M9) with fingertips on a wall surface in Massat. The surface was previously smoothed (from Gailli 2004, figure 40)

Table 2.44: Inventory of the impressed graphic units on clay recorded in the French Pyrenees.

Site name	Printed sign				Total
	With tool		With fingers		
	On wall	On floor	On wall	On floor	
Tuc d'Audoubert				4	4
Fontanet		4		3	7
Massat			1		1
Montespan	1+3 ?				4
Bois du Cantet		1			1
Total	4	5	1	7	17

Technical scheme of impressing on clay

In the sites of Montespan, Bois du Cantet and Massat, the impressed surfaces have been prepared by gathering material and smoothing. In the two remaining sites, Tuc d'Audoubert and Fontanet, the clay floors have been used in their natural state without further preparation work visible.

The making of impressed patterns on clay can therefore be a complex process that implies procuring, moving and preparing the material (Table 2.45).

The impressing of the pattern itself requires having the necessary tools available, unless the fingertips are used. Casts have been done of some of the holes in order to determine the shape of the tool used. They are usually of two types: conical (as would be made by a spearhead) or cylindrical with a flat bottom. In some instances like in Fontanet other tools have probably been used as the sign F10, made of over 50 prints, shows the use of a quadrangular implement to print some of the holes (Vialou 1986).

Impressing and the properties of clay

Impressing is a technique that can only be applied to a very soft material as it uses the property of plasticity of the medium receiving the impression. It is therefore totally suited to the working of clay.

In at least two cases in the site of Fontanet, for the impressed signs F4 and F7, the impression was done on a previously calcite-hardened clay surface, which cracked under the pressure. In these cases, it was attempted to work the clay floor surface in the same manner than it had been done in eight other locations in the cave without realising that the surface was different. The final result is different visually as the 'cracked cupules' are much less distinct than on the soft clay. The same technique, however, was carried out, even if the pressure needed was probably greater to pierce effectively the hard surface.

Table 2.45: Technical scheme of the technique for impressing on clay

Phase	Sequence
Procurement of material	Extraction
	Transport to final position
	Obtain suitable surface (artificially or naturally)
Preparation of surface	Fashioning/shaping
	Smoothing
Marking of surface	Impressing clay with finger or tool
	Repeat printing until wanted pattern obtained
Abandoning	Left in chosen final position in cave

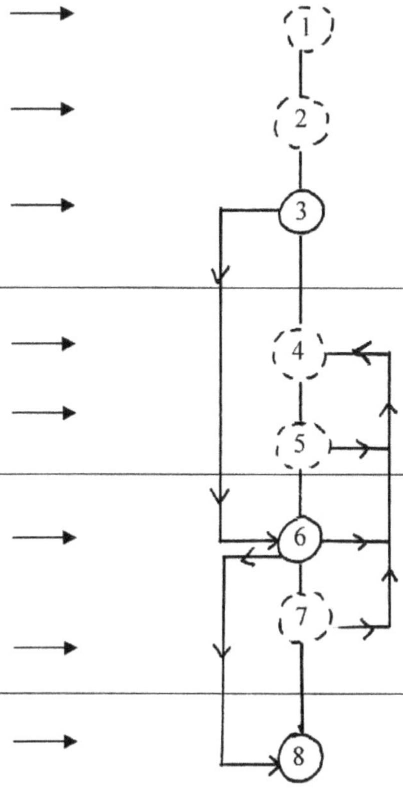

Portable objects: three dimensional sculptures

Another technique looked at is sculpture in the form of portable objects. In this category we find the most complex technical use of clay in the French Pyrénées as some objects appear to be made of fired clay. The objects looked at in this category were separated from the 3D modelage as they are made of fired clay or of well dried material and appear intended as fully portable objects.

A total of at least 128 portable objects made of clay are known in the French Pyrénées (table 2.2). They represent 43.9 % of all the recognised graphic units on clay, which is the most numerous technical category: this record however is entirely biased by the site of Bédeilhac in which at least 123 objects were found (see part 1, section on Bédeilhac in this chapter) (Table 2.46).

Table 2.46: Inventory of the portable art objects in clay found in the French Pyrénées

Site name	Dried	Fired object
Enlène	1	2
Mas d'Azil	1	1
Bédeilhac	123	
Total	125	3

The portable art objects found in this category are actually rare in the French Pyrénées, if we exclude the large series from Bédeilhac, unique to the site. Some other portable clay objects exist in Mas d'Azil but they have been recorded as objects resulting from human activity rather than as portable art as they include shapeless lumps and some possible ornaments. All these objects are described in the catalogue in appendix 1.

The technical scheme for the making of the clay plaquettes will be considered apart from the other objects as their characteristics are unique to the site of Bédeilhac.

Technical scheme of the portable clay objects

This first technical scheme (Table 2.47) considers the five objects found in the two sites of Enlène and Mas d'Azil. Three appear fired (the animal body (E1) and possible animal head (E2) in Enlène and the possible feline head (MA1) in Mas d'Azil) on the basis of their harder texture as well as the colour they have taken (dark when fired in reducing conditions for E1, bright red when fired in oxidation for MA1). The two remaining are a facetted and engraved bloc (E3) in Enlène and a small bison sculpture in Mas d'Azil (MA2). It is possible these two objects were also exposed to fire, although at a temperature below that necessary for ceramic change around 550°C, which would have hardened them enough to explain how they survived buried in the archaeological record without disintegrating.

Table 2.47: Technical scheme of the making of portable clay objects in the French Pyrénées

Phase	Sequence	
Procurement of clay	Find source	1
	Testing clay for plasticity	2
	Extraction of the clay	3
Preparation of the material	Adding ochre pigment	4
	Kneading	5
Fashioning of the bulk of the object	Give rough shape	6
Drying	Partial/full drying	7
Fashioning	Fashioning of surfaces	8
	Fashioning of details with small tool	9
Full drying	Full Drying	10
Firing	Firing in hearth	11
Reworking after firing	Fashioning with hard tool	12

Each of the objects studied is unique. The technical scheme presented implies that the final purpose of the process is to have an object of portable art. But it remains difficult to be sure that some of the manufacturing steps, especially those relating to the firing, are fully intentional. It has been argued that a human during the Magdalenian period would have known that fire could harden clay, since any fire lit on a clay surface, as is often the case in caves, would have baked the surface underneath (Bahn 1997). The fact that so few objects are made of fired clay, however, would rather indicate that this knowledge was not associated with the making of artefacts and that the cases studied here are possibly accidentally fired or a one-off trial. But it is certain that some sequences of actions have been followed in order to obtain the final result. For example, from the animal body (E1) from Enlène (Figure 2.36), which is the best preserved of all the portable objects presented here, it is possible to infer the details of this technical scheme from the traces visible on its surface.

Figure 2.36: the animal body in ceramic from Enlène (E1) (private collection Begouen Family)

The technical scheme for the objects from Bédeilhac is presented in Table 2.48 below. Out of the 123 clay plaquettes that are known to exist from Bédeilhac, 58 have been studied in the context of the present research. The resulting technical scheme presented here concerns these 58 pieces, 52 of which have not yet been published.

The silty clay used as material for the plaquettes in Bédeilhac naturally dries off and breaks into thin plaquette fragments in parts of the site (Sauvet 2006); this might have simply encouraged the Magdalenian people visiting the site to use them as plaquettes as other sites with decorated plaquettes are also found in the region at the same period, for example in Enlène (Bégouen *et al.* 1984-1985 (1999). But Bédeilhac is the only site in which a material other than stone is used for this purpose.

For some of the objects, however, it is clear the material was still soft and could still be modelled when the representations were made on the plaquettes. This has been well emphasised in the published study of a modelled plaquette (B17 in the catalogue) on which a horse is represented (Figure 2.37) (Mons and Delporte 1973; 1980; Mons 1974). The plaquette was clearly modelled with the fingers to represent the forelegs of a horse when the material was still very plastic. The material was compressed, pushed and possibly removed when still wet. Then the details of the legs of the animals, like details of the hairs were incised with a very fine tool to obtain the desired result, with a total mastery of the technique. The horse is represented in movement, legs extended towards the front, in a very dynamic attitude, recognisable even though it is fragmentary.

The conclusion can be drawn that the material was used rationally at different stages of its plasticity. This is also the case for a number of engraved plaquettes.

Sauvet (2006) points out another aspect of the plaquettes in Bédeilhac: the fact that often the representations on them appear to be inspired by the natural shape of the object. This is visible in the clay plaquettes and also in the numerous series of sandstone plaquettes of various origins that was also found in the same context that the clay ones.

Figure 2.37: plaquette B17 from Bédeilhac: front part of a horse modelled and engraved. It measures 21 cm long by 14 cm wide and 3.3 cm of maximal thickness (Photograph Musée des Antiquités Nationales).

The phase of suspension was added to the technical scheme of the making of the clay plaquettes as observations were carried out on the studied objects. 48 of the studied Bédeilhac plaquettes out of 58 have what can be interpreted as notches on their edges, which could have been used to allow suspending the objects with the help of string or rope. This number represents over 80 % of the studied clay plaquette inventory; this makes up a very consequent group. As the hypothesis of suspension of decorated plaquettes has already been put forward for other plaquette sites, such as in Labastide (Simmonet et al. 1984), it has been included here as a possibility although of course the rest of the existing objects remain to be studied to determine if similar traces exist. This would provide information on a possible use of these objects.

Table 2.48: Technical scheme for the making of portable clay plaquettes in the site of Bédeilhac

Phase	Sequence
Procurement	Find clay plaquettes naturally formed /or cut out shape from clay floor
Preparation	Preparation of the shape of the plaquettes Preparation of the surface of the plaquettes
Fashioning of the plaquette	Fashion contour
Fashioning of the surface	Model the surface to obtain figure Engrave details
Drying	Drying
Suspension for final use	Make notch(es) on contour or use natural ones Tie string/cordage Carry around for intended use
Abandonment	Plaquette abandoned on floor surface

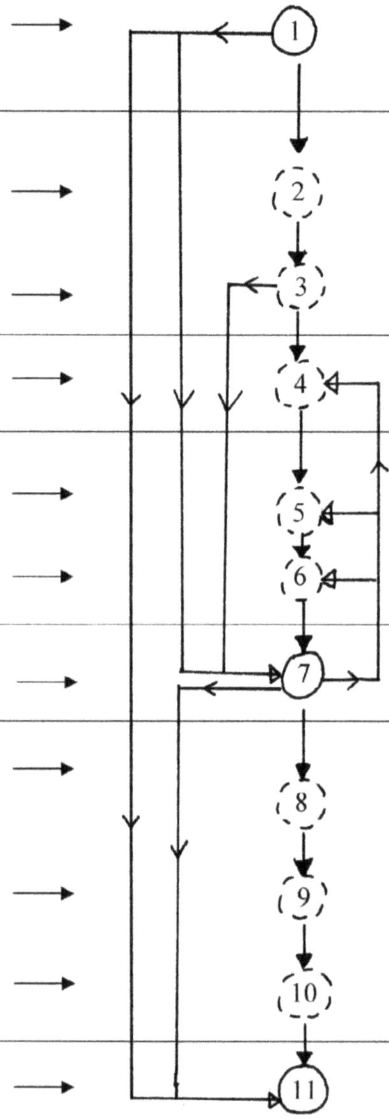

The making of Portable objects and the properties of the material used

The property of plasticity of the clay is used in the fashioning of the final shape. Also, the capacity of the material to harden by drying and then by firing have clearly been noticed and used here.

It is likely the objects would not have survived without this final hardening as they were buried in archaeological layers where they are very susceptible to re-hydration and dissolving. However, the small number of objects found with this technique implies it was not widely used, or simply the phase of hardening was not mastered and most of the objects made this way not have survived to the present day.

Clay used as Pigment

At a number of sites a lump of clay, possibly taken directly from the clay floor in the cave, was used to draw a figure on the wall nearby, such as in Bédeilhac (Octobon 1939; Vialou 1986), Labastide (Clottes 1989: 77), Oxocelhaya-Hariztoya (Laplace and Larribau 1984: 285) and Etcheberri (Laplace *et al.* 1984: 271; Clottes 1989: 77).

For this purpose, the lump of clay needed to be roughly modelled and shaped with the hand to make it a sort of 'crayon' (Couraud 1983: 106). This might account for some of the shapeless modelled portable clay objects, now dried (see the ones from Mas d'Azil mentioned above), found in a number cave sites often with finger traces on them, sometimes with traces of scraping as well. This type of object is also found in Lascaux for example (Couraud and Laming-Emperaire 1979: 156).

Some pigment, possibly ground ochre, was apparently also mixed with the soft clay in some instances such as in the so-called 'tattooing plaque' found in Mas d'Azil (Figure 2.38). This object was found together with bone needles and it was assumed that the needles would have been used to collect some pigment from the previously modelled plaque, thus piercing it with small deep holes, and then to pierce human skin in a form of tattooing process (Péquart 1960). The plausibility of such a hypothesis still remains to be proven, but it cannot be denied that the object is made of a material with colouring properties which is likely to have been used in some form of decorating process. The holes were made on a still soft material. Some of them do pierce the plaque from one side to the other but not all do so. The site of Mas d'Azil can, therefore, also be added to the list of other sites where a clay mixture was used as pigment in the French Pyrénées even if no trace remains of the application of this pigment.

Figure 2.38: the 'tattooing plaque' from Mas d'Azil a possible bloc of prepared pigment made of ochre and clay, kneaded and flattened then pierced possibly in order to take small quantities of pigment to use (Péquart Collection, Mas d'Azil Museum).

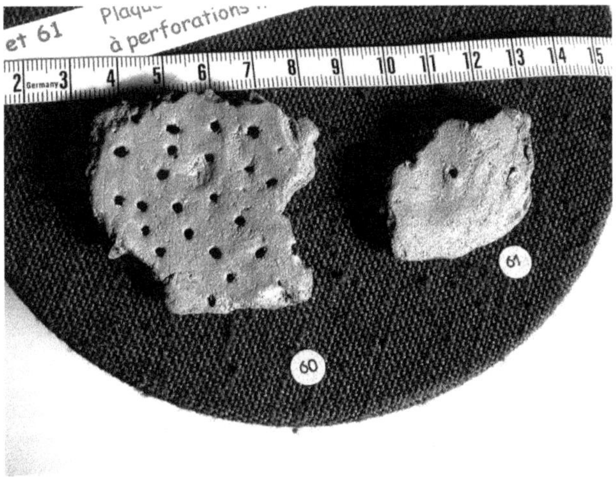

In the other sites mentioned above, Bédeilhac, Labastide, Oxocelhaya-Hariztoya and Etcheberri, figures are mentioned as drawn with the clay pigment, especially one horse in each of the site of Bédeilhac, Labastide and Etcheberri (Clottes 1989: 77). However, determining the exact nature of the pigments used can be difficult. When brown pigment has been used in the Ariège, it is described as a type of ferrous clay (Clottes *et al.* 1990: 171). Clay can also be one of the minerals added to pigments in small quantities in paint recipes as an extender. It is found for example in pigments recipes in Lascaux and Altamira (Couraud and Laming-Emperaire 1979).

In the same manner that clay can vary from source to source, according to the context it is connected to (for example for a geologist, a chemist or a potter), the definition of ochre is fairly variable: in Clottes *et al.* (1990: 174) they define it as 'a clay rich in iron oxide', naturally coloured from yellow to red, from which a more intense red can be obtained by heating up the pigment. However, they also mention that in the Ariège ochre was used in rare cases, never in parietal painting for which haematite-based paint recipes were applied. They do indicate that the brown paint, which is more likely to be clay-based, but is rarely used, has not been analysed up to now in this region as the analyses have concentrated on red and black pigments (Clottes *et al.* 1990: 179).

All this seems to underline a possible further avenue for looking at a technique for using clay in the French Pyrénées, which for the present remains rather 'invisible' in the archaeological record.

Conclusion

A combination of techniques

In many instance a combination of the techniques suitable for the working of clay have been used together for making the various parts of a figure, often with differences for the bulk of the shape and the indication of finer details. This shows a full appreciation of the properties of the material which makes it extremely versatile.

In the case of *modelage*, another technique is needed if fine details are required, since shaping a quantity of clay with the hands and fingers does not allow detailing beyond a certain size. Small tools then become useful for engraving for example.

Finger-tracing and engraving are techniques combined with *modelage* if clay is moved and worked in order to make a suitable surface. They are sometimes found next to each other on similar wall surface like in Massat for example. In Niaux, on the other hand, the technique used on clay is very consistent and unique: engraving.

Impressing is possibly the only technique that is not found combined to other although when the surface was previously prepared it can be argued that *modelage* was used.

On portable objects, similar making of details by engraving, scraping or smoothing with tools are seen after using *modelage* and carving for the bulk of the object.

The example of the large bison *modelage* at Tuc d'Audoubert illustrates well this combination of techniques. The general process for the making of the bulk of the body is reductive which, in a way, is similar to the way other materials (antler, bone or ivory) are commonly worked in Upper Palaeolithic art. However, the fashioning and smoothing of the surfaces by hand and the addition of separately fashioned elements, such as the tails, horns and ears, indicate the use of a specific quality of clay as a material: the possibility to be worked in an additive manner (Bégouen and Breuil 1958; Bégouen and Clottes 1984).

Table 2.49: inventory of clay uses with and without surface preparation. Surface preparation involves moving clay material to obtain a suitable surface. Portable objects have not been included as by their portable nature they do not offer a context informing on material preparation. As this table was made using the inventory in the catalogue in Appendix 1, only the 27 described and referenced engravings from Niaux have been included. The remaining 38 are indicated in bracket as unreferenced.

Technique used	Preparation of surface	Total	Surface used as found	Total
Modelage	TA1, TA2, TA3, TA5 a, b, c, d, e Labouiche bison B1, B5, B7 MO3, MO5, MO8, MO9, MO10, MO13 ER 1, ER2, ER3, ER4	22	TA4 F2, F3, F6 B2, B3, B4, B16 MO11, MO14, MO15, MO16 L1	13
Engraving	MO8 ET1, ET2	3	N1, N2, N3, N4, N5, N6, N7, N8, N9, N10, N11, N12, N13, N14, N15, N16, N17, N18, N19, N20, N21, N22, N23, N24, N25, N26, N27 (+38 unreferenced engravings in Niaux) TA6, TA7 F11 B6, B8, B9, B10, B11, B12, B13, B14, B15 M1, M3, M4, M5, M6, M7, M8, M14	85
Finger-tracing	MO19, MO21, MO22 O1, O2, O3	6	MO1, MO6, MO7, MO17 M2, M10, M11, M12 ER5, ER6, ER7, ER8, ER9, ER10, ER11 ER12	16
Impressing	MO2, MO4, MO12, MO20 BO1 M9	6	TA8, TA9, TA10, TA11 F1, F4, F5, F7, F8, F9, F10	11

Previous surface preparation

There is a general sense that the various techniques used on clay in the Magdalenian in the French Pyrénées were used with purpose and intention. This is visible in the numerous occasions where the wall surfaces have actually been prepared and fashioned in order to be able to support the works on clay (Table 2.49).

As clay is a commonly available material in many cave sites, one could easily have argued for a simple opportunistic use of the material at hand without planning. The analysis of the clay inventory in the French Pyrénées clearly disproves this hypothesis.

The proportion of *modelage* for which clay was moved for the making of the figure is the highest of all the techniques looked at, as is to be expected as this particular technique implies building up relief.

It is more revealing to see that for all the other techniques used, a number of cases are found in which the clay surface were clearly prepared by moving material when this was not actually a requirement of the technique used. They remain less numerous than the ones made on naturally available surfaces, but their existence is indicative of intention. Also as this type of detail is not always mentioned in the relevant descriptive literature, it is likely the examples mentioned here are a minimal quantity.

In Chapter 2, the clay uses found in the French Pyrénées were presented and analysed in order to underline the techniques used and the visual aspect taken by the art on clay in the region. A further level of analysis of the Pyrenean material will be found in Chapter 4 which looks at the levels of knowledge recognised in the working of clay and at some implications for the social contexts considered.

CHAPTER 3: THE USE OF CLAY IN CENTRAL EUROPE IN THE UPPER PALAEOLITHIC

Moravia (Czech Republic) in the Pavlovian period (29,000 to 22,000 BP) (Svoboda 1994) is mentioned as one of the five important centres existing in Europe in the Upper Palaeolithic by Soffer (1985). Very rich open-air sites are found in this region, yielding a large range of material and techniques used. These were created by people usually thought to have been mammoth hunters (Oliva 2000) because of the very large accumulations of mammoth bones found in all the major sites.

In the context of this culture, loess soil was wetted, modelled and fired into ceramic at the sites of Dolni Vestonice (I, II and III) and Pavlov (I, II and V) in what appears to be a systematic manner for over 2,000 years (Soffer *et al.* 1993). Emerging evidence also suggests that this use of clay was a regional phenomenon, spreading across Central Europe at roughly the same period of time, associated with the local Pavlovian/Gravettian culture. Thus, examples of fired clay production are known in the following Pavlovian sites (beside the sites of Dolni Vestonice and Pavlov): Predmosti, Petrkovice I, Spytihněv, Jarošov and Boršice in Moravia (Czech Republic) (Svoboda *et al.* 1996; Svoboda personal communication February 2008), Krems-Wachtberg in Austria (Neugebauer-Maresch 1999), and Moravany-Lopata in Slovakia (Valoch 1996). By radiocarbon dating, they can all be attributed to the same time period: 27,000 BP to 25,000 BP, except for Moravany-Lopata, associated by typological comparisons (Otte 1981; Valoch 1996) and Petrkovice I, where there were problems with the radiocarbon dates.

In addition, further East and South in Central Europe, in Slovakia and Hungary, three more sites are known to have yielded fired clay finds (figure 3.1). Kasov is attributed to an Upper Gravettian/Epigravettian context (22,000 to 18,000 BP) and Sagvar and Cejkov to an Epigravettian context (19,000 to 18,000 BP) (Valoch 1996; Djindjian *et al.* 1999; Svoboda (ed.) 1996). Their exact connection with the Pavlovian is certainly a topic to investigate through the presence of fired clay objects, especially bearing in mind the fact that fired clay has also been found in the Russian site of Kostenki, in Epigravettian layers (Vandiver *et al.* 1989), a site with striking similarities to the Moravian sites.

Table 3.1 summarises the cultural attributions and the dates of the sites where fired clay is present in Central Europe.

After describing the types of clay uses found in Central Europe in the Gravettian period in the first part of this chapter, the extensive material record available for the use of clay at the sites of Dolni Vestonice I and II and Pavlov I will be presented and analysed through looking at the themes of representation and the technology visible in these art objects in the second part.

Figure 3.1: Map of the Gravettian sites with fired clay in Central Europe; **Moravia**: 1 Dolni Vestonice I, 2 Dolni Vestonice II, 3 Dolni Vestonice III, 4 Pavlov I, 5 Pavlov II, 6 Pavlov V, 7 Predmosti, 8 Spytihněv, 9 Jarošov, 10 Boršice, 11 Petrkovice I; **Slovakia:** 12 Moravany-Lopata, 13 Kasov I, 14 Cejkov; **Austria:** 15 Krems-Wachtberg; **Hungary**: 16 Sagvar.

Table 3.1: Sites with fired clay in Central Europe and their cultural attributions and dates (the dates are approximate and not calibrated).

(Predmosti and Petrkovice I are indicated here as they seem to have yielded at least one ceramic for Predmosti and three for Petrkovice, but it must be noted that recent excavations at these sites did not yield any ceramic finds – personal communication Jiri Svoboda February 2008).

Site name and location	Cultural attribution	Date range using published C^{14}	References
Dolni Vestonice I, Moravia	Pavlovian	27,000-25,000 BP	Valoch 1996; Vandiver *et al.* 1989
Dolni Vestonice II, Moravia	Pavlovian	27,000-25,000 BP	Valoch 1996; Soffer and Vandiver 1997
Dolni Vestonice III			Jiri Svoboda pers. Com. Feb. 2008
Pavlov I, Moravia	Pavlovian	27,000-25,000 BP	Soffer and Vandiver 1994, 1997
Pavlov II, Moravia	Pavlovian	27,000-25,000 BP	Soffer *et al.* 1993a
Pavlov V, Moravia			Jiri Svoboda pers. Com. Feb. 2008
Predmosti, Moravia	Pavlovian	27,000-26,000 BP	Valoch 1996, Jiri Svoboda pers. Com. Feb. 2008
Spytihněv, Morava Valley			Jiri Svoboda pers. Com. Feb. 2008
Jarošov, Morava valley			Jiri Svoboda pers. Com. Feb. 2008
Boršice, Morava valley			Jiri Svoboda pers. Com. Feb. 2008
Krems-Wachtberg, Austria	Pavlovian	27,000 BP	Neugebauer-Maresch 1999
Moravany-Lopata, Slovakia	Pavlovian	Not available	Gonysevova 1999
Petrkovice I, Czech Republic Moravia	Upper Gravettian	23,000 BP	Svoboda *et al.* 1996; Jarošova *et al.* 1996, Jiri Svoboda pers. Com. Feb. 2008
Kasov, Slovakia	Upper Gravettian/ Epigravettian	22,000-18,000BP	Valoch 1996; Banesz 1996
Sagvar, Hungary	Epigravettian	19,000-18,000 BP	Gonysevova 1999; Gabori 1968
Cejkov, Slovakia	Epigravettian	19,000 BP	Djindjian *et al.* 1999; Banesz 1996

The conclusions obtained from the analysis of the clay record, especially the social context visible through elements of apprenticeship, techniques and styles of the ceramic objects will be developed in Chapter 4, as well as some elements relevant to the wider Gravettian art context.

1 The types of clay uses found in Central Europe

The earliest discovery of fired clay in Upper Palaeolithic contexts was made in the early 1920s at the site of Dolni Vestonice I in Moravia in the Czech Republic when bones, flint tools and a mammoth figurine made of ceramic were noticed in the section of a cutting near the village of Dolni Vestonice (Valoch 1996). More discoveries of clay uses were made at this site and at the neighbouring site of Pavlov, near the village of Pavlov, until recently (Vandiver *et al.* 1990; Svoboda *et al.* 1996). These discoveries exemplify the type of use of clay found in the sites of Central Europe as the firing of clay.

The evidence available for clay being used in the sites of Dolni Vestonice and Pavlov shows that this material was submitted to the action of fire, apparently in a systematic and repeated manner, between 27,000 BP and 25,000 BP (Soffer *et al.* 1993).

The material record shows two main types of artefacts: fashioned figurines (or fragments of) (figure 3.2), and fired clay fragments with no clear traces of fashioning (figure 3.3).

Figure 3.2: fragmentary female figurine DVI 7 in profile from Dolni Vestonice I (the head and feet are missing)[16]

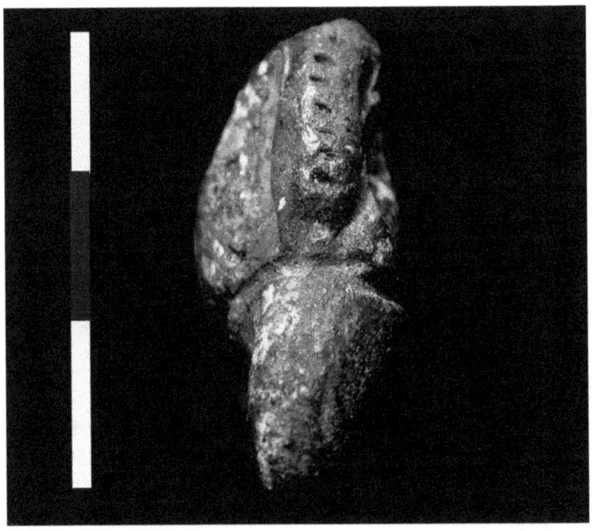

In their analysis of the ceramics of Dolni Vestonice and Pavlov, Soffer and Vandiver (Soffer and Vandiver 1994, 1997; Soffer *et al.* 1993a; Vandiver *et al.* 1989; Vandiver *et al.* 1990) describe the various stages of firing the pieces have been submitted to, deduced from careful analysis of the material and comparison with experimental replicas. They mention that some pieces are unfired, low fired or well fired with a wide range of temperature up to about 800°C (Soffer and Vandiver 1997: 387).

It is worth recalling here that clay (and any assimilated material such as the loess soil of the Moravian sites) undergoes a non-reversible structural change into ceramic at a firing temperature of 450°C to 500°C, with some variation according to the exact nature of the material and climatic conditions at the time of firing (Hodge 1964). Any material fired under this temperature can be described as unfired, even if it has been submitted to the action of fire, as the material can still revert to a soft paste and dissolve if re-hydrated (Vandiver *et al.* 1990: 49). It is likely that exposure of the material to fire makes it more likely to survive in the archaeological record even if the temperature of ceramic change is not reached as it helps hardening significantly any fashioned object.

Figure 3.3: Non-fashioned fired clay 'pellets' from Dolni Vestonice I

In the other sites of Central Europe where fired clay was found: Predmosti, Petrkovice I (Soffer *et al.* 1993a), Spytihněv, Jarošov and Boršice (Svoboda et al. 1996; Svoboda personal communication February 2008), Krems-Watchberg (Neugebauer-Maresch 1999), Moravany-Lopata, Cejkov (Valoch 1996), Kasov and Sagvar (Gonysevova 1999), only figurine fragments were recovered, in very fragmentary states, and in small numbers. This confirms that the most visible type of use of clay in Central Europe in the Gravettian period is related to the making of ceramic figurines. The large number of shapeless fragments of fired clay found is also thought to be associated with figurine-making, or as will be seen, possibly figurine destruction (Soffer and Vandiver 1993, 1997).

Another possible type of clay use found at the site of Dolni Vestonice and Pavlov also needs to be mentioned here: clay used as a 'structural ceramic', in the words of Soffer (2000:60). A small part of the ceramic record at the sites of Dolni Vestonice I and II and Pavlov I has been interpreted as accidentally fired clay-lined baskets or daub used on constructions (figure 3.4) (Soffer and Vandiver 1997). However, these pieces are few, they are difficult to recognise and interpret and they are still in the process of being studied by Soffer and her team (Soffer 2000). They will not be studied in detail in the present research.

They are of interest for the evidence they bear of an increased material complexity at the sites and will be considered as such when suitable in the discussion.

[16] The objects from Dolni Vestonice I belong to the collections of the Anthropos Institute in Brno. Those from Dolni Vestonice II and Pavlov I are from the Collections of the Archaeological Institute in Dolni Vestonice.

They also raise the question of the problems of the recognition of such objects in archaeological excavation. It has been noticed in countless excavation sites since the beginning of archaeology, such as in Enlène in the Pyrénées (see chapter 2), that hearths lit on a clay floor have baked the clay, leaving fragments sometimes resembling potsherds (See Bahn 1978, 1986 and Bahn and Otte 1985 for full discussion on this topic). After some debate at the end of the 19th century deciding whether these were actual potsherds, it has generally been accepted by archaeologists that they were not, and that they did not represent more interest than for indicating the presence of a hearth. Thus it is likely that most shapeless baked clay fragment found in excavations have not been examined or even kept in earlier excavations and that much information has been lost there about possible structures made of clay or fired clay.

Figure 3.4: possible structural ceramic fragment from Dolni Vestonice I. a) profile b) side

a)

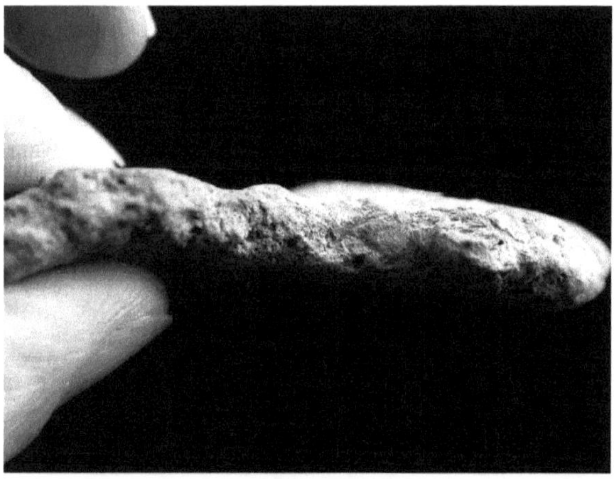

b)

Structural ceramic fragments were recovered in Dolni Vestonice and Pavlov because the excavators were aware of the presence of ceramic at the sites and kept every small and shapeless fragment they found of it. However, in most of the other Gravettian sites of Central Europe, which have been excavated very early for some of them (i.e. 1880 for Predmosti) (Valoch 1996), this was not the case, and it can be assumed that some evidence has been missed. For example, many of the small fired clay pellets (figure 3.4) from Dolni Vestonice and Pavlov were recovered by the technique of wet-sieving (Soffer and Vandiver 1997: 389) and would have been missed without this technique.

The material from Pavlov I and Dolni Vestonice I and II form most of the record available for the use of clay in central Europe and have been studied for the resent research. They will be analysed in detail below with the aim of looking at the technology used in the making of the fired clay figurines.

2 Description of the archaeological record of clay uses at Dolni Vestonice and Pavlov

Introduction

The cluster of sites forming Dolni Vestonice and Pavlov are sometimes described as one super-site or mega-site (Oliva 2000: 225) with at least six stations (Dolni Vestonice I, II and III, Pavlov I and II and V) (Svoboda *et al.* 1996) found on the foothill of the Pavlovské hills, between the villages of Dolni Vestonice and Pavlov in the Moravia region, Czech Republic (Figure 3.5 and 3.6). They are spread over 2.5 km, with the two most important sites, Dolni Vestonice I and Pavlov I being less than 500 m. apart.

In these sites, a complex and very rich Gravettian cultural sequence was deposited during a long period of time, between 30,000 BP and 20,000 BP (Svoboda 1994; 1996: 288). The specific faciès taken by this Gravettian in the region is described as the Pavlovian culture, the site of Pavlov being the eponymous site.

The sites of Dolni Vestonice and Pavlov overlook the Moravian corridor, a natural opening between the Carpathian Mountains to the South and the Moravian uplands to the north connecting the Russian plain to the Danube plain (Absolon 1949; Svoboda 1994). This would have been an important communication route in the Upper Palaeolithic, as well as the major migration route for herds. This partly explains the importance of these sites and their continuous reoccupation over a few millennia. The duration of the occupations appears to vary between the sites, but it seems that they were mainly occupied during the winter months (Svoboda 1994).

As far as ceramics are concerned, only the ones found in the sites of Dolni Vestonice I and II and Pavlov I are studied in the present research. It was not possible to access the other fragments mentioned in the literature as coming from the Pavlovské Hills sites, but it is true they are much fewer in quantity.

A research team consisting of P. Vandiver, O. Soffer, B. Klima and J. Svoboda has recently studied and published precise data which describe in great detail the ceramics of Dolni Vestonice I and Pavlov I, for which there is a large corpus of information available (Soffer *et al.* 1990, 1993; Vandiver *et al.* 1989, 1990; Vandiver 1988; 1997; Soffer and Vandiver 1994, 1997). Information on Dolni Vestonice II is more difficult to find (Valoch 1996; Vandiver *et al.* 1989; Svoboda 1994), and information on Pavlov II is scarce yet (Soffer *et al.* 1993a; Valoch 1996).

For the purpose of the present research, the collections of ceramic objects coming from Dolni Vestonice I and II and Pavlov I have been studied in the institutions where they are kept: the Moravian museum in Brno (Czech Republic) for the pre-World War II finds from Dolni Vestonice I, and the Anthropological Institute in Dolni Vestonice (Czech Republic) for the rest. It was decided to concentrate on the shaped ceramic objects coming from the sites, for which stylistic and thematic studies can complete the technical ones. Therefore, the numerous pellets and shapeless fired clay fragments yielded by each site have not been studied in the present research, and we will rely on Soffer and Vandiver's work as far as they are concerned.

All the ceramic objects seen and studied in the Czech museums are described individually in the catalogue of portable clay objects in Appendix 2. The reference numbers used in the present research correspond to the references given to the objects in the catalogue and are our own.

Not all the ceramic objects from Dolni Vestonice and Pavlov have been published yet, however, especially those from the most recent excavations (Martin Oliva and Jiri Svoboda, personal communication May 2004). Therefore the present inventory does not claim to be exhaustive as the most recent material is not included.

A general presentation of each site will be followed by a study of the ceramic objects based on their thematic and stylistic analysis.

Figure 3.5: Map of the sites of Dolni Vestonice and Pavlov (After Svoboda *et al.* 1996, p. 174)

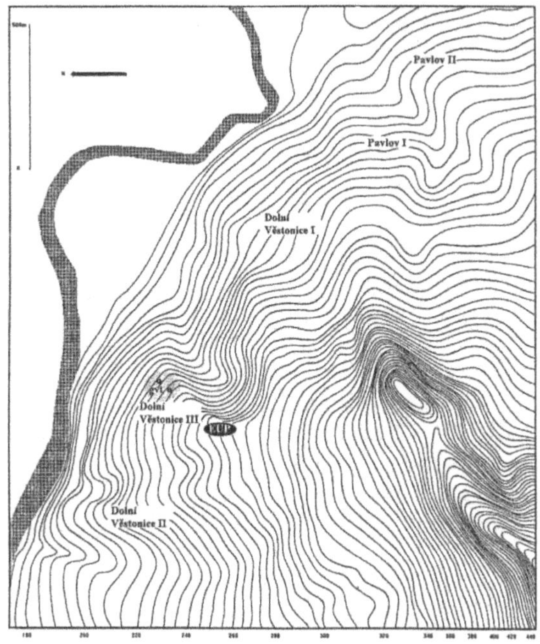

Figure 3.6: View of the Pavlovské Hills from the north (From http://donsmaps.com/dolnivphotos.html, first accessed October 2006).

The fired clay objects from Dolni Vestonice and Pavlov

The largest concentration of ceramic artefacts known for the Upper Palaeolithic has been found at the Moravian sites of Dolni Vestonice I and Pavlov I. Altogether, the two sites yielded over 10,000 ceramic objects consisting mainly of fragments of animal and human figurines, and of various sorts of pellets and diversely shaped fragments (Soffer *et al.* 1990, 1993; Vandiver *et al.* 1989, 1990).

Dolni Vestonice I

The site was discovered in the first part of the 1920s (Valoch 1996), by H. Freising who spotted bone, flint tools and a mammoth figurine made of ceramic in the section of a cutting near the village of Dolni Vestonice. Excavations by Karel Absolon started that same year (1924), and continued from 1924 until 1938. During the German occupation of the Czech Republic in World War II, A. Bohmers continued the excavations, between 1939 and 1942 (Vandiver *et al.* 1990). K. Žebera then excavated in 1945-46. B. Klima took over the excavation in 1947-1952 with further seasons in 1966 and in 1971-1979. J. Svoboda worked there in 1990 and 1993 (Svoboda *et al.* 1996). Some of the finds of the early excavations were destroyed in a fire together with the material from Predmosti during World War II, but the works of art survived (Valoch 1996).

Very thick ashy deposits were found in Dolni Vestonice I, with concentrations of fired clay and decorative objects around them. They seem to point to a long occupation, or reoccupation, of the site (Svoboda 1994).

The climatic studies indicate that the cultural layer was probably deposited during a period of changing climate, with a certain instability and a changing environment. This was apparently a period of gradual cooling that eventually possibly led to the departure of the Pavlovian people from this region (Svoboda 1991). More specifically, pollen analyses and faunal remains indicate a winter occupation of the sites of Dolni Vestonice and Pavlov (Svoboda 1994; Svoboda *et al.* 1996).

Dolni Vestonice I, together with all the Dolni Vestonice-Pavlov sites, can be attributed to the Pavlovian culture, which dates to between 29,000 and 22,000 BP (Svoboda 1991). In order to obtain more radiocarbon dates for these important sites, trenches were dug in 1990 to collect samples (Svoboda 1994). The dates obtained confirm the stratigraphic data, indicating that the lower location at Dolni Vestonice I is the oldest (31,000-27,000 BP) and partly predates the Pavlovian. The other areas (middle, upper and uppermost parts) of Dolni Vestonice I date to around 25,000 BP. (Svoboda *et al.* 1996; Svoboda 1994). The ceramic objects were found in these younger locations.

These artefact-rich upper and middle parts of the sequence also yielded structures traditionally interpreted as oval or circular-shaped dwellings with central hearths (Absolon 1949; Klima 1955; Valoch 1996). However, this is being questioned by Oliva (2000 b) who worked directly from the notes left by the excavator, K. Absolon, as the early excavations have not been published. The hearths and kilns found are associated with lithic and bone tools, portable art and ceramic figurines. Few burials were also found (a woman's and a child's) as well as several scattered human remains. A large accumulation of mammoth bones measuring 45 m by 12 m was found on the edge of the Upper part of Dolni Vestonice I as in many other Pavlovian sites in Central Europe (Svoboda *et al.* 1996).

The lithic industry for the middle location is as follows: the burins represent between 30% and 40 % of the lithic assemblage; endscrapers between 12 and 18%; there is an absence of Aurignacian types; backed microliths are rather frequent (between 8 and 26%), with numerous microsaws (microdenticulates: 5 to 7 %) representing a tool form specific to the site (Svoboda *et al.* 1996).

What is possibly more important to point out is the fact that during the Pavlovian at Dolni Vestonice I, most of the raw materials for the lithic industry were imported from long distances, between 60 and 300 km (Oliva 2000a: 221; Absolon 1949: 22; Svoboda 1994: 56). This implies the organisation of a complex network of exchanges or seasonal migration that emphasises the cultural complexity seen at the sites.

Dolni Vestonice I is the first site where ceramic-making was recognised to have existed for such an early period (27,000 BP). It yielded a total of over 5,761 fired clay objects that Vandiver *et al.* (1989) have categorised as follow in their inventory: figurines, fragments of figurines, slab-like objects, nearly spherical balls and small irregular pellets in fired clay (Table 3.2).

About 3,700 broken fragments of figurines and over 2,000 small grey pellets were found (Vandiver *et al.* 1989). Ninety percent of the pieces are grey to black in colour, which indicates firing at low temperature in a reducing atmosphere. The remaining ten percent are yellow, red or brown, often with black ash stains.

Vandiver *et al.* (1989) mention among their recognisable animal figurine inventory a series of felines, a mammoth, a bear, a rhinoceros, birds, a wolverine, bovine figures, and some herbivores. Many isolated animal legs or feet were also found, sometimes just in the shape of a stump, without any clearer determination. Most figurines are body parts, and often isolated heads or legs.

Table 3.2: inventory of fired clay pieces from Dolni Vestonice I after Vandiver *et al.* 1989.
(The probable figurines fragments are recognised because of the compound curvature of unbroken surface; slab-like objects are thin flattened fragments, slightly concave, 6 to 8 cm maximum length; the nearly spherical objects are balls 1 to 2 cm in diameter; the small pellets measure 4 to 10 mm and are irregular in shape).

Animal figurines	707 (of which 77 are nearly whole and 630 broken)
Human figurines	14 (of which one is nearly whole and 13 broken)
Probable figurines fragments	Over 3,000
slab-like objects	20
nearly spherical objects	20
Small pellets	Over 2,000
Total	Over 5,751

Vandiver *et al.* (1989, 1990) mention 77 pieces as 'nearly whole' animal figurines in the collections from Dolni Vestonice I from the Moravian Museum in Brno. However, it was not possible to find 77 of these when examining the collections (by taking 'nearly whole' as indicating that a large proportion of the original figurine is present, with only small fragments missing). The number of complete figurines found would rather be closer to 15 in my analysis, with only the following missing small fragments: DVI 1, DVI 3 and 4, DVI 16, DVI 29, DVI 30, DVI 62 and DVI 27 (the reference numbers correspond to the catalogue in Appendix 2). It is true that Vandiver *et al.* (1989, 1990) do not define what they mean by nearly whole. The large majority of the animal figurines are now extremely fragmentary.

Figure 3.7: The well-known Dolni Vestonice 'Black Venus' made of fired clay.

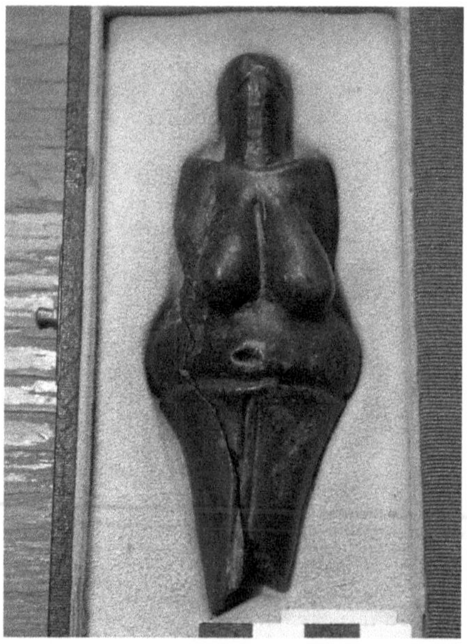

A large part of the fired clay collection (over 2,040 pieces) is made of non-figurative fragments such as the pellets, slabs or spherical objects as described in Table 3.2.

The rest is thought to come from figurative ceramics. These can represent human figures such as the well-known 'Black Venus' (figure 3.7) and animal figures.

Table 3.3 summarises our own inventory of the ceramic objects from Dolni Vestonice I by types as seen in the collections of the Moravian Museum in Brno. The reference numbers correspond to the catalogue in Appendix 2.

This inventory will be used as the base for the subsequent analysis of the objects.

Many objects, such as the Dolni Vestonice I Black Venus, have an artificial shiny surface caused by a conservation treatment consisting of consolidation with polyvinyl acetate. They would originally have had a matt surface texture (Vandiver *et al.* 1990) but the treatment has changed the appearance of the objects, mainly their shine and original colour, and has unfortunately robbed us of any possibility of gaining information from detailed surface examination of the pieces, like details of the fashioning or indication of firing temperature from the colour (M. Gonysevova, pers. com. May 2004). This is true for most of the ceramics from Dolni Vestonice and Pavlov.

The figurative ceramic inventory from Dolni Vestonice I has been divided into categories and sub-categories. The first category is made of human representations which are sub-divided into 'female' (female without doubt), 'probably female' (recognisable as female by comparison with known figurines; mostly fragmentary), 'male' (without doubt) and 'anthropomorphs' (no indication of sex). The second category gathers the animal figurines. Subdivisions have been made only for the two most numerous animal types, the felines and the mammoths. The third category of figurative pieces has been called here 'body extremities' as it is made up of legs, possible trunks or muzzles. The pieces described as 'rods or cones' in Soffer and Vandiver's inventory (table 3.5) have their place here. This category is by far the most numerous with 73 pieces in Dolni Vestonice I. The 'other' category gathers pieces which are too shapeless or broken to have a recognisable shape, or again fully shaped pieces in which one cannot recognise a human or animal shape.

Table 3.3: Descriptive inventory of the 132 ceramic pieces from Dolni Vestonice I seen at the Moravian Museum in Brno. The reference numbers correspond to the numbers given in the catalogue in Appendix 2. The numbers in brackets indicate the quantities of objects for each category (DVI 3 and DVI 4 bear different reference numbers but are in fact two fragments of the same figurine).

Humans (10)				Animals (30)		Body extremities (feet, legs, muzzles, trunks) (73)	Other (18)	
Female figurine (4)	Probably female (2)	Male figurine (1)	Anthropomorph (3)	Feline (6)	Other animals (24)		Undetermined with fashioning traces (9)	Balls (9)
DVI 1 DVI 3 and 4 DVI 7 DVI 8	DVI 2 DVI 5	DVI 10	DVI 6 DVI 9 DVI 13	DVI 15 DVI 18 DVI 20 DVI 21 DVI 23 DVI 24	DVI 14 DVI 16 DVI 17 DVI 19 DVI 22 DVI 25 DVI 26 DVI 27 DVI 28 DVI 29 DVI 30 DVI 31 DVI 32 DVI 33 DVI 34 DVI 36 DVI 37 DVI 38 DVI 39 DVI 40 DVI 41 DVI 51 DVI 62 DVI 63	DVI 11, DVI 12, DVI 35, DVI 42, DVI 43, DVI 44, DVI 46, DVI 47, DVI 48, DVI 49, DVI 52, DVI 53, DVI 55, DVI 56, DVI 57, DVI 61, DVI 64, DVI 65, DVI 66, DVI 67, DVI 68, DVI 69, DVI 70, DVI 71, DVI 72, DVI 73, DVI 74, DVI 75, DVI 85, DVI 86, DVI 88 , DVI 89, DVI 90, DVI 91, DVI 92, DVI 93, DVI 95, DVI 97, DVI 98, DVI 99, DVI 100, DVI 101, DVI 102, DVI 103, DVI 104, DVI 105, DVI 106, DVI 107, DVI 108, DVI 109, DVI 110, DVI 111, DVI 112, DVI 113, DVI 114, DVI 115, DVI 116, DVI 117, DVI 118, DVI 119, DVI 120, DVI 121, DVI 122, DVI 123, DVI 124, DVI 125, DVI 126, DVI 127, DVI 128, DVI 129, DVI 130, DVI 131, DVI 132,	DVI 45 DVI 54 DVI 58 DVI 60 DVI 76 DVI 87 DVI 94 DVI 96 DVI 133	DVI 50 DVI 77 DVI 78 DVI 79 DVI 80 DVI 81 DVI 82 DVI 83 DVI 84

Dolni Vestonice II

In the 1985, a rescue excavation was carried out about 1 km from the original site of Dolni Vestonice I, when a new industrial exploitation of the loess was initiated above the ancient brickyard near the village of Dolni Vestonice. It unearthed the remains of the site of Dolni Vestonice II (Valoch 1996, Svoboda 1991), which was excavated until 1987 (Vandiver *et al.* 1989).

The site is composed of three large concentrations of archaeological material, called agglomerations by Svoboda (ed.) (1991: 6). They comprise several isolated settlement units and a mammoth bone accumulation (Figure 3.8). The first settlement agglomeration excavated in 1986 was oval-shaped and was found in the highest part of the site on the foothills. The well-known triple burial was found in its southern part. The second agglomeration was excavated in 1987 and was oval-shaped. The third agglomeration, also excavated in 1987, was located on the western slope and yielded a male burial. Three isolated settlement units were located at the lower part and closer to the uppermost settlement agglomeration there was another settlement unit. The mammoth bone deposit extended about 150 m to the west from the western edge of the settlement (Svoboda (ed.) 1991).

A few elements seem to indicate a rather short or temporary occupation in comparison to the sites of Dolni Vestonice I or Pavlov I (Svoboda (ed.) 1991), in a changing environment. For example, the dwellings are rather insubstantial surface structures. The representational art is very limited, as well as the production of fired clay, and the ash lenses are smaller than in the neighbouring sites (Svoboda (ed.) 1994).

Figure 3.8: Dolni Vestonice II, Plan of the upper part of the site. The western slope, the mammoth bone deposit and the human burials DV XII-XV and DV XVI are indicated. The settled areas are dotted. (From Svoboda *et al.* 1996).

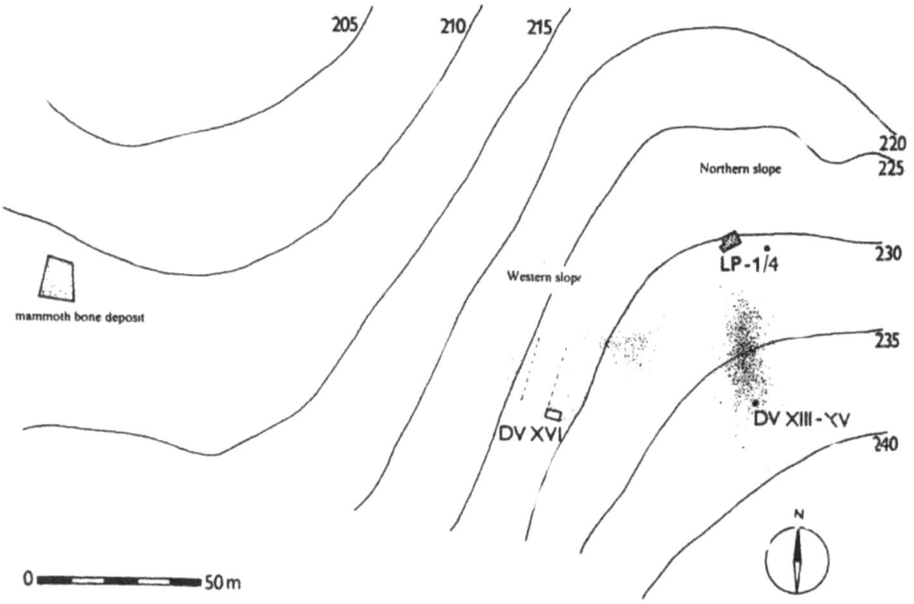

Due to the recent excavations (1985-1987) and the intensive research carried out at Dolni Vestonice II, the site has the largest series of radiocarbon dates for the Dolni Vestonice-Pavlov settlements (Svoboda (ed.) 1994). There seem, however, to be some problems as the dates coming from different laboratories were very different for similar materials: the Gröningen laboratory gave a series of dates between 27,500 and 25,000 BP; dates from Prague were between 25,000 to 22,000 BP; and dates by Illinois laboratory were between 26,690 and 22,630 BP (Svoboda (ed.) 1991). The authors (Svoboda *et al.* 1996) decided to retain only the Gröningen dates as they did show certain coherence with the other known dates and could then be applied to interpretation.

The ceramics are few in comparison with Dolni Vestonice I and Pavlov I, as is representative art. This fits in well with the generally accepted idea that Palaeolithic portable art is mainly associated with long-term settlements. (Svoboda (ed.) 1991).

The ceramics were found in the western slope excavation (Svoboda (ed.) 1994) and described as six pellets and a fragment of reindeer nose with incisions concentrated in the depressions in the first settlement unit. All the objects there seem related to the central hearth (Svoboda (ed.) 1991: 20).

It was not possible to find more literature about the ceramics from Dolni Vestonice II, but some pieces other than the ones mentioned above were seen in the Anthropological Institute at Dolni Vestonice (Table 3.4) and are clearly labelled as coming from the site. Apart from the reindeer nose (reference DVII 1) and pellets mentioned above, they are: the head of a female figurine (DVII 5), the body of a female figurine (DVII 6), a feline head (DVII 4), another possible feline head (DVII 3), a large shaped fragment (DVII 7) and a smoothed shaped fragment (DVII 2).

Table 3.4: Descriptive inventory of the 13 ceramic pieces from Dolni Vestonice II seen in the Anthropological Institute in Dolni Vestonice, with their reference numbers from the catalogue in Appendix 2. The numbers in brackets indicate the quantities of objects for each category.

Animal figurine (3)			Female figurine (1)	Probably female figurine (1)	Rounded fragments (6)	Other (2)
horse (1)	Felines (2)	other	DVII 6 body	DVII 5 head	6 rounded fragments, under 1.5 cm in size	DVII 7 large worked fragment DVII 2 smoothed fashioned fragment
DVII 1 muzzle of	DVII 4 feline head	DVII 3 fragmentary head with small pointy ears				

Pavlov I

B. Klima discovered the site from surface finds while he was working in Dolni Vestonice I. He excavated here between 1952 and 1965 and again between 1971 and 1972 (Valoch 1996). So far three volumes of studies have been published on the different areas excavated at Pavlov I: the 1952-1953 excavation in the south-eastern area (Svoboda ed. 1994); the North-western part excavated in 1957-1958 (Svoboda ed. 1997) and the South east part excavated in 1954 and 1956 (Svoboda ed. 2005). The middle part still remains to be published (Svoboda 2005)

Pavlov I is an extremely rich site that is the eponymous site for the Pavlovian culture. Various radiocarbon dates from the site seem to all fit between 26,700 and 25,000 BP (Svoboda *et al.* 1996).

Some habitation structures were recognised. They seem to be semi-subterranean (Svoboda 1994) with a central hearth or hearths, with kettle-shaped pits on the surrounding floors and sometimes with large stones or bones along the external margin of the dwellings (Svoboda *et al.* 1996). In the 1952-53 excavation area, the structures were interpreted as two huts next to each other, 4 to 5 m in diameter, surrounded by activity areas with hearths and mammoth bones deposits (Svoboda 1994). It is the only site in Moravia where a storage pit comparable to the ones known from Eastern Europe has been found. Thick ash deposits are also found in Pavlov I with concentrations of fired clay and decorative objects around hearths. The duration of the occupation seems quite long (Svoboda 1994).

The site produced a large record of the earliest known ceramics, the earliest evidence for textile or basketry (impressions in fired clay), evidence for early ground stone technology (Svoboda 1997), many pieces of portable art, a male burial, antler, bone and stone tools (Svoboda *et al.* 1996).

The lithic industry is represented by over 100,000 pieces, among which a few thousands retouched tools are found that overall correspond to other known Gravettian assemblages of the Austrian-Moravian-Silesian territory by the dominance of burins over endscrapers. Some aspects of the lithics however, such as the large number of microlithic backed implements and other microliths is unique to Pavlov I (Svoboda 1994: 54; Klima 1997c).

The faunal remains are mainly hare, reindeer, polar fox, mammoth, wolves, horse, birds, as well as some rare bears, felines and rhinoceros (Musil 1997). The quantity of figurative ceramic art found at Pavlov I will allow comparisons with the hunted fauna at the site: this will be developed further in relation to the study of the themes of animal representations.

The bone tool industry is varied with borers, baguettes, spades made of mammoth bone, pierced batons and axes made of antler (Valoch 1996).

Portable art at the site is very rich with ornaments in ivory and shells, pierced teeth, ivory rings and numerous pendants. Many statuettes in ivory, bone and fired clay have been found, mainly representing animal and female figurines. An engraved mammoth tusk has been interpreted as a possible map representing the topography of the site and could be the oldest known map (Valoch 1996).

The burial of an adult male in flexed position on the right side was also found under an incised mammoth shoulder blade (Klima 1997a).

As for the fired clay inventory, 4,445 ceramics objects were found in Pavlov I (1,332 in the 1952-1953 excavation (Soffer and Vandiver 1994: 167) and 3,113 in the 1957 excavation) (Soffer and Vandiver 1997). They are mainly figurines, figurine fragments and pellets similar to Dolni Vestonice' s, but different sorts of flat fragments, usually unfired or low fired, are also found in Pavlov. Soffer and Vandiver (1997) have interpreted them as possible structural ceramics or accidentally fired architectural elements, pointing out that unfired clay could have had a wider utilitarian use in Pavlov I than in Dolni Vestonice I.

The inventory from Pavlov I is slightly different from Dolni Vestonice I as it consists of smaller ceramic pieces (Soffer and Vandiver 1994) and seems to include both fired and unfired objects, grey to yellow in colour (Soffer *et al.* 1993a).

The objects from Pavlov I are the most studied of all the ceramic evidence from the Czech sites, again by Soffer and Vandiver. As they had previous experience of the study of the Dolni Vestonice I ceramics, they remarked that they felt more capable of recognising the objects (Soffer and Vandiver 1997).

Table 3.5 summarises the ceramic inventory from Pavlov I published by Soffer and Vandiver (1994, 1997,

2005), which totals at 5,716 pieces. In this inventory, only the material excavated between 1952 and 1958 at Pavlov I is taken into account. Ceramic objects have also been found in the later excavations of the site but they are not published yet (Svoboda pers. com. May 2004).

The unworked pieces of their inventory are fragments of fired clay that do not show any sign of being shaped or smoothed. All the worked pieces have at least one side bearing traces of smoothing and shaping (Soffer and Vandiver 1994).

Table 3.5: Published Inventory of the ceramic objects from Pavlov I compiled after Soffer and Vandiver 1994, 1997, 2005).

Unworked pieces	2000		
Worked pieces 3716	Worked non-figurative 3572	Pellets	3000
		Flat fragments	301
		Slab-like fragments	129
		Spalls	12
		Rods or cones	30
	Worked figurative 144	Anthropomorph	18
		Zoomorph	30
		Anthropomorph or zoomorph	96

Soffer and Vandiver (1997, 2005: 420) define the descriptive terms used in their worked non-figurative categories in the table above as follows:

- Pellets are fragments that show smoothing on more than two of their surfaces, even if they add that they suspect this smoothing could be the result of the method of wet sieving on metal screens used to recover them for the low fired or unfired pieces (1997:389); part of the pellets however, the ones black in colour, are fired at high temperature and thus were originally intentionally fashioned and fired as 'pellets' (2005).
- Flat fragments are rectangular in shape, with smoothing on just one surface; likely fragments of larger clay constructions.
- Slab-like fragments are thinner in profile than the flat fragments, slightly bowed, and have evidence of working or smoothing on at least two of the flat surfaces. They are small in size and range in thickness from 0.6 to 1.5 cm with most measuring 1 cm or less. Some could be interpreted as daub applied on various supports (Soffer and Vandiver 1994, 1997, 2005: 420).
- Spall fragments have one flat or slightly concave unworked side and a convex and well smoothed opposite side. They are broken off a larger body (2005: 421).
- Rods or cones have elongated rounded contours with one end larger in diameter than the other, and they are broken only at one end by mechanical breakage.

As for the nature of the figurines made of ceramic, Soffer and Vandiver (1994, 1997) describe 18 anthropomorphs, 30 zoomorphic figures, and 96 pieces that could be anthropomorphic or zoomorphic. This makes a total of 144 figurative figurines from PI.

Again, as for the Dolni Vestonice material, our own inventory based on the pieces seen in the Institute in Dolni Vestonice does not match the one published by Soffer and Vandiver (Table 3.6). This time, however, our inventory of figurative pieces is actually larger that the one they publish with 178 pieces described, out of which 156 represent humans or animals. One can only assume that this has to do with the fact that all the ceramic pieces from the various years of excavation of the site of Pavlov I are kept together in the collections whether they have been published or not.

Our figurative ceramic inventory from Pavlov I has been divided into categories and sub-categories (Table 3.6). The first category is made of human representations which are sub-divided into 'female' (female without doubt), 'probably female' (recognisable as female by comparison with known figurines; mostly fragmentary), 'male' (without doubt) and 'anthropomorphs' (no indication of sex). The second category gathers the animal figurines. Subdivisions have been made only for the two most numerous animal types, the felines and the mammoths. The third category of figurative pieces has been called here 'body extremities' as it is made up of legs, possible trunks or muzzles. The pieces described as 'rods or cones' in Soffer and Vandiver's inventory (table 3.5) have been added here. This category is by far the most numerous with 117 pieces in PI. The 'other' category gathers pieces which are too shapeless or broken to have a recognisable shape, or again fully shaped pieces in which one cannot recognise a human or animal shape.

Table 3.6: Descriptive Inventory of the 178 ceramic pieces from Pavlov I seen in the Anthropological Institute in Dolni Vestonice. The reference numbers correspond to the numbers from the catalogue in Appendix 2. The numbers in brackets indicate the quantities of objects for each category.

Humans (10)				Animals (29)			Body extremities (feet, legs, muzzles, trunks) (117)	Other (22)
Female (1)	Probably female (5)	Male (1)	Anthropomorph (3)	Feline (4)	Mammoth (4)	Other animals (21)		
PI 6	PI 1	PI 9	PI 4	PI 17	PI 22	PI 11	PI 10, PI 28, PI 32, PI 33, PI 35, PI 36,	PI 40
	PI 2		PI 8	PI 19	PI 23	PI 12	PI 37, PI 38, PI 39, PI 44, PI 51, PI 61,	PI 41
	PI 3		PI 115	PI 20	PI 24	PI 13	PI 62, PI 63, PI 68, PI 69, PI 70, PI 71,	PI 42
	PI 5			PI 21	PI 26	PI 14	PI 73, PI 74, PI 75 PI 76, PI 77, PI 78,	PI 43
	PI 7					PI 15	PI 79, PI 80, PI 81, PI 82, PI 83, PI 84,	PI 46
						PI 16	PI 85, PI 86, PI 87, PI 88, PI 89, PI 90,	PI 48
						PI 18	PI 91, PI 92, PI 93, PI 94, PI 95, PI 96,	PI 49
						PI 25	PI 97, PI 98, PI 99, PI 100, PI 101, PI	PI 52
						PI 27	102, PI 103, PI 104, PI 105, PI 106, PI	PI 54
						PI 29	107, PI 108, PI 109, PI 110, PI 111, PI	PI 55
						PI 30	112, PI 113, PI 116, PI 117, PI 118, PI	PI 56
						PI 31	119, PI 120, PI 121, PI 122, PI 123, PI	PI 57
						PI 34	124, PI 126, PI 127, PI 128, PI 129, PI	PI 58
						PI 45	130, PI 131, PI 132, PI 133, PI 134, PI	PI 59
						PI 47	135, PI 136, PI 137, PI 138, PI 139, PI	PI 60
						PI 50	140, PI 141, PI 142, PI 143, PI 145, PI	PI 64
						PI 53	146, PI 147, PI 148, PI 149, PI 150, PI	PI 65
						PI 114	151, PI 152, PI 154, PI 155, PI 156, PI	PI 66
						PI 125	157, PI 158, PI 159, PI 160, PI 162 PI	PI 67
						PI 144	163, PI 164, PI 165, PI 166, PI 167, PI	PI 72
						PI 169	168, PI 170, PI 171, PI 172, PI 173, PI	PI 153
							174, PI 175, PI 176, PI 177, PI 178	PI 161

Pavlov II

Fired clay was also found in Pavlov II (Soffer *et al.* 1993). The site which is located 350 m to the south of Pavlov I was excavated in 1966-67 after some test pits were dug in 1960. The living surface seems restricted to a 40 square metre area (Djindjian *et al.* 1999).

The site is smaller than Pavlov I, but culturally comparable, with a concentration of artefacts and animal bones around five hearths, axially arranged (Svoboda *et al.* 1996). As at Dolni Vestonice II, the presumed habitations are rather light surface structures. The site lacks representational art. The production of fired clay was limited and the ash deposits are small. This seems to indicate a short or temporary occupation (Svoboda 1994).

The lithic industry shows many burins, the presence of scrapers, backed pieces, microliths, and has similar lithics to Dolni Vestonice. Faunal remains consist mainly of mammoth, reindeer, horse, hare and carnivores. A complete lion leg in articulation suggests the presence of an animal hide. There were also some fragments of ivory points (Valoch 1996), 33 pendants in schist, 67 pierced tertiary shells, some pierced teeth, and 275 blocs of iron oxide ochre (Djindjian *et al.* 1999).

As for fired clay, some fragments were found but in much smaller numbers than in Pavlov I and Dolni Vestonice I (Valoch 1996).

Valoch mentions that 135 small fragments of fired clay were found in Pavlov II, nine of which were modelled (Valoch 1996). Soffer *et al.* (1993) confirm that there are ceramics in Pavlov II, without giving any more details.

The themes represented in the ceramic art

The interpretation of the figurines described in this section sometimes follows published descriptions, sometimes disagrees with them. When no published description was available, as for the majority of the ceramic objects, the interpretation given is our own and as often in the field of Upper Palaeolithic art, can sometimes be debated.

The themes found represented in the ceramic art are summarised in the tables below with the corresponding percentage of the total pieces they express (tables 3.7, 3.8, 3.9 and 3.10).

Table 3.7: Quantitative inventory of fired clay objects with the themes represented from Dolni Vestonice I (DVI), Dolni Vestonice II (DVII) and Pavlov I (PI). This inventory is drawn from our own database.

Theme		Dolni Vestonice I	Dolni Vestonice II	Pavlov I	Total	% of total
Human	Female figurine	4	1	1	6	1.9
	Probably female	2	1	5	8	2.5
	Male	1		1	2	0.6
	Anthropomorph	3		3	6	1.9
Animals	Feline	6	1	4	11	3.5
	Mammoth	1		4	5	1.6
	Rhinoceros	1		1	2	0.6
	Bear	1			1	0.3
	Bird	3			3	0.9
	Bovid	1			1	0.3
	Horse		1	3	4	1.3
	Reindeer	1		1	2	0.6
	Ibex			1	1	0.3
	Mustelid	1			1	0.3
	Undetermined	15	1	15	31	9.8
Body extremities		73		117	190	60.3
Other	Balls	9			9	2.8
	Other worked fragments	9	2	22	33	10.5
Total		131	7	177	315	
% of total		41.6	2.2	56.2		100

The following analysis is based on our own observations of the figurines in the museums where they are kept and the data found in tables 3.7, 3.8, 3.9 and 3.10 come from our own study of the objects. As described earlier in this chapter, these inventories do not match the ones published by Soffer and Vandiver. The various recognised themes of representation are studied for the ceramic record considered here from Dolni Vestonice I and II and Pavlov I. The main categories of theme represented are: human representations, animal representations, body extremities and others, among which some elements that could be described as abstract or schematic.

The human category is made up mostly of female figurine fragments (14 altogether), 2 male figurine and six undetermined anthropomorphs. We will see that some of the body extremities can also be considered as human as they have a heel and a flat sole.

The most represented animals found in the ceramic art in Dolni Vestonice and Pavlov are felines and mammoths. There are also some recognisable rhinoceroses, horses, birds (owls?), caprine, reindeer, bear and at least one bovid.

The body extremities category describes modelled fragments of what appears to be limbs, with or without feet, or possibly of muzzles. They make up by far the most numerous category of representations at the Dolni Vestonice-Pavlov site with 190 pieces. It has been possible to determine some sub-groups among the numerous feet and legs according to their shape and size.

The objects gathered under the heading 'Others' do not fit in any of the previous categories. A number of them could be described as abstract or schematic. This is a category that has not been considered in Soffer or Vandiver's studies as they only look at figurative representations of animals or humans. This will be developed further here.

Table 3.8: Quantitative inventory of fired clay objects with the themes represented from Dolni Vestonice I (DVI)

Theme		DV I	% of total
Human	Female figurine	4	3
	Probably female	2	1.5
	Male	1	0.8
	Anthropomorph	3	2.2
Animals	Feline	6	4.6
	Mammoth	1	0.8
	Rhinoceros	1	0.8
	Bear	1	0.8
	Bird	3	2.2
	Bovid	1	0.8
	Horse		
	Reindeer	1	0.8
	Ibex		
	Mustelid	1	0.8
	Undetermined	15	11.5
Body extremities		73	55.7
Other	Balls	9	6.8
	Other worked fragments	9	6.8
Total		131	100

Table 3.9: Quantitative inventory of fired clay objects with the themes represented from Pavlov I (PI)

Theme		PI	% of total
Human	Female figurine	1	0.6
	Probably female	5	2.8
	Male	1	0.6
	Anthropomorph	3	1.7
Animals	Feline	4	2.2
	Mammoth	4	2.2
	Rhinoceros	1	0.6
	Bear		
	Bird		
	Bovid		
	Horse	3	1.7
	Reindeer	1	0.6
	Ibex	1	0.6
	Mustelid		
	Undetermined	15	8.5
Body extremities		117	65.1
Other	Balls		
	Other worked fragments	22	12.4
Total		177	100

Table 3.10: Quantitative inventory of fired clay objects with the themes represented from Dolni Vestonice II (DVII)

Theme		DVII	% of total
Human	Female figurine	1	14.3
	Probably female	1	14.3
Animals	Feline	1	14.3
	Horse	1	14.3
	Other	1	14.3
Other worked fragments		2	28.5
Total		7	100

The human representations

Between the sites of Dolni Vestonice I, Dolni Vestonice II and Pavlov I, a total of 22 human representations have been found. They make up 7 % of all the ceramic pieces studied for the three sites (Table 3.7). Among them, 1.9 % are female figurines, 2.5 % are probably female, 0.6 % are male and 1.9 % are non-sexed anthropomorphs.

A distinction has been made here between the 'female' figurines which are without doubt female as they show specific attributes such as breast or sex, and 'probably female' figurines, which are interpreted as female by comparison with other known female figurines as they are too fragmentary to show any sexual attribute. As the interpretation of some of these pieces can be subject to debate, it has seemed useful to create these categories.

Also, it will be seen further that some pieces in the body extremities category can be interpreted as human, which changes the statistics of human representations. These pieces will be considered in the body extremities section.

Table 3.11: Inventory of the fired clay human figurines from Dolni Vestonice I (DVI), Dolni Vestonice II (DVII) and Pavlov I (PI).

	DVI	DVII	PI	Total	% of total
Female	4	1	1	6	36.4
Probably female	2	1	5	8	27.3
Male	1		1	2	9
Anthropomorph	3		3	6	27.3
Total of human figurines	10	2	10	22	100

- **The female figurines**

The majority of the human representations from the sites of the Pavlovské Hills are female or probably female with 14 pieces out of 22 known human figures, representing 63.7 % of this category (table 3.12). Among the individually recognisable pieces (i.e. to which a precise species or gender can be attributed), this is the category with the highest number, just before the felines.

The domination of the female theme of representation echoes perfectly the Gravettian cultural background of the ceramic finds at Dolni Vestonice and Pavlov: the representation of human females is one of the main characteristic of this culture, after the lithic industry (White 2003: 82).

The figurines were identified as female mostly by comparison with already known pieces as many are fragmentary. Indeed, some 'models' seem to have been copied repeatedly. This is particularly true for the 'Black Venus' type at the Dolni Vestonice locale. Some specific elements (belts and 'hats') also seem to be repeated on a number of figures.

Table 3.12: The female figurines in ceramic from the sites of Dolni Vestonice I, Dolni Vestonice II and Pavlov I.

Female figurines	DVI 1 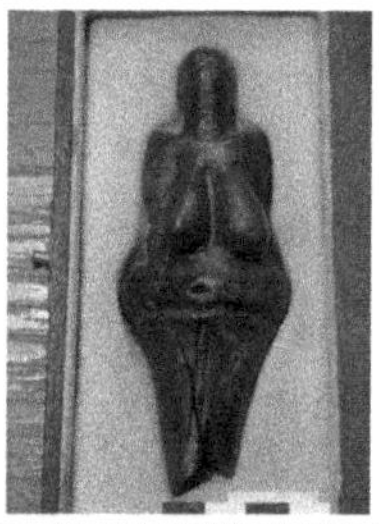	DVI 3 and 4 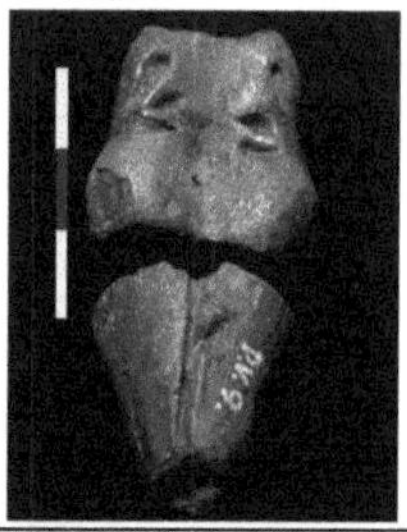	DVI 7 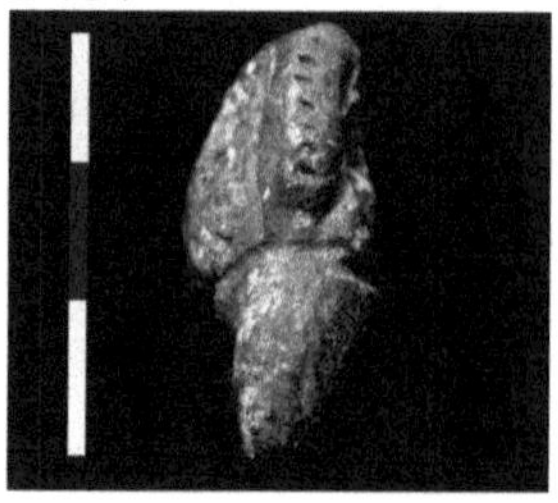	DVI 8 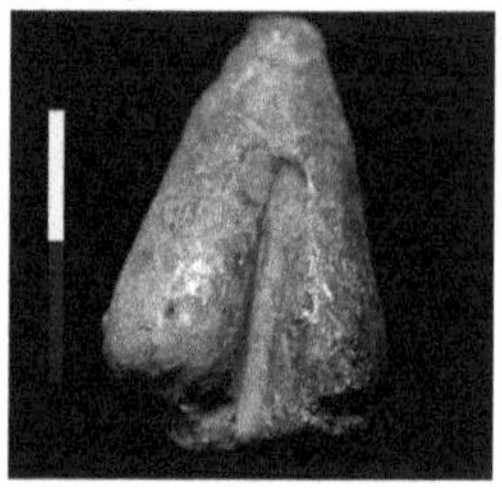
	DV II 6 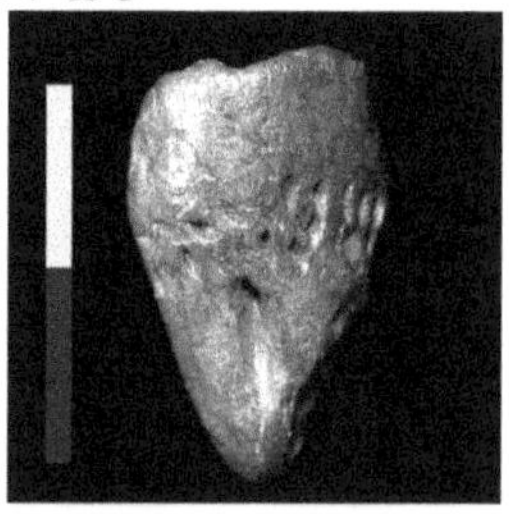	PI 6 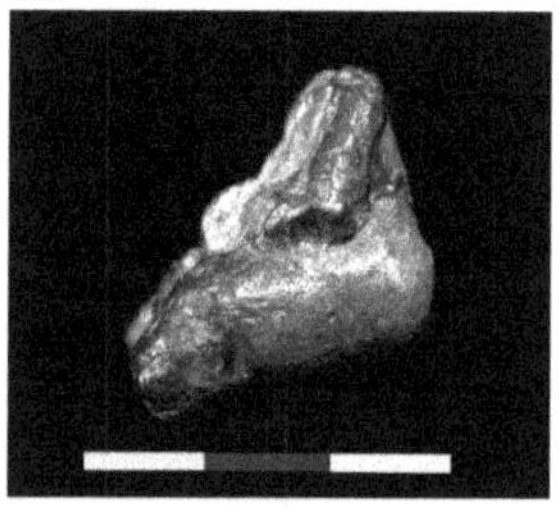		
Probably female figurines	DVI 2 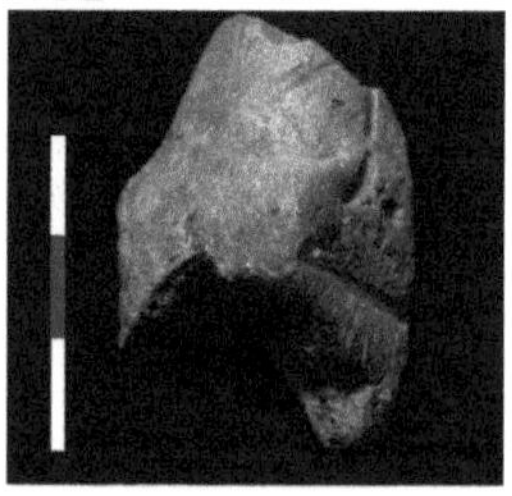	DVI 5 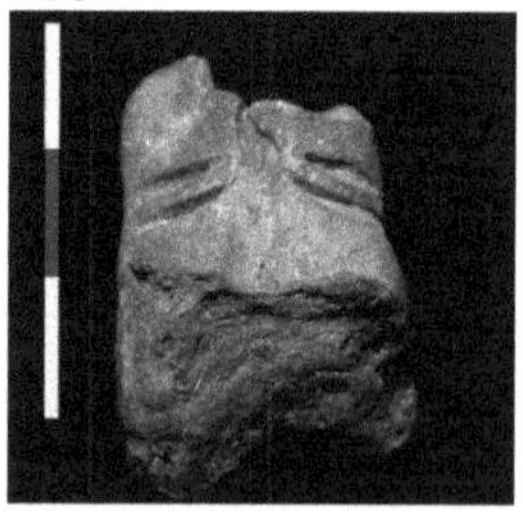	DVII 5 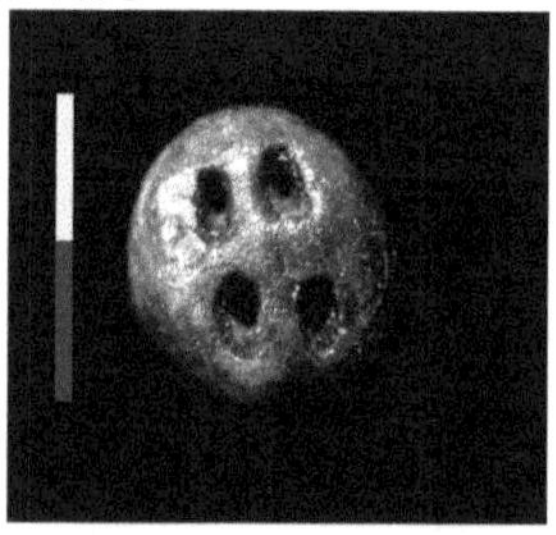	PI 1 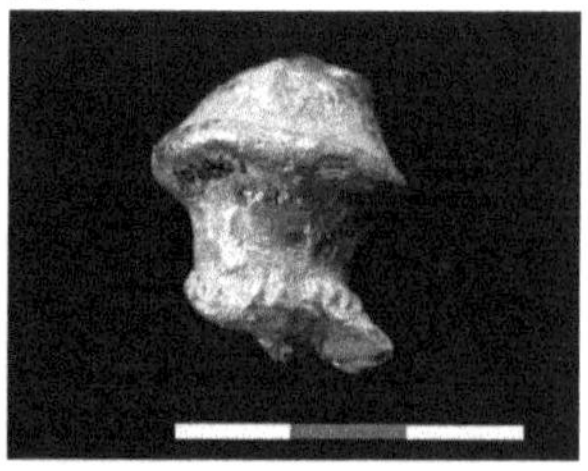
	PI 2 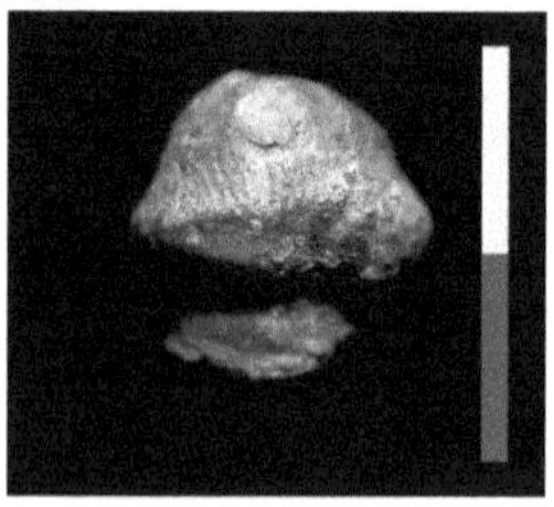	PI 3 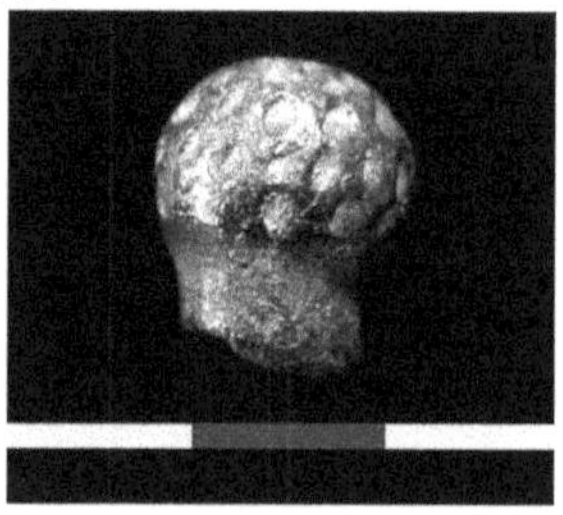	PI 5 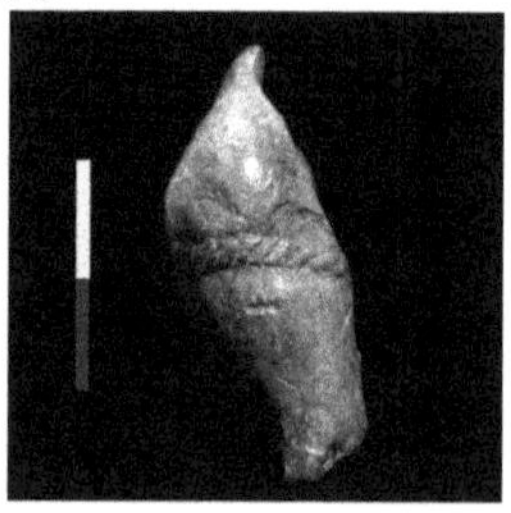	PI 7 

A 'Black Venus' model

The well-known 'Black Venus' (DVI 1; figure 3.7 and 3.12) from Dolni Vestonice is the most complete female figurine of the inventory even if a small area at the bottom of the legs is missing. It is striking to see that five other female/probably female figurines show very similar characteristics as if all were following a similar model.

The back of the Black Venus (figure 3.12) has been fashioned by making a deep groove to represent the spine. This groove then continues without interruption to mark the separation of the legs On the upper part of the torso, on each side of this groove, two parallel short lines are found in a symmetrical position, possibly indicating some form of body decoration (tattoos or body painting), unless they simply represent the folds of the skin of this rather plump lady.

Figure 3.12: the back of the Black Venus (after Bosinski 1990: 126).

These short parallel lines are also found on the back of two fragmentary figurines: DVI 3 and 4[17], and DVI 5 (figure 3.13) show similar parallel marks on each side of their spine. Some variation is clearly visible however: on DVI 5, although more fragmentary (only part of the bust has survived) the markings are exactly parallel and at an angle, on a very similar manner to the ones visible on the Black Venus. DVI 3 and 4, although more complete (only the head is missing) feels more like a variation on the Black Venus model: it shows similar traits but they seem done by another 'hand'. The markings on the side of the spine are more approximate and not strictly parallel; the groove that marks the separation of the legs is also deeply incised but it is not strictly straight. There is a general feel that it was done copying the main model but quickly and without as much care.

Figure 3.13: similar manners to represent the back of female figurines:

a) The back of figurine DVI 3 and 4

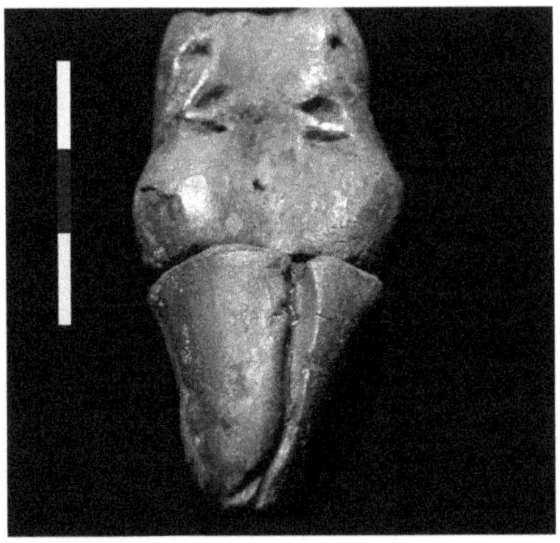

b) The back of figurine DVI 5

The head of the Black Venus is cylindrical, elongated, rounded on top, and is directly attached to the body with hardly no indication of the neck, although the collarbones area is beautifully fashioned. On the face two long slits seem to indicate the eyes, but there is no other realistic detail. An area starting at the junction between the 'eyes' until the base of the 'neck' seem to have been scraped with a microdenticulate tool that left very small parallel incisions. It is not possible, however, to determine if this was intended as the representation of some meaningful detail on the head, especially as it is very lightly marked. The top of the head shows four deep holes, oval in form, placed as the corners of a rectangular shape (figure 3.14).

One nearly identical head (in its remaining fragment) is DVII 5 (figure 3.15). It shows exactly the same shape and disposition for the holes, as well as a similar colour and surface texture. It is also similar in size. One could argue here that this piece was made by the same 'hand' that the Black Venus, even if in this case one side of the head is damaged and does not allow seeing if the eyes were similar.

Figure 3.14: the top of the head of the black Venus (DVI 1) with the four holes.

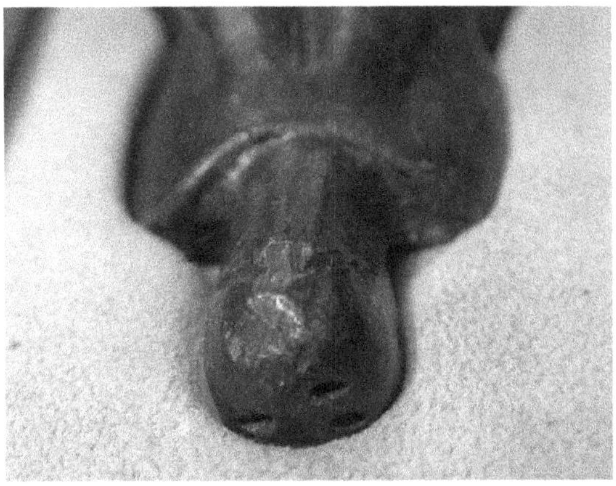

Figure 3.15: DVII 5: holes on top of the head of a Black Venus type.

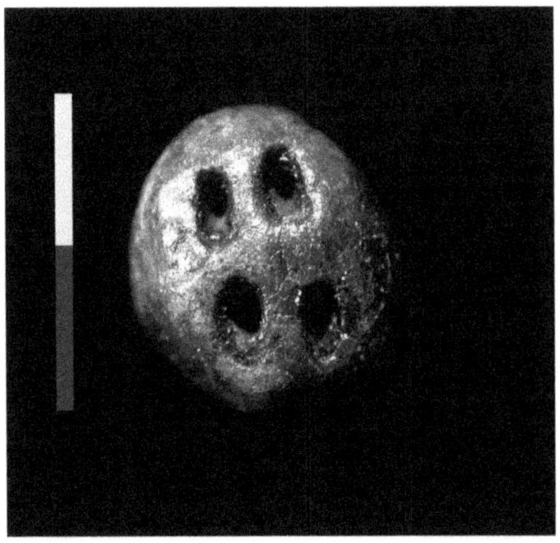

Another figurine fragment shows strong similarity with the Black Venus (figure 3.17); DVI 8 (figure 3.16) represents the breasts of a female figurine that have become detached from the body. Their overall oblong and elongated shape as well as the nature of the groove made to separate them follow a similar model as the Black Venus.

[17] This figurine has been given two numbers in the original inventory as it is broken into two at the belt. There is no doubt however that it is just one figurine.

Figure 3.16: DVI 8 breasts of a female figurine.

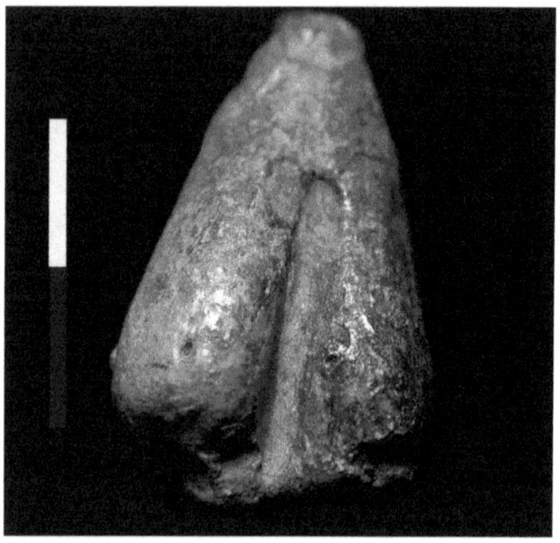

Figure 3.17: Bust of the Black Venus.

Finally the fragmentary figurine DVII 6 (figure 3.18:) seems to be a variation on the Black Venus model with the presence of characteristic elements such as a similar conical shape of the leg with deep grooves used to mark the separation both at the front and on the reverse. For the back, the groove of the leg continues towards the top to indicate the spine. However, there are also some marked differences especially for the belt which is not indicated here by a groove, but by the juxtaposition of small holes visible only at the front. This piece is also extremely small, measuring only just over two centimetres in length.

Jiri Svoboda thinks DVII 6 actually represents a vulva (personal communication February 2007), which is indeed arguable from the front. The fact that the back is fashioned in a similar way to other figurines, even if it is fragmentary, connects it, in our opinion, to the other female figurines represented at the Pavlovské Hills sites. It is completely possible, however, that a dual representation was intended on the same object, as has been argued for other portable art ivory objects of the Pavlovian.

It can be concluded, therefore, that by their size as well as by their strong similarities with the Black Venus model, the fragmentary figurines DVI 8 and DVII 5 probably belonged to extremely similar figurines that were made by the same person or by people working together according to a same pattern. The other figurines DVI 3 et 4, DVI 5 and DVII 6 follow a similar model but with some variations that makes them more individual. They are also much smaller in size.

Figure 3.18: DVII 6, female figurine body: front on top picture, back on the bottom one.

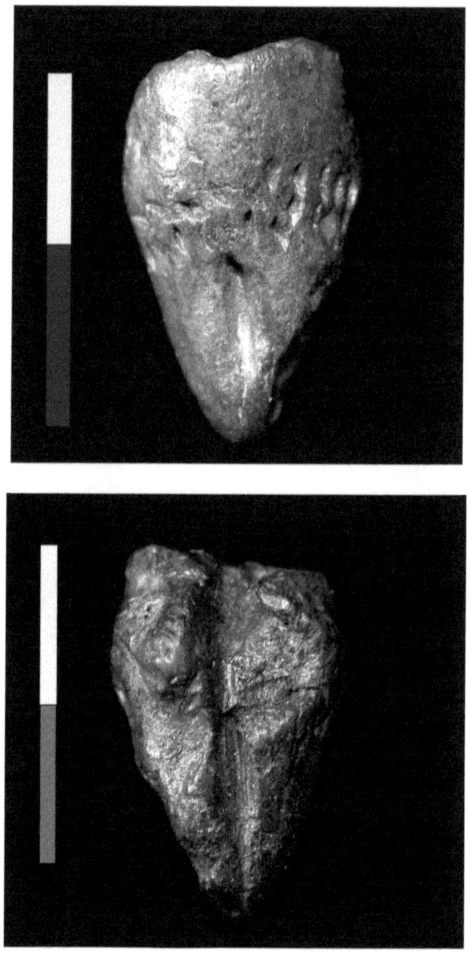

It is interesting to note that the use of this 'Black Venus' model is only found in Dolni Vestonice I and Dolni Vestonice II, and that indeed the only two human figurines found in Dolni Vestonice II are female figurines fragments made on this model. It is absent from Pavlov where other types of female figurines are encountered. This is obviously indicative of the transmission of some form of tradition in the appearance given to female figurines at these sites. Issues concerning the social context surrounding these 'cannons' of representation will be looked at in more detail in Chapter 4.

Another criterion for recognising female figurines as such among the fired clay fragments is by comparison with known elements usually found represented on other Gravettian female figurines.

Belts

One such criterion is the presence of a marked belt, which is actually also found on the Black Venus. It has been accepted here as an indicator of a female representation, even when no other elements indicating gender (breasts, pubic triangle or genitalia) were present.

DVI 7 (Figure 3.19) is one of the reasons for accepting this fact as there a belt is marked by an incision, and the figurine's sex is clearly indicated even if the breasts are not depicted. The belly button is also clearly marked as well as elements of body decoration (a circular shaped made of juxtaposed dots around the belly button, ornaments on the arms).

Figure 3.19: DVI 7 fragmentary female figurine from Dolni Vestonice I.

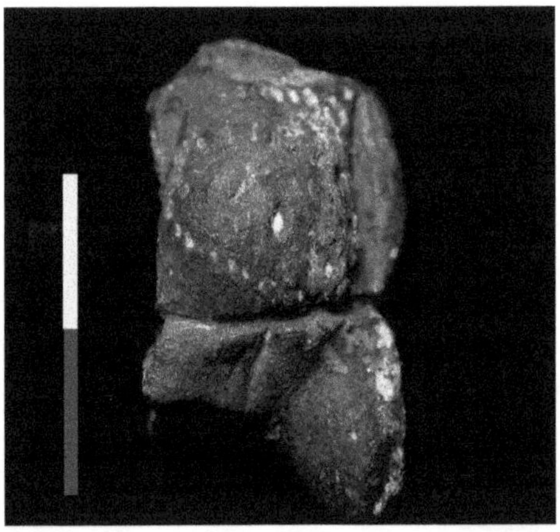

The fragmentary figurine DVI 2 (figure 3.20) has been interpreted as possibly female because of the presence of a belt and of a marked belly button. However, it is much broken.

Figure 3.20: fragmentary female figurine DVI 2

The figurine PI 5 (figure 3.21) has been identified as a fragmentary female figurine by its general profile as well as by the belt indicated. In this instance, however, the belt is made of a series of short parallel incisions on a ribbon in relief, located low on the hips as is found on a number of female figurines.

Figure 3.21: fragmentary female figurine from Pavlov I: PI 5

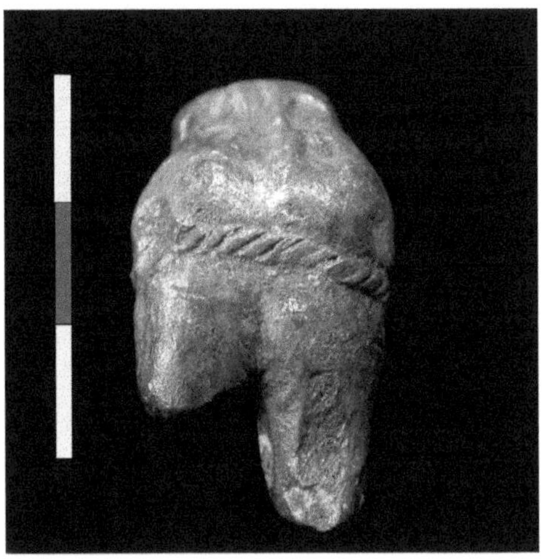

Figure 3.22: fragmentary female figurine from Pavlov I: PI 7

Figurine PI 7 (figure 3.22) has been interpreted as a fragmentary female representation as it shows a similar belt to PI 5. This is the only element indicating it could be a female figurine; the general shape is that of the lower part of a body (the top part is broken off) but no detail is fashioned, it forms a bloc and the separation of the legs is not indicated. The surface is also fairly damaged. However, the belt clearly follows the same model as the belt of PI 5. These are the only two figurines in the Pavlovské Hills sites with such elaborate belts. They do remind of the ornaments visible on some Gravettian figurines of Eastern Europe, for example at Kostenki I (figure 3.26) (Abramova 1995).

Heads

Three heads have also been identified as coming from female representations.

PI 1 (figure 3.23) can be interpreted as a head wearing a hat. It is detailed enough to see an ornament in relief at the base of the neck and the large rim of a cone-shaped hat. This object has been interpreted in the past as the extremity of a penis (Klima 1997b: 247). However, the presence of the necklace as well as the width of the rim of the hat which forms a marked angle with the 'face' area are elements that, in our opinion, go against this interpretation.

Figure 3.23: head of female figurine from Pavlov I: PI 1

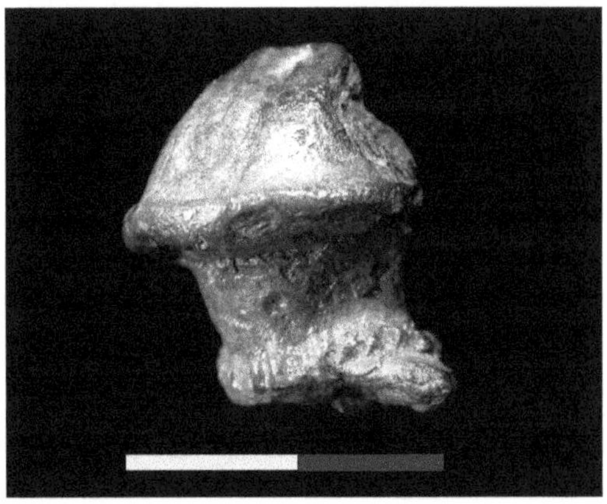

Figure 3.24: head of female figurine from Pavlov I: PI 2 head with hat.

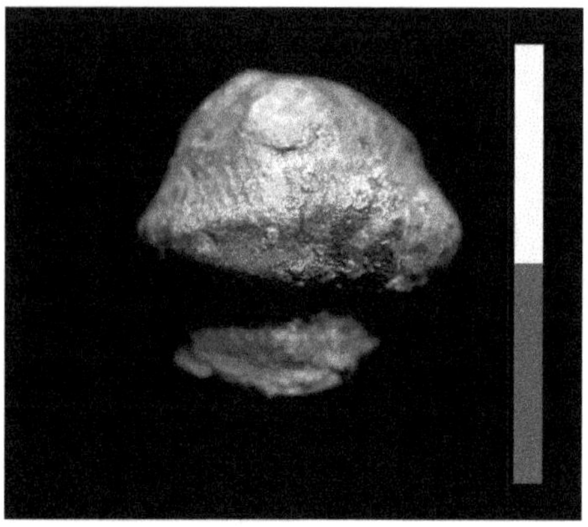

The object PI 2 (figure 3.24) is interpreted entirely by comparison with PI 1 as representing also a head with a hat as it shows a similar overall shape, although much less detailed. One could argue the two pieces were copied from a similar model but with less skill for PI 2. This is another case in which similar stylistic norms could be visible in the making of the Pavlovian ceramic figurines.

Figure 3.25: PI 3 head of a female figurine of the 'golf-ball' type from Pavlov I.

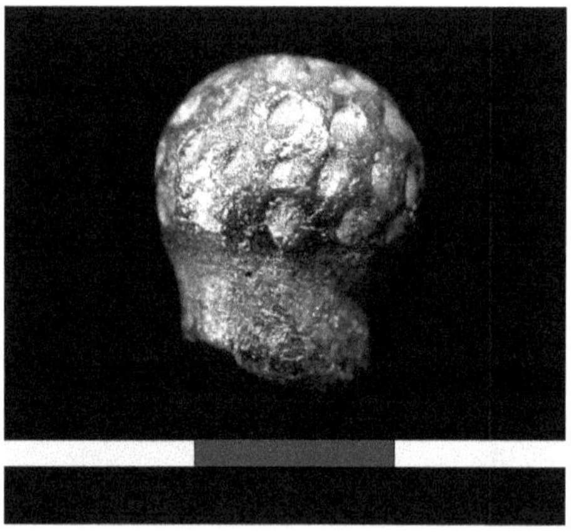

The third head, PI 3 (figure 3.25) is of another type. It is unique at the Pavlovské hills sites, but found elsewhere in Eastern Europe, notably at the site of Kostenki I in Russia (figure 3.26), although there the material used is ivory. This is just one of a number of striking common points between the Moravian sites and the sites of the Kostenki-Avdeevo culture in Eastern Europe.

The head is described as a 'golf-ball' head. The head PI 3 is round, like a small ball, planted on a smoothed neck. It is decorated all around with regularly placed circular depressions. There is no realistic indication of a face. It is broken at the neck. The surface colour is dark.

We have seen that out of the six representations of female figurines found in Dolni Vestonice I, four follow the 'Black Venus model'. The two other show as attributes a belt and a bellybutton. In Dolni Vestonice II, out of two figurines one can be described as a close copy of the Black Venus' head (DVII 5), the other (DVII 6) as a variation on its model. Six figurines out of eight do follow a similar model in Dolni Vestonice I and Dolni Vestonice II.

Figure 3.26: 'golf-ball' headed Gravettian female figurine from Kostenki I (from White 2003: fig. 110A and B).

On the other hand, in Pavlov I, out of the seven female figurines present, none resembles the 'Black Venus'. There seems to be much more variability in the types of female figurines represented with five different types found. Three of the pieces (PI 5, PI 7 and PI 3) show traits that have been found on female figurines from other Eastern European Gravettian sites. Obviously, information on the makers of the figurines can be seen through the study of theses objects with some knowledge being passed on for the models reproduced.

The high variety of types of female figurines found represented in Pavlov I also needs to be seen in the light of the usual consideration that the site was an aggregation site where human groups from different backgrounds gathered on regular occasions (Soffer 2000: 59). A first statement visible through the study of the female figurines in ceramic is that of a marked stylistic difference between the two sites of Dolni Vestonice I and Pavlov I.

- **The male figurines**

There are only two male figurines found among the record of ceramic figurative figurines at the sites of Dolni Vestonice and Pavlov.

One is the figurine DVI 10 (Figure 3.27) from Dolni Vestonice I. Only the pelvis area and the top of the legs remain. The sex is clearly indicated and realistic with the representation of the testicles. The penis is broken off at the base. The fragment measures about 3 cm high.

The second one, PI 9 (figure 3.28), comes from Pavlov I. It is a roughly shaped human form, with the top of the body and the bottom of the legs missing. The top of the legs connecting to the body is recognisable. The swollen area between the legs seems to indicate the sex. The thin legs are also an element found on the other male figurine. This figurine seems to have been modelled rather quickly.

Figure 3.27: Ceramic figurine DVI 10 male figurine.

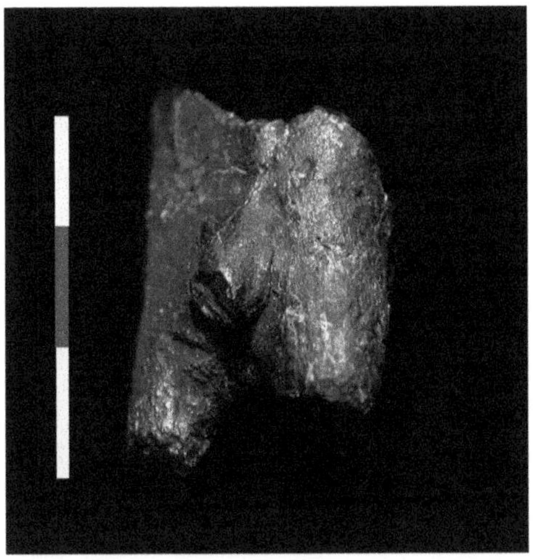

Figure 3.28: Fragmentary male ceramic figurine PI 9

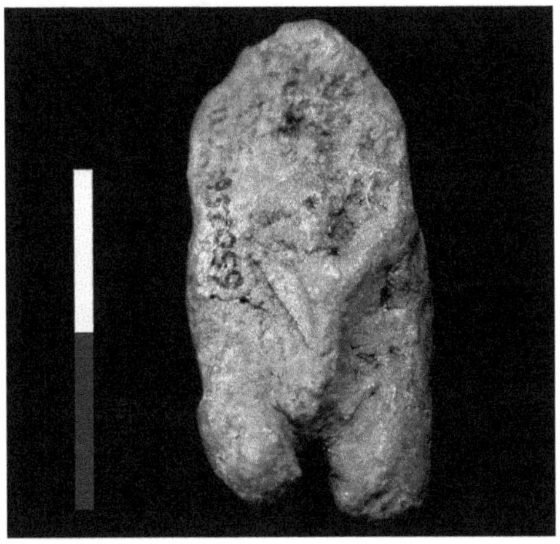

The representation of male figures in the Gravettian is very rare. The only other known male human representation is the famous ivory articulated 'puppet' found in the Brno II burial (Oliva 2000c). It is interesting to see that although rare (one recognised piece for each major site), the theme of male human representation is present in the Pavlovian ceramic art.

- **Anthropomorphs**

This category is made up of six figurines (1.9 % of the studied inventory) that represent undoubted human beings, or part of, which cannot be identified as male or female. They are shown in table 3.13.

Some human feet have also been recognised in the ceramic record but it has been decided to study them in the body extremity category further in this chapter as this type of representation in itself appears to be a characteristic of the Pavlovské Hills sites.

Three figurines have been recognised for Dolni Vestonice I and three for Pavlov I.

In the site of Dolni Vestonice I, the figurine DVI 6 seems to represent the top part of the crossed legs of a figure attached to the beginning of the pelvis area, the rest being broken off. The piece shows traces of distortion, after the fashioning of the figure but before its firing, which could have resulted into giving this impression of crossed legs. There is no indication of the sex of the figure. The legs are thin and elongated and the pelvis remains fairly narrow, which would rather fit with a male representation.

The figurine DVI 9 is not broken. It is a small human representation with the legs and the unique visible arm reduced to their simplest expression as small short stumps. Similarly, the neck is represented as a small stump with no head indicated. The middle part of the body is rather large and plump in comparison with the extremities, but no anatomical detail is indicated. The only detail seems to be a belt figured by a deep incision that goes around the figure at the hips level. One side of the figurine is completely flattened and smooth and has been strongly deformed when still wet and plastic, possibly by throwing it onto a flat surface. The presence of a belt and the wide belly would tend to indicate a female representation, but the sex is not clearly indicated.

The last figurine from Dolni Vestonice I is DVI 13 which is also nearly whole (only slightly broken at the legs) but which is interpreted here as a much stylised human figure that possibly doubles as a sexual symbol. The piece is cylindrical, elongated and curved. It is fully rounded at one end, which would be the head, and the whole surface is very smooth and has been carefully worked. The legs' end is broken off, but here is just enough of it left to see the beginning of the top of the legs separating. The curved parts are placed where the chest and the buttocks would be, encouraging interpretation as a human figure, even if there is absolutely no detail on all the upper part of what would be the body. The general shape is also obviously rather phallic if it was not for the remains of legs visible. Perhaps we have here again an example of intentional duality in the representation as has been noticed for some of the portable art objects in ivory found at the Pavlovké Hills sites (Svoboda 1996).

In Pavlov I, PI 4 is also a much stylised human figurine. It is flattened and one can see a rough human shape with a rounded head, a narrower part indicating the neck, then a cylinder with straight edges that end up by two small stumps indicating the legs (one is broken underneath). An incision across the waist area could indicate a belt. The surface of the piece is damaged in pieces but altogether is seems fairly whole and does not present other detail except for a short vertical incision on the face. Again it is noticeable that the overall shape is rather phallic, even if a belt has been considered as a female attribute.

The next figurine, PI 8, is much broken. It appears to be a fragment of the thighs of figurine that would be rather large (probably about the size of the Black Venus) in this portable art context. The separation between the conical legs is well marked. The large size of the legs would match a female representation.

The last human figurine from Pavlov I, PI 115, is also fragmentary. The torso and the left leg only remain. The torso is a simple cylinder, slightly flattened to which the thin leg is attached at the base. The other leg is broken as well as the top part. There is no indication of other detail such as arms, sex or belt. The overall figurine is thin and long which would match a male representation.

Table 3.13: The anthropomorphs in ceramic from Dolni Vestonice I and Pavlov I.

DVI 6	DVI 9	DVI 13
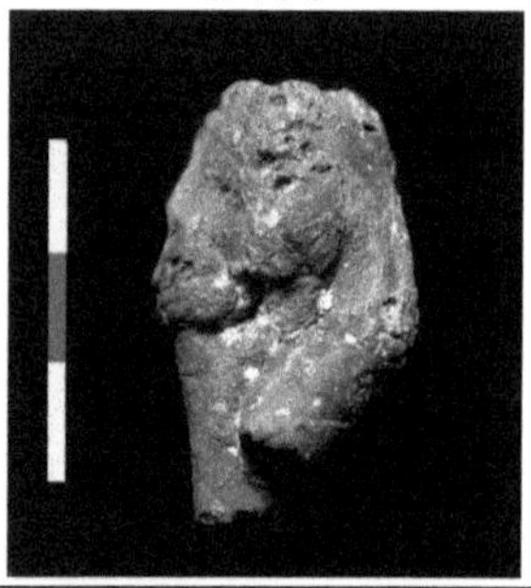	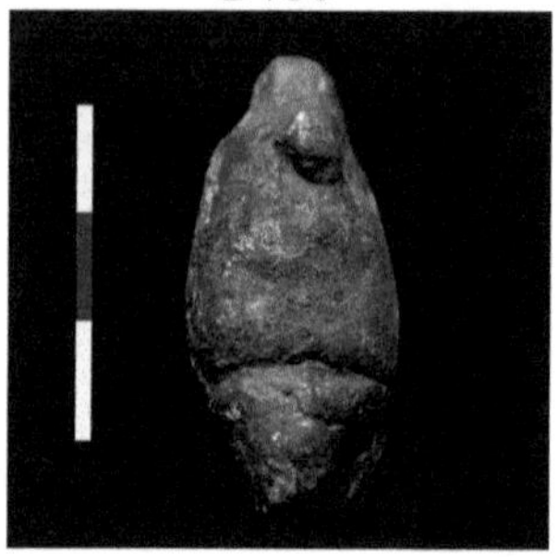	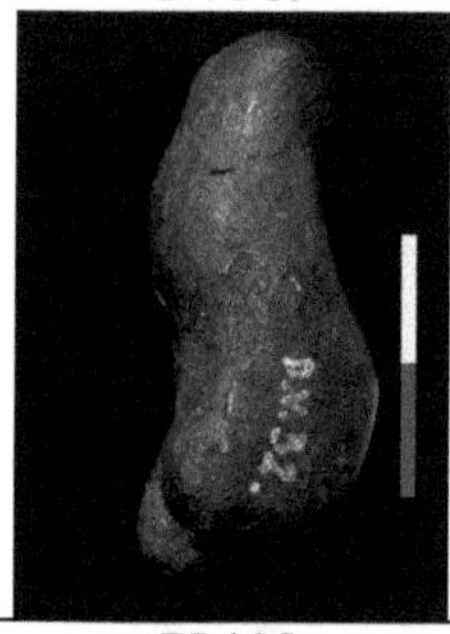
PI 4	PI 8	PI 115
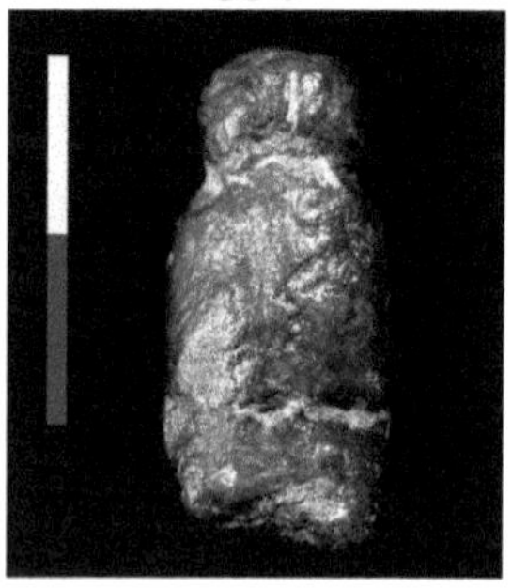		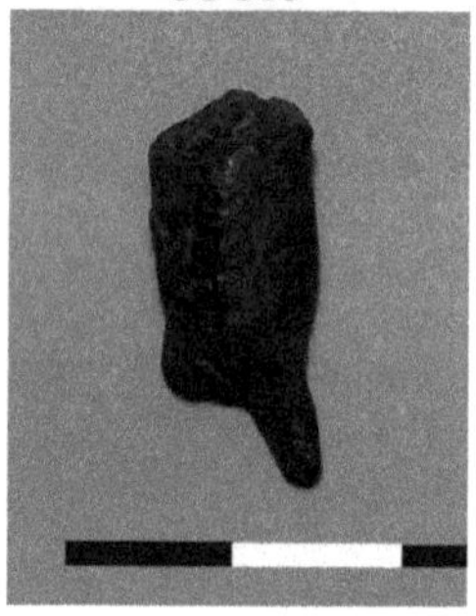

- **Conclusion on the human figures**

The female figurines dominate the record of the human representations in fired clay as is perfectly fitting for a Gravettian art context. Their number, however, is very high (14 between the three sites).

The presence of two male figures and of other anthropomorphs is more original and shows the wealth of material actually available in fired clay in the Pavlovian sites.

Also, the wide variety of styles and types of these representations is striking, showing the participation of a large number of different persons in the making of the figurines.

We will see further that some human feet also exist in the art record in ceramic which can be added to the quantities seen above.

The animal representations

Among the animals represented in fired clay in Dolni Vestonice I and II and Pavlov I, the largest number of animals whose species can be identified is made up of representations of felines: 11 pieces making up for 3.5 % of the general inventory (Table 3.6 and 3.14). The representation of this animal is clearly dominant in Dolni Vestonice I with 6 pieces. Then bird representations come in second place with 3 pieces. All the other recognisable animal species there are then found in single examples at this site. The situation is different in Pavlov I where felines are in equal quantity with mammoths with 4 pieces of each, followed then by horses with 3 pieces, and then by single representations of other animals. Among the small inventory of Dolni Vestonice II, one feline representation can also be found as well as one horse and one undetermined.

It must be pointed out, however, that undetermined animal representations neatly dominate the whole record with 31 pieces: this is often due to the high fragmentation of the objects which prevent determination; but some animals, although whole or nearly whole, are also simply too vaguely shaped to be determined other than by their overall category as herbivore or carnivore for example.

We are aware of certain anciently published inventories of the types of animals found in the art of the Pavlovian that mention a large number of bear representations and that are still referred to today in some publications (Clottes 2008) Such an inventory however seems totally different from what was seen while studying the collections and is not found again in more recent publications about Pavlovian art. This seems to raise questions about the initial interpretation given to some of the Pavlovian art objects.

The animal representations considered here are of whole (or fragments of whole) figurines. Some of the body extremities looked at further in this chapter can possibly be attributed to animal species which modifies the statistics presented below. However, this will be discussed in the relevant part concerning the body extremities as we will see that interpretation plays an important part there.

Table 3.14: inventory of animal representations

Animal species	DVI	DVII	PI	Total	% of total
Feline	6	1	4	11	17.7
Mammoth	1		4	5	8
Rhinoceros	1		1	2	3.3
Bear	1			1	1.6
Bird	3			3	4.8
Bovid	1			1	1.6
Horse		1	3	4	6.5
Reindeer	1		1	2	3.3
Ibex			1	1	1.6
Mustelid	1			1	1.6
Undetermined	15	1	15	31	50
Total	30	3	29	62	100

The examples of recognised animal species will be presented in order of quantitative importance, starting with felines, mammoths, birds and horses.

- **Felines**

Among the animals represented in the ceramic art, felines constitute the largest category, with 11 pieces, representing 17.7 % of the animal inventory. They are found at the three studied sites. It is remarkable that all these representations all heads of felines, or fragments of heads (table 3.15).

They can be recognised as felines as most of them are well modelled, especially for the detail of the muzzle (nose, lips and chin). It is true that some of these representations have sometimes been published as bears, sometimes as felines (For example two of the very similar heads following the type most commonly found in DVI are found described one as bear and one as a feline in the same page illustrated with drawings reproduced from Klima (in Valoch 1996: Fig. 117, No 11 and 16) when a view of the top of the two heads is identical.)

Bear muzzles, however, are more elongated and thinner that the ones of felines. We will also argue here that the high consistency of the manner of fashioning the details of the muzzle is an indicator of the same species being represented.

Table 3.15: The feline representations in ceramic from Dolni Vestonice I, Dolni Vestonice II and Pavlov I.

DVI 15 	DVI 18 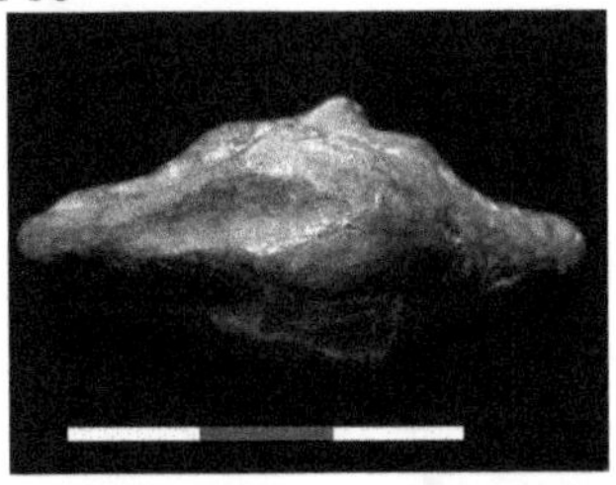	DVI 20 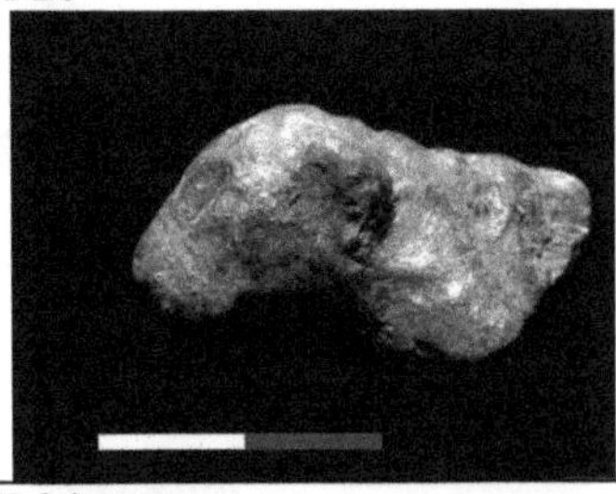
DVI 21 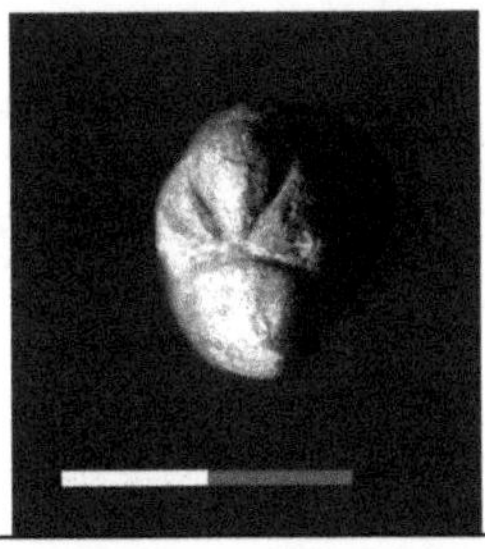	DVI 23 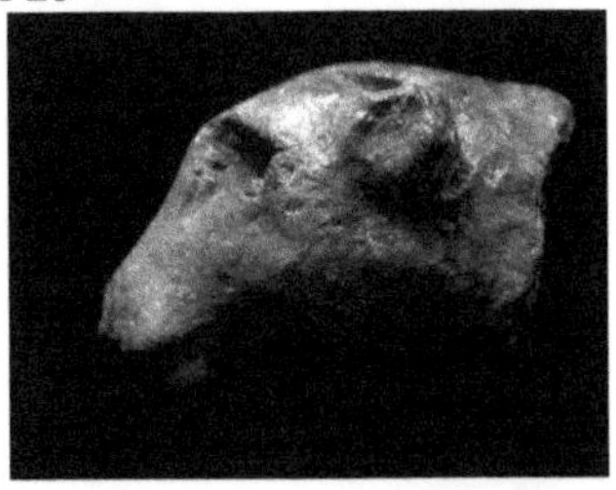	DVI 24 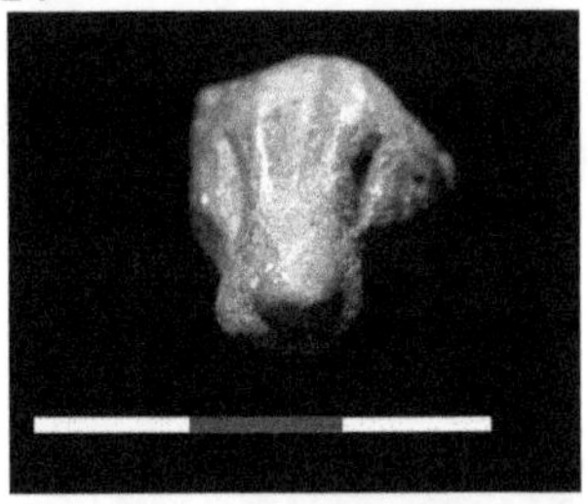
DVII 4 	PI 17 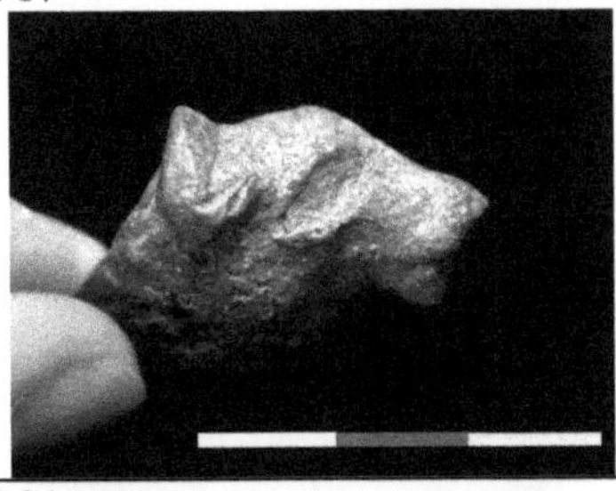	PI 19 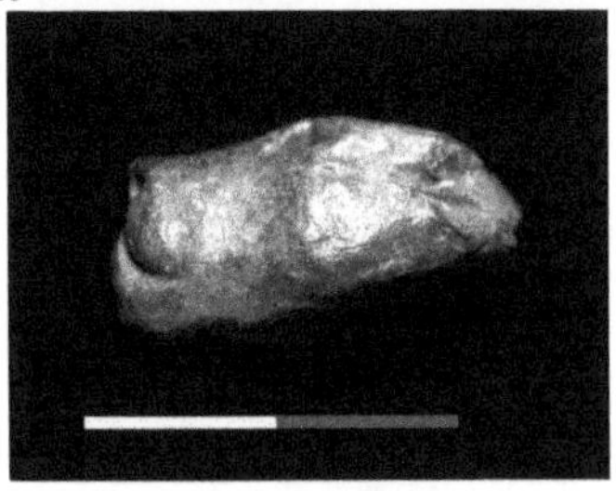
PI 20 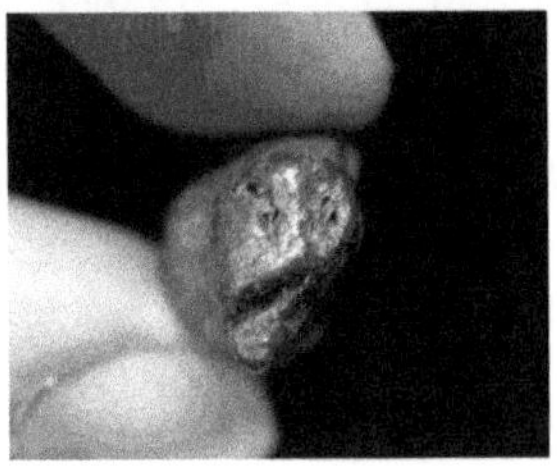	PI 21 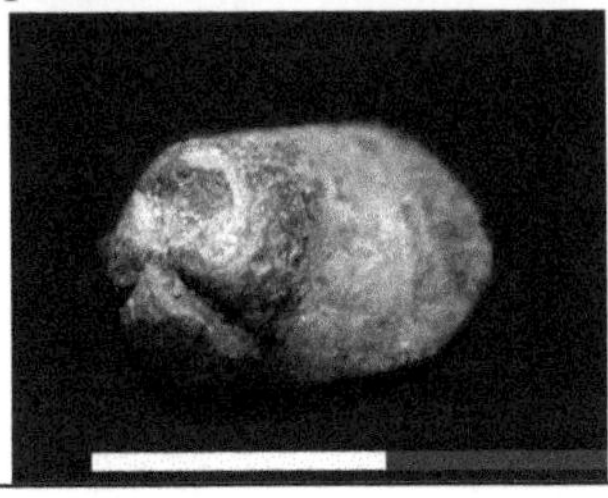	

The heads are small in size, between two and four centimetres-long, and are very carefully detailed to shape the muzzle, the eyes, ears and volume of the head in general. The profile of the head, the way the eyes are indicated, the position of the ears and the way the detail of the muzzle allows distinguishing nose, lips and chin are major stylistic elements that allow comparison between the objects.

All the profiles of the animals except one depict realistic profiles with the nose protruding more or less comparing to the mouth. The head PI 19 (see table 3.15) however is different in this aspect as it shows the mouth at the same level that the nose in a rather flattened profile. This impression is strengthened by the fact that the angle formed by the line of the muzzle is close to a 90° angle with that of the ridge of the nose. This stylisation of feline profiles is found in other known felines representations from the Early Upper Palaeolithic, for example in Vogelherd in Germany (Hahn 1993) or in the famous human-lion figure from Holenstein-stadel. This has been interpreted by some as a humanised profile.

The eyes are consistently represented by two incisions, varying in length, that run parallel to the ridge of the nose. This appears to be a widespread convention in the feline representations. These incisions vary in length between 2 and 8 millimetres.

The ears of the felines are represented on 5 of them (DVI 18, DVI 20, DVI 23, DVI 24 and PI17) with some details, often as rounded protuberances. They are all placed rather low on the sides of the head and all of them have the particularity of not being symmetrical: could this express some special posture of the animals with a certain meaning?

The way the nose of the feline heads is shaped seems to be following four different types in which 10 of the 11 feline heads can be categorised, the last one (DVI 15) being too fragmentary. The stylisation looked at considers the manner the separation of the different parts forming the muzzle (nose, lips and chin) is indicated (table 3.16). Thus it has been possible to determine types which are schematised in table 3.17.

Table 3.16: Detail of the muzzles of the felines representations in ceramic from Dolni Vestonice I; Dolni Vestonice II and Pavlov I.

DVI 18	DVI 20	DVI 21	DVI 23
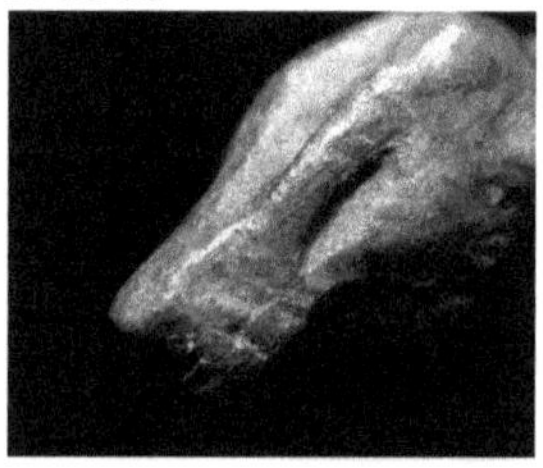	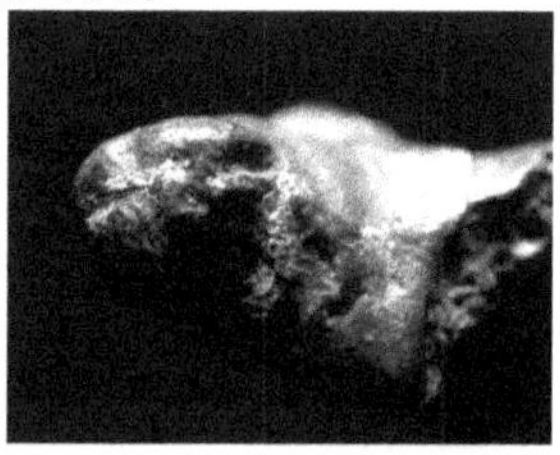	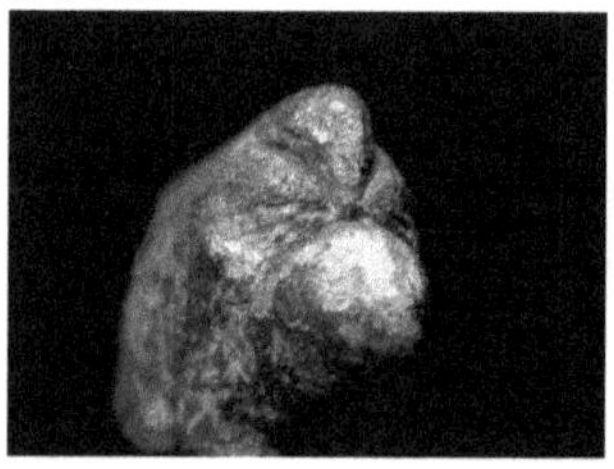	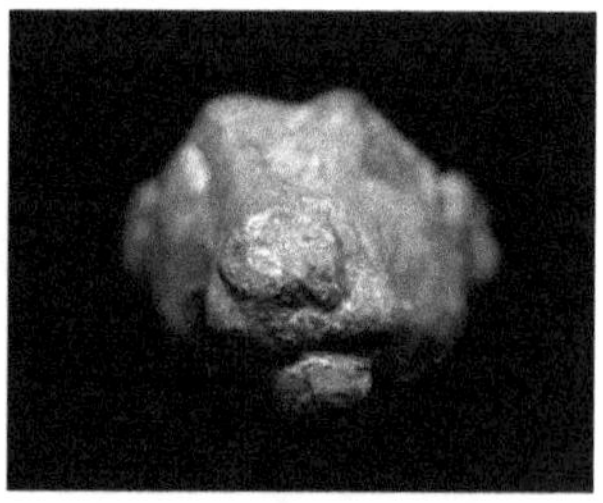
DVI 24	DVII 4	PI 17	PI 19
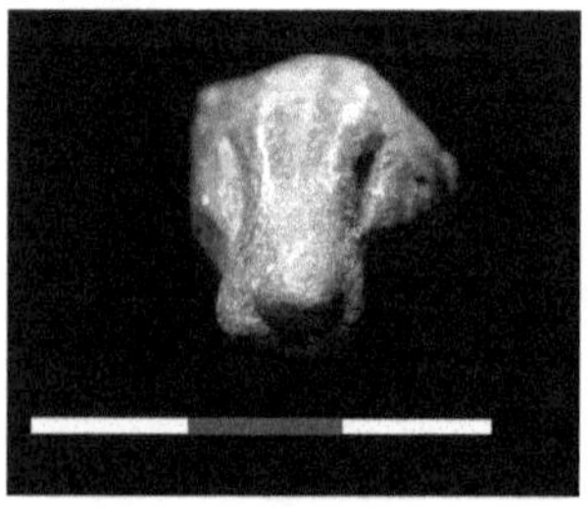	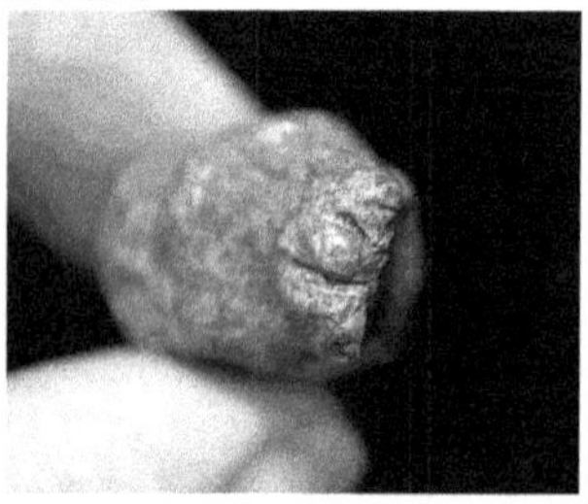	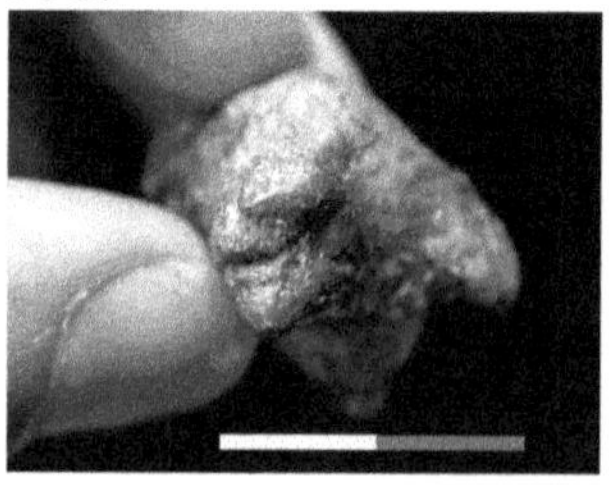	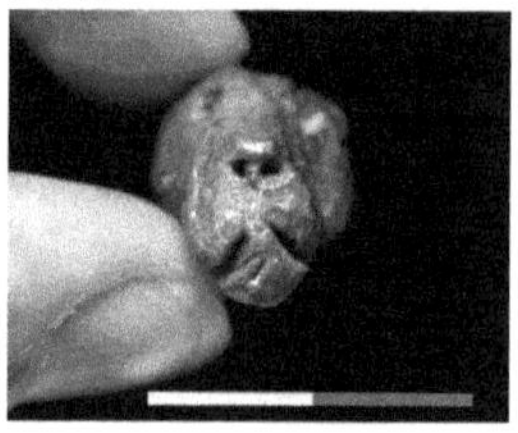
PI 20	PI 21		

These types are as follows (Table 3.17):

Type 1: the nose is shaped by a roughly semicircular incision underlining it. Two incisions or a continuous one indicate the mouth by a horizontal line which does not touch the line of the nose. This type is associated with a modelling of the lips, nose and chin which brings it into three dimensions and is very naturalistic.

Type 2: 4 incisions are placed in a sort of cross shape, radiating from a central point, sometimes nearly touching in the centre.

Type 3: the muzzle is also shaped by 4 incisions, the two for the nose being oblique and the ones for the mouth horizontal. Only one object, DVII 4, is in this category.

Type 4: the nose is indicated by two holes. Then the mouth is marked by two oblique incisions. Only one piece, PI 19 is in this category. This is also an object with strong stylistic differences from the other feline heads with a much flattened profile.

Only two styles (types 1 and 2) are seen among the five feline representations of Dolni Vestonice I showing therefore a stronger stylistic unity. Three styles (types 1, 2 and 4) are found in Pavlov I possibly indicating more variability in this site, a fact that has already been seen in the female representations. In Dolni Vestonice II, only one muzzle is recognisable that follows type 3. It can be argued that type 3 is a variation on type 2 (the cross shape), that could denote links with the other sites as type 2 is found in equal quantities in Dolni Vestonice I and Pavlov I.

Analysing the detail of the muzzles also allows confirming the similarity between four feline heads which are visually very much alike: DVI 18, DVI 23, DVI 24 and PI 17. They all belong to the type 1. DVI 18, DVI 23 and DVI 24 are so alike that it can be argued they were made by the same person or by persons working closely together according to strictly similar standards. They have the same pointy nose and they are carefully shaped with many details, notably in the fashioning of the muzzle. The eyes are represented by parallel slits on each side of the forehead. The ears are small and rounded. PI 17 is also following this description, but varying slightly.

Type 2 is found equally in Dolni Vestonice I and Pavlov I. The repartition of these types clearly indicates connections between the sites of Dolni Vestonice I and Pavlov I regarding the making of the ceramic figurines with some form of stylistic transmission.

Table 3.17: Typology of the feline muzzles indicating the shape of the incisions or holes used to mark the separations between the chin, lip and nose of the felines.

DVI 18	DVI 20	DVI 21	DVI 23	DVI 24
Type 1	Type 2	Type 2	Type 1	Type 1
DVII 4	PI 17	PI 19	PI 20	PI 21
Type 3	Type 1	Type 4	Type 2	Type 2

- **Mammoths**

Mammoth are the second most important category of recognisable animal figurines by their number after the felines. They make up 8 % of the inventory of fired clay animal with 5 pieces (Table 3.18). We have seen that for feline representations only heads were found; for the mammoths, on the opposite, the inventory only contains bodies, possibly because of the characteristic profile of this animal, which is often seen in Upper Palaeolithic art. It will be seen further that some of the body extremities could also be interpreted as representing mammoths.

The repartition of the mammoth figurines between the sites of Dolni Vestonice I and Pavlov I is very unequal: in Dolni Vestonice I only one representation is found; in Pavlov I there are four of them and they are more numerous than the feline representations at this site; in Dolni Vestonice II, they are absent. This is another marked difference between the two major sites of Dolni Vestonice I and Pavlov I.

Table 3.18: The mammoth representations in ceramic from Dolni Vestonice I and Pavlov I.

DVI 62	PI 22	PI 23
	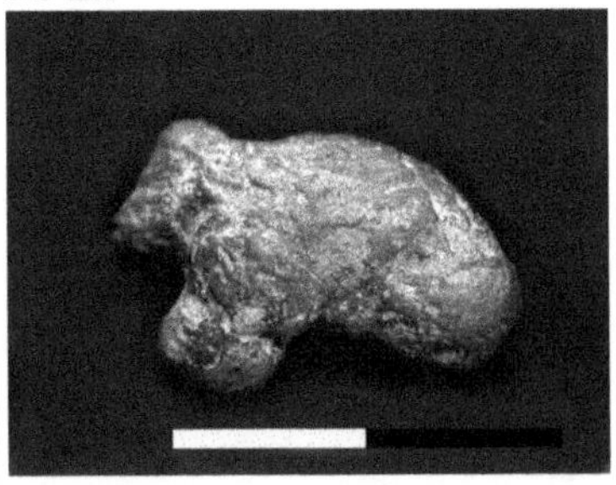	
PI 24	PI 26	
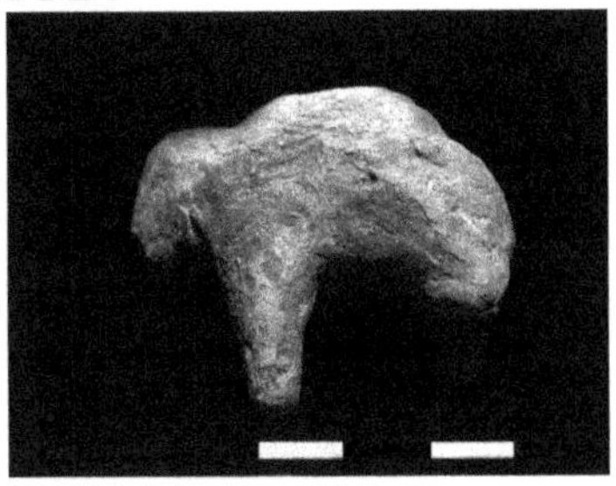	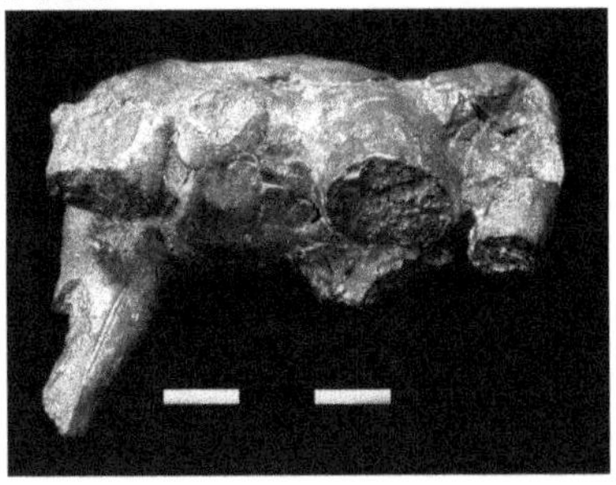	

There seem to be an acute importance of the mammoth in the economy of the Pavlovian sites where huge accumulations of mammoth bones are found, even if the debate about whether the animal was actually hunted or simply scavenged by the Pavlovian people is still ongoing (Musil 1997; Soffer 1985; Roebroeks *et al.* 2000). It is striking that the animal is also one of the major theme of representation at the Pavlovské Hills sites; in fact, in Upper Palaeolithic art in general, there is usually a great difference between the proportions of animals represented and the actual faunal remains found at the sites (Bahn and Vertut 1997; Lorblanchet 1995). This would underline that, once more, the situation at these sites was different from most Upper Palaeolithic contexts.

The pieces are recognised as mammoths because the curve of the back with the bump of the head is very distinctive. The large bulk of the animal is also very typical.

There are little descriptive details on the figures except on the figurine PI 26 (figure 3.29) which appears to have been carefully worked and smoothed all around with details such as the eyes indicated by a small bump (figure 3.30). Its legs also appear to have been shaped with the intention of rendering the real volume of the various parts. Unfortunately, the figurine is much damaged and large areas of the surface are deteriorated.

Figure 3.29: Mammoth figurine in ceramic PI 26 from Pavlov I view underneath the belly.

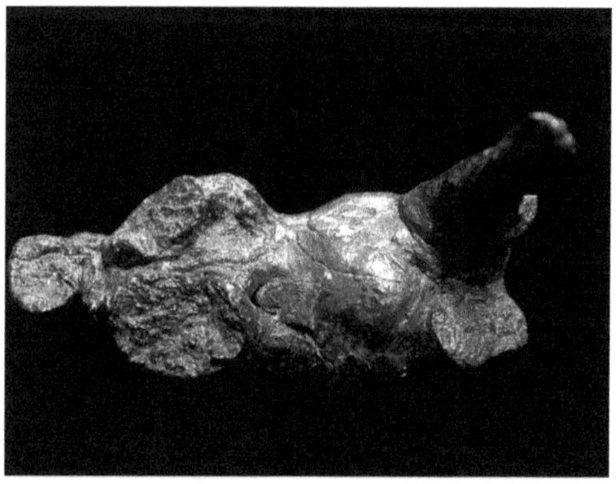

Figure 3.30: frontal view of PI 26 mammoth figurine: the detail of the left eye is still visible as a protuberance.

As for the style of the mammoth figures, a simple look at table 3.18 shows how different they all are from each other. Although the mammoth representation is recognisable without doubt in all cases, the manner of fashioning each piece would indicate they were possibly made by different persons that were not working together or not copying each other.

Out of the five figurines in Table 3.18, three appear whole or nearly whole (DVI 62, PI 22 and PI 23), one has only some of its body extremity missing (trunk and two legs for PI 24), and the last one, PI 26, is fragmentary by its extremities as well as the side of the head.

DVI 62 is schematic and lacks the typical mammoth profile with a mark bump for the head. Instead the outline of the back continues towards the head and the trunk in a semi-circular line. The legs are represented by two chunky stumps, one at the front, and one at the back, separated by a marked arch for the belly. Only the fact that the trunk is clearly whole indicates it is a mammoth. It is represented detached from the body at its based, by the mouth, and then it attaches to the legs. This figurine is the only mammoth representation found in Dolni Vestonice I.

PI 22 can be interpreted as a schematised young mammoth. The figurine is actually whole; the legs are only indicated by a rough stump at the front, quite oversized in comparison with the rest of the body; at the rear of the animal the body ends up in a rounded shape without indication of the legs. The trunk is indicated by a tiny pointy protuberance. The figure gives the impression of a very young mammoth being represented because of its general proportions.

PI 23 shows the outline profile of the back of a mammoth. The base of the legs and the trunk are visible on one side but these are the only details. The figure is rather flat. It appears incomplete, with all the lower part missing but all the remaining surfaces do not show traces of breakage. They have a worn-out feeling, however, which makes it difficult to determine if we are considering the result of taphonimic processes or not.

PI 24 is a mammoth represented with its full length of legs originally, even if only two remains unbroken today: one at the front and one at the back. It also had the trunk represented. The whole figure however is done without any other detailing and the surfaces are roughly smoothed.

The larger number of mammoth representations at the site of Pavlov I (4) in comparison to the ones at Dolni Vestonice I (1), and the great stylistic difference visible between all these figurines are the striking elements appearing through their study. They inform us on inter and intra-site variability.

- **Horses**

Four horse representations (Table 3.19) were found in the record of ceramic figurines representing 6.5 % of the total of the animal representation. It is the third most represented recognisable animal among the ceramics of the Pavlovké Hills sites.

All the pieces come from Pavlov I and Dolni Vestonice II and none from Dolni Vestonice I. They are all very different in style. Three are heads (or fragments of): PI 11, PI 16 and DVII 1. The last one is a fragment of hind side.

PI 11 is a fragmentary head with the lower part (jaw and neck) broken off. The remaining part is much detailed, with ears, eyes and nostrils indicated and show an emphasis on the mane that is high in relief and fashioned with the detail of the hair.

PI 16 is another horse head interpreted as such because of the length of the muzzle, its shape and the way it connects with the widening neck. It is fragmentary with the top of the head broken off, and the surface is very uneven. Details such as the eyes, represented by a simple hole, are still visible, however.

DVII 1 is a fragment of a head with only the muzzle remaining. It is extremely well rendered with volume given to the surfaces; the nostrils are indicated by two small protuberances; the chin is well shaped. The surface of the object is pierced by 3 deep circular holes (Table 3.19). One of them is located where one of the nostrils would be, possibly to indicate it, but there is only one. The two other holes are placed on the cheeks, in symmetry, but they do not completely pierce the object. They have been made at an angle, which is possible to see because of the marks left on the edges of the holes. It is possible to see on the edge of the broken part that at least three more holes existed there. They are rather enigmatic. It has also been interpreted in the past as a possible reindeer (Svoboda (ed.) 1991: 51)

PI 12 represents the hind side of a horse clearly recognisable by the typical tail and the way it connect to the rest of the body. There is little detail on this piece other than the general form of the animal represented.

Table 3.19: horse representations in ceramic from Dolni Vestonice II and Pavlov I.

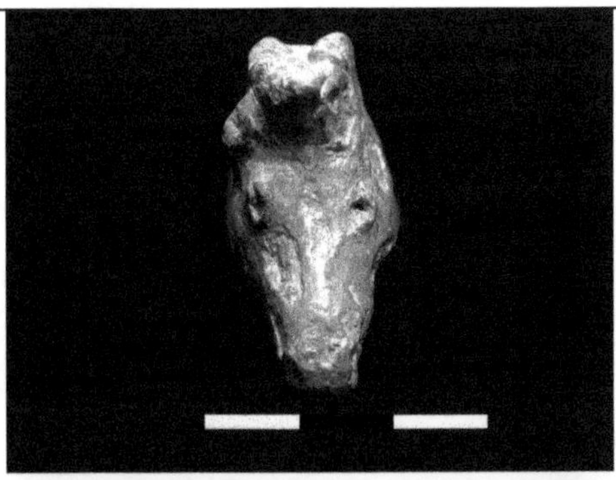
PI 11

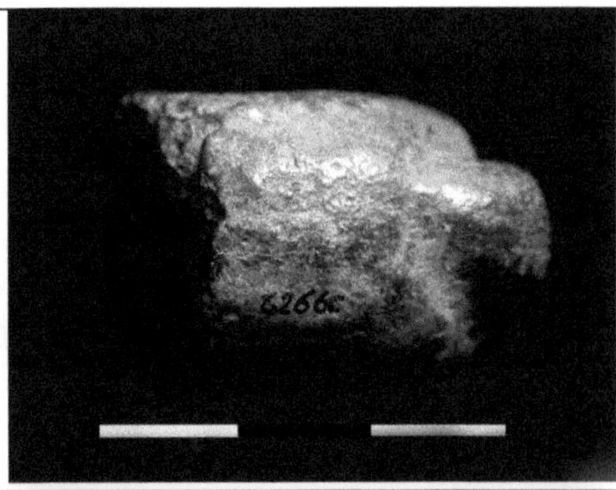
PI 12

PI 16

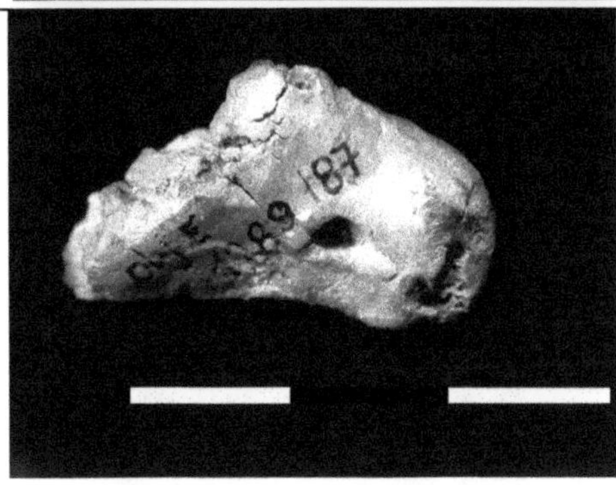
DVII 1

- **Birds**

Three bird representations have been recognised from the site of Dolni Vestonice I, representing 4.8 % of the animal total. Bird representations are rather rare in Upper Palaeolithic art, so it is striking to find this theme as the fourth most numerous animal type represented. It is also remarkable that bird figurines are found only in one site, Dolni Vestonice I.

Two of the figurines, DVI 28 and DVI 29, are similar in shape, although one is much larger and fragmentary; the third one, DVI 30, is different (Table 3.20).

DVI 29 shows the head of a bird of prey at the top of a body simply shaped like a conical cylinder, without details. The piece is whole and measures 3.8 centimetres high. The bird of prey is recognisable by the shape of its pointy and slightly curved beak combined with two 'ear-like' appendices on top of the head, probably indicating the shape of the feathers as in some species of owls (i.e. Bubo bubo). No other details are indicated but these are enough to recognise the animal type without much doubt.

The figurine DVI 28 has been interpreted as a bird representation by comparison with DVI 29 as it is fragmentary with much of the front of the piece missing (figure 3.31). This object is the largest of the ceramics from Dolni Vestonice, together with the Black Venus, as it measures 11.8 centimetres high. It appears to have a similar 'body' to DVI 29: a cylinder ending up as a cone. The top part is much broken and only the two pointed 'ears' and the back of the head are left. At the difference of DVI 29, the remaining surfaces are all well smoothed. A groove also appears to run all around the object at the base of what would be the head. Just at the base of the 'ears', traces remain of two holes drilled at the ends of what would have made a rectilinear axis crossing the head from side to side. The front of the head broke off along this axis, and one half of the holes can still be seen. They were at least 0.5 cm deep, but they did not completely pierce through the object. The rest is broken, with no real possibility of knowing if the making of the holes would have resulted in the breakage of the object, or if it happened at a later date.

The figurine DVI 30 (Table 3.20) also appears to represent an owl but of a different type as this time the head is rounded and without ears. The eyes are indicated by two holes. The figure is animated as the head is at an angle with the body as if the animal was turning its head in a natural position. The beak is indicated by a slight protuberance without more detail.

Table 3.20: Bird representations in ceramic from Dolni Vestonice I

DVI 28	DVI 29	DVI 30
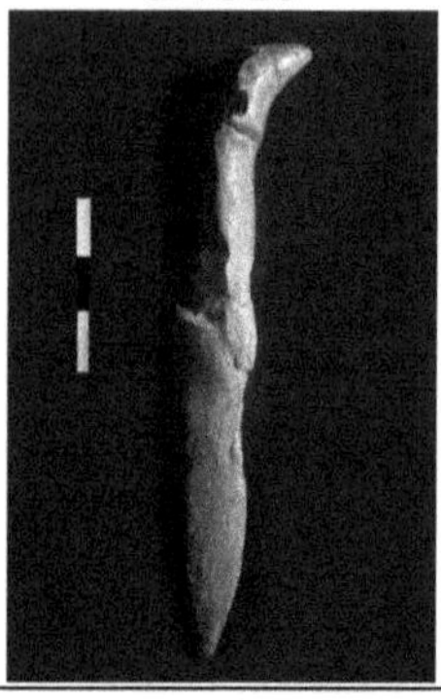	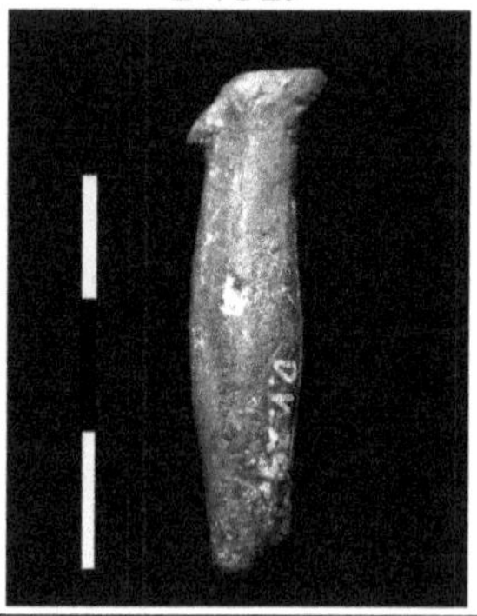	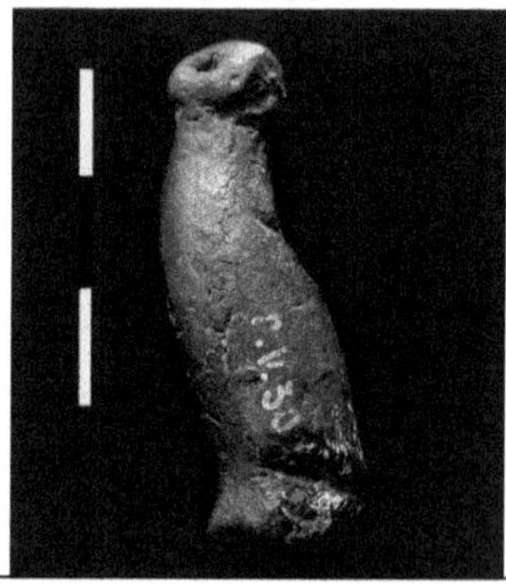

Figure 3.31: comparison of a back view of the bird representations DVI 29 and DVI 28

DVI 29 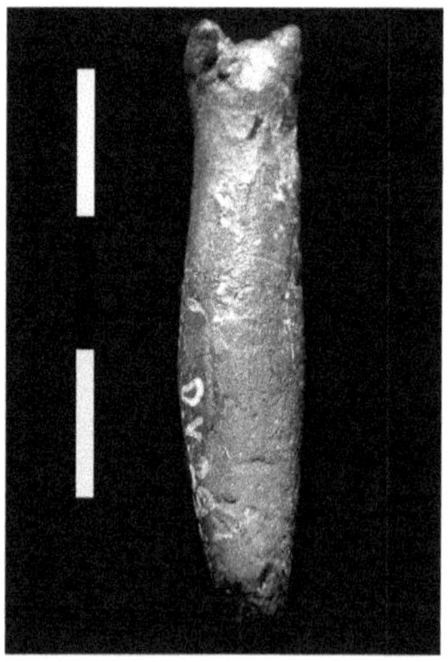

DVI 28 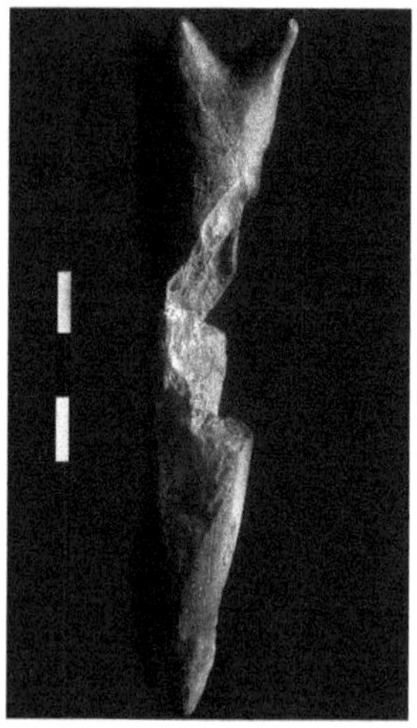

- **Rhinoceroses**

Two animal heads can be attributed to woolly rhinoceroses. They make up 3.3 % of the animal total (table 3.21).

One comes from Dolni Vestonice I (DVI 14) and one from Pavlov I (PI 13). They are recognisable by the presence of a horn for DV 14[18] or of the base of two horns for PI 13.

The head DVI 14 has only one horn sprouting from the very end of the muzzle, vertical in comparison with the angle of the chin. It is rather thin and long. The top of the head is shaped in a rather conical form. One hole on the left side of the head possibly represents an eye, but it is not seen on the other side of the head. There are no other details on this figure. The surface is very uneven.

This head appears rather stylised and the single horn represented raises questions as woolly rhinoceroses (coelodonta antiquus) found over Europe during the Upper Palaeolithic had two horns as is visible in the known representations in cave art in Chauvet (Clottes ed. 2003) or Rouffignac for example (Plassard 1999). One can argue for this being a result of the reconstitution process of the piece (see footnote 17), for some form of artistic licence or again that a species of rhinoceros with a unique horn did exist in the late Pleistocene.

It is true that the second head, PI 13, does have the bases of two horns, which rather invalidates this latest hypothesis. It is much more realistic than the first head with the details of the eyes, the nostrils, the open mouth and well-smoothed surfaces. The overall shape of the head is also well rendered. The neck area is very wide but it is broken off.

These two heads are stylistically extremely different: DVI 14 is much more stylised with the only indicative details being the single horn and the overall shape of the head; PI 13 on the other hand is realistically rendered even if it is partly broken.

Table 3.21: Rhinoceros heads in ceramic from DVI and PI.

DVI 14	PI 13
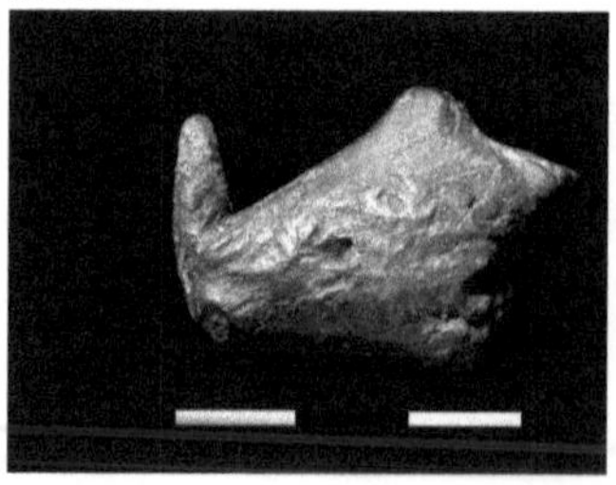	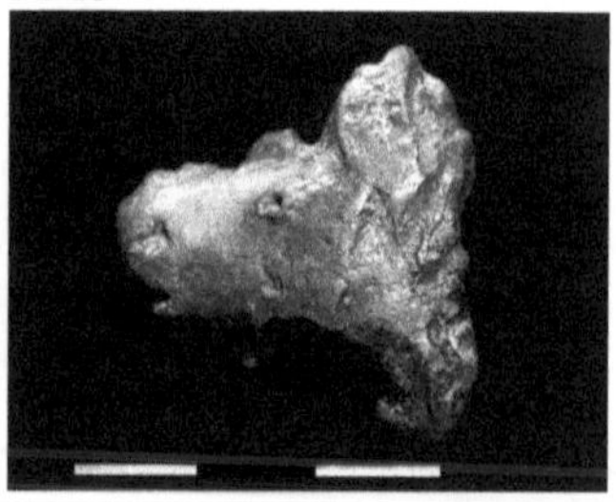

[18] Although it is likely the horn of DVI 14 was reconstituted by Absolon after excavation – Martin Oliva, personal communication May 2004

- **Reindeers**

Two figurations of reindeer are recognisable in the ceramic inventory. They represent 3.3 % of the animal total. One comes from Dolni Vestonice I, the other from Pavlov I (table 3.22).

DVI 19 is the rather schematic head of an animal interpreted as a reindeer because of its proportions. The head has a long muzzle and quite a broad neck. The mouth and the nostrils are indicated by incisions and the eyes by protuberances, which are not symmetrical on each side of the head. Between the ears on top of the head, a damaged surface is visible with two incisions at the bottom of it, which indicate it could have been made intentionally. The surface is rather smooth, with some accidents in the material. The head is broken at the neck.

PI 18 is finely modelled. It is broken at the back of the head. The eyes are indicated by two protuberances with incisions on them. The ears are also protuberances. The head seems stretched towards the front. The detail of the muzzle is well visible on one side with incisions for the mouth. The surfaces are smooth.

Some of the body extremities have also been interpreted as belonging to reindeer. This will be seen further in this chapter.

Table 3.22: the reindeer representations in ceramic from Dolni Vestonice I and Pavlov I.

DVI 19	PI 18
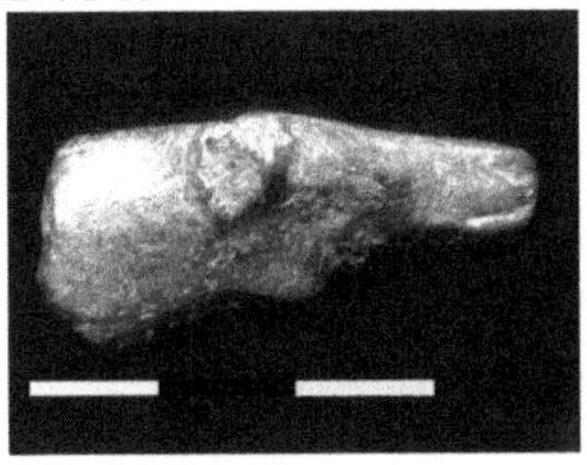	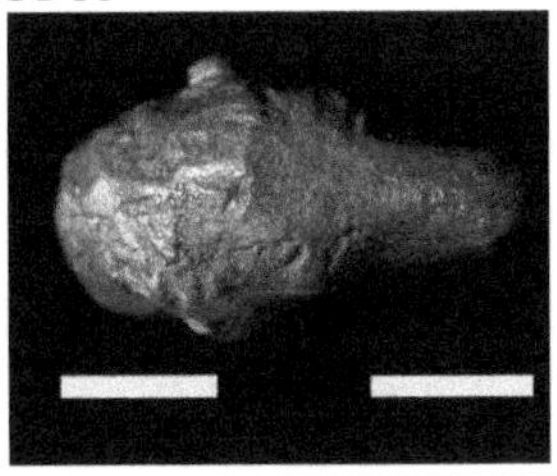

- **The bear**

One nearly whole bear figure is found in Dolni Vestonice I (DVI 16) representing 1.6 % of the recognisable animal total (figure 3.32).

The bear figure DVI 16 is a nearly whole figure from Dolni Vestonice I. Only the extremities of the legs and the surface of the rear of the animal appear broken off, a fact interpreted by Absolon as resulting from magic practices on the figurine in order to ensure it would not run away. In fact the front legs do seem to have been shaped and broken off, but the rear legs are much less detailed: they seem to form a wide stump. There are no details of the face of the animal and part of the surface is damaged as on the right side of the head. The general volume of the animal, however, is well rendered, with emphasis on determining elements such as the specific profile of the back with the relief of the shoulders indicated and the thin elongated muzzle.

Among the fragmentary indeterminate animal pieces it is possible some could belong to bears, but it has not been possible to be certain. Also, it is possible that among the body extremities some can be attributed to bears. Thus, the original quantity of bear representations in ceramic at the sites is likely to be higher than just this one certain piece.

Figure 3.32: bear figurine DVI 16 from Dolni Vestonice I

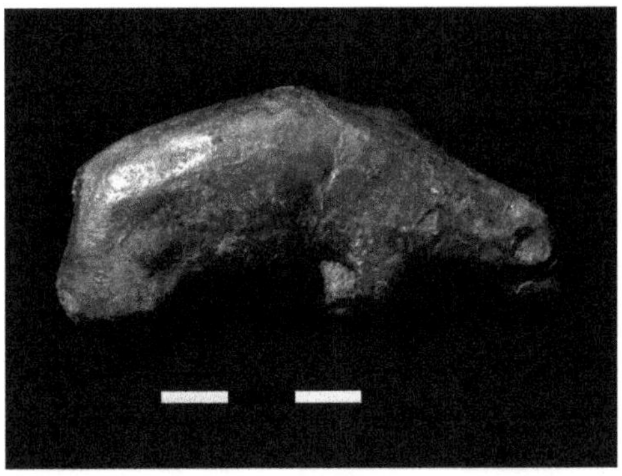

However, it was not possible to find as many bear representations among the ceramic animals as the quantities published in the past by Klima and still repeated today (Clottes 2008). It is true that some of the feline heads have sometimes been interpreted as being bears. This is not, however, the current dominant view.

- **The caprine**

One caprine figure was recognised during the study of the inventory representing 1.6 % of the animal total.

This is the figurine PI 15 from Pavlov I (figure 3.33). Its horns are clearly represented, even if they are not symmetrical. The figure is not detailed enough however to allow determining which type of caprine is depicted, possibly an ibex. The muzzle is rounded with possibly two very small holes representing the nostrils. It is broken at the neck.

This representation raises questions as there is no mention of caprine at all among the faunal remains found at the sites at the Pavlovské hills (Musil 1997).

Figure 3.33: the caprine figurine PI 15 from Pavlov I.

- **The bovid**

One bovid figure was recognised during the study, representing 1.6 % of the total animal inventory. The figure (DVI 27: figure 3.34) evokes a young bovine, possibly an aurochs calf. It comes from Dolni Vestonice I. The object is nearly complete except for the rear legs and the right ear. Two holes indicate the nostrils on the muzzle. The eyes are indicated by small holes on protuberances. The left ear is indicated. The two front legs appear as superimposed today but it seems they were simply distorted by pressure applied before the firing. The material is full of inclusions and is very irregular.

This figurine is different by its theme and style from the rest of the ceramic objects. M. Oliva (pers. com. May 2004) mentioned that it is indeed very similar to the numerous Neolithic clay figurines found in the same region.

Figure 3.34: the bovid figure in ceramic DVI 27.

- **A weasel?**

One figurine was determined as representing an animal from the mustelid family, possibly a type of ferret of weasel (figure 3.35). It makes 1.6 % of the total animal inventory.

The piece is very small as it measures just less than one centimetre in length but it is very expressive. The muzzle is pointed and elongated. The eyes are indicated by very small slits under the ears, which are rendered by two round flattened surfaces. The beginning of the neck is clearly formed, and then it is broken. The surfaces are quite rough.

Figure 3.35: The mustelid head DVI 63 from Dolni Vestonice I

- **Indeterminate animals**

31 objects have been categorised as indeterminate animals representing 50 % of the studied animal inventory (Table 3.14). They represent body parts that are too fragmentary or imprecise to recognise the animal species, but that can be attributed to animals without doubt (i.e. body with legs, isolated head).

This is the case for example of the piece PI 25. It is a fragment of an animal body with all the front missing as well as the legs. When viewed from the back (figure 3.36) and in profile this animal could be interpreted as a bear, as is published in Klima 1994. However, the overall profile also resembles some of the mammoth figurines, even if the front part is missing.

Figure 3.36: indeterminate animal PI 25 from Pavlov I

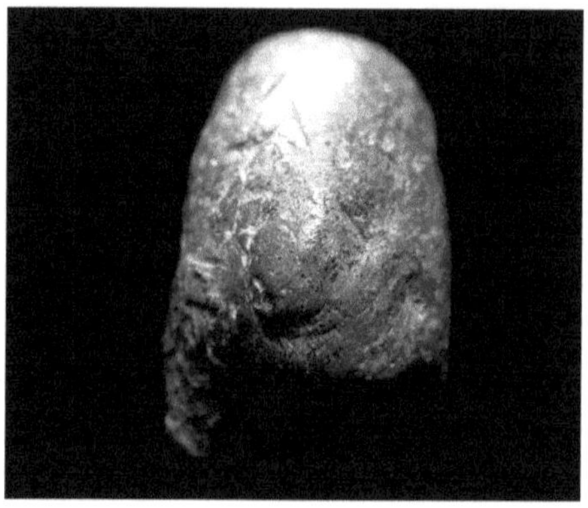

The figurine PI 14 (figure 3.37) is an animal head which can only be interpreted as a large herbivore without more precision. The only details are the two holes for the eyes. The general shape reminds of the rhinoceros head from Pavlov I, but here the forehead is completely smoothed with no trace of horns. The back of the head is pinched between the fingers and narrows to form the neck, which is only broken on a very small area. The piece appears to have been made and fired just as a head, without having been attached to a body.

Figure 3.37: indeterminate animal PI 14 from Pavlov I

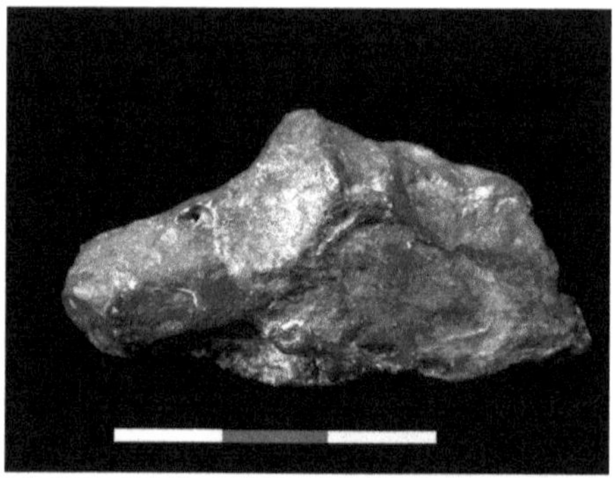

The figurine DVI 17 is quite unique in the ceramic assemblage as we will see it can be interpreted as a fantastic or composite animal (Table 3.23), a theme of representation that exist in some Upper Palaeolithic art contexts, notably in parietal art.

K. Absolon (1949:25) points out that this animal head was first described as a crocodile at the time of its discovery as it looks 'somewhat exotic', but that is was later identified 'beyond doubt as the head of reindeer'. This interpretation has remained since, but a close examination of the object reveals some features that are not fitting with it. These are notably a sort of ridge on the muzzle, two small protuberances under the chin and a well marked mane.

The 'ridge' (Table 3.23: b) is in relief and runs on the whole forehead from the nose to the beginning of the mane. The nose is a triangular flattened surface.

It is quite surprising to see two small protuberances neatly marked on each side of the chin (Table 3.23: c). They remind of long fangs or teeth. On each side of the nose, there is also a swollen area that reminds of the lips of a feline or a canine.

Table 3.23: Fantastic animal DVI 17 from Dolni Vestonice I

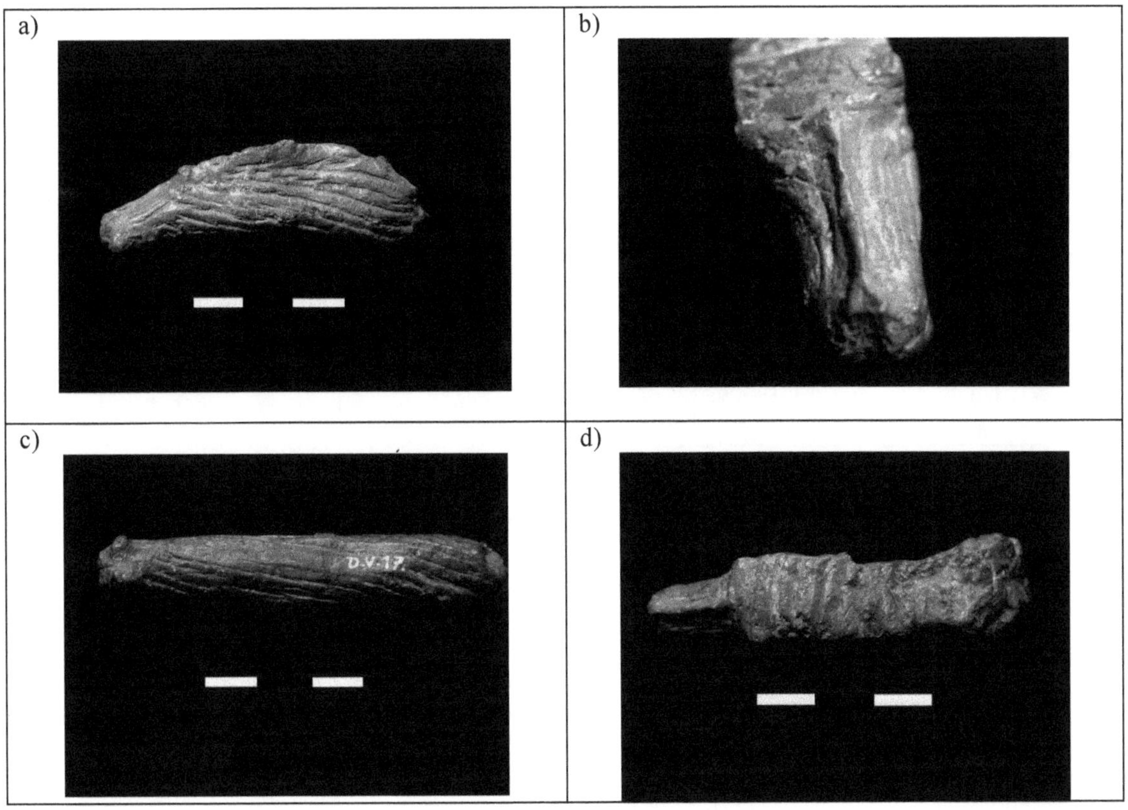

Then, above the very elongated neck, there is a 1.5 cm wide flattened surface that apparently represents a mane (Table 3.23: d). It is covered with parallel incisions running across the width, in a different direction than for the rest of the surfaces which are entirely covered with incisions forming parallel lines. This surface decoration is unique in the figurative ceramic record of Dolni Vestonice and Pavlov.

The back of the neck has been worked and smoothed; this suggests that it was not attached to a body, but that the piece was originally intended to be a head only.

Thus if some animal representations are not recognisable due to their fragmentary state or again their very schematic appearance, it is clear that for some other such as the head DVI 17 the theme itself is not one we are able to recognise.

- **Conclusion on animal representations**

The recognisable representations of animals in the ceramic art at Dolni Vestonice I, Dolni Vestonice II and Pavlov I show that certain species were deliberately given a special importance: felines and mammoths dominate the record with respectively 11 and 5 pieces. Horses come in third position with 4 pieces and then birds in fourth position with 3 pieces. The importance of this last theme is quite a surprise as it is unusual. Rhinoceroses and reindeer are present with two pieces each. And finally the species of bear, bovid, ibex and mustelid are also present with one piece each. It is noticeable that the recognised variability in the animal themes is larger at Dolni Vestonice I (with 8 different types of animals) than at Pavlov I (with 6 different types of animals).

The relation between animal species actually hunted and the ones represented in Palaeolithic art has long been a topic of research (Rice and Patterson 1986; Clottes and Delporte 2003) with the main conclusion that, except for some exceptions, the species represented normally differ significantly in percentages from those actually hunted. This fact has also been noted in Dolni Vestonice I and Pavlov I (Svoboda *et al.* 1996).

Musil (table 3.24) provides statistics regarding the faunal record found in the parts of Pavlov excavated in 1952-1953 and 1957-1958. Hare is consistently the most hunted species in both parts of the sites, together with foxes, wolves and reindeer. Mammoth remains are found in great number in the 1957-1958 part but in much smaller quantities in the other part. It can be assumed that some of these species were specially hunted for their fur (foxes and wolves) rather than for their meat.

Similar proportions of hunted animals were found in Dolni Vestonice I also where hare, mammoth and reindeer were very important in the faunal record (Svoboda *et al.* 1996:167).

Some of the most represented animal species in the ceramic art such as the felines, bears and rhinoceros are very rare in the faunal record, although they are not totally absent. Some of the most hunted animals, hares, fox and wolves appear totally absent from the figurative record up to date. Mammoth are possibly both very present in the faunal record of Pavlov I and in the art found at this site. Some species such as reindeers, horses, bovine and birds are present in the faunal record and in small number in the art record. One animal species represented in ceramic, a caprid figurine, is totally absent from the faunal record.

Table 3.24: percentage of representation of animal species in the faunal record of Pavlov I 1957-1958 and the 1952-1953 excavated areas (after Musil 1994 and 1997).

Animal species	1957-1958 excavation	1952-1953 excavation
Mammoth	18.9	7.5
Wolf	14.6	12.5
Reindeer	15.1	10.1
Horse	9.0	4.6
Hare	19.2	18.5
Polar fox	13.9	16.9
Fox	3.2	12.7
Bear	0.6	1.6
Wolverine	2.3	4.4
Lion	0.3	0.5
Rhinoceros	0.9	0
Birds	1.7	8.3
Felids	0	0.7
Bovids	0	0.5
Red deer	0.3	0.2

These results emphasise the symbolic importance given to some animal species, especially the feline and mammoth.

The presence of animal species that have been described as dangerous or very large animals is also a major characteristic of the animal representation in fired clay at the Pavlovian sites. This corresponds to a wider tradition seen in the art of the early Upper Palaeolithic in Europe in which these dangerous or very large animals dominate the record for representations for example in the site of Vogelherd in Germany (Hahn 1993), in the Chauvet cave (Clottes (ed.) 2003) or in the Grande Grotte in Arcy-sur-Cure in Burgundy (Baffier 1998). The wider cultural implications of this statement are developed in Chapter 4.

It is also interesting to see an association between some types of representations and the fact that they are represented as whole bodies or simply as heads (Table 3.25). Thus, out of all the recognised categories, only the horse representations and the indeterminate animals do have pieces in both 'head only' and 'body' categories. All the feline, rhinoceros, reindeer, ibex and mustelid are found as heads only. The mammoths, bear, birds and bovid are only found as bodies. These tendencies connected to the themes of representations seem to hold across the Dolni Vestonice-Pavlov locale, even if some themes are specifically associated to a site.

For the indeterminate animals, there is a tendency to find more bodies or body fragments than heads, which is probably the result of an important fragmentation that prevents recognising the pieces. It is interesting, however, to see that when looking at the proportions of heads and bodies in each of the site (excluding Dolni Vestonice II for which the record is too small) the proportions are very similar with 6 heads and 9 bodies of indeterminate animals in Dolni Vestonice I and 5 heads and 10 bodies in Pavlov I. This would possibly indicate that altogether, it has been possible to determine heads more easily than bodies, probably because they are less fragmented.

Table 3.25: percentage of each animal category found as head only or as whole bodies or body fragments

Animal type	Head only	Whole body
Feline	100	
Mammoth		100
Rhinoceros	100	
Bear		100
Bird		100
Bovid		100
Horse	75	25
Reindeer	100	
Ibex	100	
Mustelid	100	
Indeterminate animals	38.7	61.3

The consistency of these statistics, however, indicates that the fact of finding a type of animal represented only by its head, without a body, is not the result of a fragmentation of the piece. It seems that the Pavlovian people were purposefully using a process of metonymic representation (a head to represent a whole animal here) in association with some animal species. It will be seen further that this process is especially important in the body extremities category to the point that it is proposed here to see metonymic representations as a characteristic of the Pavlovian ceramic art.

Body extremities

In the global inventory of the studied ceramic pieces the dominant category by far consists of what is recognised as body extremities: isolated feet and/or legs of animals or human, as well as some possible muzzles or trunks: 190 pieces were found representing 60.3 % of the total (table 3.7). 73 come from Dolni Vestonice I and 117 from Pavlov I. None have been found in Dolni Vestonice II. This theme of representation is not unknown (at least one other example is known from Kostenki I -Abramova 1995- and some legs have been represented in the Magdalenian of southern France) but it is unique in such numbers in portable art in the Upper Palaeolithic of Europe.

These representations in the ceramic art from Dolni Vestonice and Pavlov greatly surprised the first people who found them (Valoch 1996) as they are a very unusual theme of representation in Upper Palaeolithic art. Some of them are fragmentary and thus could possibly be seen as broken off a main body, but sometimes they bear no traces of having been attached at all. Soffer and Vandiver (1994, 1997) think that they result from the additive manner in which the Moravian ceramics were fashioned. They were able to determine through radiography that the whole or nearly whole figurines were made by fashioning each part separately and then by sticking them together. The figurines would have broken or separated at the joint during the firing process. As for Valoch (1996: 140) he points out that such 'little feet' are hardly ever seen on whole figurines, questioning the fact that they could have been attached at all.

We are rather following Valoch in the present study arguing here that we can see in the ceramic art of Moravia a form of systematic metonymic representation in which the body extremities were made as symbols of a whole figure, animal or human. The following study of the pieces is supporting this view.

Among the body extremities, it has been possible to determine various precise types; they form the basis of the following analysis. The size range of the pieces was also considered as a possible indicator of the making of isolated body extremities without intention of attaching them to a larger body. Looking at the degree of fragmentation of the pieces is also an argument in that sense.

- **Various types**

The existence of these types does support the view that the body extremities were made as precise shapes, recognisable by their makers as specific animal types. It has been possible to classify them as various categories.

The various types recognised are detailed in Table 3.26 below and illustrated in Table 3.27.

For three of these types, an interpretation is put forward: Type 1a seems to correspond to human feet, naked or shooed; Types 2d and 4 are interpreted here as mammoth extremities.

Table 3.26: Recognised types of body extremities from Dolni Vestonice I and Pavlov I

Types			DVI	PI	Total
1 With a marked foot	1a	Boot-like	DVI 12/DVI 52	PI 127/PI 128/ PI 145/PI 149/PI 170/PI 171/ PI 176/	9
		Human	DVI 11/ DVI 64	PI 10	3
	1b Thin		DVI 70/ DVI 73/ DVI 86/ DVI 99/	PI 81/ PI 82/ PI 111/ PI 130/ PI 134/ PI 139/ PI 154/ PI 167	12
2 With a worked foot extremity	2a Thin with a curved extremity		DVI 48/ DVI 67/ DVI 69/ DVI 74/ DVI 101/ DVI 103/ DVI 117/ DVI 118/ DVI 119 / DVI 120/ DVI 123/ DVI 127	PI 32/ PI 37/ PI 70/ PI 73/ PI 78/ PI 80/ PI 89/ PI 92/ PI 98/ PI 96/PI 120/ PI 121/ PI 122/ PI 124/ PI 140/ PI 143/ PI 146	29
	2b Thin with a straight extremity		DVI 43/ DVI 66/ DVI 68/ DVI 71/ DVI 102/ DVI 106/ DVI 107/DVI 116/DVI 125 DVI 128/ DVI 131/ DVI 132	PI 68/ PI 71/ PI 76/ PI 77/ PI 79/ PI 84/ PI 101/PI 106/ PI 107/ PI 109/ PI 110/ PI 112/ PI 119/ PI 126/ PI 129/ PI 141/ PI 148/ PI 162/ PI 164/ PI 172/ PI 175	33
	2c Short and large		DVI 44/ DVI 47/ DVI 53	PI 142/ PI 165/ PI 173	6
	2d Mammoth type		DVI 89/ DVI 90/ DVI 97/ DVI 110/ DVI 112	PI 33/ PI 36/ PI 38/ PI 39/ PI 85/ PI 93/ PI 95/ PI 104/ PI 118/ PI 131/ PI 132/ PI 136/ PI 151/ PI 156/ PI 158/ PI 160/ PI 178/ FLATTENED PI 44/ PI 94/ PI 135/ PI 155/ PI 166	27
3 Pointed extremity	3a Straight		DVI 75/ DVI 95/ DVI 100/ DVI 122	PI 62/ PI 87/ PI 123/ PI 137/ PI 138	9
	3b Curved		DVI 35/ DVI 42/ DVI 46/ DVI 115	PI 63/ PI 74	6
4 Trunks			DVI 61	PI 28	2
5 Broken at both extremities	5a With straight parallel edges		DVI 88/ DVI 91	PI 90/ PI 177	4
	5b Conical shape with straight Edges		DVI 72/ DVI 85/ DVI 98/ DVI 104/ DVI 105/ DVI 108/ DVI 109/ DVI 111/ DVI 113/ DVI 124/ DVI 129	PI 69/ PI 86/ PI 97/ PI 99/ PI 100/ PI 174	17
	5c Conical shape with other edges		DVI 56/ DVI 57/ DVI 114	PI 35/ PI 105/ PI 108/ PI 150/PI 152/ PI 157/ PI 163	10
6 Very fragmentary			DVI 55/DVI 93/ DVI 126/	PI 51/ PI 83/PI 88/ PI 91/ PI 116/ PI 113/PI 117/ PI 133/ PI 147/ PI 159/ PI 168	14
7 Others			DVI 49/ DVI 65/ DVI 92/DVI 121/ DVI 130	PI 61/ PI 75/ PI 102/ PI 103	9
Total			73	117	190

Table 3.27. Illustration of the determined types of body extremities in ceramic.

Types	Illustration of the various categories			
	1a		1b	
Type 1	'Boot-like' PI 149 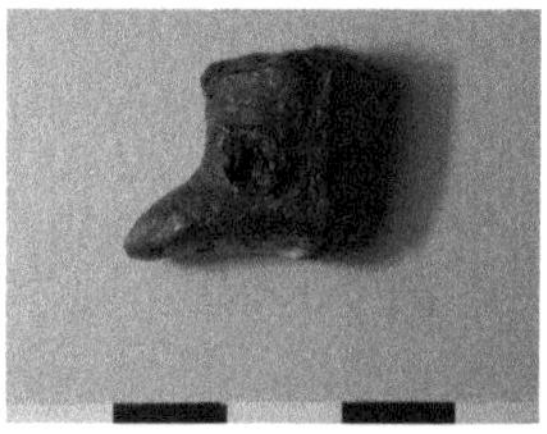	Naked foot DVI 11 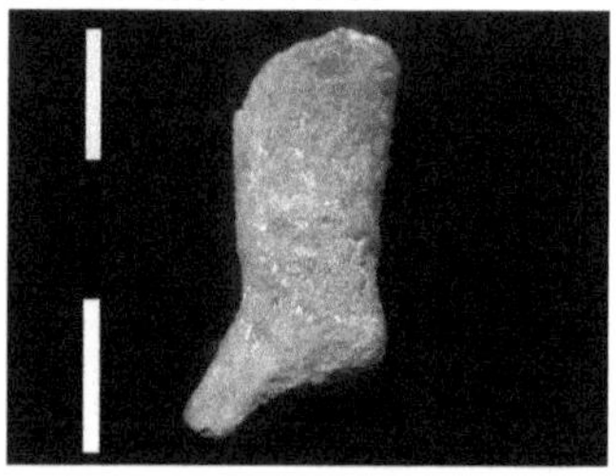	PI 154 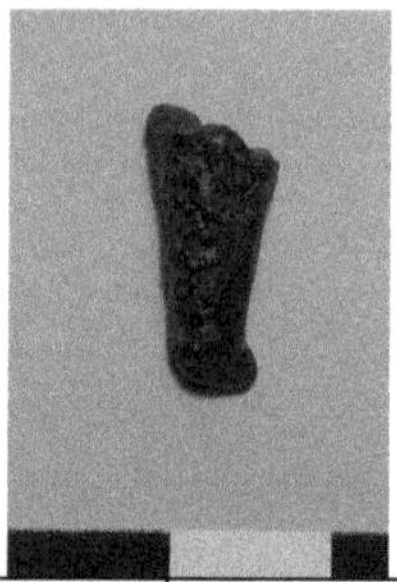	
Type 2	2a DVI 69 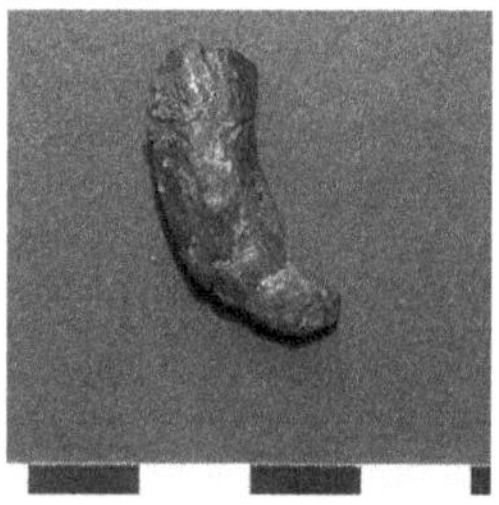	2b PI 107 	2c DVI 44 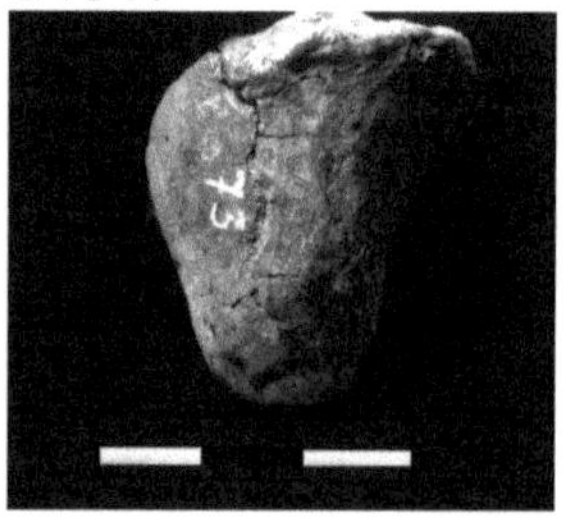	2d PI 39 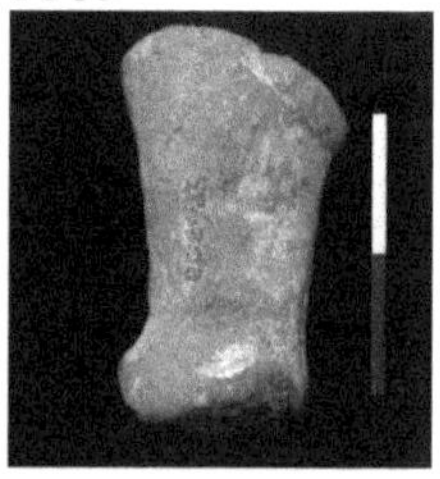
Type 3	3a DVI 75 	3b DVI 46 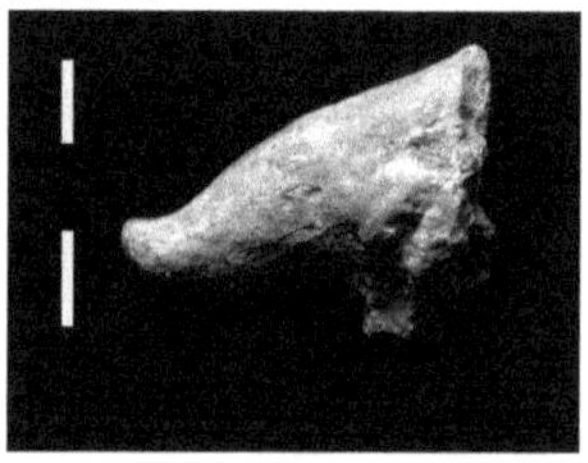	Type 4	PI 28 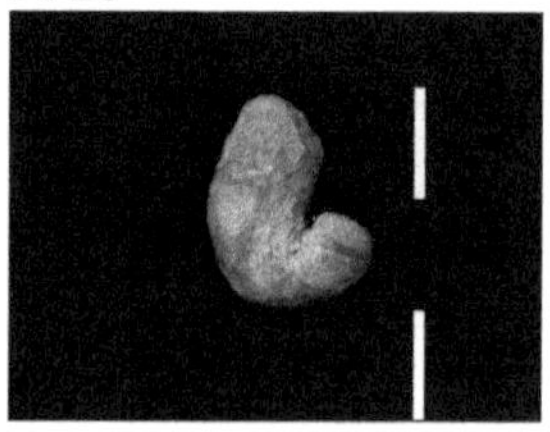
Type 5	5a PI 90 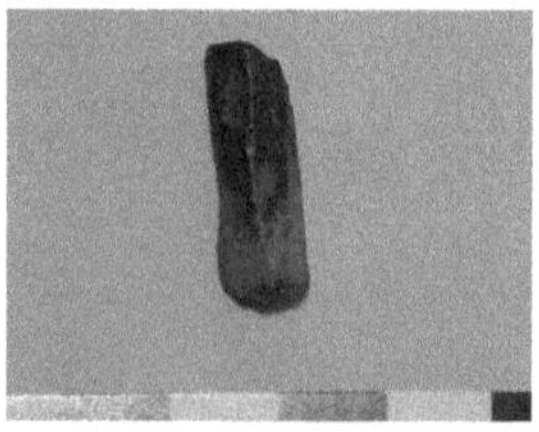	5b PI 97 	5c PI 152 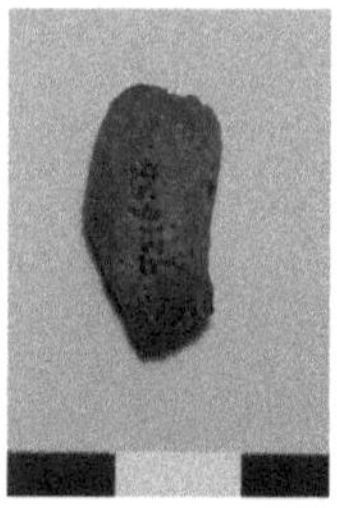	
Type 6	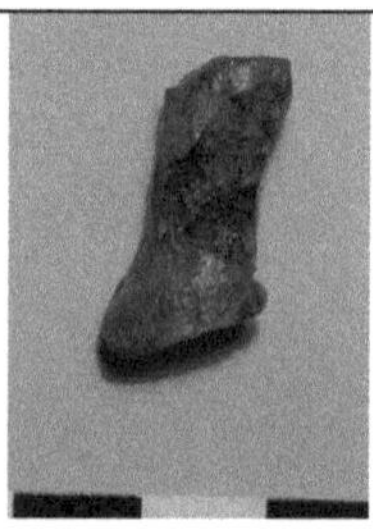 PI 159	Type 7	DVI 92 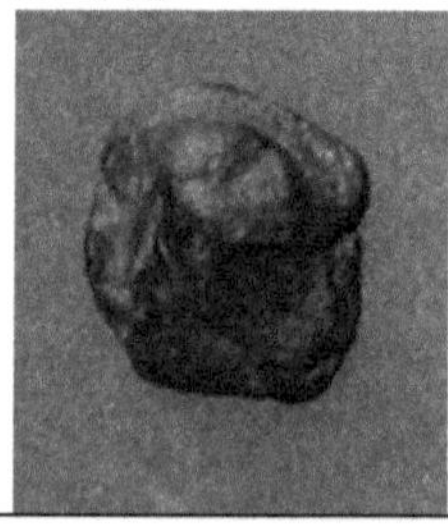	

Mammoths legs

27 pieces are interpreted as possible mammoth legs as they are wide, cylindrical and with straight sides. Most are flattened at one end to represent the foot which is sometimes fashioned in relief. 2 pieces are also clearly mammoth trunks.

This interpretation is based on the similarity of some pieces (for example PI 39 in Table 3.27, type 2d) with actual animal legs (here with elephant legs on figure 3.38) showing a typical wide leg and a flattened foot. Also, one of the representations of mammoth in fired clay is the only animal representation at the Moravian sites with two whole legs and feet present (PI 24, figure 3.39). The rear leg corresponds to the mammoth type legs. They can be used for comparison with the assemblage of body extremities.

Figure 3.38: modern elephant legs.

Figure 3.39: mammoth figurine PI 24 in fired clay with legs from Pavlov I

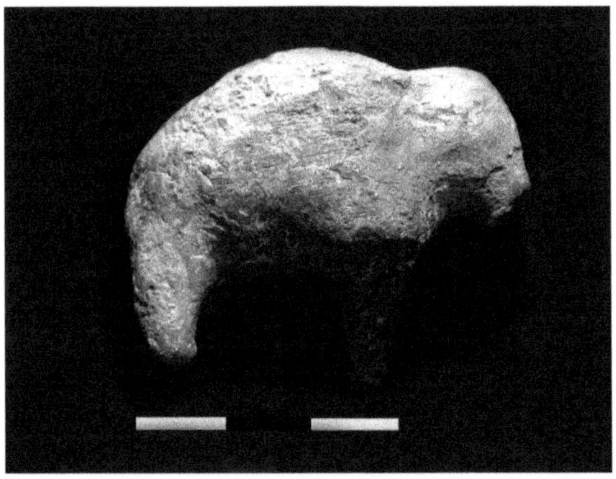

If this interpretation as mammoth legs is accepted, the mammoth would then become the most represented animal by far at the site of Pavlov I as 22 legs could be added to the four mammoths already recognised at the site. On the opposite, only five mammoth legs were found in Dolni Vestonice I and only one mammoth figurine was recognised.

The mammoth would then become the most represented animal in the ceramic art of the Pavlovské Hills, with however a major difference: in Dolni Vestonice I the number of mammoth figurines then becomes equal to the number of feline heads found; In Pavlov I, the mammoth becomes neatly dominant in the representation. These differences might imply variations in the subsequent symbolic implications.

The connections between this theme of representation and the faunal remains actually present at the sites also need to be emphasised as the mammoth was one of the most hunted (or scavenged in the opinion of some, see Soffer 1993) animal at the Pavlovské Hills in the Pavlovian period (Musil 1994, 1997). Materials issued from mammoths (bones and ivory) were widely used at the sites. It is therefore not surprising that this animal would have a strong place in the symbolic bestiary of the sites.

Human feet

Among the body extremities, 12 pieces are recognisable as human feet or legs with feet. Three pieces are represented naked (Table 3.28), with a marked heel and a flat sole, a combination not seen in the animal world. They are all very different in style from each other. 9 other pieces are interpreted as booted human feet.

These representations of isolated human feet are very rare in Upper Palaeolithic art in general. The fact of having shooed feet represented is even rarer. One could argue they were broken off a larger body during firing. However, the large majority of the known human statuettes, especially the Gravettian female figurines, do not have the feet detailed. At least one well fashioned human foot is known from the site of Kostenki 1 in Russia (Abramova 1995), carved in limestone, indicating once again a common point between the art of the two sites.

Table 3.28: the naked human feet in fired clay from Dolni Vestonice I and Pavlov

DVI 11	DVI 64	PI 10
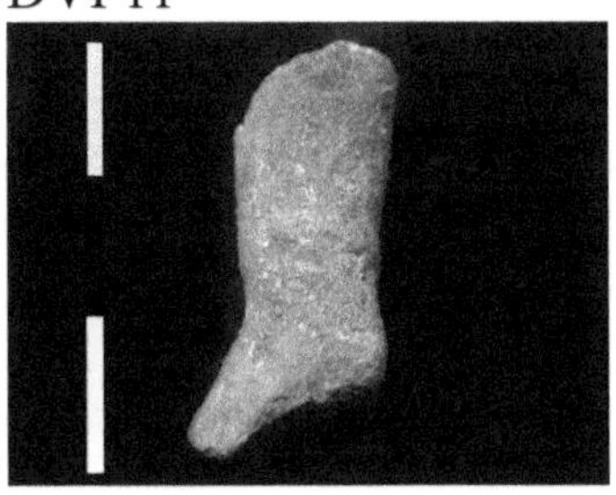	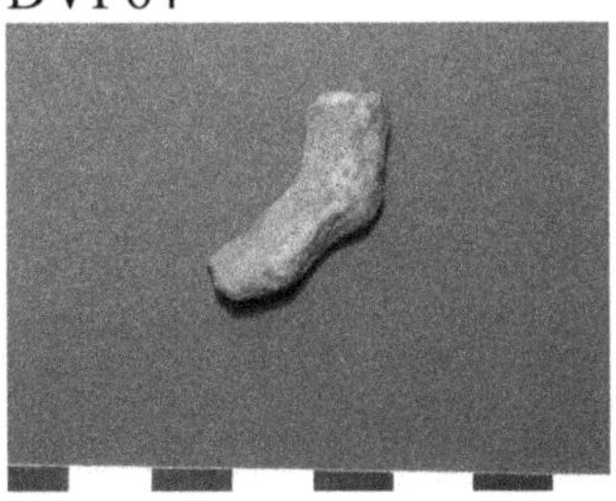	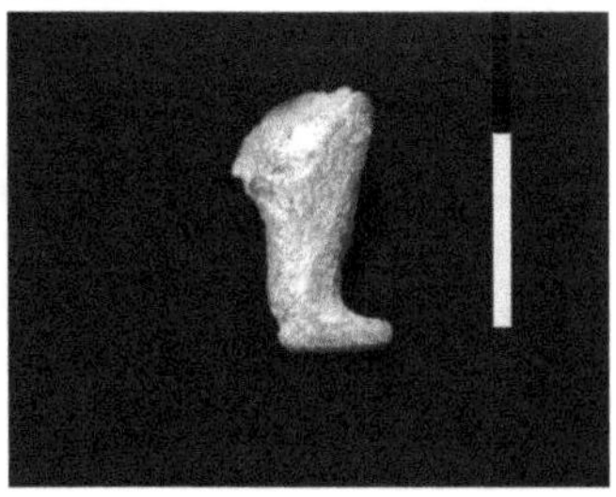

Other recognisable animals

Some of the body extremities are very finely shaped which indicates they were most probably modelled with the intention of representing precise animals. For most of them, it remains difficult to propose an exact interpretation. Two pieces, however, are so carefully shaped that we propose here an interpretation.

PI 32 (Figure 3.40) is an extremely well fashioned leg; the foot is indicated, the angle of the knee is well-marked and the volume of the muscles of the thigh is rendered accurately. The piece is quite large among the ceramic statuettes with about 5 centimetres long. It is broken at the top of the thigh. There is no doubt it represent the leg of a member of the deer family, whether a deer or a reindeer, as is illustrated in figure 3.40

PI 73 (figure 3.41) is a very thin and long black piece (1.6 centimetres long). It is slightly curved at one end to indicate the foot, and then it widens at the thigh. This extremity is whole: the thigh end is cut off rather than broken. It can be compared to a young artic fox leg as is illustrated in figure 3.41

Figure 3.40: comparison between reindeer legs and the body extremity PI 32, most likely also a reindeer.

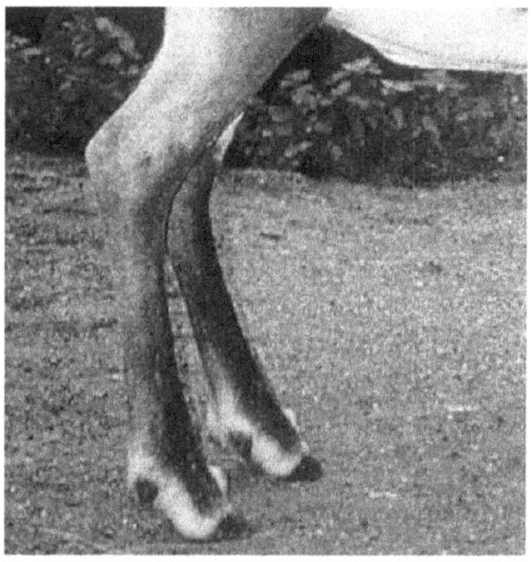

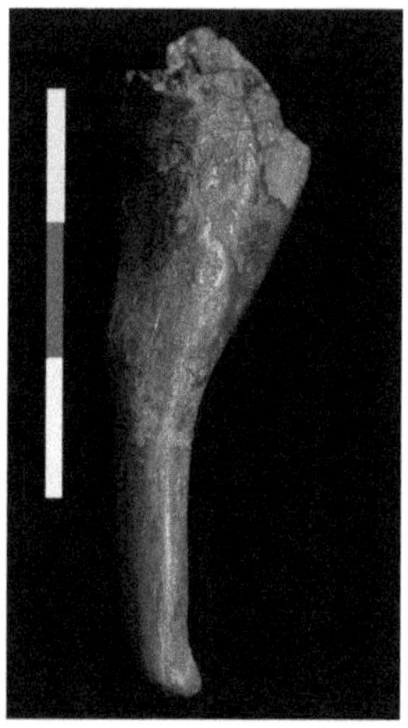

Figure 3.41: comparison between the front legs of a young artic fox and the body extremity PI 73

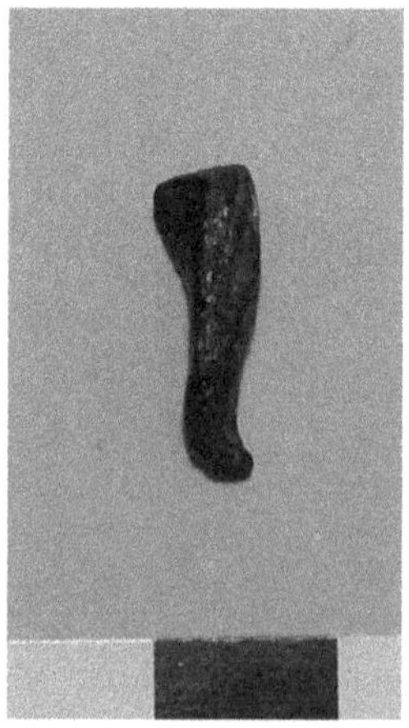

- **Size range of the body extremities**

In order to test the idea that the body extremities in ceramic were made originally as metonymic representations (as symbols of a whole animal), statistics on the size range of the studied inventory of body extremities (190 pieces) were done. The purpose was to determine possible patterns as well as if all or parts of these body extremities were indeed part of whole figurines (i.e. if the estimated figurine size would fit into the size bracket normally seen for figurines at the sites) or fashioned separately and intentionally as limbs or muzzles (i.e. if the estimated figurine size was too big). This size range is visible in figure 3.42 below.

At first glance, it is possible to notice that in Dolni Vestonice I the pieces tend to be larger in size than in Pavlov I where the vast majority is below 2.5 cm in their maximal dimension. A simple conclusion can be drawn: the body extremities are larger in Dolni Vestonice I than in Pavlov.

One could question whether this is due to excavation methods as Dolni Vestonice I was excavated earlier than Pavlov I (Vandiver *et al.* 1990). This is possibly partly the case. However, even for Dolni Vestonice I, there are over 2,000 objects classified as pellets (Vandiver *et al.* 1989) in the archaeological record that measure between 4 and 10 mm and that were nonetheless recovered. This would indicate that the excavation was carried out carefully enough to recover even small pieces.

The statistics on the size range of the body extremities for Pavlov I also seem to confirm the general tendency previously noticed of a wider variability at this site than in Dolni Vestonice I.

In order to assess the possible figurine height one can expect to have if these body extremities were indeed part of a whole figurine, a ratio consisting of the actual height of the few available animal figurines with legs divided by the height of the leg (from the extremity to the junction with the belly) was calculated (table 3.29). It should be pointed out that some of the figurines legs are fragmentary and would have been longer originally.

Figure 3.42: Comparative diagram of the size range (in centimetres) of the body extremities of the sites of Dolni Vestonice I and Pavlov I in relation to the percentage of the inventory of body extremities in each site (73 pieces in Dolni Vestonice I and 117 in Pavlov I).

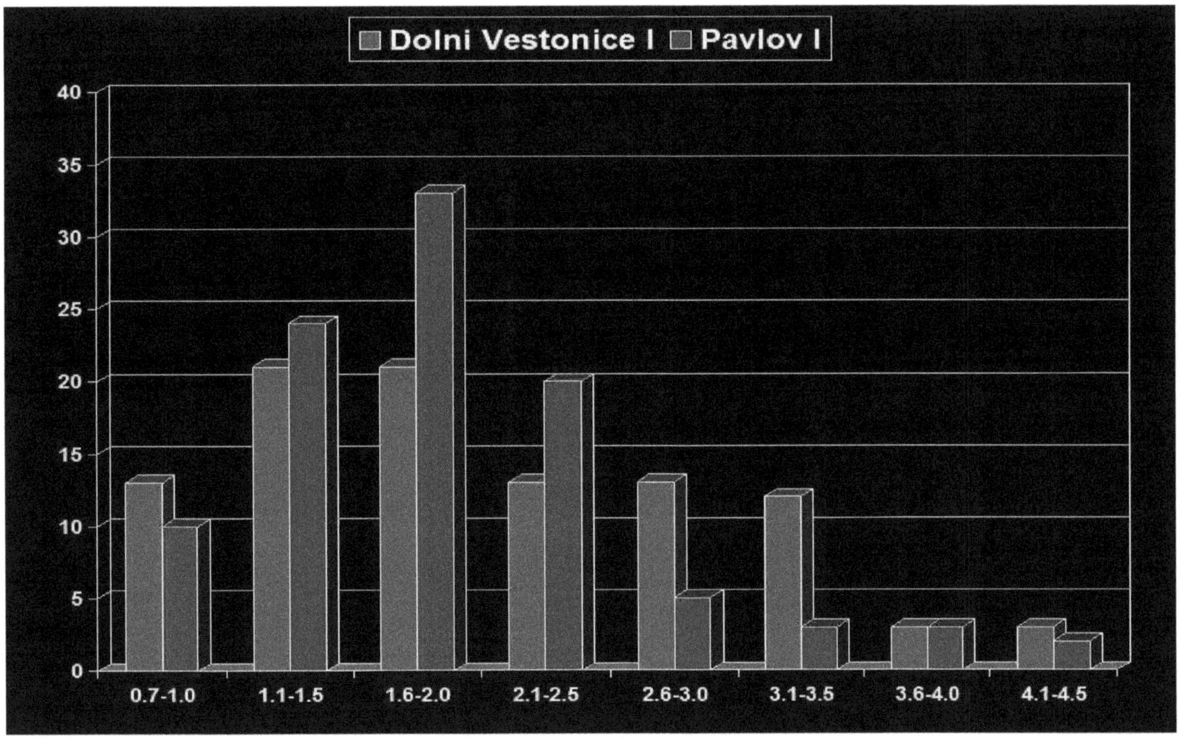

Whole or nearly whole fired clay animal figurines are few; only five examples are known from Dolni Vestonice I and four from Pavlov I, among which one mammoth representation (PI 24) has two legs with very different ratio for each. It can be pointed out that out of the five animals with legs in DVI, three are undetermined animal bodies, and out of the four animals with legs from PI, three are mammoths.

From this table, it appears the ratio variations are fairly consistent between the two sites with the smallest ratio being around 2, thus corresponding approximately to a leg half the size of total height, and the larger ratios being around 4, which corresponds approximately to a leg measuring a quarter of the total height.

Table 3.29: Ratio of the body height divided by the height of the leg (from the extremity to the junction with the belly) of the animal figurines with legs from DVI and PI. The smaller the ratio is, the tallest is the leg in comparison to the body.

Dolni Vestonice I		Pavlov I	
DVI 27 (bovine)	2.5	PI 26 (large mammoth)	1.96
DVI 22 (undetermined animal)	3.5	4 (for the front f a mammoth)	2.3
DVI 51 (undetermined animal)	3.66	PI 31 (flattened undermined animal body)	3
DVI 16 (bear)	4	PI 22 (small mammoth)	3
DVI 32 (undetermined animal)	4.3	PI 24 (for rear leg of mammoth)	4.2

Therefore, it will be assumed that the legs could measure in their extreme dimensions between a quarter and a half of the height of the body they would have been attached to. This will allow estimating the range of body sizes one could expect in relation with the size range obtained for the body extremities (table 3.30).

Table 3.30: estimate height of animal bodies in fired clay in relation with the size range of the body extremities found all in centimetres.

Height of body extremity	Body height if leg = 1/4 of body	Body height if leg = ½ of body
0.7 cm	2.8	1.4
1 cm	4	2
1.5 cm	6	3
2 cm	8	4
2.5 cm	10	5
3 cm	12	6
3.5 cm	14	7
4 cm	16	8
4.5 cm	18	9

Looking at the maximal height seen for animals in the fired clay figurine record at Dolni Vestonice I and Pavlov I, one can see the tallest figurine in height is the mammoth PI 26 that measures 5.5 cm in height. The next tallest figurine found is also a mammoth from Pavlov I (PI 24) that measures 4.2 cm in height. In Dolni Vestonice I, the highest animal figurine (DVI 16) measures 3.2 cm high.

Considering these sizes, it appears unlikely that any body extremities measuring over 2.75 cm in their maximal dimension could have been attached to an animal body as the resulting figurine would be very large, larger than what is actually seen in the archaeological record for fired clay objects, as well as in Palaeolithic art standards in general. The longest known fired clay object in all categories for the Moravian sites, DVI 28 measures 11.8 cm but most objects are much smaller. The rest of the art made of other materials at these sites is also rather small; for example a mammoth ivory found in Pavlov I measures 4.2 cm in height (Valoch 1996: fig. 114) respecting a similar size range than the art made of fired clay.

If one accepts that part of the pieces found between 2.1 and 3 cm could indeed be part of whole figurines, the large size of the rest makes it unlikely they were ever part of whole figurines. It could be argued that the body extremities at Dolni Vestonice I were fashioned as such without intention of attaching them to an animal body.

As for Pavlov I, the very large majority of the body extremities measures less than 2 cm (figure 3.42) and could indeed have been part of larger whole figurines as far as size is concerned. A group of pieces is much larger in size (up to 4.5 cm) and could indicate the presence of some body extremities fashioned as such without intention of attaching them to a body, unless one argues for the presence of much larger fired clay figurines that have not survived in the archaeological record.

- **Level of fragmentation**

It is also interesting to look at the visible level of fragmentation of the body extremities inventory. A simple inventory (Table 3.31) shows that 11.6 % of the pieces are actually whole: they are fashioned at both ends and do not show any sign of breakage.

Table 3.31: Inventory of the body extremities in ceramic from Dolni Vestonice I and Pavlov I according to their level of fragmentation (whole, broken at one end, broken at both ends).

	DVI	PI	Total	% of total
Whole or nearly whole pieces	11	11	22	11.6
With one broken extremity	43	78	121	63.7
With two broken extremities/fragmentary	19	28	47	24.7
Total	73	117	190	100

If we consider the size range of the whole body extremities (Table 3.32), we can notice that for Dolni Vestonice I the whole pieces are indeed among the larger ones in the inventory, which would confirm they were not attached to a body. For Pavlov I, however, the picture is different: the whole pieces are all below 2 cm in size, which shows that even small pieces could be made for themselves without ever being attached to an animal body.

It is also true that the majority (63.7%) of the inventory of body extremities is broken at one end, with also 24.7 % being broken at both ends. The ceramic inventory as a whole is extremely fragmentary, which can of course bias observations.

Table 3.32: Quantities of whole body extremities recognised for each size range on each site.

Size range in cm	0.7-1.0	1.1-1.5	1.6-2.0	2.1-2.5	2.6-3.0	3.1-3.5
Dolni Vestonice I		2	2	1	1	5
Pavlov I	2	5	4			
Total	2	7	6	1	1	5

Systematic metonymic representations in the ceramic art

Following the various observations above, it seems that the idea of a systematic use of metonymic representation is fully acceptable for the ceramic art of the Pavlovian sites.

This aspect has been developed above for the case of the body extremities, which neatly dominate the record, but this can be extended without risk to a large part of the animal representations: all the ones where heads only are found. This is the case for example of the feline figurines, which are only represented as heads instead of whole animals. It must be emphasised also that for some of these heads the neck area, where the figurine should be broken off, is actually also fashioned and smoothed. This clearly indicates that these objects are not broken off a body.

This systematic use of metonymic representations should be seen, in our opinion, as a major characteristic of the Pavlovian ceramic art with the corresponding symbolic implications it carries, notably as a form of shared language between the Pavlovian people. These symbolic implications do deserve further research that will be the topic of future publications but are not looked at in detail here.

Non-figurative objects

Some ceramic objects found in Dolni Vestonice I, Dolni Vestonice II and Pavlov I have been shaped intentionally in the form they have today but cannot be described as human or animal.

This is the case for a series of nine spherical objects from Dolni Vestonice I (figure 3.43).

Figure 3.43: Spherical objects made of fired clay from Dolni Vestonice I.

Some fired clay fragments also bear traces of possible surface decoration, as is commonly found in decorated objects made in other materials at the Pavlovské Hills sites (Valoch 1996). It can be questioned whether the clay fragments were a simple support for geometrical/ patterned decoration (figure 3.44), or whether they formed part of an object which had decorated surface but which is not recognisable because of its fragmentary state. Two flat small fired clay fragments bearing parallel incisions were found in Dolni Vestonice I and two in Pavlov I.

Figure 3.44: PI 42 ceramic fragment with parallel incisions.

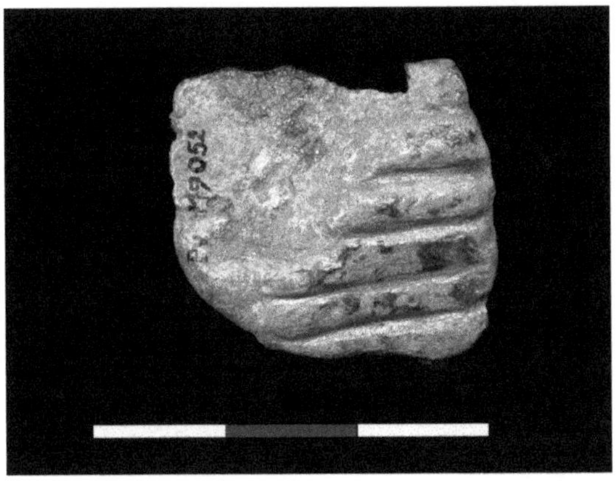

The fact that some elements in the ceramic art can be described as abstract or surface decoration connects the art on clay of the Moravian sites to the wider context of the art of the Gravettian period in Central and Eastern Europe, and indeed to the art of the Upper Palaeolithic in general as abstraction is consistently found during this period (i.e. the signs found in cave art) together with animal and some human representations.

Conclusion

The study of the themes of representation found in the ceramic art in the sites of Dolni Vestonice I, Dolni Vestonice II and Pavlov I allows underlining similarities and variability between these sites. A general tendency shows more uniformity in Dolni Vestonice I and much more variability in Pavlov I. The very small number of fired clay objects found in Dolni Vestonice II makes it difficult to draw general conclusions for this site.

We have seen that the female figurines found in Dolni Vestonice I and II can be separated into two types only, when five different types are found in Pavlov I. The major type of female figurine found in Dolni Vestonice I, the Black Venus type (DVI 1), is completely absent from Pavlov I.

The feline heads also show more variability in Pavlov I with three types present, for only two in Dolni Vestonice I. However it can also be pointed out that one of the type found shows exactly similar heads both in Pavlov I and in Dolni Vestonice I. By the quantity of feline representations found in both site (7 in Dolni Vestonice I and 5 in Pavlov I) it appears that this animal was treated as symbolically important in both places.

The mammoth representations underline a very different situation between the two sites with only one figurine found in Dolni Vestonice I and seven in Pavlov I with 27 more possible mammoth legs to add to this record. One other point to emphasise is that among the five mammoth figurines from Pavlov I, there is very little uniformity in style, underlying various makers.

As for the three known bird representations, they come from Dolni Vestonice I only.

There is much variability visible in the body extremities both in the quantities found in the two sites and in their size range. A large number of pieces (117) are found in Pavlov I, a much smaller in Dolni Vestonice I (73). The pieces tend to be much smaller in size in Pavlov I.

As for the non-figurative pieces, similar objects are found in Pavlov I and Dolni Vestonice I in the shape of flat fragments with parallel incisions but other objects such as the clay balls are absent in Pavlov I.

The elements listed above will provide the basis on which to build an analysis of the elements of a transmission of knowledge and skills visible in the fired clay record, taking into account inter-site and intra-site variability and their social application. This analysis is developed in Chapter 4.

3 Technology of the use of clay in Moravia: the making of the objects

Thanks to the large record of fired clay objects available at the sites of Dolni Vestonice and Pavlov, much research has been carried out (Vandiver *et al.* 1989, 1990; Soffer and Vandiver 1994, 1997; Soffer *et al.* 1993) that made it possible to determine fairly precisely the way the Pavlovian ceramics were made from the procurement of the material used, to the fashioning methods and the firing technology applied. These various processes allowed reconstructing the technological scheme visible in the making of the objects: this is presented in Table 3.33 below. The analysis of the social implications of these technological processes is treated in Chapter 4.

Procurement of the material

Chemical analysis was undertaken in 1924 and 1949 (Absolon 1949) to determine the nature of the material used to make the Dolni Vestonice ceramics. A sample of the Black Venus was notably taken inside the break at its thigh. It was thought then that the figurine was made of the local loess soil mixed with bone ash, mammoth bone and perhaps fat, leading the archaeologists to believe that the ceramic objects of Dolni Vestonice were made of well-prepared and tempered clay (Vandiver *et al.* 1990).

Vandiver *et al.* (1989, 1990) did more analysis by X-ray diffraction, radiography, amino-acid determination and liquid chromatography to try and detect the presence of bone or other organic remains, but failed to do so. They sampled again the Black Venus, and also analysed different other fragments. Only traces of fatty acids and ammonia (0.01% to 0.1%) were present in the surface layer and interior of the samples, the same amount as found in loess taken from occupation levels at the site (Vandiver *et al.* 1989, 1990).

They interpret the presence of fatty acids and ammonia as showing that the porous ceramic samples are in equilibrium with the surrounding soil, ammonia and fatty acids being normal breakdown products of surface plant remains. They mention that this could have misled the first analyses, being interpreted as the presence of bone. But they can certify that no bone or organic material was present in the Dolni Vestonice figurines (Vandiver *et al.* 1989, 1990).

Table 3.33: Technological scheme of the making of ceramic figurines in Dolni Vestonice and Pavlov

Phase	Sequence	
Procurement	Dig loess from habitation soil	1
Preparation	Adjust plasticity by adding water	2
Fashioning	Shape different parts of the figurines separately (bodies, heads, tails ears, limbs)	3
	Stick them together to form whole figure	4
	Fashion some details	5
Drying	Partial drying	6
	Full drying	7
Fashioning	Fashion details and surfaces	8
Drying	Full drying	9
Re-wetting	Wet the fully dried figurines just before firing	10
Firing	Preparation of firing structure	11
	Procurement of fuel	12
	Ignite fire	13
	Expose figurines to flames	14
	Watch and hear explosion	15
Obtainment of figurines	Obtainment of whole or fragmentary figurines	16
	Possible reshaping after firing	17

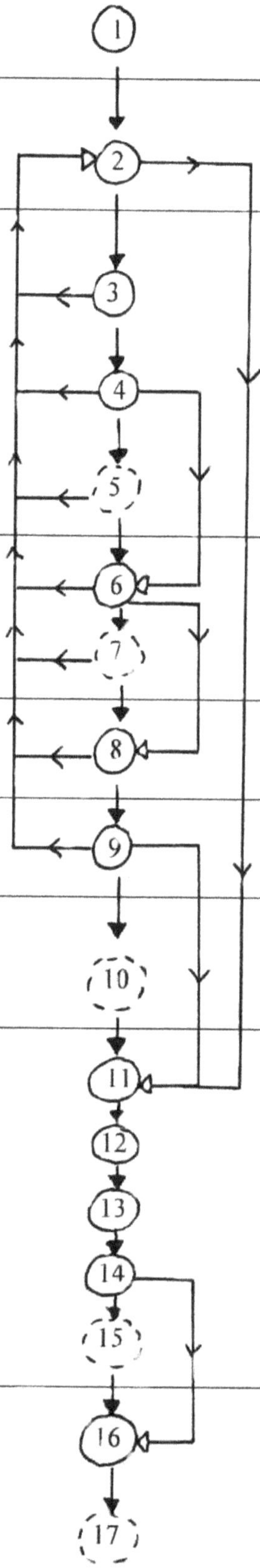

Analysis indicated that the ceramic samples are primarily composed of rock fragments, quartz and feldspar with only a minor amount of clay (Soffer and Vandiver 1997). This corresponds to the nature of the local loess soil. An important difference between the loess and the ceramics is the relative density of particle packing: the loess is about twice as porous as the objects. This indicates that local loess was not selected in bulk form and carved into shape but rather mixed with water and modelled (Soffer *et al.* 1993a). This is also seen in the micro structural and macro structural observations of kneading and of the joining of wetted and semi-plastic parts to form the figurines (Vandiver *et al.* 1990).

Because of the large size and number of the inclusions in the loess, the material would not have been plastic in the sense of modern potter's clay. It actually contains only about 20% of illitic clay. It would have been a gritty and sandy paste, not very malleable and easily cracked (Vandiver *et al.* 1990).

Soffer and Vandiver (1994) also note that there seems to be a difference between the material from Dolni Vestonice and Pavlov, with a higher proportion of the Dolni Vestonice pieces being actually fired (over the point of ceramic change around a firing temperature of 500° C). Looking at the material composition, they think the reason could be that the nature of the loess at Pavlov was slightly different, with a higher proportion of calcium carbonate. This would have made it less suitable for firing.

Fashioning

The different parts of the figurines (bodies, legs, feet, tails, heads and ears) appear to have been hand-shaped and preformed separately out of the material, and then stuck together to form the whole figurine (Vandiver *et al.* 1989). There is evidence for hand modelling of the pieces in the traces of fingerprints or scraping marks. Xeroradiography confirmed the practice of this additive manufacturing process (Soffer *et al.* 1993, Soffer and Vandiver 1997) by showing the internal joints in the objects. The many isolated body parts found, such as legs or heads, have been interpreted by some as having broken at the joint, which would be a weak point, during the firing process thus demonstrating this additive process (Soffer *et al.* 1993). We have seen, however, that this does not work for part of the body extremities which were fully modelled without being attached to a body. And this certainly does not explain the large quantity of body extremities found when full figurines with fashioned extremities are in fact very rare.

There is evidence that the modelled figurative pieces were smoothed with tools, wiped or scraped. Many exterior surfaces seem carefully shaped. A point that seems confirmed by the study of Kralik and Novotný (2005) about the fingerprints on the figurines: they are actually quite rare in relation to the large corpus of pieces. Convex surfaces were usually not reworked but many concave surfaces show traces of impression from small implements (finger nails, pointed tools and microliths), which were used to incise details of bodies or faces (Vandiver *et al.*1990).

Soffer and Vandiver determined that the technique used for fashioning was mainly additive, with some possible reductive activity (removal of material) in the making of small details, when incisions are made for example. One aspect they do not take into consideration is the notion of drying in relation to the fashioning as they consider drying only in relation to the firing, in order to determine a possible role for the thermal shocking of the objects (see below). However, some traces are still visible on the objects that tell us at which stage of dryness of the material they were fashioned. This bears interesting implications as to the time investment necessary or chosen to make the objects as the material needs to be left to dry for a certain time in order to obtain certain results.

In spite of the alteration of the surfaces by the conservation agent used, there remains evidence on some pieces that allow determining if they were fashioned when still wet or when dried. This is the case on the rhinoceros' head PI 13 from Pavlov I (Figure 3.45a): it is clear here that the eye was made by impressing a thin stick-like object into the material when wet, as a rim of excess material created by this action formed and is still visible around the eye (Figure 3.45b). This only happens if the material is worked when wet.

Figure 3.45: Rhinoceros head PI 13 from Pavlov I. It measures c. 3.5 cm high.

a) Profile

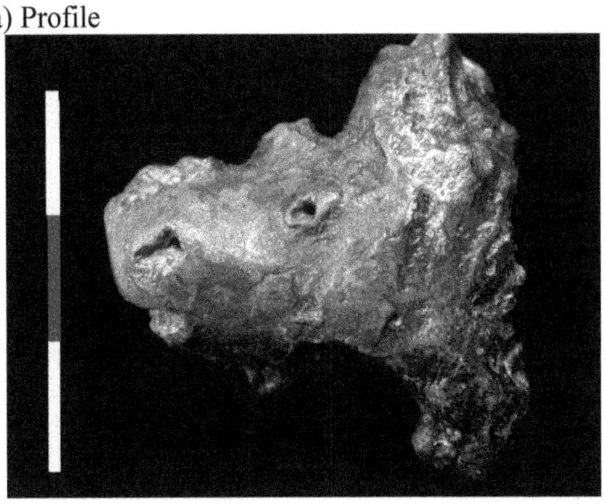

b) Close view of the eye, with the rim of excess material visible.

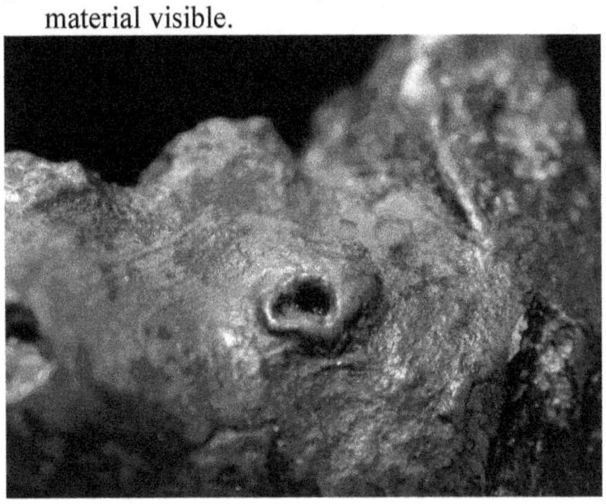

Many of the ceramic objects also bear traces of scraping, which seem to indicate a reworking of the surfaces after drying (complete or partial), as scraping on a wet material would be a messy process that would not serve a useful purpose in the fashioning of a small figurine. It is much easier to work wet clay with the fingers, keeping the help of implements to indicate details. Scraping, on the other hand, is extremely suitable for use on partially or fully hardened clay. It allows erasing some other traces of fashioning left on the wet material and smoothing the surfaces. In some instances, this scraping action leaves a series of thin parallel incisions.

This is also seen in the rhinoceros head PI 13. Scraping has reshaped the end of the muzzle (Figure 3.46). It is likely there was also originally a rim of excess material around the hole of the nostril, if it was done on wet material. This was taken off during the scraping process.

Figure 3.46: Rhinoceros head PI 13 from Pavlov I with visible traces of scraping on the muzzle indicated by the arrow. It measures c. 3.5 cm high.

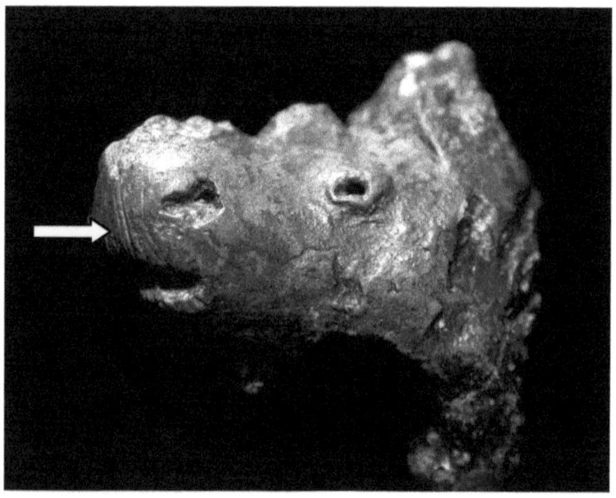

The scraping action is especially visible when it has been carried out with what looks like a micro-denticulate tool that leaves small parallel grooves on the surfaces. This type of tool was indeed rather frequent in the site of Pavlov and Dolni Vestonice (Svoboda 1997; Klima 1997c). A micro-denticulate tool was thus possibly used on the face of the Black Venus from Dolni Vestonice I as it is marked by a band of thin vertical parallel grooves (Figure 3.47).

Figure 3.47: Detail of the face of the Black Venus (DVI 1) with vertical parallel grooves.

Scraping with a smooth edge tool would not necessarily leave very visible marks on the surfaces of the figurines, especially if one remembers that they have all been treated with polyvinyl acetate, which clearly affects the reading of the surfaces. The partially/fully-dried clay can also be re-worked in many other manners, such as carving, fashioning or engraving as one would work a very soft stone.

Whatever the technique used on the figurines, another element indicating that a number of the figurines were reworked after partial or full drying is the very small average size of these representations. Most of them measure between 2 and 5 cm in their maximal dimension and are very finely detailed. Anybody who has worked clay will know how difficult it is to fashion a very small and fine sculpture from wet clay since the whole of the object is subject to being deformed by the slightest pressure applied. Working on a soft material could not be done all around the object, and would leave specific marks. On many of the figurines from Dolni Vestonice and Pavlov the surfaces are very well smoothed, and the objects are worked in the round. This suggests they were re-worked after some form of drying, as is the case for a well crafted reindeer leg from Pavlov I (Figure 3.40).

The fact that reworking of the figurines after drying took place is important in indicating that they were carefully made in a way that took time. Drying necessarily introduces the idea that a longer time is needed to make the objects than when working directly on the soft material. This means the whole process could not be instantaneous, but had to be done in separate steps. This emphasises the investment in time put into these objects by the people who made them. Whether they were destined to be exploded in a subsequent firing, as Soffer *et al.* (1993) argue, or not, it confirms their symbolic importance. The fact that they were dried also fits in well with Soffer and Vandiver's (1997) idea that the best way to obtain an explosion during firing would be to dry the objects, and then re-wet them before throwing them into the fire.

Finally, one object from Pavlov I (PI 35) also raises the possibility that some re-working could take place after the firing. It appears to be a limb fragment, broken at each end. On the whole length of the object, a scraped area can be seen (Figure 3.48). This scraped area differs from scraping done on a dry material in a manner that looks as if it was done after the firing. This observation is based on personal experience of how clay reacts to being worked at various stages of the manufacturing process. The surface of a ceramic undergoes changes during the firing process. Scraping the ceramic after firing produces a surface that reflects light in a different manner than the surrounding area, appearing lighter in colour, with a different texture. Scraping done on dried clay before firing affects the surfaces in a mechanical way, by flattening it for example, but does not produce this change in colour and texture. This is because the internal structure of the clay object is affected by the firing process in a different way than the external surfaces are.

Figure 3.48: Body extremity from Pavlov I (PI 35) with traces of post-firing scraping indicated by the arrow.

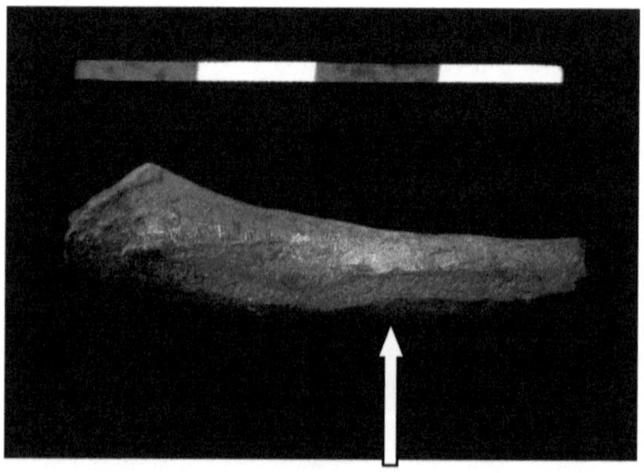

Fashioning a ceramic after it has been fired can be compared to the working of stone or osseous materials as the material has hardened. Fired clay is then treated as another material and worked with techniques used for other material.

Firing

Firing ceramic is a complex process involving a succession of steps that need to be followed, from the gathering of fuel, the preparation of firing structures and objects to burn, the obtaining of the necessary firing temperature to the retrieving of the objects after the firing. Wood was apparently the fuel used in fires in Dolni Vestonice and Pavlov (Damblon 1997: 439).

Firing temperatures

Vandiver *et al.* (1989, 1990) carried out an extensive range of analyses in order to determine the firing temperatures at Dolni Vestonice and Pavlov. These included differential thermal analysis, scanning electron microscopy, x-ray diffraction and a durability test, which simply consists of boiling the fragments (a durable ceramic would survive intact). They sampled between eight and twenty fragments for these analyses for each of the sites of Dolni Vestonice I and II and Pavlov I, using the unshaped fragments found and choosing from a range of different colours of the material. This determined the firing temperatures as ranging between 500°C and 800°C, which is consistent with using a bonfire, a hearth or a banked walled kiln and with the firing structures found (Soffer and Vandiver 1994).

They also looked at the colour of the fragments as being indicative of the firing temperature (Soffer and Vandiver 1997). For example, they determine that hard black and red ceramics usually corresponded to a firing temperature of about 800°C, the black and red ceramics softer in texture and the harder orange and grey would have been fired to about 700°C, and the softer orange and grey ceramics to around 600°C.

But they also note that the duration of the firing has an effect on the final appearance of the ceramic, but it is not possible to be certain about how long the firing lasted. However, some elements can give a rough idea of the length of the firing. Only a small amount of glass formed at the contact points between the particles, which mean that the vitrification process that occurs in the clay during firing was not high. This process, which leads to the hardness of the ceramic, depends both on the temperature and on the duration of the firing. As we know the upper limit of the temperature to be around 800°C, notably because of the nature of the firing structures that could not have produced higher temperatures, Soffer *et al.* (1993a) say the firing time was not long, probably not more than a few hours.

The firing structures

In general the spatial distribution of ceramics in both Pavlov and Dolni Vestonice seem to correspond to the hearth or kiln structures, in which they were probably fired (Klima 1997a).

A very interesting point about the ceramics from Dolni Vestonice I is the existence of two firing structures, kiln-like, that seem to have been especially manufactured for ceramic firing. They seem to have been used several times, and many ceramic objects were found inside the structures. In Dolni Vestonice I, a large number of fired clay objects were concentrated in a small hut about 80 m away from the other dwellings, up the slope, with a kiln structure. A special, possibly sacred nature of behaviour was argued as being associated with this dwelling. Another firing structure was also found, this time among the dwellings (Vandiver *et al.* 1989, 1990).

The firing structures of Dolni Vestonice I were identified as kilns on the basis of the abundance of ceramics present; one (the up-slope hut) contained over 2,300 ceramic fragments (Vandiver *et al.* 1989). This kiln consisted of an oval depression measuring 130 cm by 40 cm and 40 cm deep. It was domed on the north-eastern and north-western sides (Figure 3.49). The second kiln was excavated in 1979, some 40 m west of the first one and measured about 1m in outside diameter and 60 cm deep (Figure 3.50). Nearly three quarters of its perimeter was a horseshoe-shaped loess wall, and figurine fragments were found inside the ash lens at the base (Vandiver *et al.* 1989). As this kiln had been recovered as a block, it was micro excavated by Vandiver *et al.* (1990). This allowed them, with the aid of comparative experimental work, to show that it must have been fired a few times and/or over a long period, as the structure itself was fired to 600°C to 700°C.

Figure 3.49: The domed kiln structure in the upslope dwelling at Dolni Vestonice I.
a) Plan of the upslope dwelling, b) Reconstruction of the dwelling, c) The domed kiln (a and c from Vandiver *et al.* 1990, 60; b from Soffer *et al.* 1993, 270)

a)

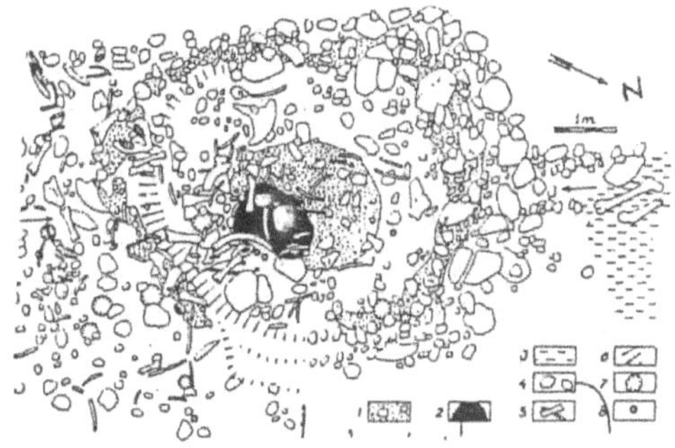

b)

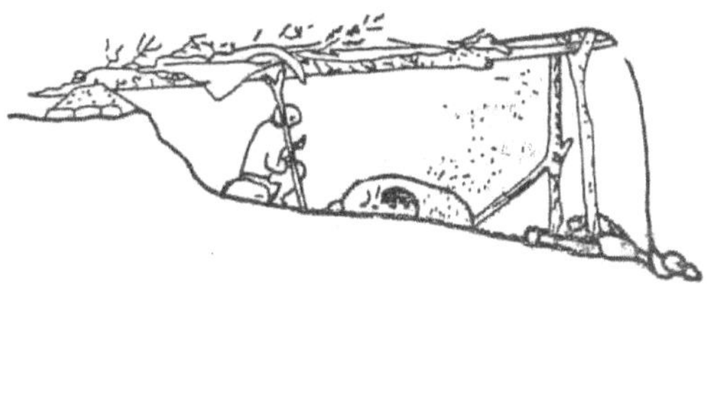

c)

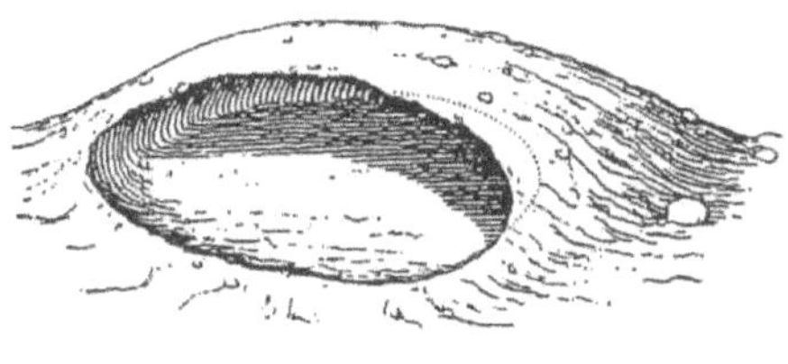

The two kilns at Dolni Vestonice I had a refractory wall with an opening on or towards one side that would allow some air in (Vandiver *et al.* 1990). Evidence of ash on the figurines indicates that they were probably set on the fuel, which then burnt to ash. The fact that many pieces were burnt in reduction to a dark grey-black colour is caused by this ash layer, which kept the oxygen out, creating a reduced atmosphere that produces dark ceramics.

In the 1957 excavations at Pavlov I (Soffer *et al.* 1997), a few hundred ceramic pieces were recovered with quite precise provenience data; all seem associated with one of the dwellings, also with a hearth. There is no special firing structure there. It was noticed that out of 448 pieces, 34 were burnt in oxidising and 385 in reducing conditions. Soffer and Vandiver (1997) think that an accidental burning of the dwelling itself could have produced the oxidised pieces, whilst the other reduced pieces could have been fired purposefully. The idea of an accidental burning of a dwelling is confirmed by Klima who noticed a lot of charcoal in the same dwelling, especially in small holes that must have been post-holes. This could have accidentally preserved some 'structural' ceramics among the fired clay fragment as Soffer (2000) commented in view of the special nature of some larger fragments.

Gonysevova (1999) did some experimental replication of the kilns from Dolni Vestonice I. She interpreted them as simple structures with a pit dug in the soil in which she placed the objects to be fired, then above them sticks of wood placed upright, with the whole thing covered by a dome of wetted loess, leaving some ventilation holes at its base (Figure 3.51). The structure used was very efficient for the firing of ceramics, reaching temperatures of about 1000 °C. The dome of loess had to be removed at the end to retrieve the objects, but was rebuilt with more wet loess for another firing.

Figure 3.50: Horseshoe-shaped kiln from Dolni Vestonice I. a) Plan of the excavation in 1978: 1 ash lens, 2 kiln banking, 3 dug out hearth. b) The kiln (a from Soffer et al. 1993, 271; b from Vandiver et al. 1990, 60).

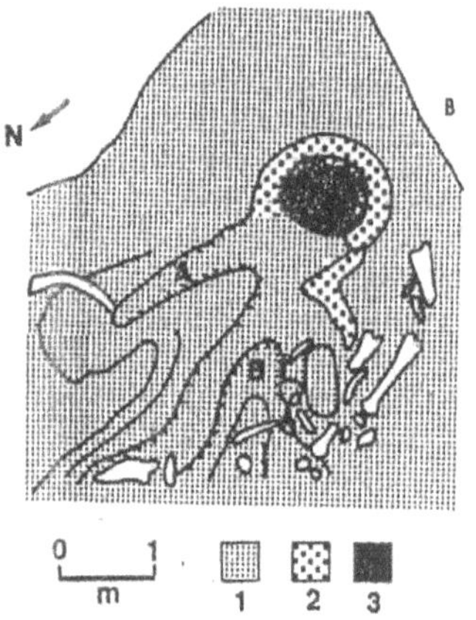

a)

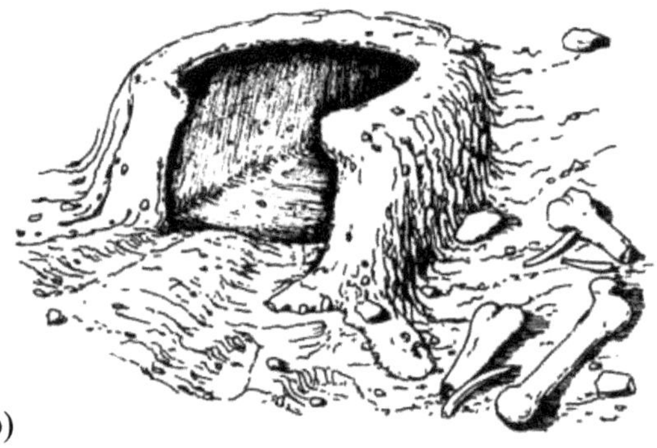

b)

Figure 3.51: Diagrammatical illustration of the experimental replica of a Dolni Vestonice kiln by Gonysevova and photograph (From Gonysevova 1999, p. 524-525).

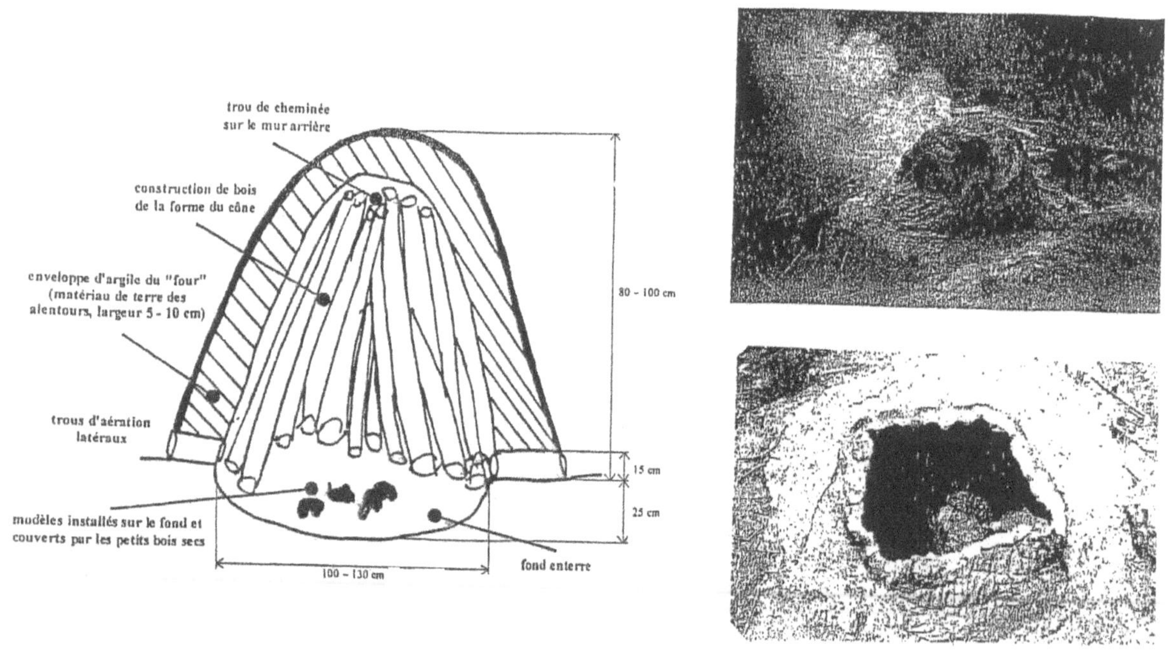

The very high firing temperature reached (1000 °C) is problematical as there is no evidence that any of the Pavlovian ceramics were ever fired at temperatures above 800°C. The factors responsible for reaching such a temperature are probably the large quantity of fuel accumulated in proportion to the small clay figurines, and the presence of two opposite ventilation holes at the base of the dome. These would ensure that oxygen feeds the firing enough to burn the wood intensely, as seen much later in updraft kilns (Hodge 1964). The fire would also be lit by one of the ventilation holes. It seems unlikely, in view of this difference in firing temperature, that the Pavlov and Dolni Vestonice figurines would have been fired in this manner.

Gonysevova (1999) based her experimental work on the functioning of the kilns as means of producing finished objects of fired clay, and on knowledge we have today about controlling airflows and fuel in order to fire ceramics. A more likely interpretation of the excavated remains of the firing structures from Dolni Vestonice I would consist in a pit firing, possibly banked with walls of loess on all or some of the sides, with an opening left at the top to allow air into the firing. This would work well for firing ceramic objects and would produce the range of temperature expected.

However, the idea that the ceramics from Dolni Vestonice and Pavlov were intended as finished ceramic objects has been seriously questioned by Soffer and Vandiver in view of the fragmentary nature of most of the objects and of the nature of the material used. The hypothesis that intentional thermal shocking of the objects in the fire, in a form of ritual use of pyrotechnology, was the pattern of their use has been developed instead (Soffer *et al.* 1993; Vandiver *et al.* 1997). It will be presented in Chapter 4.

Conclusion

In this chapter, we reviewed the factual data obtained from the study of the Pavlovian ceramics including the existing quantity of objects, the shape they take and the technology behind their making. The little known importance of the figurative inventory was emphasised and time was spent on describing the detail of the themes of representations found as many of these objects had not yet been published. This allowed underlining the dominance of the body extremities as an unusual theme of representation. As for the way the ceramic objects were made, it has been possible to reconstitute the associated technical scheme that shows a much standardised way of obtaining these figurines during all the corresponding occupation period.

The context of knowledge seen through the study of clay in this region is analysed below in Chapter 4.

It can also be added in this conclusion that recent research shows that the use of fired clay seen in the sites of Pavlov and Dolni Vestonice is in fact part of a regional tradition of making ceramic during the Pavlovian and Epigravettian. Archaeological evidence has now been found in ten sites outside the Pavlovské Hills (see table 3.1 for details). For each of these sites, the fired clay objects found are very fragmentary and in small quantity. They appear to be fragments of animal figurines or have a geometrical shape (in Kasov), the only whole figurine up to date being the so-called glutton from Predmosti (figure 3.52), discovered anciently but isolated. They have been described as very similar to the ones found in Pavlov and Dolni Vestonice (Gonysevova 1999).

Investigation on these new finds and their link with the ceramic from Pavlov and Dolni Vestonice is a new avenue of research opening up that will need to be looked at in detail in the future.

Figure 3.52: The ceramic wolverine figurine from Predmosti, Czech Republic. It measures about 5 cm long.

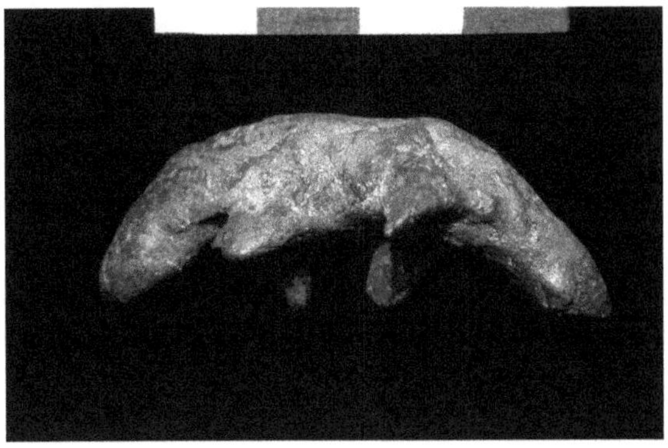

CHAPTER 4: THE CONTEXT OF KNOWLEDGE FOR THE WORKING OF CLAY IN THE UPPER PALAEOLITHIC OF EUROPE

By presenting and analysing the collected data about the use of clay in the Upper Palaeolithic of Europe in the previous chapters, it has been possible to underline the various technical processes of the working of clay. In this chapter, the elements of knowledge visible in this working are gathered and presented as a constellation of knowledge related to the use of clay. The elements of a constellation of knowledge interact together in the application of a technology (Sinclair 2000: 200; Dougherty and Keller 1982). Aspects of the social contexts considered can then be emphasised by looking at the elements possibly indicating a transmission of this knowledge, at formal or informal levels.

Four elements constitute the basis of any constellation of knowledge as they are recognised as being present in all technological processes; they are the raw material(s), the implements used to work the raw material(s), the technical processes involved and the final end-points. All these components are put in relation with each other, notably through monitoring criteria applied by the individual carrying out the action that take into account stylistic, aesthetic, procedural and functional considerations. These elements properly represent a constellation of knowledge as all the components depend upon the knowledge that an individual has acquired of them (Dougherty and Keller 1982; Sinclair 2000: 200). In the case of the present study, the constellation of knowledge approach is applied to the production of symbolic expression from clay material. In a technological study of a material, the emphasis is placed on the act of making and studying it; the fact that the resulting 'production' is art does not erase the importance of the technical process; quite the opposite; it is thought a further understanding of Palaeolithic symbolic expression can be gained from its technological study (White 1997; Fritz 1999).

Reconstructing the constellation of knowledge used for the working of clay should allow understanding the manufacture and use of this technology (Sinclair 2000: 200) although Sinclair also points out (2000: 201) that the specific elements considered in the constellation (materials, techniques, end-points and implements) are limited by the current state of our archaeological knowledge. They can be implied or hypothesised, however, through the reconstruction of the working processes, even if the archaeological record did not preserve all traces. The potential elements for the reconstruction of the constellation of knowledge of the working of clay are listed in Table 4.1.

Table 4.1: Potential elements for the reconstruction of constellations of knowledge of the working of clay in the Upper Palaeolithic

Material	Implements	Processes	End-points
Clay:	Osseous tools: Polisher, Spade,	*Modelage*	Figurative representations:
Soft	Spatula, Spear, Needle, Awl	Engraving	3-D : Animal
Partly dried		Finger-tracing	Human
Hardened	Stone tools: Blade, Flake, Point,	Shape Impressing	Other
Fired	Scraper, Burin, Denticulate,	Carving	2-D: Animal
Silt	Piercer, Knife, Grinding stone,	Drawing	Human
Mondmilch	Polisher	Firing	Other
Loess		Scraping	
	Wooden implements	Smoothing	Ornament
Water		Rubbing	
	textile	Polishing	Other:
	Skin (for storing prepared clay)	Mixing (with pigment)	Pigment bloc
		Piercing	Unfired clay
	Kiln, hearth	Softening (with water)	Containers
		Cutting	Structural (specific hearth
	The human body	Exploding	structure for cooking, boiling,
		Preparation of fuel	containing, floor levelling,
			daub...)

The different parts of the present chapter will follow each of the categories of the components of the constellation of knowledge: the raw material, the implements, the technical processes and the desired end-points obtained. The knowledge gained from the study of clay for each of these elements will be summarised and analysed. A fifth part will then concentrate on the aspect of social context visible through the study of the use of clay in the upper Palaeolithic.

1 Knowledge of the raw material

It was seen in Chapter 1 that the raw material considered in the present study was clay, and by extension, also other materials such as silt, mondmilch and loess which have similar properties to clay and have been used in a similar manner by Upper Palaeolithic people.

The various states the clay (and assimilated materials) can take can be in themselves considered separately as for each state different properties of the material can or cannot be used. These states are: soft, partly dried, hardened and fired clay. Related to these different states, water can be added in the list of raw materials involved in the working of clay as it is essential in controlling the plasticity of the clay.

That these 'variations' of the clay material were recognised and used implies specific knowledge of the raw material notably how to recognise and procure it for the purpose one had in mind; how to prepare it, if necessary, to make it usable; and how to store it to plan future uses. It will also be seen first that the exact variations in the nature of the raw materials do correspond to distinct regional and environmental contexts.

The different contexts

As was seen from the material presented in Chapters 2 and 3, the two regions considered for the technological study of clay in the Upper Palaeolithic of Europe present very different contexts chronologically and geographically and also in the very nature of the sites studied: in the Magdalenian of the French Pyrénées, they are all cave sites; in Central Europe, the Gravettian sites are open-air locations. This has implications on the exact nature of 'clay' used as in all cases the raw material appears to come from the site itself or its close vicinity.

The French Pyrénées: cave sites

In the French Pyrénées, all the sites in which clay uses have been found are cave sites. There, the raw materials used are real clay, silt and mondmilch, which are all found naturally in cave contexts. There is no evidence to date for any use of clay outside a cave site, but open-air sites are crucially lacking in this region in spite of the progresses in the regional archaeology. Cave and shelters do appear privileged habitation sites in the Magdalenian of the French Pyrénées (Clottes 1989: 38).

The uses of clay in the region are all found in deep parts of the caves, in environments secluded from everyday life and away from daylight and from the glacial climate outside. The habitation sites are frequently found at the mouth of the cave, but clay was mostly used away from them, where other forms of symbolic expression were also made.

There is some evidence for the raw material being transported within the caves (Larribau and Prudhomme 1984, 1989 ; Rivenq 1984) (see tables 2.38 and 2.49), but again this appears to be rather short distances: in Tuc d'Audoubert, a 20 metres distance separate the bison modelages and the hole thought to be left by the extraction of a clay plaque used as raw material in the nearby Chamber of the Heels (Bégouen and Breuil 1958); in Montespan, a large hole a few metres away in a corner of the chamber where the bear sculpture is found is thought to be the source from which the clay was extracted (Trombe and Dubuc 1947). It can also be said, therefore, that clay appears to be used in the vicinity of its location of procurement in the French Pyrénées.

The Gravettian in Central Europe: open-air sites

The major Moravian sites for the use of clay in Central Europe, Dolni Vestonice I and II and Pavlov I and II, are large open-air settlements. There the exact nature of the raw material used for the making of the fired clay has been firmly determined by the analysis of Vandiver *et al.* (1989, 1990) as being the loess soil on which the sites are situated. This loess is large grained with rock fragments, quartz and feldspars and a small proportion of clay in it (Gonysevova 1999; Soffer and Vandiver 1997).

Furthermore, there appears to be enough evidence in the material record from Dolni Vestonice I and at least two separated locations of Pavlov I to imply that this loess soil was obtained very locally, probably from the near vicinity of the dwellings. It appears that the loess soil from Pavlov I is less suitable for firing than the one in Dolni Vestonice I, situated only about 500 metres away as its content in calcium carbonate is higher. This resulted in a higher rate of unfired pieces (under the point of ceramic change at 500°C) found in Pavlov I than in Dolni Vestonice I (Soffer and Vandiver 1994: 164; 1997: 400).

This evidence would indicate that the raw material in Central Europe was found directly in the context of the living habitation site in the open. It is also useful to consider the environmental conditions at these sites: it was a cold climate with low precipitation ratio (Opravil 1994 : 177) in which the ground was possibly frozen at the time of the occupations, as several factors seem to indicate a winter occupation of the sites (Musil 1994 ; Opravil 1994). The implications for the procurement and the working of the material are looked at below.

Recognition and procurement of the raw material

The recognition and the procurement of the raw material imply some previous knowledge from the people involved. It is necessary for them to know where to find it and how to recognise it. This process involves learning to identify the properties sought after, and also learning to test these properties. The methods of procurement of the material must also be known: how to extract it, with what appropriate tools.

The knowledge required for recognising and procuring the material could be acquired in two ways: in a passive manner, simply by watching others in action, or through a purposeful form of teaching.

In the French Pyrénées

In the French Pyrénées, the raw material was found and used inside deep caves. This implies that the individuals involved in the process of recognising and procuring clay were away from the living areas in the majority of cases. Exceptions are found with the sites of Mas d'Azil and Enlène in the Ariège, which are habitation sites located deep inside caves (Clottes 1989).

A natural source of raw material was found when human beings ventured away from daylight, probably with the purpose of leaving their marks in the cave through the means of art. All the sites in which clay has been used in the French Pyrénées are sites where other techniques of art have also been used, mostly parietal art but also portable art as at Mas d'Azil (Péquart 1960), Enlène (Clottes and Bégouen 1984; Bégouen and Bégouen 1936) and Labouiche (Méroc 1959). It can reasonably be inferred that prospecting for clay sources was not a systematic occurrence, but that finding sources might have been an opportunistic event within the wider framework of Magdalenian symbolic expression.

The knowledge of what could be done with this material, however, must have been there to use when the opportunity arose. This implies being able to recognise the material. For this, testing was most likely carried out to decide if it was suitable. In caves, clay is found as a plastic material, whose degree of plasticity can be tested straight away by a simple touch of the finger, or by extracting a very small amount and shaping it. This might account for the various mentions of 'clay balls' found in the literature (for example Méroc 1959; Clottes *et al.* 1984; Trombe and Dubuc 1947), although there is no way to be sure of the age of most of these finds.

If testing took place, this implies that certain properties of the material were sought; this might be certain states of plasticity that the people involved must have had the knowledge to recognise as suitable for the making of *modelage* or for shaping on the spot by finger-tracing, engraving or impressing. It is also possible here to hypothesise that anybody living in a hunter-gatherer context, in close contact with nature, would have had some experience of plasticity through contact with mud. This is a material children are likely to have played with, and thus experienced. There might have been some rather widespread knowledge about plasticity among Upper Palaeolithic people which could have been transferred to the working of clay.

One can argue for an immediate use of the material found. This is most likely in caves in which the art appears to have been made during a single visit, and then abandoned and not visited again. This could be the case for the caves of Massat or Fontanet in the Ariège. In Fontanet the art on clay is found on floors which have been very well preserved and which also preserve numerous footprints resulting from the Palaeolithic intrusion in the deep part of the cave. A habitation site is found in the first part of the site, close to the original cave mouth. The deeper part of the cave becomes fairly dangerous with a steep well at one point. This did not stop Palaeolithic people who have marked the clay floors with a series of impressed signs around this well and beyond (Clottes *et al.* 1984). This was possibly a unique visit to this part of the cave. Similarly in Massat, the engravings, finger-tracings and the impressed signs on clay were all found in a deep part of the cave, accessible only through two very narrow successive crawling passages. There are very few traces left behind by the Palaeolithic visitors beside the art which was possibly made in a single visit (Gailli 2004). In these cases, it could be implied the source of clay was found during a chance exploration, the opportunity of using it was taken but the knowledge of the existence of this source did not result in further use of the material in this cave and was not transmitted.

In other cases, it is of course also possible that the source of clay was memorised for further use, or at least that the art on clay was done in a series of separate events, especially in sites where large quantities of clay were used and cave art in general was found. This can be seen at the sites of Niaux, Montespan and Bédeilhac for example.

In Niaux, the art on clay is found as engravings on clay floors, most of which are situated at the foot of the main panels of wall art in the Salon Noir. There is clearly a connection between the two art techniques, stylistically as well as by location. The style of the art of the Salon Noir is very uniform; for this reason it had been suggested that it was done in a short period of time by the same person(s) or by individuals belonging to the same group (Leroi-Gourhan 1965; Vialou 1986). Advances in AMS dating techniques have recently permitted the direct dating of the art on the walls of the Salon Noir, and this has seriously questioned the former implication as it appears that the images were painted over approximately a thousand-year period (Clottes *et al.* 1992; Clottes 1995). There is no way to date directly the art done on the clay floor; however, by its association with the parietal art and also by its quantity (at least 65 recorded separate graphic units), it seems reasonable to assume it was also possibly done over a long time, in a few separated events. This has implications for the transmission of the knowledge that art could also be done on the floors in this cave which would have been passed on during generations.

In Montespan, it has been noted that the Casteret-Godin Gallery where most of the art, engraved and sculpted on the limestone walls or made on clay, is found, has been subjected to large scale reshaping of the clay masses and clay banks available in the site (Rivenq 1984). This has been a laborious and time consuming process and it is likely it was done during successive visits to the site. The same thing can be said of part of the engravings found there: many actually combine engraving and low relief shaping, which also requires much work on a rock surface. This implies that knowledge of where the site was, how to access it and what could be done in it with the clay or rock surfaces must have been kept by the individuals making the art with some possible transmission if the time period involved were long. Unfortunately, dating engravings is as difficult as dating works on clay; little chronological information is available in the context of Montespan that would provide an idea of the time scale considered.

Bédeilhac is possibly the only site in which information about the procurement of raw material for making clay portable objects can be obtained in a Magdalenian context. This is because the objects have been found in large numbers very close to their raw material source in the Terminal chamber of the cave. It seems the clay floor of this part of the site naturally breaks into plaquette shapes when drying slightly (Sauvet 2006). It is likely this phenomenon was noticed and inspired individuals who had the previous knowledge of using stone plaquettes as a form of symbolic expression, as is found in other sites in the region (Bégouen and Clottes 1981b). Also in the site of Bédeilhac, engraved and fashioned sandstone plaquettes have been found in quantities together with the clay ones (Sauvet 2006). A point to note is that, if some clay plaquettes were naturally found as plaquettes, there are also traces of clay being extracted in the same part of the site probably to shape as a plaquette and use in a form of symbolic expression. The fact that some plaquettes show clear traces of being worked when the material was still soft (Mons 1974; Mons and Delporte 1973) also indicates that soft material was extracted from the floor and worked as well as some drier naturally formed plaquettes. The knowledge obtained from the observation of a natural phenomenon was used in an active behaviour for recognising and procuring the clay plaquettes.

There are few elements for dating exactly the plaquettes in order to establish the duration of their manufacture. However, data concerning the exact types of sandstone or stone used for the stone plaquettes in Bédeilhac indicates at least eight different types of raw material (Sauvet 2006), most of which were brought to the site from various external sources. This fact combined with the richness and variety of the portable and parietal art, is indicative of important ritual activity, and the large size of the cave could be made sense of by considering Bédeilhac as an important aggregation site (Conkey 1980). The current lack of publication of the material and excavations from Bédeilhac might be the reason why this site is not usually found mentioned as such in any relevant literature. However, it is likely the current research carried out about this site (Georges Sauvet, personal communication August 2004) will shed light on this topic in the short term.

The knowledge related to the recognition and procurement of the raw material and its passing-on needs to be considered in this particular context. The knowledge that needed to be transmitted in the French Pyrénées was mostly how to recognise the raw material as the procurement, which only took place when extracting it, could be done with the hands or with any tools without any difficulty thanks to the softness of the plastic clay. This transmission could take place in a passive informal manner by observing those using the clay. In the cave context of the Pyrénées this implies being present inside the cave where the cave art was made. Footprints found indicate that all age groups could be present deep in the caves, even young children (Clottes and Simmonet 1984; Bégouen and Clottes 1984b; Clottes 1989: 85) who could then have integrated the knowledge that clay was a suitable material for the making of art.

It is of course also possible that a more active teaching of this knowledge existed, but it certainly needed to be done inside the cave sites. This question refers back to the question of the exact nature of the rituals carried out in the cave, which form the context for cave art in general. This is one for which at present there is no obvious answer.

In central Europe

The sites of the Pavlovské Hills in Moravia are open-air sites, but they are made of built dwellings that structure space, creating private and enclosed areas surrounded by open and 'public' ones. The degree to which this distinction is visible from the archaeological record is of course a subjective matter as much depends on how easy the access to the enclosed areas was both physically and socially; this is difficult to assess from the remaining material traces (Oliva 2000b).

However, regarding the recognition and procurement of the material at the sites of Dolni Vestonice I and II and Pavlov I and II, some assumptions can be made.

First, the extraction of the loess soil used to make the ceramics could have been done inside a dwelling, and thus only in view of the individuals allowed in this dwelling. If one considers a normal habitation context, individuals from all ages from an extended family group could have been present and could have seen others extracting the loess soil by digging it out, probably with some appropriate tool such as a 'spade'. They would have learnt from witnessing these actions that the soil could be extracted for some further use and how to do it. This passive knowledge would have been available to any individual, from all ages and sexes.

If one hypothesises the existence of a habitation structure dedicated to a specific ritual as has been widely done for the upslope dwelling at Dolni Vestonice I (the 'Sorcerer's hut'; Valoch 1996), the implications do change. Inside this structure, a kiln dedicated to the firing (and possibly the exploding – see further in this chapter) of ceramic figurines has been excavated. As the structure seems spatially separated (it is 80 metres apart) from the group of dwellings in the upper part of Dolni Vestonice I, it has been interpreted as having had a possible 'special' purpose (Vandiver *et al.* 1990), a statement encouraged by the accumulation of broken fired clay animal and human figurines found there. It must be noted however that another kiln structure was found at Dolni Vestonice I, this time to the north-west (Oliva 2000b: 287). The knowledge of how to recognise and extract the material could then have been controlled and its passing-on restricted to the chosen persons by carrying out these actions away from everyone's eye, inside the structure and by purposeful teaching. There do not seem to be any pits excavated inside the structures around the two kilns (see plans in Chapter 3), however, from which the clay could have been procured except if one considered the kiln pit itself of course.

One can argue for the impracticality of having holes on the floor inside a dwelling but small pits around hearths do exist at the site (Soffer 1989: 725). They have been interpreted as possible boiling pits. Therefore this is not impossible. However, the quantities of raw material requested need to be taken into account: altogether over 2,000 fragments of fired clay were found in the upslope dwelling in Dolni Vestonice I alone, and over 3,000 in the north-west structure (Vandiver *et al.* 1989). Even if the fragments themselves are rather small, it seems large quantities of clay were fired over time, especially if we assume a reoccupation of the sites over a long period (Svoboda 1994) and if one considers that the remaining archaeological record is likely to be only a small portion of the original production. This would have produced large holes which are not suitable to have inside a dwelling, unless one argues that the material excavated to make the kiln pit structure was kept, prepared and used.

Also, it can be noted that the interpretation of the hearths of Dolni Vestonice I and their surroundings as dwellings is questioned by Oliva (2000) who points out that the reconstruction of the dwellings has been based on the concentration of artefacts around the hearths rather than on real physical structures. This is not, however, the most widely accepted view.

Secondly, the recognition and procurement of the raw material could have been done outside the enclosed dwelling structure, but still in its close vicinity as indicated by the variations in the nature of the loess across the Dolni Vestonice –Pavlov locale (Soffer and Vandiver 1994: 164; 1997: 400). Some pit structures are found around the site (Soffer 1989). Some of them could result from extracting raw material for ceramic making but any archaeological evidence of this is missing. This suggests that the activity of extracting the clay could have been seen by everyone and the knowledge of recognising and procuring the material could have been widespread.

But of course the fact that some steps of the technological process of the working of clay such as the recognition and procurement could have been carried out in full view of all at the site does not mean this was the case for the whole working process.

To the opposite of the situation in the French Pyrénées, the specific properties of the loess soil used in the working were not recognisable from the rough material extracted: the fact that the loess could become plastic if mixed with water and kneaded, and the fact that it would harden into ceramic (or explode) under the action of fire, are the elements of a complex knowledge that would have needed to be passed on.

The glacial environment at the time of the occupation of the Moravian sites (27,000 to 25,000 BP) is also a factor to take into account when considering the procurement of the raw material for the ceramics, as was noted by Soffer (2000: 68). She indicated that loess for ceramic production was likely far more easily accessible from spring to autumn when the ground was not frozen. However, there seems to be evidence for a winter occupation of the sites at Dolni Vestonice and Pavlov (Opravil 1994; Musil 1994). This has implications for the methods of extraction; suitably strong tools and possibly some way of softening the soil would have been needed. Another alternative might imply long term storage of the raw material with an extraction before the heart of winter sets in, or again an association of seasonality with the use made of clay.

Preparation of the raw material

The preparation of the raw material involves the knowledge of any special treatment needed to apply to the material in order to be able to use it as wanted.

In the French Pyrénées

In the French Pyrénées, any preparation involved does not seem to concern the nature of the raw material itself, with the possible exception of the making of portable objects.

For the parietal art in clay, the material is used as found. The only preparation involves selecting from the naturally available plasticity of clay. There is no work done on transforming the nature of the material. Work can be done, however, on preparing the chosen surface for the final location of the art. This can imply gathering the selected clay by transporting it. This clay can be used for preparing a smooth flat surface on which drawings will be made like in the Hunting Frieze in Montespan (Rivenq 1984); or it can be gathered to form the volume of *modelages*. The knowledge is concentrated in recognising the right plasticity, probably by testing it, of the available material.

For the portable art in clay, the material needs to be extracted, although in Bédeilhac there is evidence that the silt floor of the cave could naturally break in plaquette shapes when drying (Sauvet 2006). It is likely some kneading would have been needed when one wanted to shape the material for the other cases of portable objects found. In Mas d'Azil, it even seems that iron oxide pigment was added to the clay and mixed with it (Péquart 1960). These processes probably implied adjusting the water content of the material by adding water or by drying. However, no traces remain in the archaeological record and the portable objects are too few, with the exception of the Bédeilhac plaquettes, to be able to generalise.

In Central Europe

In the sites of Moravia, Soffer *et al.* (1993a) found evidence in the analysis of the density of the ceramic particle packing of an important difference with the loess: the loess is about twice as porous as the fired clay objects. This indicates that it was mixed with water and kneaded to achieve a suitable plasticity to be fashioned before firing. Vandiver *et al.* (1990) indicate that the kneaded and wetted loess would have remained a gritty and sandy paste, not very malleable and easily cracked. But Gonysevova mentions the result of replication done with loess soil from the brick yard site in Dolni Vestonice, in the vicinity of the site of Dolni Vestonice I; in two cases out of three the texture of the wetted loess was good, and in the third case the material would break easily but the plasticity was sufficient to allow fashioning figurines (Gonysevova 1999: 522).

Thus another knowledge included in the process of making fired clay figurines in Moravia was that the material needed to be prepared by mixing it with water and kneading it, a process which also implies the presence of water, and of containers suitable to carry it as part of the process of working of the clay.

A transformation of a material which is not originally plastic takes place with a total change in its nature through its working; this is a complex process that would point towards the recognition of a transmission of knowledge.

Storage of the material

If one argues for a purposeful preparation of the raw material, the question of storage of the prepared clay also needs to be addressed.

In the French Pyrénées

It can be argued that the cave sites of the French Pyrénées constituted natural stores of plastic clay, without any need for human involvement in the action of storing. The knowledge that it was available in a given environment was the only requirement.

In Central Europe

As noted above, a special preparation was needed to obtain the material for making figurines in the sites of the Pavlovské Hills which involved much effort. Over 10,000 fired clay fragments have been recovered from the excavations and it is likely much more material was originally worked which has not survived (if it was low fired for example). This implies that large quantities of material were used and therefore prepared. As the preparation of the material was only one step in a much longer complex process (see Chapter 3), it is possible the preparation could have taken place in advance. If so, some form of storage of the prepared material would have been needed.

The basic knowledge necessary for successfully storing the prepared material would be how to keep its water content stable; otherwise it would dry out and would need to be rewetted and re-prepared before use. For successful storage the prepared clay needs to be wrapped in an insulating material that will prevent water from evaporating, and that needs to be flexible to fit closely to the prepared mass; waterproof animal skins could be used. Plastic clay stored this way remains plastic indefinitely if kept in stable, preferably cool, conditions. In view of the cold environment of the Pavlovian period in Moravia, it could also be argued that prepared plastic clay could simply be frozen, which would keep the water content of the material stable. If storage by freezing was already applied to other resources, such as meat, the notion of applying it to other substances could have existed. However, up to date only one find of a permanent storage pit has been discovered at the sites of the Pavlovské Hills at Pavlov I (Svoboda 1994: 220). Such storage pits are frequent in the Palaeolithic record of Eastern Europe and are interpreted as being for the storage of meat in the ground through freezing (Soffer 1989). Soffer (1989: 725) indicates that in Dolni Vestonice II small pits situated about one metre away from the hearths have been interpreted as possible boiling pits; this could imply that meat storage took the form of drying or smoking which then would have needed boiling before consumption. There is therefore little evidence for storage through freezing in Dolni Vestonice and Pavlov.

If one assumes the storage of quantities of prepared clay, it also needs to be assumed that this storage would have been in a permanent location as plastic clay is a very heavy material.

Conclusion

In all the cases reviewed in this section, a striking point is the local origin of the raw material used as if the use of clay in the Upper Palaeolithic was closely associated to certain places and to the specific forms of symbolic expression taking place there. Since clay appears to be used principally for symbolic expression in this time period, this could be taken to indicate that symbolic expression was then closely connected to places, more than is usually visible from the study of other art techniques for the period, especially for portable art. This is possibly an important contribution Upper Palaeolithic clay studies can make towards understanding better the forms of symbolic expression associated with this period, and will be considered again later.

2 Knowledge of the implements used to work clay

All the possible implements known in the Upper Palaeolithic tool-kit that could have been used to work clay are listed in Table 4.1. It should be noted, however, that there is presently a lack of any tool specifically dedicated to the working of clay. This could be because of a general lack of awareness among archaeologists of the existence of clay as one of the materials used in the Upper Palaeolithic; this could have prevented recognising specific traces left by use-wear analysis for example (Cahen and Caspar 1984). But it is also true that clay is an extremely versatile material that can be worked by most existing tools and even without tools. One is here reminded of the words of Ingold (1990: 7): 'you do not necessarily have to use a tool to implement a technique'. In the case of clay, the human body itself can be the implement used. The possible implements used for the working of clay are considered below according to their suitability to its different states of texture: soft, partly dried, dried and fired clay. The knowledge of the suitability of certain implements for working the various states of the raw material is also a knowledge for which a certain degree of transmission can be implied, especially in specific cases such as for the making of the figurines in Central Europe.

Implements for working soft clay

Soft clay can be worked simply with hands or fingers thanks to its property of plasticity. This makes it a material that can be used in all circumstances, even when one has no tool available. This might have been an important factor in its use inside the caves of the French Pyrénées. One draw back of using hands and fingers is that they do not allow the creation of very fine details, for example the details of the faces in the Moravian ceramic figurines (although there is also the possibility of using fingernails).

The plasticity of the soft clay also implies it can be worked with most forms implement, from wooden tools to osseous material or stone tools. For example flint burins normally used to engrave the rock surfaces in caves have also been used to engrave clay surfaces, as can be recognised from the shape of the marks left on the clay. In the case of small *modelages* or for the Moravian figurines, it can be noted that it is difficult to obtain small details if the material is very soft as it deforms easily.

The fact that prints of textile have been recognised on fired clay fragments in the Moravian sites (Adovasio *et al.* 1996, 1997; Soffer *et al.* 2000) might also imply that textile was one of the implements involved in the working of the material, possibly as part of the storage process for example, as a bag to contain the prepared material. Animal skin can also be inferred as an implement for similar reasons.

In the French Pyrénées, two cases of possible implements used for clay working and then left stuck in the soft wall surface have been found. One was found in the cave of Montespan, in the Casteret-Godin Gallery; there, three natural fissures have been filled with clay in different locations in the gallery and then decorated with a line of impressed cupules made with fingertips. Next to the third fissure (MO12 in the catalogue) from the entrance of the gallery an implement described as a large bone spatula was left stuck inside the clay. It has been assumed it was used to smooth the artificially filled clay surfaces (Trombe and Dubuc 1947: 96). In the cave of Fontanet, a reindeer antler was found stuck in the clay wall by the sign F11, composed of 11 parallel vertical incisions. It could have been used in some way of working the surface although it has also been interpreted as possibly resulting from some sort of ritual, as a kind of *ex-voto* (Clottes 1989: 87; Clottes *et al.* 1984: 434).

Implements for working partly dried clay

Partly dried clay is still possible to work with the hands, depending on the degree of dryness, but more strength would need to be applied, possibly with a less controlled result, or the use of a harder part of the hand such as the finger nails.

Any wooden implement would be suitable to work partly dried clay, as well as any osseous or stone tools. These tools would, however, need to have fine working edges as they were mostly used to make small details or thin line drawings.

For Pavlov I and Dolni Vestonice I, the surfaces of the object mostly seem very carefully shaped but there is very little evidence of their being worked with a tool. Small implements however are recognisable as having been used for marking details on faces and bodies such as microliths which were made in quantity at the Moravian sites (Klima 1997c). These were used for example to incise the details on the Black Venus of Dolni Vestonice I (see figure 3.47 in Chapter 3): three short parallel lines marking the middle of its face are recognisable as having been made by a microdenticulate tool (Vandiver *et al.* 1990: 45; Soffer and Vandiver 1994: 168, 1997: 388).

Implements for working dried clay

Fully dried clay is a fairly hard material that can be the equivalent of soft stone. Wooden implements are not suitable. Hard tools are needed to work it such as osseous material and flint tools, the same ones that can be used to work semi dried clay. It is often difficult to determine from the archaeological record visible if details were made on partly dried or on fully dried material as the traces left on the objects do not differ much. Also, in the Moravian sites, the surface of the objects was altered by the firing process and also by the treatment of the figurines with a conservation agent (polyvinyl acetate) after their recovery (Martina Gonysevova, personal communication May 2004).

Implements for making fired clay

The necessary implement for firing clay is the hearth or the kiln in which the clay will be fired.

Data from excavations at Dolni Vestonice I and experimental replica carried out by Gonysevova (1999) have produced information about the two kiln structures used in Dolni Vestonice I. They were built by digging out a pit, 30 to 40 cm deep that was then covered, at least partially, by a wall of clay inclosing both the fuel and the objects to fire. Gonysevova' s experimental replication (1999) was based on the kiln structure found in the second habitation structure of Dolni Vestonice I, also called the upslope dwelling or the 'Sorcerer's hut' for the possible special association to ritual that has been given to it (Valoch 1996). She interpreted traces of grooves and holes in the original fired clay bank as air vents used to regulate the quantity of air allowed inside the kiln and thus, to an extent, the temperature of the firing (1999: 523). Clay figurines were placed at the bottom of the kiln with the fuel piled on top. In Gonysevova' s experiment, wood was piled up 80 to 100 cm above the objects and the whole thing was covered with a wall of soft clay, except for a air vent at the top in addition of the ones placed at its base. This structure was very successful in firing the figurines, producing temperatures between 700°C at the bottom of the pit to 1,100°C at the top, and firing during five and a half hours, slightly less when the kiln was reused for a second firing

This experiment raises a few issues, however, with regard to the possibility of firing such a structure inside a dwelling as has been the usual interpretation for the data from the excavation. Gonysevova mentions that flames up to 1.5 metre high were rising from her kiln at one time with temperatures as high as 1,500°C at the top of the flames (1999: 525). Clearly this would pose problems inside a low roofed dwelling such as the usual reconstitutions of the upslope hut normally assumed (Figure 4.1). The temperature in the close surroundings of the kiln would also rise greatly and it can be seen from its plan that the distance between the edges if the kiln and the walls of the dwelling is only between 1 to 2 metres. This would also make it difficult for someone to remain in the close vicinity of the kiln during the firing process as the air temperature would rise dangerously. So we can suggest that either smaller quantities of fuel were used in order to produce a smaller and more controlled fire, or the structure, interpreted as a dwelling, did not actually have a roof. This questioning of the interpretation of the habitation structures of Dolni Vestonice I is found in Oliva (2000) who points out that beside artefacts concentrations, actual structuring elements are rather missing.

Figure 4.1: on top, drawing of the reconstituted upslope dwelling from Dolni Vestonice I (from Soffer *et al.* 1993: 270); lower, plan of the excavated area (from Vandiver *et al.* 1990: 60)

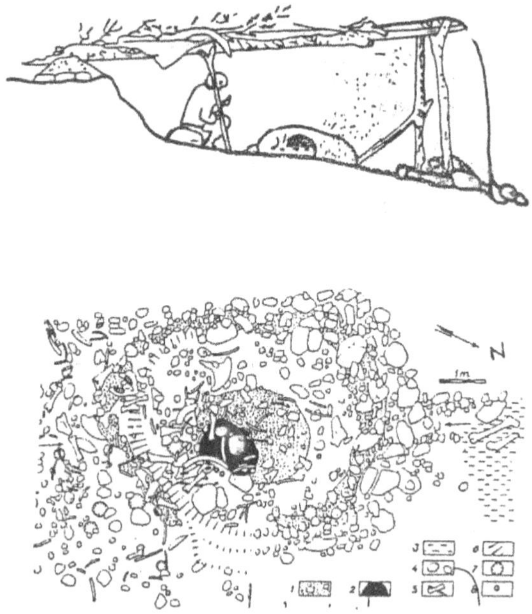

The construction of firing structures implies a full understanding of the whole process of firing clay. One can talk here of specialist knowledge of a complex technology. With regard to the kiln, its successful working would have required the person(s) making and using it to have skill acquired by practice as knowledge from a simple observation of others in action would not have been sufficient. But it is true that in the sites of Dolni Vestonice II, Pavlov I and Pavlov II, all the ceramics seem to have been fired in simple hearths as no kilns structures are found. Some of the hearths, however, appear to have produced higher temperature fires for a longer duration than is visible in other hearths at the same sites, possibly indicating a difference in their use (Vandiver *et al.* 1990; Soffer and Vandiver 1994, 1997).

It is also true that the two kiln structures were found in Dolni Vestonice I only. There might be some temporal differences with the site of Pavlov I through which the knowledge of making ceramic figurines was still passed on but with variations. Similarities and differences in the styles of the figurines themselves also give some indications on this question (see further in this chapter).

3 The processes of the working of clay in the Upper Palaeolithic: Knowledge and know-how

In the context of *chaîne opératoire* studies, it has been widely recognised that the processes of the technical action involve two types of knowledge in order to obtain a successful result: knowledge, or *connaissance*, and know-how, or *savoir-faire* (Pélégrin 1991; Ingold 1990; Schlanger 1994, 2004; Karlin and Julien 1994).

Knowledge, or *connaissance*, can be defined as regrouping the concepts of ideal forms and raw materials and the gesture associated to the action (Pélégrin 1991). It is a discursive knowledge, the 'recipe' for making things. It is 'explicit rather than tacit, objective rather than subjective, context-independent rather than context-dependent, discursive rather than practical, 'knowledge that' rather than 'knowledge how'' (Ingold 1990: 8). It is a conceptual, abstract knowledge (Schlanger 1994: 148). Knowledge can be acquired by simple observation of others in action (Pélégrin 1991), or transmitted by teaching, even in contexts outside those of its practical application as it is encoded in words or artificial symbols (Ingold 1990: 8)

Know-how or *savoir-faire* is a physical knowledge, a skill that is at once a form of knowledge and a form of practice; it is tacit and subjective (Ingold 1990: 8), practical and procedural (Schlanger 1994: 148). It implies the capacity to apply abstract concepts and to assess the result of the technical operations (Pélégrin 1991). Know-how involves dexterity and judgement; it could be seen as 'an overall capacity to succeed in an activity through personal engagement and interpretation' (Schlanger 1994: 148)

Know-how is strongly connected to the practical experience of the individual and is not really transmissible as it exists and gets acquired essentially through practicing, through observation and imitation (Pélégrin 1991; Ingold 1990: 8). The acquisition of know-how implies the comprehension by the individual of the technical fact itself, that is to say a consciousness of the parameters involved and the technical relations in term of cause and consequence. This is badly suited to systemisation or a transmission (Pélégrin 1991). However, Schlanger (1994: 148) points out that know-how is also what allows one to transfer and generalise knowledge, implying that if know-how itself is not transmissible, it is necessary to have know-how in order to transmit knowledge to others. One should keep in mind that know-how and knowledge are very much interconnected although different by definition.

Assessing the levels of knowledge and know-how visible in the working of clay in the Upper Palaeolithic of Europe will be carried first through establishing to what extent the techniques employed are actually adapted to the properties of the raw material and use them. In a second part, the norms and variations visible in a *chaîne opératoire* analysis of the techniques will be used to evaluate the degree of recognisable transmission of knowledge in the two cultural contexts studied in this research, and thus possibly the extent to which the working of clay was related to knowledge or to know-how. For this purpose a comparison with the working of osseous materials will also be drawn in each of the studied region, in order to replace the working of clay in its wider technological context within each culture. Schlanger (1994) reminds us that the *chaîne opératoire* is made of both variable and fixed strategic events that need to adapt constantly to a raw material that is never standard in shape or composition (Pélégrin 1990). The implications of both these 'variable' and 'fixed' events will be looked at with regard to the type of knowledge involved.

The adaptation of techniques to the properties of the material: what knowledge is visible

The degree to which the properties of clay were made use of in the technical processes is important as indicating to what extent the material was understood, and therefore its properties were known. As Sinclair reminds us, 'the knowledge of the techniques that may be employed is appropriate to knowledge of the raw material's properties' (Sinclair 2000: 200). Recognising this knowledge is the first step towards considering its transmission.

The processes of the working of clay are found in Table 4.1 relating to the elements of the constellation of knowledge surrounding it. They can be listed as follow: kneading, softening (with water), *modelage,* engraving, finger-tracing, shape impressing, carving, drawing, mixing (with pigment), waiting, scraping, smoothing, rubbing, polishing, piercing, cutting, firing, exploding and preparing fuel.

The properties of clay as a raw material, both unique to it and shared with other materials are summarised in Table 4.2. From this, one can divide up the processes according to the main property of the material used. For example, the property of plasticity is central to the processes of kneading, softening, modelage, finger-tracing, impressing, drawing and mixing. The property of hardening by drying implies a process of waiting as a first step before being able to engrave, carve, scrape, rub or polish. The property of hardness, caused by firing, is central to the process of firing. The capacity to explode in fire was used in the way figurines were fired.

Some of the properties that clay acquires through changing nature artificially during the manufacture process can make it akin to a soft or a hard stone and it can then be worked according to similar processes that these materials. Similarly clay has variable colouring properties according to its iron oxide content: this has allowed clay to be used in painting, like other pigments.

That the properties of the clay material are especially suitable to some of the techniques used or are properly used by some of these techniques is certain in the Upper Palaeolithic contexts considered. However, it has also been seen that some of the techniques applied to clay do not make specific use of the typical properties of the raw material. For example, engraving, done on plastic soft surfaces, has been found in the caves of the French Pyrénées; there is no technical reason for engraving on a plastic surface: it could be simply marked with a finger. However, a choice was made that resulted in engraving with a hard tool on a soft clay surface; it can be implied that the reasons for this choice were not resulting from material constraints: other motivations must be sought, possibly social in nature.

The extent of the use of the properties of clay will be looked at separately for the French Pyrénées and Central Europe.

Table 4.2: The properties of clay, shared and unique.

Properties of clay	Unique to the material	**Hardening by drying (reversible)** **Hardening by firing (non-reversible)**
		Capacity to burst in fire
		Plasticity (capacity to retain a shape, possibility to change any shape; reversible)
		Reversibility
	Shared with other materials	**Soft stone/osseous material hardness**
		Colouring properties
		Hard stone hardness

In the French Pyrénées

It has been seen in Chapter 2 that in the French Pyrénées the following technical processes were commonly found in relation with the working of clay: *modelage*, engraving, finger-tracing, shape impressing, carving and drawing. To this can possibly be added kneading, mixing (with pigment), waiting, scraping, smoothing, rubbing, polishing, piercing, cutting, firing although with less evidence from the record and as minor parts of the processes that allow combining techniques for a final result.

The technique of *modelage* has produced the most impressive of the works made in the region by using the property of plasticity of the material. Plasticity has made it possible to produce large and elaborately fashioned sculptures such as the two Tuc d'Audoubert bison and the Montespan bear statue (figure 4.2). It transforms the volumes of the surfaces found in the caves, shaping them into three-dimensional animal representations, with much less effort than would have been required for obtaining three-dimensional volumes in stone for example. Modelage is found in seven out of the fourteen recorded sites with clay uses, in the sites of Tuc d'Audoubert, Labouiche, Fontanet, Bédeilhac, Montespan, Labastide and Erbérua with a total of 35 graphic units representing 12% of the total of the inventoried graphic units on clay in the French Pyrénées (Table 2.2 in Chapter 2).

The plasticity of clay was also fully used in the making of finger-tracing, whether as recognisable graphic units or as 'macaronis'. They are found in four sites (Massat, Montespan, Erbérua, Oxocelhaya-Hariztoya) and represent 7.5% of the total of the inventoried graphic units on clay in the region with 22 graphic units. In all these sites 'macaronis' tracings are also found, as is also the case in the cave of Tuc d'Audoubert.

Figure 4.2: The headless bear statue (MO13) in clay from Montespan: a) photograph of profile (Trombe and Dubuc 1947, figure 61); b) explicative drawing (Bégouen and Clottes 1988: figure 3).

a)

b)

The plasticity of the clay is the property that permits its working by impressing: it keeps the shape of any implements - tool or finger- pressed into the material and thus allows forming patterns. 17 printed graphic units are made by impressing in the French Pyrénées in the sites of Tuc d'Audoubert, Fontanet, Massat, Montespan and Bois du Cantet, representing 5.8 % of the total of the inventoried graphic units on clay.

The process of engraving on clay created a total of 88 graphic units found in seven sites (Tuc d'Audoubert, Niaux, Fontanet, Bédeilhac, Massat, Montespan and Etcheberri) out of the fourteen sites of the French Pyrénées in which clay was used. They represent 30.3% of all the graphic units on clay. This makes a large part of the art on clay in the region. However, it must be pointed out that the technique of engraving is not adapted to the working of clay. Engraving incises a surface with a suitable hard tool, such as a burin, in order to mark it. It is adapted to hard surfaces. On soft clay surfaces, it works exactly as it would on any other surfaces and does not use the plasticity of the soft material. It can be argued that some clay-covered surfaces may have hardened naturally and become unsuitable to work simply with fingers. However, even on the inventoried clay floors used for engraving such as in Niaux or Bédeilhac, it is clear the surfaces remain soft even today as has unfortunately been noticed by the recorded cases of accidental destruction of the art (Clottes 1995; René Gailli, personal communication July 2002). It is likely reasons other than technical constraints motivated this choice of technique.

In the making of portable clay objects in the French Pyrénées, the combination of a number of processes was used in which the various properties of clay were applied successively as the material underwent successive changes in its nature. This is the most complex clay technology found in this region. The sites where this type of object is found, however, are few: Bédeilhac, Enlène, Mas d'Azil. The number of objects is also very small: three in Enlène, two sculptures in Mas d'Azil, plus seven possible ornaments and fashioned pigment lumps. But in Bédeilhac, there is a series of at least 122 clay objects that have been recorded in this study, and certainly more that have not been published or studied yet (Marie-Sylvie Larguèze, personal communication March 2007). This single site make the portable clay object the most numerous category of the total inventory with 43.9 % of the recorded graphic units.

The making of unfired portable objects can use the properties of plasticity and the capacity to harden through drying to various levels of hardness. Among the processes used in the making of portable objects, kneading and mixing (with pigment) use the property of plasticity. The processes of scraping, smoothing, rubbing, polishing, piercing, cutting are all techniques that can be used in order to add details on a portable sculpture for example. They are mostly adapted to a working on a hard/non-plastic surface such as one obtained by the drying of the clay. However they have also sometimes been found on plastic clay like in Montespan where the soft wall surfaces was sometimes smoothed and also prepared by scraping in places (Trombe and Dubuc 1947).

Waiting is a process only visible if intentional changes of nature of the clay are carried out; for example, in an intentional firing, air-drying is a step of the making that will ensure the objects will survive the firing. Waiting for the objects to dry naturally then becomes an integral part of the process.

Making fired portable objects uses the unique property of clay of being able to harden permanently though firing, normally after having used the plasticity of the material in the fashioning. Only three fired clay objects are recorded in the French Pyrénées among the graphic units studied: the headless body of an animal (E1) and a possible animal head (E2) in Enlène; an animal head interpreted as a feline from Mas d'Azil (MA1). Two more objects from Mas d'Azil also appear fired: they are two beads.

Figure 4.3: Traces of scraping done after the firing around the tail of the animal figurine E1 from Enlène (Private Collection Begouen).

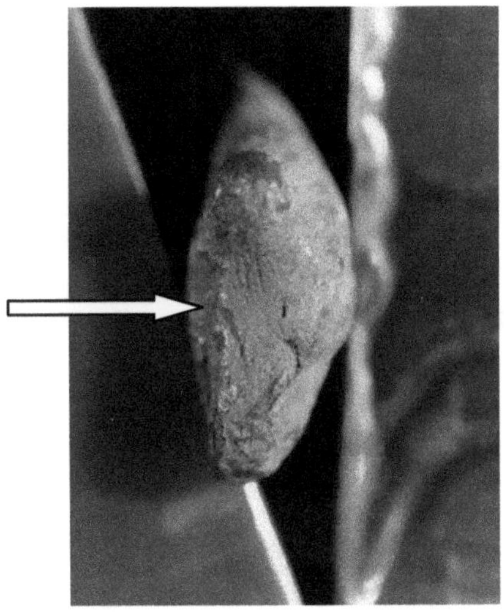

It is true the question of the intentionality of the firing of these pieces can be posed in the French Pyrénées: the few examples recorded show little uniformity, little that could indicate an intentional firing as this could also have happened accidentally in an environment in which open hearths were constant features. Only the animal figurine E1 from Enlène really shows all characteristic of a well-fired well-fashioned fired clay object which has been worked by first using the plasticity of the clay, then its property to harden by air-drying and finally its property of hardening through firing. Only one element could bring some doubt on the intentionality of the firing process: it seems that some details around the tail of the animal (figure 4.3) were added after the firing was completed by scraping, thus using a stone-working technique to work the fired clay. The two other objects are made of fired clay but their surfaces are much worn and it is difficult to get much information from examining it: it is entirely possible they were fully carved out of an accidentally fired lump of clay used as a hard stone. However, in the case of the possible animal head E2, one small untouched concave area on top of the head (figure 4.4) seems intact and resembles well fired ceramic.

Figure 4.4: Intact area on top of the possible animal head E2 in fired clay from Enlène, indicated by the arrow (Private collection Begouen).

Similarly, it is hard to know if indeed the two beads found in Mas d'Azil were fashioned from a wet lump of clay, using the quality of plasticity of the material, or if they were fully carved with a tool from a dry lump of clay, or again if it combined both, fashioning a clay ball on wet material, and piercing it when dry, as would be the most practical approach. Detail is especially difficult to see on the objects because a conservation agent strongly affects the appearance of the surfaces.

The making of fired clay objects would be the most complex technology visible in the French Pyrénées with a true application and understanding of all the properties of the clay material if it could be proven to be intentional. The rare examples encountered, however, are not presently enough evidence to confirm it was the case.

The process of drawing with clay, which was only used in three certain occasions in the French Pyrénées in the sites of Bédeilhac, Oxocelhaya-Hariztoya and Etcheberri, also made use of the capacity of the material for plasticity.

Overall in the context of Magdalenian art in the French Pyrénées the property of clay that was dominantly used in the working of the material is its plasticity; this is especially visible in the *modelage* that form an impressive and unique part in the art of this region. The capacity of clay to harden by drying and firing are also seen to be used occasionally, but in a less systematic manner across the region. This seems more a site-specific occurrence, in the case of the clay plaquettes from Bédeilhac for example. Also, it can be said that the techniques of the working of clay are not usually found in isolation: often combinations of techniques are seen for the working of the material itself, some of which use the properties of clay, some of which do not. The uses of clay are found mostly in the vicinity of other art techniques such as engravings and paintings. Much variation is clearly visible in the way clay was worked in this region. It can be said than in the French Pyrénées the quality of plasticity is the material property on which the regional phenomenon of use of clay is based as the large majority of the data shows clay was found in plastic form inside the caves, used as a plastic material, and abandoned still a plastic material in the locations where the working took place. Little change in the nature of the material by the application of technological processes is seen in the French Pyrénées.

In Central Europe

In Central Europe, the visible use made of clay involves fired figurines or objects: this implies the properties of clay related to the action of fire are the ones that were dominantly used in this context. There are two properties of clay that relate to fire: the capacity to harden permanently under the action of fire and the capacity to burst in the firing process. This last property of clay is not one commonly known as there is no practical use for it; on the opposite, it is a property that is normally seen as a potential problem in the working of clay and that requires controlling and avoiding, as is done in all the forms of later pottery-making. The tradition of adding a temper (a mineral or organic element) to the clay during the phase of preparation is aimed primarily at avoiding explosion during firing (Gibson and Woods 1990).

The fact that the Czech ceramics have hardened permanently under the action of fire is something clearly visible simply by considering the fired clay record and its importance (over 10,000 pieces) (Soffer *et al.* 1993). The possibility that these same ceramics would have been intentionally made to explode in fire by using this unique property of clay is less obvious but has been put forward by Soffer and Vandiver' s research after they noticed some patterns that questioned previous interpretations. They proposed the hypothesis that the firing of the ceramics of Dolni Vestonice and Pavlov was in fact a 'pyrotechnology' in which the element of intentional thermal shocking of figurines was central (Vandiver *et al.* 1990; Soffer *et al.* 1993a; Soffer and Vandiver 1994, 1997).

They noticed that the clay material obtained from the loess of the sites of Dolni Vestonice and Pavlov is naturally resistant to thermal changes, as it has a very low shrinkage ratio, both during drying and firing, and would normally stand a firing process well, without breaking, even with no prior episode of drying for pieces under 1 cm thick. With this in mind, they considered the huge number of fragments found on the sites - out of 10,000 pieces they counted only 77 nearly whole objects – and started wondering if this fragmentary production was intentional (Vandiver *et al.* 1990; Soffer *et al.* 1993).

They tested the effects of percussion and pressure flaking, of weathering and trampling on the ceramics by experimental replication, but concluded this did not produce the most common type of fracture found (Vandiver *et al.* 1990). They examined the ceramic assemblage and noticed that over half of the fragments showed traces of thermal shock: this consists of a rough branching structure of the fractured surfaces. This means that two fragments of the same objects could not be fitted back together, due to the nature of the broken surface. This shows that these objects exploded during the firing. This phenomenon was noticeable both in Dolni Vestonice I and Pavlov I (Soffer and Vandiver 1994).

As for the fragments not showing traces of thermal shock, the majority was broken, and traces of mechanical breakage are also visible on many (Vandiver *et al.* 1990).

Because the loess is naturally resistant to thermal changes, an intentional effort was needed to obtain an explosion, as it would normally withstand a firing in most conditions. A manner of making the loess figurines explode with certainty was replicated: they need to be left to dry after they have been modelled, and then to be wetted before throwing them into an open fire. The figurines would then explode loudly, sending bits and pieces flying in all directions. Soffer *et al.* (1993) point out that the walls found around the pit of the firing structure from Dolni Vestonice I could then have been built to protect people from the flying pieces of ceramic.

If one accepts this interpretation, one seems to have a situation in which the process of breaking and exploding ceramic objects in fire was a common pattern in their use, possibly in some sort of ritual, the resulting terracotta figurines, broken or not, not being intended as finished objects. However, it is true that not all the objects show the torn broken surfaces typical of thermal shock. Traces of mechanical breakage could also indicate a possible destruction of the objects by other means than exploding (Soffer *et al.* 1993a).

This form of pyrotechnology implies the use of the unique property of clay to be able to explode under the action of fire. The use of this property can be associated with some form of ritual involving an intentional destruction of figurines as there is no practical application of this property.

In the sites of Moravia, there is no doubt that, before the firing process took place, elaborate figurines were made with processes that involved using the other properties of the clay material. These were fashioned with techniques such as kneading, softening (with water), modelage, engraving, impressing (of details), carving, scraping, smoothing, rubbing, polishing, piercing or cutting. Waiting would also have been a necessary process to apply at various times to change the nature of the clay surfaces and to allow the application of some of the techniques. For example, the very fine details visible on some of the feline heads could not have been made on a soft material as it would have deformed; at least partial drying was necessary. These processes all draw on the qualities of the clay of plasticity and of hardening to different stages of hardness through air-drying.

The whole process of ceramic-making or ceramic-exploding in the sites of Moravia shows a successive mastery and use of all the properties unique to clay. The material was fully understood and thus what was made of it was something unique in relation to the rest of the material culture found at the sites.

Some rare elements also indicate a possible further integration in the wider technological context: the possible re-working of surfaces of fired clay objects after the firing. This is visible for example in the body extremity PI 35 from Pavlov (see figure 3.48 in Chapter 3). There, fired clay has been treated just like stone. The clay has changed its nature through firing, and even if the unique properties of clay were exploited in the making before the firing, it appears that after firing the material is similar to some sort of stone, and is, therefore, susceptible to be treated in the same manner, even if this is not done systematically.

This is what can be made of the visible part of the record for Dolni Vestonice and Pavlov. However, it must not be forgotten that enough 'hints' exist to indicate further uses of clay material in Dolni Vestonice I and Pavlov I that have left very few traces in the archaeological record but that would have drawn mostly on the property of plasticity of the material: 'structural ceramics' such as the use as daub over the dwelling structures or some fragments that appear coil or strip-built (Soffer and Vandiver 1994, 1997).

Norm and variation in the working of clay: transmission of knowledge and levels of skills

Through *Chaîne opératoire* studies, it has been possible to recognise various levels of norm or standardisation in the manner clay was worked in the Upper Palaeolithic. In this process, the individuals carrying out the action, and to a certain extent their relationships, come into light and with them the possibility of seeing hints of how knowledge and know-how related to the techniques was possibly transmitted (Dobres 2000).

Recognising the knowledge related to the technique of clay working that were visibly passed on, through a certain norm in the resulting processes or final artefacts, will allow drawing conclusion notably on the social context for the making of clay things. Recognising the variation visible from a certain norm, possibly resulting from the application of personal know-how, will allow drawing further conclusions about the individuals working the clay in the Upper Palaeolithic.

The regional contexts will be looked at separately and for each, a comparison will be drawn with the working of osseous material in order to determine if the patterns of the working of clay associated with symbolic expression fit into a wider technological context.

In the French Pyrénées

- **Transmission of knowledge and know-how**

The technical schemes drawn for the working of clay in the French Pyrénées have been presented in 'flow charts' in Chapter 2 for each of the recognized techniques in the region: *modelage*, engraving, finger-tracing, impressing and the making of portable objects.

These flow charts, by summarising the linearity of the use of a material across sites (Dobres 2000), allow seeing if patterns are visible, and from these patterns inferring the knowledge and know-how necessary for a successful process. From this base, one can elaborate on the degree of transmission of knowledge visible or hypothesised

It has been seen above that the property of plasticity of clay was the common element used in the region. But from looking at the various technical 'flow charts', a striking point was that it has not been possible to draw a unique technical scheme for the working of clay in the French Pyrénées as the techniques applied are too varied.

The techniques of engraving, finger-tracing and impressing on clay (respectively Tables 2.40, 2.42 and 2.45 in Chapter 2) as they have been applied in the various sites are the exception as they could easily be combined in a single flow chart. They show the same phases and sequences up to the point of marking the surface, which is where they differ. They all consist in marking a clay surface, artificially prepared or naturally existing, with different implements: hard pointed tool, fingers or other tools. The knowledge recognisable here is that these implements have the capacity to mark the clay surface. But there is a difference in the case of engraving: the knowledge involved is one related to the wider cave art context: knowing that a stone tool can leave marks on a rock surface. It can be used on clay without one knowing anything about the properties of the material. Using finger-tracing or impressing, on the other hand, imply the knowledge that clay is plastic and can be worked with these implements.

In the case of modelage or portable objects, the whole process is aimed at making a different surface altogether, one that will shape a representation by using three-dimensional relief. It is a much more complex process, as is visible from its technical scheme which includes 19 possible sequences in the working for modelage (table 2.40) and 12 for portable objects (table 2.45) (as opposed to a number of between 3 to 8 sequences for the techniques mentioned above). However, very different results are visible from case to cases that imply varying understandings of the properties of the material, beyond the recognized plasticity.

In many cases in the French Pyrénées, it is possible to wonder if the Magdalenian people working the clay did not simply improvised and innovated by an opportunistic use of a material found by chance when being present in the caves for reasons connected to the rituals that resulted in Palaeolithic cave art. There is possibly some of that around. However, the repetition of patterns that becomes visible across sites indicates the sharing of a wider base of knowledge behind the use of clay. This is especially seen when the surface used has been previously prepared, since the intentionality of using clay is then underlined: the cases in which this is found are summarised in Table 2.49 in Chapter 2. For the *modelages*, 22 cases out of a total of 35 (63%) indicate previous surface preparation; for the engravings this is 3 out of 88 cases (3.4%); for the finger-tracing 6 out of 20 (30%) and for the process of impressing, 6 out of 11 (55%). As expected, it is the engraving technique that shows the least evidence of surface preparation; this can be taken as confirming that the soft or hard nature of the surface used was not normally considered in the choice of the technique, even if it is most likely that in both cases a fine surface was looked for.

The knowledge visible in the working of clay in the French Pyrénées in the Magdalenian is fairly general: it relates mostly to the possibilities offered by the clay material and its plasticity, the notion that plasticity can be used in cave art representations, notably in relation with the representation of volume. The knowledge is that for this purpose the material could be extracted, transported and accumulated to obtain the wanted result. Also, in relation with the environment in which the works on clay are found in the region, there would have been a transmission of the knowledge that plastic clay was found in deep caves, where it could be used for making art. It is also possible the knowledge related to plasticity would have formed part of basic knowledge shared in a hunter-gatherer society notably through familiarity with soil and mud, as hypothesized above when looking at the raw materials.

Related know-how visible through the working of clay is very variable in the region. A wide range of skills is present, from site to site, but also from graphic unit to graphic unit even in the same site. The case of the group of bison representations from Tuc d'Audoubert can be raised here: the two large bison sculptures in clay (TA1 And TA2) show a full mastery of the material used and the possibilities offered, with a high degree of skill in the representation (Beasley 1986). They are the best known figures at the site but it must also be pointed out that they are in fact part of a group of four bison representations (Figure 4.5). The two other bison are a partly engraved-partly modelled figure (TA4) on the clay floor and a small sculpture (TA3) of a headless bison, unfired, that was placed on the floor as well.

The partly engraved/ partly modelled bison (TA4) is represented in a very schematic manner, with a much lower degree of skill in the making than the large bison. But on it, some detail is added that remind of the large bison: it appears small horns were modelled in the round and then added to this rather flat figure (Bégouen *et al.* 1977: 117).These horns are similar to the ones visible on the large bison. As for the small bison sculpture, it show all the characteristics of the animal with the large hump and a respect of the general proportions even if the head is missing, with an attempt to use three dimensional volume. Some details are indicated such as the tail. But it is not as precisely detailed as the large bison and it is much smaller in size.

It could be argued that three different individuals with varying levels of skills are recognisable as working side by side to represent these bison figures, each of them making something different from the material, except for the added horn that could indicate a copy of a detail of the large bison. Of course it is not possible to be certain that the making of all these bison was contemporaneous, but their grouping was certainly meaningful and intentional, even if the figures were made at different time periods. Besides this, it can also be said that the maker(s) of the large bison sculptures (TA1 and TA2) in Tuc d'Audoubert possessed a very high level of know-how in the working of clay, that they certainly had much previous experience of using the material in the making of sculptures (Beasley 1986).

Figure 4.5: Schema of the position of the group of four bison in Tuc d'Audoubert: the small sculpture is found in 2 between the other bison representations (From Bégouen *et al.* 1977: 113)

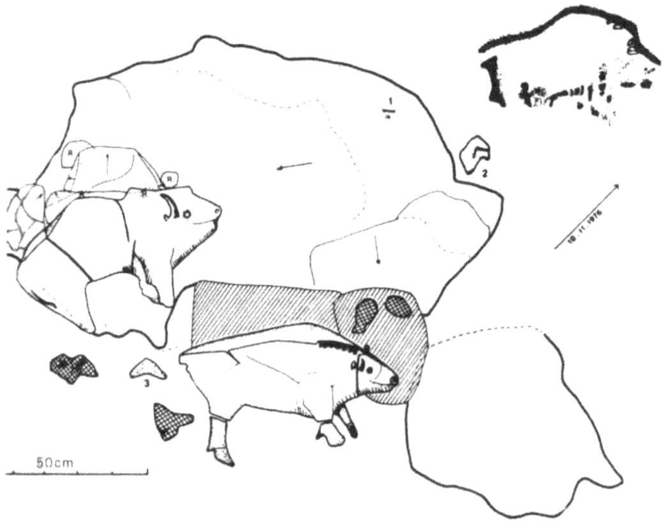

It is clear that in the Magdalenian period in the French Pyrénées, the knowledge of what could be done with clay by using the plasticity of the material and of the suitable context in which it could be done was transmitted as this seem the basis for this regional tradition. This is a simple knowledge, one that could easily have been learned by simple observation in an informal transmission of knowledge. The detail of the working of the material, on the other hand, seems to vary almost case to case.

- **Comparison with the working of bone and antler in the Magdalenian in the French Pyrénées**

Dobres has applied the chaîne opératoire concept to the study of bone and antler technologies in the French Pyrénées in the Magdalenian period (Dobres 1996, 1999, 2000). Her analysis and the resulting conclusions will be developed here with the purpose of comparison with the working of clay, which can then be considered in its wider Magdalenian technological context.

Dobres' concern is mostly the human agency visible behind technology, especially retracing individualities and through them, by means of *chaîne opératoire* research, 'an understanding of precisely when, how and to what extent a technical procedure was faithfully practiced and when, how and to what extent technicians deviated from it in a particular manner' (2000: 175). She emphasises the idea of looking at the shared norm and procedures visible in the technological processes and also at the variability from this norm as a manner to identify practice, negotiation and/or contestation, and therefore social agency (2000: 176).

Dobres chose to look across classes of artefacts in her study rather than sticking with the traditional approach of concentrating on specific classes of artefacts as she postulated that 'people usually engage in more than one activity at a time' (2000: 193). She considered artefacts made of osseous materials, mostly tools including needles, 'baguettes', harpoons, polishers, awls, spear-points and others, (Dobres 196: 154) with the idea of determining how knowledge and skills were applied to the making, use or repair of these objects and if these were transferred between artefact types (Dobres 2000: 195). She recorded observations along a number of different axes such as raw material, artefact morphology, class-specific work chain, use and breakage patterns, as well as details of repair and recycling strategies (Dobres 2000: 193).

This allowed recording an important variability in the technological treatments at the same site and across the region (French Pyrénées) (Dobres 2000: 200). For example, looking at the piercing technique applied to the eye of bone needles from four sites in the Ariège allowed seeing that four different techniques were used across the region, with two to four of them being used at the same time in each site (Table 4.3) (Dobres 1996: 186). Also, very different levels of skills were visible in the piercing of the eyes, from a perfectly circular perforation to a roughly done uneven hole.

Table 4.3: Techniques for piercing the eye of bone needles in the Ariège (after Dobres 1996: figure 13).

Site name	Circular rotation	Deep groove	Alternate rotation	Pressure
Mas d'Azil	X	X		
Bédeilhac		X	X	
Les Eglises	X	X	X	X
Montfort			X	X

Dobres concludes that along every measure of technical activity studied for the French Pyrénées she found 'evidence not of a pan-Magdalenian template of what to do, how to do it and when, but rather a situated application of knowledge and skill' (Dobres 2000: 203). The technical decisions seen were site specific (Dobres 1999: 136). She deduces from this that in the Magdalenian in the French Pyrénées the technical rules of procedure and comportment related to the cultural context were little more than guidelines (Dobres 2000: 208). Dobres explains this by reference to social factors that will be developed later in this chapter.

Looking at the variability visible in the working of clay in the French Pyrénées as is done below allows seeing that very similar conclusions can be drawn for this material. The fact that it has been used only for art-making when Dobres' study concentrated on tools in osseous material does not change the outcome of the technological studies. This can be taken as an important argument to show Upper Palaeolithic art also needs to be studied from a technological perspective besides notions of style.

- **Variability of the working of clay**

It has been seen that much variability is also visible in the technical processes of the working of clay in the Magdalenian in the French Pyrénées. Very different applications are found from site to site, and also even within the same site, as was seen above for the Tuc d'Audoubert for example.

The number of clay techniques found in each site and their types show much variability as is shown in Table 4.4 below.

Three sites show four different techniques used: Tuc d'Audoubert, Bédeilhac and Montespan. In Tuc d'Audoubert and Montespan, it is the same techniques that are used: all except portable sculpture and use as pigment. This means that all the working of clay at these sites is consistent with using the plasticity of the raw material in order to produce cave art intended to stay in the position it was left in. In Bédeilhac, the techniques used are modelage, engraving, portable sculpture and use as pigment, with the portable objects being the most common.

At the sites of Fontanet and Massat, three techniques are used. It can be pointed out that the three techniques found in Massat are engraving, finger-tracing and impressing, all techniques that consist in marking a clay surface. It does not seem that in Massat the concept of volume in representation was associated to the use of clay in the art, nor to the rest of the art found on its walls.

Table 4.4: the clay techniques present in the sites of the French Pyrénées

Site name	Modelage	Engraving	Finger-tracing	Impressing	Portable sculpture	Use as pigment
Tuc d'Audoubert	X	X	X	X		
Enlène					X	
Labouiche	X					
Niaux		X				
Mas d'Azil					X	
Fontanet	X	X		X		
Bédeilhac	X	X			X	X
Massat		X	X	X		
Montespan	X	X	X	X		
Labastide	X					X
Bois du Cantet				X		
Erbérua	X		X			
Oxocelhaya-Hariztoya			X			X
Etcheberri		X				X

Four sites show evidence for the use of two techniques: Labastide, Erbérua, Oxocelhaya-Hariztoya and Etcheberri. In three of these (Labastide, Oxocelhaya-Hariztoya and Etcheberri), the second technique found is the use as pigment.

Finally, five sites show only one clay technique used in them: Enlène, Labouiche, Niaux, Mas d'Azil, Bois du Cantet. In all these sites, except Niaux, the number of graphic units found on clay is extremely small, between one and three. In Niaux, the technique used to make the art was engraving on the clay floors. It was probably made over a long time-period as it seems related to the painted art on the walls of the Salon Noir for which AMS radiocarbon dates have shown it was made over a thousand year period (Clottes 1995). It is also stylistically close in style to the parietal art. It seems reasonable to imply a form of transmission of knowledge for making this art.

There is also much variability in the manner one technique can be used, even in the same site. The example of *modelage* will be considered here. In all cases a quantity of plastic clay is used to shape the volume of a figure. But this has been done in many different manners. For example, a clay plaque could be extracted, transported and shaped by a reductive process as would have been done in sculpture using other materials. This is seen in the two large bison at Tuc d'Audoubert. Clay could also be accumulated on a surface by an additive process before being shaped. This could be done in varying quantities thus resulting in the obtaining of low relief, high relief or three dimensional reliefs. Low relief is found in Bédeilhac for example and high relief in Montespan. Accumulating large quantities of plastic clay resulted in a monumental form of three dimensional sculpture unique to the cave of Montespan with truly large figures such as the three felines (Trombe and Dubuc 1947) (also interpreted as horses by Rivenq 1984) that each measured about 1.5 metre long before their partial destruction or the headless bear sculpture. The Tuc d'Audoubert bison can also be described as monumental but for them the bulk of their shape was produced by reduction.

Modelage in the French Pyrénées has also been obtained by a technique close to *champlevé* as it also lowered the level of the surrounding surface to emphasise the volume given to the figure. This is seen in the partly engraved-partly modelled bison figure in Tuc d'Audoubert (Bégouen and Clottes 1977).

Modelage was also obtained by modelling a flat contour découpé figure in clay and then sticking it onto a rock surface by simply pressing it on or possibly by wetting the back of the figure so it would adhere. In the case of the horse head MO3 in Montespan, it seems such a figure became detached from its wall surface probably when it dried: it has been removed from the site and sometimes confused with a portable clay sculpture. But the early discoverers of the site clearly state its position (Bégouen and Clottes 1988). The four horse *modelages* from Erbérua were also made this way ((Larribau and Prudhomme 1989: 480).

Thus a common knowledge related to the plasticity of clay and its capacity to give volume is visible in the Magdalenian region in the French Pyrénées. But the exact manner to apply this plasticity to obtain volume shows no widespread uniformity across the region. In the exact detail, there is much variability. It must also be pointed out that the various techniques for using clay are often combined, for example engraving can be used on clay *modelage* to draw details.

- **Uniformity of style**

If the techniques of the working of clay show much variability, the visual aspect of what is made with clay, on the other hand, is the unifying factor in this regional tradition of clay working.

There is no doubt that the art of the Magdalenian period, especially in the Middle Magdalenian, shows much uniformity and consistency (Sieveking 1976). This has been interpreted as the sign of a strong unity of the Pyrenean Magdalenian, with a likely community of concepts possibly resulting from complex contacts and exchanges alongside the Pyrenean range (Clottes 1989: 89). The use of clay in the French Pyrénées, as a material applied to the making of art, does fit in this wider regional context. It follows similar conventions in style and theme of representation.

This visual uniformity and consistency in the art of the Magdalenian, however, is not seen in the techniques used for the making of the art. In term of implications for the transmission of knowledge visible in the region, it would mean the Magdalenian people shared knowledge of the visual aspect things were supposed to have. This is striking in art, but can also be seen to a degree in tool-making; for example, in the case of the bone needles studied by Dobres (1996, 2000), it seems the knowledge of what the final object was supposed to look like was shared amongst the makers. All obtained a thin pointed object with a pierced eye. But deciding on the manner of piercing the eye, and possibly also of shaping the rest of the object, seemed to be down to the individual, or possibly the group of individuals carrying out the action.

In the art on clay, the case of Fontanet is a perfect example of this same phenomenon (Vialou 1986): the graphic units are made of various patterns each made of an association of cupules: visually it is very uniform. But if ones looks at the technique used, one can see that these cupules were obtained by impressing with fingers, impressing with a with a tool and by modelage: three different techniques as if three different persons had made them each with their own interpretation of how to make a cupule on a clay surface.

So, it can be said that the transmission of technical knowledge in the Magdalenian in the French Pyrénées, as has been seen in relation to osseous material and clay techniques, does not appear to be the subject of any form of formal learning. Knowledge related to the visual aspect of the end-points resulting from the technological process, on the other hand, was widespread. Could it be implied for example that knowledge was generated in the context of people making things together or next to each other, by a passive learning by observation that would not involve specific discursive knowledge? The resulting technological applications would then be built up individually by personal know-how with a great degree of choice left to the individual in decisions.

Lithic technology in the same cultural context would need to be studied with the aim of comparison to see if similar conclusions could be drawn. However, this remains a topic for further research.

This appears to be the widespread situation across the region. In the case of the working of clay, it can also be considered that thanks to the versatility of the material and the possibilities it offers of reversibility of large parts of the working process, the knowledge transmissible in an apprenticeship did not have to be very precise, and the skills involved would not be very difficult to learn nor require long practice. In comparison, some flint knapping techniques require years of practice to be mastered (Pigeot 1987; Karlin and Julien 1994).

But it is also true that the use of clay found in some specific sites might imply a more precise transmission of technological knowledge. The case of the clay plaquettes found in Bédeilhac for example shows very similar objects in the pattern of their making. There is certainly a transmission there of specific technical knowledge about the plaquettes as they are unique to the site and found in large quantities. The possibility that they were made in the context of aggregation might explain a different situation.

The fact that the 'end-point' of the use of clay is the making of art form, implies that more needs to be said about style and themes of representations in the wider Magdalenian art context. This will be looked at in part 4 further in this Chapter.

The fact that, doubtlessly, it is the social context of the use of clay that determines the visible aspects of the transmission of related knowledge also implies that this aspect needs developing; this is the topic of part 5 in this chapter.

In Central Europe

In Central Europe in the Gravettian period, clay was worked in a very different manner that in the French Pyrénées, with perhaps one exception; clay was also used in the making of art.

- **Transmission of knowledge and know-how**

The working of clay visible in the sites of Dolni Vestonice and Pavlov can be summarised by its technical scheme in the flow chart in table 3.33 in Chapter 3. It can be seen that the whole chain of sequences is linear with few options for variation (as indicated by the arrows skipping sequences), except for the fact that every step is reversible before the firing permanently changes the nature of the raw material.

A few sequences, indicated by broken circles around the sequence numbers, can be skipped such as sequences 5 and 7: these two relate to possible variation in the order of the fashioning of detail on the figures. Sequences 10 and 15 relate to accepting the interpretation of the firing of the Moravian ceramics as a ritual destruction involving pyrotechnology (Soffer *et al.* 1993). Also, the process of intentionally exploding non-fashioned clay – if the thousands of clay pellets found are interpreted as resulting from such process- could be done by jumping directly from sequence 2 to sequence 11 on the flow chart.

Sequence 17, a possible reshaping of a figurine after firing has been added as a few examples indicate this was sometimes done. However, this remain anecdotic in relation to the bulk of the worked ceramic found and cannot be implied as a frequent technological pattern.

From this 'flow chart', it is possible to say that the pattern of the making of ceramic figurines for Dolni Vestonice and Pavlov is a very linear process in the sense that one sequence follows the other in order. This fact, combined with the complexity of the whole technological process visible, implies that at the Moravian sites, the process of working clay was a standardised process in which complex knowledge was involved.

It was seen above that the various properties of the loess material used as clay appeared as understood and mastered in the technical application made of them. The state of plasticity would first be used for fashioning; a controlled hardening by air-drying would then allow the making of some details; finally the capacities of hardening in firing and/or exploding in firing were exploited.

The use made of clay in the Moravian sites is a complex technological process that was applied in a systematic manner probably in a form of symbolic expression connected to a specific ritual. The fact that the ceramic inventory is important in quantity and spans over two thousand years suggests that the cultural practices carried out over time involved repetition and transmission and the learning of a specific patterned behaviour of ceramic technology (Soffer *et al.* 1993).

The highly standardised nature of the working process would point out towards a formal teaching in order to master all the knowledge required by the process. Simple observation would probably not be enough to understand and control all the aspects of this working and to keep this specific use of clay unchanged over long time-periods. For example, the specific properties of the loess soil used were not recognisable from the rough material extracted: the fact it could become plastic if mixed with water and kneaded and the fact that it would harden into ceramic (or explode) under the action of fire are knowledge that needed to be passed on. Knowledge related to the effect of heat on the clay also needed to be transmitted.

But it is true that little skill is needed then to apply the learned knowledge successfully: the techniques themselves would not require long practice in comparison with some flint-working techniques for example. Various levels of know-how are not really possible to see in the application of the technological process, but they are clearly there in the visual aspect of the figurines: some appear fashioned with much talent; other could have been done by any young child. These elements could suggest a technological process which might have been controlled by (an) individual(s) possessing the knowledge of the whole process, but in which a wider range of participants could take part, notably in the shaping and fashioning of the figures. The stylistic variations visible in the figurines seem to indicate that a number of different individuals did the fashioning. These stylistic elements will be considered further in this section.

- **Comparison with the working of osseous materials**

The processes used for working osseous materials at Dolni Vestonice and Pavlov will be reviewed here in order to see how the treatment given to clay material fits into the wider technological tradition at the sites.

Svoboda (1996: 291) reminds us that the industry of organic materials (bone, antler and ivory) and art (ivory carvings, stone carvings, pendants, engravings, and ceramics) are documented mainly from the Evolved Pavlovian stage (27,000 -24,000 BP) in Moravia. The use of clay at Dolni Vestonice and Pavlov is dated between 27,000 and 25,000 BP (Svoboda 1996). It appears that the osseous industry of the Pavlovian, although it did not preserve everywhere, is very rich. However, this is not a widespread aspect in the wider Gravettian context in Europe; in most other regions, there is rather a rupture with the rich osseous industries of the preceding Aurignacian: these materials seem less used in the Gravettian (Goutas 2004: 53-54).

The range of objects made from osseous materials is very varied; utilitarian implements, ornaments (beads, rings), portable art (carvings, decorated objects), hunting weaponry (spears and lances), processing implements (digging tools, spades, pestle, pickaxes, shovels, hammers, clubs, tent pegs, polishers, handles, piercers, awls) (Soffer 2000: 62; Goutas 2004: 55).

In the Gravettian in general, all the hard animal materials (bone, antler and ivory) were worked. In Moravia, ivory was used more than bone. Mammoth bones also seemed to be systematically exploited during the Pavlovian period (Goutas 2004: 55). In Dolni Vestonice, artefacts made of antler are few and are mostly pierced batons and axes. In Pavlov, on the other hand, tools in reindeer and deer antler are more numerous than bone ones with varied types of tool. Both attached and shed antlers were used (Goutas 2004: 55, 57).

Various authors have underlined the wide variety of techniques used for the working of the osseous industry in the Pavlovian sites (Soffer 2000; Oliva 2000: 219; Svoboda 1996). They can be listed as follows: breaking, sawing in to chunks (by notching, sawing and bifacial grooving), splitting, longitudinal grooving, bipartition by opposed grooving, extraction of baguettes by double longitudinal grooving (Goutas 2004; Svoboda 1996).

The double grooving technique was a novelty in the early Upper Palaeolithic: it allowed predetermining the dimensions and morphology of the support and thus was the base for much improvement and development of the working of osseous materials. The initial investment in time at the debitage stage is important, but the rest of the process is made much easier. It does seem to originate in the Gravettian period and was found at the Moravian sites (Goutas 2004: 58, 66).

In Dolni Vestonice and Pavlov, the techniques for the debitage of the raw material apparently varied according to its nature. In Pavlov, it seems some mammoth bones were worked by percussion, just like stone (Valoch 1996), thus obtaining flakes sometimes used as such or retouched (Goutas 2004: 57). For antler, the technique of extraction by double grooving is sometimes visible. Chunk of cut antler are also found. Some antler show many traces of working such as cuts, blow, abrasion and various wears and some might have been used as tool (ref to Valoch 1996). For ivory, the tusks were cut out into chunks by cutting and percussion associated to breakage, and then the chunks were split to obtain long flakes (Otte 1981; Goutas 2004: 58). Apparently the double grooving technique was not used in the debitage of ivory in Moravia (Goutas 2004: 58).

After the debitage of the osseous raw material into rough-out shapes, their fashioning could be continued by the application of the following techniques: double grooving (for obtaining the exact shape required); scraping (to refine the final shape); abrasion by friction; polishing (with finer abrasives than for the abrasion) which makes the objects surface shiny and which was systematically used on ivory in Moravia; perforation; incisions with functional, decorative and/or symbolic role (Goutas 2004: 64).

The variety of the techniques used in the working of osseous materials in the Pavlovian or Moravia has been underlined previously as well as the variety of the range of objects made with them. Unfortunately, *chaîne opératoire* studies of the working of these materials do not seem available at present in Pavlovian context. The lack of description of the technical acts in references and the great differences in terminologies encountered are also a problem in trying to reconstitute the details of the technological working in Central Europe in general (Goutas 2004: 57)

However, some elements can be pointed out. The techniques for the working of osseous material in the Pavlovian period in Central Europe show a full adaptation and understanding of the materials used; the processes of the working seem adapted both to the type of material worked and to the intended final result. The variety seen in the techniques and in the artefacts they produce indicates the extreme mastery of the whole technological process.

Also, the fact that certain types of identical objects were found in number, would indicate that a specific knowledge related to these objects was transmitted, both from a technological point of view and for the visual aspect they were supposed to have. Valoch (1996: 121) underlines that some specific types of tools in osseous material are found in the Pavlovian; this is the case of small axes made of antler found in Predmosti, Dolni Vestonice and Pavlov that are normally made with the same part of the antler, the basal part, by bevelling and polishing it in order to obtain a sharp edge. The function of such axes with a polished edge is not clear. Valoch hypothesises on a use for working soil (Valoch 196: 121). These objects appear similar in typology as well as by the manner of their making.

The processes involved in the working of osseous material are complex processes; to obtain a desired artefact a few techniques need to be combined into a chaîne opératoire.

The working of ivory shows a clear pattern of working; the debitage of ivory was done by cutting out and percussion to obtain chunks that were then split to obtain long flakes from which the artefacts were made. It has also been underlined that the double grooving technique was not used on ivory in Moravia, and that ivory objects were systematically polished. These elements are consistently found and they relate to knowledge which is shared.

Although more research is needed, it would seem that a certain level of technological standardisation is visible from the study of the osseous material used in the Pavlovian of Moravia implying that some specific knowledge was transmitted. Whether this transmission was the result of simple observation or of formal learning is difficult to determine. However, the fixed elements visible in the working of ivory for example seem to indicate that specific patterns of working associated with this material were consistently followed, thus ignoring other possible ways of working it, such as by double grooving for example. It appears there were certain determined ways of doing this at technological level in Moravia, with little variability.

These elements, combined with the much standardised manner of working clay seen above, carry implications for the Pavlovian social context in which these technologies were used. These will be looked at further in this Chapter.

- **Some stylistic elements**

From a stylistic point of view, there is little in common between the art found made of ivory and the ceramic art. The art on ivory is made of some carved figurines which tend to be much stylised, and of many decorated surfaces as in Pavlov I for example (Klima 1997b: 267; Svoboda *et al.* 1996: 160). In Dolni Vestonice I, some figures are found carefully carved in the round in ivory, but what they represent is not usually figurative: some are considered as some form of sexual symbol (Figure 4.6) (Svoboda *et al.* 1996: 159). The art on clay is much more naturalistic and varied in its themes of representation, as well as being much more abundant than in any other medium.

The knowledge and know-how related to the visual aspect of the art in the making of the ceramic objects in Dolni Vestonice and Pavlov must also be considered. The aesthetic and expressive quality of some of the ceramic figurines is very high (Absolon 1949: 27) and would require talent or long practice, possibly both. In the case of two themes of representation, the female figurines and the felines heads, it is even possible to argue that some of the recorded figurines could have been made by the same person as they are exactly similar or could have been made from the same model with various levels of exactitude and success, suggesting they were made at least from similar models, possibly involving copying the work of an individual in a learning process for example.

Figure 4.6: Portable art in ivory from Dolni Vestonice I (from White 2003:136)

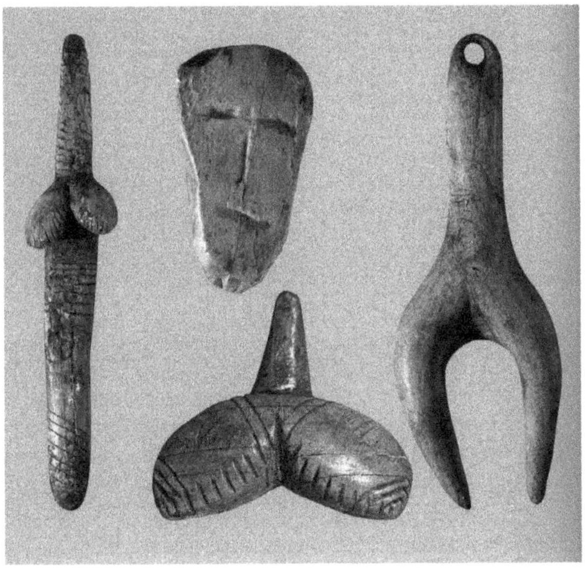

For the female figurines in ceramic, it has been seen in Chapter 3 that the best-known of them all, the Black Venus from Dolni Vestonice, was actually part of a group of six similar figurines and could have been used as their model. All these were found in Dolni Vestonice I and II. This type is absent from Pavlov.

Two of the figurines, DVI 8 and DVII 5 are so similar it could be argued they were made by the same hand or at least by two persons working with exactly the same manner, possibly copying each other closely. They are both fragmentary but details are well visible. DVI 8 (Figure 3.16 in chapter 3) shows the breast of a female figurine: the groove that marks the separation between them shows the same width and general aspect as the one on the Black Venus and could have been made with the same tool. DVII 5 (Figure 3.15) is a fragment of the head of a female figurine: the top part of the head is visible and just as for the Black Venus; it has been marked by impressing four deep holes arranged in a square. These holes are oval in shape and are exactly similar on both pieces: they seem made with the same tool. DVI 5 (Figure 3.13) is the fragment of a torso, intact on the back. It shows two incised parallel grooves on each side of the spine, just like the Black Venus' one.

Then a figurine formed by two fragments is DVI 3 and 4 (Figure 3.13). The model of the Black Venus is still clearly recognisable but it cannot be shown that the same hand did this figurine. Some individual variation is visible, such as in the two parallel grooves on the back which are not strictly parallel nor well marked anymore. The groove separating the legs and the groove making the belt have been done with less care and are approximate compared to the model. One could argue here for an imitation of the Black Venus model, but with less skill in the making: could this indicate a learning process?

Finally, the figurine DVII 6 (Figure 3.18) could be taken as an even further interpretation from the same template. The legs have a similar conical shape with a deep groove marking their separation and indicating the spine but the overall shape is shorter, and seems more quickly done. A belt is indicated, but instead of being made by a deep groove, it has been made by juxtaposing small impressed holes around the waist of the figurine. The Black Venus template was followed, but with less skill and some personal variation.

It must be pointed out that the 'Black Venus' type of female figurine is not the only type of female figurine found in Dolni Vestonice I. The figurines DVI 7 (Figure 3.19) and DVI 9 (Figure 3.20) are fragmentary headless bodies, different in style, but with the same characteristics of having a belt represented by incision and of having the breasts missing.

As for Pavlov I, out of the seven female figurines found there, five different types could be found indicating less standardisation in the physical appearance given to them. It must also be noted that the 'Black Venus' type is absent in the Pavlov I female representation.

The 14 feline representations found in Dolni Vestonice I and II and Pavlov I are all heads or fragments of heads. It has been seen in Chapter 3 that it is possible to determine a typology of the various aspects given to the muzzles through the indication of nose, lips and chin, which are finely detailed on ten of the heads (Table 3.16). These are subtle variations in the making of the figures, important for possibly differentiating the makers of the various feline heads. In Dolni Vestonice I, only two types of muzzles (types 1 and 2) are found for five recognisable heads. In Pavlov I, three types (types 1, 2 and 4) are visible for four heads. In Dolni Vestonice II, only one muzzle is recognisable and it presents another type called type 3 here. It is different from what is found in the other sites, but not to a great extent, and it could be argued that it is a variation on the type 2 which is found in the two other sites. These stylistic elements emphasise connections between the sites seen in the making of the feline representations on ceramic, in a stronger manner than for the female figurine representations. Among the many possible ways to represent the muzzle of a feline, a few selected patterns were used and repeated. This would imply the fashioning of these figurines was done by people following similar procedural criteria.

Beside the pattern of the muzzle, the overall aspect of some feline heads from Dolni Vestonice I makes them so similar that they could have been done by the same hand; this fact was already noticed by Absolon (1949: 25) who said that there was no doubt the artist duplicated here his/her own work to make what he then interpreted as bear heads rather than felines. DVI 18 (Figure 4.7) and DVI 24 (Figure 4.8) are two very similar figures. They have the same pointy nose and they are carefully shaped with a lot of details, notably in the fashioning of the muzzle. The eyes are represented by long parallel slits on each side of the forehead; they are the only two feline heads found with this characteristic (see Table 4.6 below), which again emphasises their similarity. The ears are small and rounded.

Figure 4.7: Feline head DVI 18

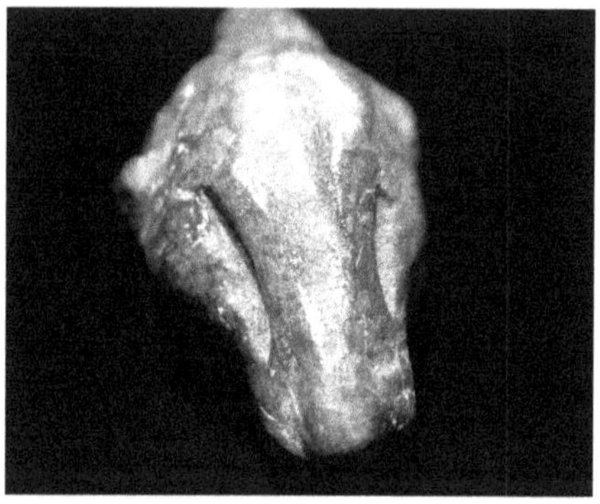

The heads DVI 23 and PI 17 are visually very similar to them in their general aspect and also share the same type of muzzle. However, details do vary slightly such as the eyes which are indicated by very short slits parallel to the ridge of the nose. This indicates that the people making them might not be exactly the same person that made DVI 18 and DVI 24 but that they did follow the same cannons of representation in the rendering of their figurines in such a close way that it is likely they learned how to do it together or from one another.

Figure 4.8: Feline head DVI 24

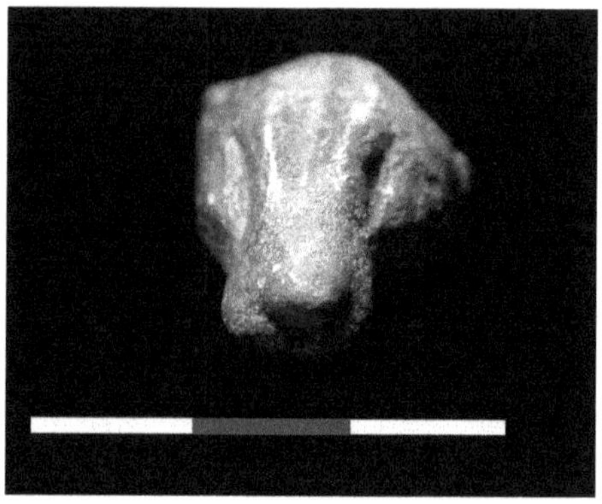

Thus stylistic elements complement the technological study of the use of clay in Moravia by helping to inform us about the makers of the fired clay figurines and the possible relations visible between them through the study of the objects.

Table 4.6: Characteristics of the feline heads from Dolni Vestonice I, Dolni Vestonice II and Pavlov I (when the information was available).

Reference		DVI						DVII	PI			
		15	18	20	21	23	24	4	17	19	20	21
Profile type	Flat									X		
	Pointy	X	X	X	X	X	X	X			?	?
Eyes	Short slits	X		X		X			X	X		
	Long slits		X				X					
Surface decoration: parallel incisions				X								
Non-symmetrical ears			X	X		X	X		X			
Pierced						X						
Muzzle type	Type 1		X			X	X		X			
	Type 2			X	X						X	X
	Type 3							X				
	Type 4									X		

Conclusion

By looking for norms and variations visible in the processes used to work clay in the two regions of Europe studied, it has been possible to underline the potential of clay studies to inform us about two very different cultural contexts.

In the French Pyrénées, an extreme variability in the techniques used for working clay and the application of these techniques has emphasized that a common general knowledge about the plasticity of clay and its potential use was passed on in the Magdalenian societies of the region. An exact transmission of techniques, however, was not recognised, which might imply that the obtaining of skill and know-how of the working of clay was not transmitted but rather was down to individual experience. On the other hand, the visual aspect of the works of art done on clay in the region followed shared cannons of representation associated with the Magdalenian culture.

In Central Europe, the working of clay appears to be much more standardised. It follows a linear technical scheme with little room for variation in a process that recognises and exploits all the properties unique to the clay material in a succession of sequences. It can be assumed that a formal transmission of knowledge existed for working clay. Stylistic details of the figurines allow implying further relationships between the actors of clay working in Dolni Vestonice and Pavlov.

Since clay was used for artistic and thus symbolic creation in each region, more needs to be said about the visual aspect taken by the works on clay and their respective wider art context. This will be considered below. Also, the study of the processes of clay working appears to have a potential for increasing our knowledge of the social contexts the techniques are used in, notably elements about intra/inter-site variability across the two regions considered, some involvement in recognising aggregation sites as well as contribution towards social agency. This will also be considered below.

4 The desired end-points: symbolic expression and clay

The end-point of the technological process of working clay in the Upper Palaeolithic is consistently, as we have seen throughout this research, the creation of forms of symbolic expression, figurative such as animals or humans, or non-figurative in the case of abstract elements. The knowledge involved in the making of these end-points is one able to combine material and form, the realisation that the properties found in clay as a raw material can be used in wider art contexts considered to result into meaningful shapes whose styles and technologies can inform us about the people who made them.

In the discussion below, the first part will consider the possible relations between the use made of the properties of the clay material and the themes of representation encountered in the two cultural contexts studied in the research. Then the remaining two sections will look at the wider art contexts visible in Central Europe and in the French Pyrénées with regard to a number of research topics toward which clay studies can contribute.

The properties of clay material and the themes of representations

Determining the properties of the raw material that have been used in the making of the art has been made possible by the application of a *chaîne opératoire* technological study. This has allowed precise observations about the making of the art, an issue that Lorblanchet (1977: 44) notes should logically be considered before dealing with the question of the potential meaning of art by simply concentrating on its content.

The unique material properties of clay are its plasticity, its capacity to harden to various levels of hardness by drying and by firing, and also its capacity to explode under the action of fire (Table 4.2). By looking at possible connections between the themes represented and the unique properties of the material used, it is hoped to gain an insight on any possible contribution of the use of clay to symbolic expression in the Upper Palaeolithic, especially with the view that the two appear strongly connected during this period.

In the French Pyrénées: modelage and naturalism

In Chapter 2, the techniques used in relation with the themes represented in the clay art of the French Pyrénées were looked at separately for each main theme of representation (animal, human and signs). Among the techniques used in this region for working clay, as was seen in the previous section, modelage, finger-tracing, impressing and portable sculpture are those that make use the most of the properties of plasticity of the clay, to varying degrees. The properties of hardening by drying and firing were also sometimes exploited.

With regard to the themes represented, some patterns of relation between them and the properties of the clay used are indeed noticeable.

It was noted from the analysis of the data from the French Pyrénées that the use of plasticity through the technique of modelage was used primarily to represent figures of animals and vulvas, except for two cases (Table 4.8); in Tuc d'Audoubert it was decided to include the five modelled 'clay sausages' (Bégouen and Breuil 1958: 98) found with the abstract graphic units as their interpretation is not straightforward and as they are the only example of this type of modelage known in the French Pyrénées; the other case of modelled signs is found in Fontanet (Vialou 1986) where in three occasions the cupules forming the signs found on the clay floors in the cave were shaped by modelling instead of impressing. This might be motivated by the fact of wanting to obtain larger cupules than was possible with impression.

For the animal and human themes of representation it is possible to see from Table 4.7 that out of 29 horse representations made on clay in the French Pyrénées, 14 are modelled, 10 are engraved and 5 are finger-traced. Out of 26 bison representations, 12 are modelled, 14 are engraved, none are found in finger-tracing. No other animal representation, with the exception of the large bear *modelage* in Montespan, is made by *modelage*. That is to say that *modelage*, which uses the property of plasticity of the clay, has been used only for representing the most commonly found animal species in the cave art of the French Pyrénées: the horse and the bison (Clottes 1989). This process allows the representation of a three-dimensional volume, thus increasing naturalism. The other animal figures are drawn in two dimensions with lines made with a tool or a finger. This treatment seems to emphasise the importance of these two animals in the belief system of the Palaeolithic people of the region and the importance put on the obtaining of a naturalistic figure by the Magdalenian (Clottes 1989 (1999): 80).

The fact that many horse and bison are also found engraved is entirely down to the considerable number of representations of these animals found engraved on the floors of two sites, Niaux and Bédeilhac.

It should also be noted that the technique of impressing, which uses the plasticity of clay but produces visual results close to two-dimensions, has only been used for non-figurative representations in the French Pyrénées in the Magdalenian, never for animals. The visual result obtained with this technique seems to echo signs also found in the cave art of the region (Clottes 1995) made by juxtaposing painted dots.

Signs, on the other hand, mostly remain two-dimensional which might confirm their abstract and highly symbolic nature: they are products of the artist's imagination or cultural background and only seen in two dimensions. 60.2% of the signs are made by the technique of engraving; 17.3 % are impressed; 14.3 % are finger-traced and 8.2 % are modelled (Table 4.8). The use of techniques that only exploit the property of plasticity of clay in a reduced manner so as to obtain two dimension representations clearly dominates the record for the non-figurative representations.

It must also be pointed out that the level of adaptation of the represented themes to the properties of the raw material used varies enormously from site to site.

Table 4.7: The recognised animal and human themes found in clay in the sites of the French Pyrénées in relation with the properties used in the techniques applied.

Site name	Animal theme	Human	Property of plasticity used			Plasticity, air-hardening/ fire-hardening	Uses clay as soft stone
			Modelage	Finger tracing	Printing	Sculpture	Engraving
Tuc d'Audoubert	4 Bison		4				
Enlène	2 Indeterminate					2	
Labouiche	1 Bison		1				
Niaux	9 Bison						9
	4 Horse						4
	2 Aurochs						2
	3 caprid						3
	4 Fish						4
	1 Bear						1
	1 Indeterminate						1
Mas d'Azil	1 Bison		1				
	1 Feline					1	
Bédeilhac	10 Bison		5				5
	5 Horse		2				3
	1 Aurochs						1
	1 Cervid						1
	1 Feline						1
	1 Indeterminate						1
		1 Vulva	1				
Massat	2 Aurochs						2
	3 caprid						3
	1 Fish			1			
Montespan	11 Horse		8	2			1
	1 Bear		1				
	1 Mammoth			1			
		1 Vulva	1				
Labastide	1 Bison		1				
Erbérua	5 Horse		4	1			
	1 Cervid			1			
Oxocelhaya	2 Horse			2			
Etcheberri	2 Horse						2

Table 4.8: Inventory of the non-figurative representations in relation with the properties of the clay techniques used to make them in the French Pyrénées

Site Name	Non-Figurative					
	Clay used as soft stone	Property of plasticity used				
	Engraved sign	Finger-traced sign	Modelled	Impressed signs		
				Circular	Linear	Other
Tuc d'Audoubert	2		5	1	3	
Enlène	1					
Labouiche						
Niaux	41					
Mas d'Azil						
Fontanet	1		3	1	2	4
Bédeilhac	13					
Massat	1	3				1
Montespan		4			3	1
Labastide						
Bois du Cantet						1
Erbérua		6				
Oxocelhaya-Hariztoya		1				
Etcheberri						
Total	59	14	8	2	8	7
% of total of signs	60.2	14.3	8.2	2	8.2	7.1

Finally, there is also a tendency in two of the caves with the greatest quantity of clay uses, Tuc D'Audoubert and Montespan, towards a monumental form of sculpture. This only finds an equivalent in Palaeolithic art in the sculpted frieze of some shelters and caves of Dordogne and Central France (such as Roc-de-Sers in Charente, Cap Blanc in Dordogne, Angles-sur-l'Anglin in Vienne for example) (White 2003; Sieveking 1976; Breuil 1952) and the topic of the possible connections between these two forms of art is certainly one that will need further research. The two bison sculptures from Tuc D'Audoubert respectively measure 61 and 63 centimetres long (Bégouen *et al.* 1977), much larger than any of the sculptures made in other materials in the region. In Montespan, the headless bear statue itself measures about 1.1 metre long, and what is left of the three large feline *modelages* is enough to determine they were over 1.5 metre long (Trombe and Dubuc 1947).

It is clear that this monumental form of sculpture is entirely based on the property of plasticity of the raw material used, which allowed the making of large figures with much less involvement in time and effort than the equivalent in stone. The fact that most of the parietal art in the French Pyrénées is found located in deep caves might possibly have contributed to this as less time could be spent in the cave to make the art, simply the duration of the fuel supply one could bring along. The examples from other regions mentioned above are all found in day-lit areas. It must also be mentioned that some works of art verging on the edge of sculpture do exist in the French Pyrénées in the cave of Montespan (Rivenq 1984) and in Isturitz (Clottes 1989). In Montespan, two types of engravings are found in the Casteret-Godin Gallery which houses most of the work on clay in the cave; one consists of engravings done with a fine sharp tracing; the other has been described by Rivenq as closer to being a low relief technique that tends to shape the volume around the outer lines of the drawing (1984). This is found all along the gallery together with the numerous works on clay for which considerable quantities of material have been shifted and fashioned ; it could be argued that this whole gallery puts emphasis on the representation of figures in volume. In Isturitz, on the other hand, a few figures in low relief are mentioned as found at the back of a massive of stalagmite by the living areas (Clottes 1989 (1999): 81).

In Central Europe, realistic representations and clay

The techniques used in Central Europe for the making of fired clay figurines involve *modelage*, using the property of plasticity of clay; waiting using the capacity of clay to harden by drying; and firing using the capacity of the material to harden permanently under the action of fire or the capacity of the material to explode in fire.

In all cases, the results obtained are three-dimensional representations. This material record in fired clay is largely dominated by realistic representations of animals (whole or limbs) and of humans. It might be assumed that this is because the properties of the material used result in fashioning a volume, which is suitable for naturalistic representation. It is true that four fragments of fired clay have been found (two from Dolni Vestonice I and two from Pavlov I) (Table 3.6 in Chapter 3). That show a rather flat surface with a decoration of parallel incisions (Figure 4.9); this reminds of some other forms of art found in the Pavlovian sites of Moravia that consists in decorating surfaces of ivory with geometrical and/or abstract patterns (Svoboda *et al.* 1996). However, the fired clay fragments found are usually small and broken, thus making it difficult to be certain of whether they were part of the outer surface of an object fashioned in three-dimensions or whether the clay support was only used for abstract decoration. Rare figurative objects with decorations of parallel incisions have been found such as the feline head DVI 20 (Figure 4.10).

Figure 4.9: Fragment of fired clay DVI 59 from Dolni Vestonice I with decoration of parallel incisions

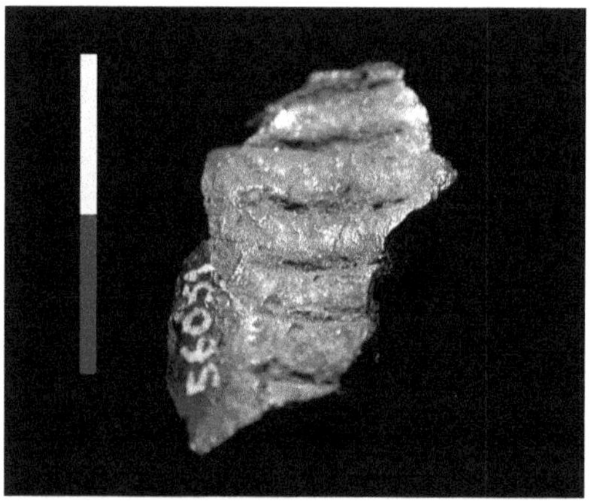

Figure 4.10: Feline head DVI 20 from Dolni Vestonice I decorated with parallel incisions on the back of the head. The two non-symmetrical ears are visible at the bottom of the picture.

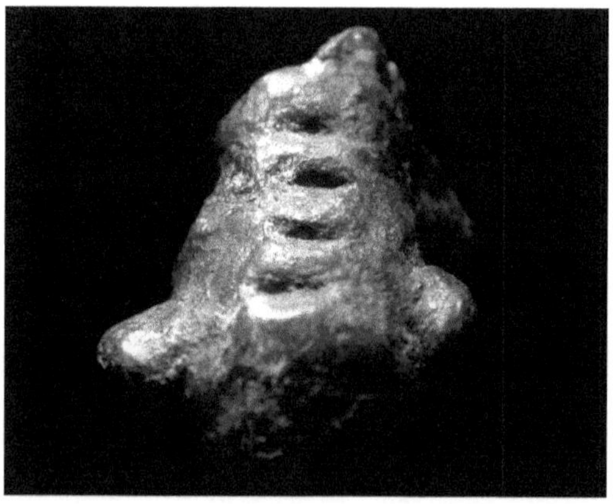

It appears, however, that in the art of Moravia, clay was used mostly for producing three-dimensional figures representing realistic themes: animals and humans, possibly a result of the great suitability of the material used for such representations. There seem to be an association between the material used, clay, and the naturalistic shapes given to the fashioned objects in the Pavlovské Hills sites. The rest of the art found at the site is mostly made of surface decoration or of rather abstract shapes. It is true some animal shapes in ivory are found, such as a well-known feline or a mammoth for example, but they remain much stylised. The argument that this would be because of the difficulty of working the ivory material does not hold much ground when one considers the elaborately fashioned objects illustrated in figure 4.6. At these sites, clay appears clearly as the preferential material used in association with the fashioning of realistic shapes.

The wider Magdalenian art context in the French Pyrénées

The Magdalenian art of the French Pyrénées shows some stylistic conventions that seem spread across the region during certain periods. Most of the dated cave art seems to belong to the Middle Magdalenian (Clottes 1995), which has been associated with the stylistic criteria determining Leroi-Gourhan's Style IV. They include the presence of claviforms; the flattened M-shape used to mark the belly of horses; shoulder lines drawn on the horses; a dorsal triangle and a double line under the mane for bison; the filling of the bodies with short lines; the naturalism of anatomical details, with horns and legs generally in perspective (Clottes 1989: 83).

The art done on clay, realistic or abstract, by *modelage*, engraving, finger-tracings or impressing show similar patterns and conventions than those used in other art techniques, such as painting and engraving, in the Magdalenian of the French Pyrénées.

The representational themes also appear to follow a similar pattern throughout all the art with two animal species dominating, the horse and the bison (Clottes 1989: 83). However there are some regional variations. In the east of the French Pyrénées, bison representations are more numerous than the horses, like in Tuc d'Audoubert for example (Bégouen and Clottes 1958); as one goes west, the horse theme starts to dominate bison representations: in Montespan in Haute-Garonne for example only horses are found in the art on clay (Trombe and Dubuc 1947).

The art on clay is often closely associated with other forms of parietal art and indeed sometimes appear to complete it like in the case of the floor engravings found at the foot of the painting on the walls of the Salon Noir in Niaux (Clottes 1989 (1999): 80). The art on clay of the French Pyrénées needs to be looked at in the wider Pyrenean Magdalenian art context as it cannot be disassociated from it.

At the same time, this art also bears specificities that are likely to contribute to certain research topics connected to Magdalenian cave art, notably the relationship between the internal space of the cave and the location of the art in it. As for portable art in clay, the only significant ensemble found is the one of the Bédeilhac clay plaquettes which can also be replaced in a wider context of decorated stone plaquettes found in the region in the Magdalenian, notably in the sites of Enlène (Bégouen and Clottes 1981a and 1981b) and Labastide (Simonnet *et al.* 1984).

Representations in clay and their locations in the caves

In recent cave art studies, some emphasis has been drawn on the need to look at the relation of parietal art with its immediate underground environment considering the way the cave itself might have been integrated in the art through the use of natural relief or through the topographical position of the representations (Lorblanchet 1995: 167). The use of clay in parietal art can also be looked at with this in mind, with taking the term 'parietal' as indicating art that was not meant to be transported rather than art strictly made on walls as the word originally implies.

Table 4.9 summarises the detail of the locations of the graphic units on clay inside the caves they were found in as was reviewed in the inventory tables for each site in Chapter 2. It becomes very clear that most of the parietal art made on clay was found on clay floors, in terminal galleries or terminal parts of side galleries and in relations with other forms of art.

The engraving of floors is very rare in the Upper Palaeolithic in general but is one of the particularities of the use of clay in the French Pyrénées (Vialou 1986).

Table 4.9: Position of the parietal art on clay inside the caves. The numbers correspond to the quantities of recognised graphic units. Sites with portable art were not included, except Bédeilhac for which only the parietal art is counted.

Site Name	Position in relation to topography			Position on cave map		Relation with other art forms			
	On wall	On floor	Other	Passage area	Terminal part	Apart	Near painting	Near engravings	Near portable art
Tuc d'Audoubert		13	2 (on boulder)		15	15			
Labouiche			1 (Clay bank)		1				1
Niaux		65		1	64		64		
Fontanet	1	10		10	1	9		2	
Bédeilhac	1	15			16			16	7
Massat	13			13				13	
Montespan	13	4	5 (Clay bank)	22		16	1	6	
Labastide			1 (on boulder)		1	1			
Bois du Cantet			1 (on boulder)		1	1			
Erbérua	12			6	6			12	
Oxocelhaya-Hariztoya		2	1 (on boulder)	1	2	1		2	
Etcheberri			2 (Clay bank)		2	2			
Total	40	109	13	53	109	45	65	51	8

Niaux is the site where this technique is better represented with eight preserved groups of representations depicting at least 65 graphic units, mostly found in the Salon Noir (Clottes 1995: 143). There, six distinct groups of engravings are found; two are outside the painted zone, four are next to the painted walls. The choices of the animal themes and even the relative proportions of represented animals are close to what is seen in the parietal art with similar species dominating: bison, hose, and ibex. Those animals are represented with the same conventions including details such as the lines on the shoulder (horse 87, between the panel 4 and 5) or the separation of the belly and the side with a line more or less parallel to the one of the belly (ibex 136 in the left side gallery). The angular or barbed signs that are so common on the painted animals in the Salon Noir are also found in six cases on the floor. The differences between the art on clay and the parietal art are limited to some animal themes that are not found in both and to the fact that many more signs seem to exist on the floor than on the wall (Clottes 1995: 124). It must be pointed out that in Niaux, the technique of engraving is hardly found on the walls at all: only one small figure exists on the walls of the Salon Noir, and five others are found in the rest of the cave. This is a very small number in comparison with the art found on the floors (Clottes 1995: 143).

Amongst the floor engravings, it can also be noted that at least two cases of engraved floor done on hard surfaces do exist in Upper Palaeolithic art: one is in the Colombier rock shelter in the Ardèche (Combier *et al.* 1984, figure 3); the other one was discovered in the cave of Trois-Frères in the Ariège in 1986 when it was decided to wash off the fine layer of mud covering some wall surfaces in the 'Salle du Faisan' to see if it could have been hiding engravings. Some engravings on the floor by these walls were also spotted then and the mud washed away. It came as a surprise to find the floor surface had been entirely engraved as this type of discovery has not been made previously in the French Pyrénées (Bégouen and Clottes 1987: 182).

As for the position of the art on clay within the cave itself, it appears that the majority of the art on clay is found in terminal parts of the caves, which are sometimes difficult to reach such as the Jauze-Mandement gallery in Bédeilhac. Clottes (1989: 80) notices that modelage in clay are often found in remote galleries in the region. For him, this might imply these figures had a very special and exceptional role in some ritual which might explain why they are so rare. But it is also true that the art on clay is very fragile and that the remaining record might be much biased by conservation factors as they would have had more chance of surviving in terminal galleries where circulation of air and people would have been limited.

Finally, it must be reminded that altogether, the art on clay is mostly found in sites in which other art techniques have also been used. It is true there are some discrepancies from site to site; if one looks at the inventory by quantity of graphic units found (Table 4.8), the associations with painted art seems to dominate. However, it is also true that in six out of the 12 sites found in this table the art on clay is found isolated from other forms of art. This might imply a slightly different meaning given to this form of art. But it is true the themes of representation found are the same for all the art techniques used in the Magdalenian in the French Pyrénées in general. Any variation seems rather site-specific.

Plaquettes in the Magdalenian in the French Pyrénées

The presence of a large number of clay plaquettes in Bédeilhac (over 123), although unique by the nature of the material used, echoes other finds of decorated stone plaquettes in the region, as well as in the same site. These plaquettes consist of thin stone slabs that have been decorated by engraving and also painting. Painted ones appear rarer but this is possibly due to conditions of conservation and recovery of the objects as recent re-examination of the few hundred engraved plaquettes from La Marche (Vienne) for example revealed minute traces of pigment still on them (Nicolas Mellars, personal communication July 2006).

In the Magdalenian of the French Pyrénées, Enlène is indeed one of the major sites known in Europe for the quantity of decorated stone plaquettes found there: over a thousand (Bégouen *et al.* 1984-1984). Plaquettes were also found in numbers in Labastide (Simmonet *et al* 1984: 529), in Isturitz (Davidson 1997 : 130) and in smaller quantities in Mas-d'Azil, Lortet, Gourdan (Clottes 1989: 70) or in Labouiche (Méroc 1959) in the Pyrenean region. Stone plaquettes were apparently also found in Bédeilhac although the age of the excavations/prospection in the site and the lack of information about the condition they were found in do not allow determining if they were indeed made and used together with the clay ones in similar contexts (Sauvet 2006).

A number of portable art objects in the French Pyrénées seem to have been intentionally broken by Magdalenian people (Bégouen *et al.* 1987; Clottes 1989: 68); stone plaquettes often appear to have had a similar treatment.

It can be pointed out that in the sites of Enlène and Isturitz stone plaquettes have been found in quantities and often broken and that in these two sites some sandstone sculptures were also found intentionally broken (Clottes 1989: 68). These two sites have also been pointed out as possible aggregation sites in the region, a fact that refers back to a possible association between decorated plaquettes and aggregation sites (Davidson 1997: 130).

The apparent intentional destruction of this category of portable objects has also been assumed to be a part of some ritual which would have formed the pattern for using the objects, and after which participants lost their interest. It is true that some of these stone have been found re-used as tools, in hearths or paving the habitation floors (Bégouen and Clottes 1979a, quoted in Clottes 1989: 71). As traces of fire are often found on them, it has even been implied that fire could be associated with the ritual and that the stone plaquettes could have been intentionally thrown in the fire in order to make them explode before scattering their remains (Omnes 1982 and Péquart 1963 both quoted in Clottes 1989: 71). Although Clottes (1989) invites our caution by pointing out that the association with fire could also result from a utilitarian re-use of the stone plaquettes for structuring hearths, the association of the ritual destruction of objects with fire reminds of what is seen in the sites of Moravia with fired clay objects, although the cultural, temporal and geographical contexts between the two regions are extremely different. The only possible common point of these sites could be their nature as aggregation sites, widely argued for the Moravian sites (Soffer 2000: 66) and assumed for sites such as Mas d'Azil (Alteirac and Vialou 1984: 394), Enlène, Isturitz and also possibly Bédeilhac (Clottes 1989; Conkey 1980: 611).

As for Bédeilhac, some of these plaquettes could indeed be fragmentary and intentionally broken, but there are no clear traces of any exposure to fire on the ones examined in the context of this research (see Appendix 1 in the catalogue). However, another aspect noticed on some stone plaquettes seems more widely found on the clay plaquettes: the idea of a relation between this type of art and the outside of the cave site possibly visible through traces of a means of suspension of the plaquettes to make them truly portable (Clottes 1989: 71). Clottes points out that in Enlène a small number of broken plaquettes have been reconstituted within the site but that most of the missing fragments were not found and could have been taken away from the cave. The recent find of an engraved plaquette in Labastide with notches on its edges and clear rubbing traces from suspension (Simmonet *et al.* 1984) could indicate a further use for the plaquette: it would have been transported and possibly worn. In Bédeilhac, as was seen in Chapter 2, 83% of the studied clay plaquettes show some form of notch, more or less marked, on the edges of the objects which could have been used for suspension (Figure 4.11). This appears to be a high percentage of the clay plaquettes; however more research is needed in order to support it. A comparison with the stone plaquettes from the same site would also be needed.

Figure 4.11: Fashioned clay plaquettes B32 from Bédeilhac with possible suspension notches indicated by the arrows (Péquart collection, Musée de Préhistoire de Menton).

The fact of taking away objects from the site of aggregation could be inferred to be part of the separation/dispersion rituals evoked for this type of gathering (Conkey 1980). It is true that many objects are still found in the sites: this could indicate this practice was possibly not widespread, unless simply the quantities of material made during various rituals was so large than much would remain behind anyway. And of course, other uses for a suspended stone or clay plaquette can by hypothesised such as its use as a bullroarer for example. The fact that some of the fired clay objects found in Dolni Vestonice and Pavlov in the Czech Republic have also been pierced, possibly for suspension, can also be correlated to the remarks on the clay plaquettes seen above.

Conclusion:

The general themes of the Magdalenian art of the Pyrénées are respected in the art on clay. These are mainly animal representations and signs. Bison and horse representations dominate the record, just like in the rest of the art of the period (Clottes 1989). In Tuc d'Audoubert (Bégouen and Clottes 1977), in Labouiche (Méroc 1959) and in Bédeilhac (not including the plaquettes) (Gailli *et al.* 1984), the representations in clay are only bison. In Montespan, however, the bison is absent completely, but the bestiary is more varied with eight horses, one bear, two felines, and one possible mammoth (Rivenq 1984). Some human elements are also found in the art on clay with the vulva representation from Bédeilhac (Gailli *et al.* 1984).

Another common point is the existence of isolated heads (as at Mas d'Azil) or of headless bodies (like the animal figurine from Enlène, the bear 'dummy' in Montespan) in the representations. This is a phenomenon that has been noticed in many other instances in Upper Palaeolithic art (Péquart 1960), even if an explanation is far from commonly agreed.

The existence of engraved plaquettes in clay in Bédeilhac can also be placed in a wider context by comparing them with the known engraved stone plaquettes bearing similar animal representations from sites like Enlène (Bégouen and Clottes 1981) or Labastide (Simonnet *et al.* 1984).

The use of the natural relief and forms of the cave and their inclusion in figurative representations is a widely found phenomenon in Upper Palaeolithic cave art, as at Bédeilhac where is has been done extensively (Sacco and Sauvet 1998). This can also be extended to the works on clay. For example, some natural cupules on the clay floor were used as the starting point of a bison engraving in Niaux (Malvesin-Fabre 1953).

But there are also some aspects of the art on clay from the French Pyrénées that are unique to the use of this material. These are the making of finger-tracings, the floor engravings and a tendency towards a monumental form of sculpture.

The wider art context in the Gravettian

The Gravettian regional culture of Moravia, the Pavlovian, indicates a real centre for creativity in the region with the decoration of weaponry and tools, personal ornament and representative art. Decorations at the sites of Dolni Vestonice and Pavlov is characterised by geometric ornamentation composed of rows of incisions, sometimes arranged in herringbone patterns and other complex patterns. Realistic figures are rare. The ceramics are also typical of the art at theses sites (Svoboda *et al.* 1996: 157, 160)

The question of whether the Pavlovian people actually hunted mammoths or rather scavenged them remains (Svoboda 1996: 29), but it is clear that in the Pavlovian sites the mammoth was an important animal, whose bones and tusks were much used (Oliva 2000a) and which was represented in the art as well.

The characteristics of the art in fired clay found in Moravia echoes some aspects of the wider regional art context in the Pavlovian/Gravettian and also possibly in the contemporaneous late Aurignacian. A few points related to this wider context will be discussed here: the question of the representation of 'dangerous' animals in the Early Upper Palaeolithic, the wider context for the representations of female figurines and the question of intentional destruction of portable art.

Representation of 'dangerous' animals

The themes found in the animal representations in fired clay are elements of connection with the wider art context. In Moravia, many of the recognisable animals made in fired clay can be described as 'dangerous' animals, a category in which carnivores (such as bears and felines) and very large herbivores (such as rhinoceros and mammoths) have been put (Clottes 1996b). In the case of the clay figurines, it can be noted that the most common animal representations in the record are, after the indeterminate animals, felines (17.7% of the recognisable representations), mammoths (8%), horses (6.5%), birds (4.8%), and then rhinoceros and reindeer with 3.% each (Table 3. 14 in Chapter 3). The felines, mammoths and rhinoceroses are recognised as 'dangerous' animals.

The dominance of 'dangerous' animals representations in the art appears to be a characteristic of the Early Upper Palaeolithic (Clottes 1996b; Hahn 1993), both in portable art and cave art. They have been identified as a characteristic feature of the portable art of the Aurignacian in Southwest Germany (Porr 2004). As for cave art, felines, rhinoceroses, bears and mammoths are also the most commonly found animal representations in the Chauvet cave in the Ardèche, a site precisely dated to the Aurignacian period around 32,000 BP (Clottes 1996b; 2003).

This tradition of themes of representation also seems to have survived at least into the beginning of the Gravettian period (Svoboda 1994): the fired clay figurines of Moravia are thought to date to the Evolved Pavlovian, around 27,000 BP to 25,000 BP (Svoboda *et al.* 1996); the cave art of La Grande Grotte in Arcy-sur-Cure (Yonne) is dominated by figures of mammoths, bears, rhinoceroses and one feline head is also found: it is associated to a Gravettian cultural level and the art is securely dated to around 28,000 BP (Baffier and Girard 1998).

This preference for 'dangerous animals' is also found in the early periods of the Upper Palaeolithic in the materials chosen to be transformed into pendants for personal ornamentation: Taborin (2004: 44) mentions that in the Aurignacian wolf, lion and bear teeth are pierced, together with fox and deer canines. This phenomenon is also still found in the Gravettian period, together with some new tendencies. This is not the case in the rest of the Upper Palaeolithic, when the dominant types of pierced teeth found are bovine incisive, fox canines and deer canines. The fact that very large teeth such as lion and cave bear canines are difficult to pierce seem to have been overrun by more important symbolic considerations (Taborin 2004: 46).

Later in the Upper Palaeolithic, the animal themes change: they are dominated by the horse and bison association (Leroi-Gourhan 1965) and the 'dangerous' animals become rare in the art.

In Moravia, it seems that a late Aurignacian culture coexisted with the apogee of the Pavlovian (Oliva 2000a: 227) in the same region. There would have been possible forms of intergroup competition with the Pavlovian culture and possibly contacts or exchanges (Svoboda 1996: 291). Some form of influence might be echoed in the themes of animal representation visible in the art, as well as in the style of some of the figures.

Among the ivory figurines of the Aurignacian in southern Germany, felines and mammoths are the most represented animals (Porr 2004: 262), just as in Dolni Vestonice and Pavlov. For the German figurines, it has been argued that these animals indicated symbols of power and aggression (Hahn 1993). More recently, Porr (2004) emphasised the very different social nature of the two animal species: mammoths probably lived in herds dominated by a mature female (by association with present day elephant behaviour); lions in prides with one dominant male and a group of females. Beside the possible social implications, Porr underlines that a male/female distinction might have been associated with each animal (lions as male and mammoths as female) thus connecting each statuette with a precise individual, possibly the person making the object and also the one wearing and/or using it as all the Swabian Jura statuettes appear worn by handling for very long periods of time (Porr 2004: 266). These figurines appear made in a perspective of long-term use.

The ceramic art of Moravia, on the other hand, is thought of as short-term (Svoboda 1996) as the pattern of its use would have implied the destruction of the figurines. Also in the Pavlovian, the actual faunal remains of mammoths are found in very large quantities in all the sites. It has been seen that the feline representations in fired clay are apparently the most numerous animal category. However, it has been seen in Chapter 3 that 27 of the loose animal body extremities inventoried could possibly also be interpreted as mammoth legs (22 from Pavlov I and 5 from Dolni Vestonice I). If this is accepted, the mammoth representations would then dominate the animal representation. But it is true that this animal representation would be the most important in Pavlov I but not in Dolni Vestonice I or II, thus underlining differences between the sites. These elements might point out to a special meaning given to the mammoth representations in this particular context, especially if one accepts the disputed possibility that the Pavlovian people were mammoth hunters (Oliva 2000a).

Some stylistic elements in the animal representations also show similarities between the arts of the Swabian Jura and Moravia. For example, a feline head from Vogelherd (Figure 4. 12) share similar details with the feline head PI 19 from Pavlov I (Figure 4.13). This manner of representation flattens the profile of the muzzle in what has been seen as a possible anthropomorphisation of the figure (Porr 2004). It is true this characteristic is also found in the head of the lion-man statue from Hohlenstein-Stadel in Germany (White 2003), in some lions from Chauvet cave (Clottes 2003) or in the Lion head from La Grande Grotte in Arcy-sur-Cure (Baffier and Girard 1998). PI 19 is the only feline head with these characteristics at the Pavlovské Hills sites as all the other show a more pointy profile.

Animal themes, other than the 'dangerous' ones, are also seen in the fired clay art; reindeer, horses and birds for example. These apparently correspond to actually hunted species at the difference of the felines, bears or rhinoceroses whose bones are only found in an anecdotic manner (Svoboda *et al.* 1996). But other much hunted species, especially hares and foxes, are apparently absent from the art record. As is consistently seen during the Upper Palaeolithic, major differences are seen in the art between the species represented and the ones actually hunted or eaten (Lorblanchet 1995) thus underlining that the motivation for the art is probably to be found away from subsistence concerns.

Figure 4.12: Feline head in ivory from Vogelherd, Southwest Germany. It measures about 3 cm long (From Lorblanchet 1999).

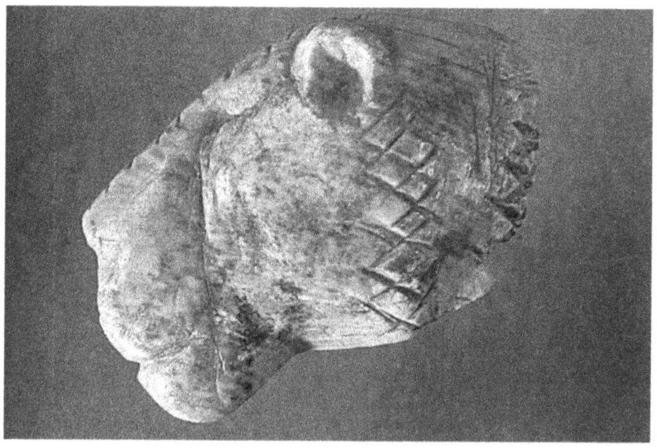

Figure 4.13: Feline head PI 19 in fired clay from Pavlov I.

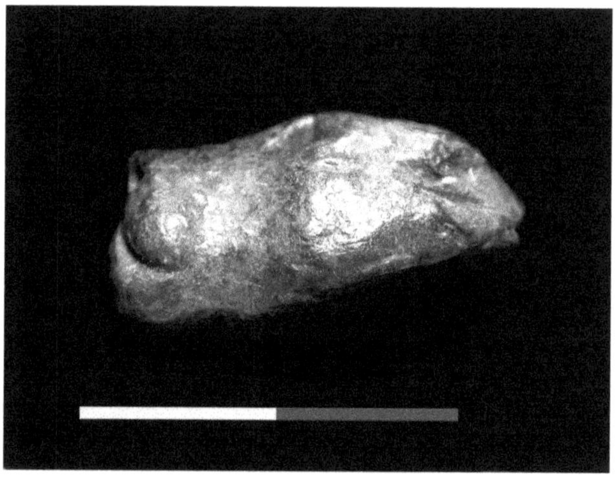

The case of the fired clay 'Venuses'

The human female figurines made of ceramic echo the widespread phenomenon of female representations, also called 'Venuses', seen in the Gravettian period across Europe, from Western Europe to Central Russia. There are regional stylistic differences in these representations, but also certain common points; the women are usually represented naked, with large breasts, buttocks and bellies, often pregnant. The extremities of the body (legs and heads) tend to be neglected, being very little detailed or simply hardly shaped at all. They sometimes wear ornaments (White 2003; Bahn and Vertut 1997; Praslov 1985; Bisson and Bolduc 1994; McDermott 1996).

These figurines have often been associated with an idea of fertility and reproduction, whether to use in some form of ritual, or again as possible amulets connected to women and pregnancy (White 2003). They are often found, especially in Central and Eastern Europe, to be made with care and much detailing out of ivory or soft stone. Some are even pierced for suspension. Effort and time seems invested in making them and they appear treated specially; for example in Kostenki I in Russia female figurines, whole and fragmentary, have been found carefully placed inside pits found in the habitation surfaces (White 2003; Abramova 1995).

In Dolni Vestonice I and II and Pavlov I, female figurines have been found made out of fired clay. If the general aspect of their representation fits into the wider Gravettian art context, the context they were found in is rather different. They were discovered with the rest of the fragmentary art in fired clay which appears closely associated with fire as most of the finds come from hearths or their close vicinity (Jarošova 1997). They did not seem to have been given a different treatment than the rest of the clay art, and indeed appear to have also been made to explode in fire. This implies a short-term use of the fired clay figurines in contrast with what is mostly seen in the Gravettian. It is also true that there might have been a difference of treatment between the fragmentary female figurines found and the Black Venus figure which is nearly whole, even if the fact that it has been used as a model for figurines that were themselves damaged in fire would rather indicate a common manner of treatment of all fired clay figurines.

The question of intentional destruction of portable art objects

The idea of intentional breakage of objects is an element bearing strong symbolic and social implications that may possibly be associated with clay uses. The act of bursting ceramics in the firing process can be linked to this process in the Czech Republic sites. It has been seen above that intentional destruction of art objects was also seen in the Pyrénées with the case of decorated plaquettes or some broken animal figures.

In Moravia, the theory of hunting magic to explain the art has also been applied to fired clay objects. Thus Absolon (1949) interpreted in this perspective the holes visible in some figures (Figure 4.14); they would have been made in ritual destruction of the animal. But it is true that some of these holes might simply have been made in order to be able to suspend the objects, possibly in order to wear them as ornaments.

When Svoboda (1994) looks at the art of the Moravian sites, he divides the portable art into long-term and short-term use objects, with the idea that some manufacturing are time-consuming and will serve a purpose for a long time (the art objects in ivory and harder materials), when for others, such as the ceramics, it seems that the moment of the making itself is the important part. For the short-term objects, he argues they would have been part of some form of ritual.

Svoboda (1994) associates the Gravettian ceramics with the short-term art and compares them to the later engraved Magdalenian plaquettes, which are usually found broken. For these objects, the moment of their creation and destruction seems to be the important aspect. Svoboda considers this sort of art constitute the non-perishable relicts of a more complex ritual creation. He does consider the possibility that it could be hunting magic rituals, but points out that there is a clear difference between the bestiary chosen for representation and the actual evidence from faunal remains from the sites. This distinction seems true in general in Upper Palaeolithic art, and has been discussed at length in the case of cave art (Lorblanchet 1995; Bahn and Vertut 1997; White 2003).

Figure 4.14: Feline head DVI 23 from Dolni Vestonice I: it has been pierced with two holes (a), but only one perforates the reverse of the object (b).

a)

b)

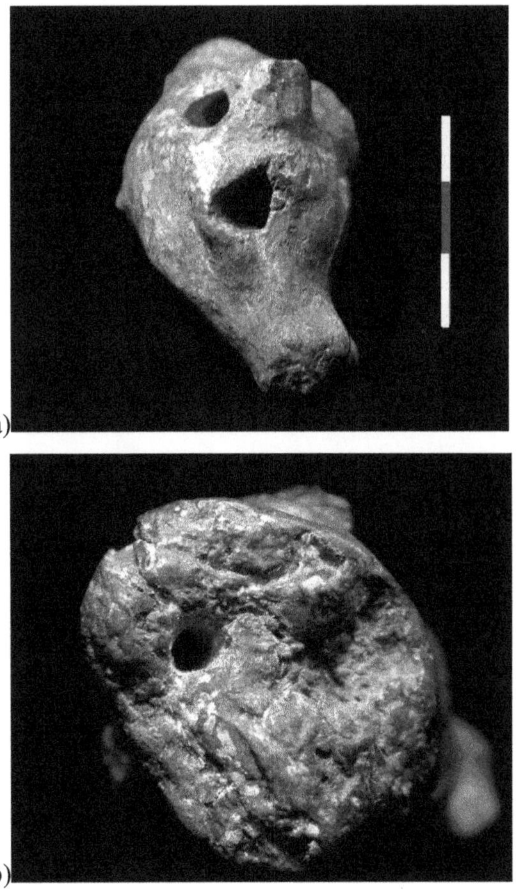

The question of intentional destruction of portable art is one that points out some incredible resemblance between the fired clay figurine from Dolni Vestonice and Pavlov and the portable art from Kostenki I in Russia (Abramova 1995). Broken fragments of animals and human figurines are extremely numerous there, with many isolated parts such as heads or feet found just as in Moravia. This connected with the fact that some fired clay figurines[19] were also made at this site (Absolon 1949; Soffer *et al.* 1993; Bruce Bradley, personal communication January 2003; Vandiver *et al.* 1990; Valoch 1996) would indicate possible

[19] It must also be added that G. Bosinski (1990) published many portable art objects from Kostenki I, 1-Poljakov and from Kostenki XIII-Kelsiev as being made of fired clay. After more enquiries, it appears that a mistake was made in the translation of the Russian word for fired clay, and that the various objects are rather made of calcareous marl as is found in other publications (Gerhard Bosinski personal communication May 2004). Some fragments of fashioned fired clay have however been found in Kostenki 1 (Soffer *et al.* 1993; Bruce Bradley, personal communication January 2003; Vandiver *et al.* 1990; Valoch 1996).

connections with the Pavlovian of Central Europe, which remain to be investigated more closely, especially with regard to the use of clay.

Conclusion

The art made of fired clay in the Pavlovian appears clearly connected to some elements of the wider Aurignacian and Gravettian art contexts. However, it can also be said that because of the unique pattern of use of the fired clay found at the site, one that implies a systematic destruction through exploding in fire in a form of pyrotechnics display, some specificities related to the sites of Dolni Vestonice I and II and Pavlov I are to be expected.

5 The Social context for clay working in the Upper Palaeolithic

The potential of *chaîne opératoire* studies to inform us on social aspects of past societies has been underlined by many authors (Audouze 2002; Dobres 1996, 1999, 2000; Dobres and Hoffman 1994; Gamble 1999; Karlin and Julien 1994; Lemonnier 1990; Leroi-Gourhan 1943; Pigeot 1987; Pélégrin 1990, 1991; Schlanger 1994, 2004; Sinclair 1995 among others).

Leroi-Gourhan was the first person to point out that the *chaîne opératoire* could fuse social and technical activities: some sequences of the technical process are fixed and depend on material constraints; but others are flexible and result from choices made among the many possible ways to do one thing. It has been shown that these choices stem from the social and cultural contexts of the individuals carrying out the action (Gamble 1999: 82; Pigeot 1987). Thus techniques can be seen as primarily social productions (Gamble 1999: 29 quoting Lemonnier 1993; Schlanger 1990: 23) and the study of techniques through the *chaîne opératoire* approach can help identifying strategic and tactical choices at every stage of the manufacturing process (Audouze 2002: 287). This approach offers the possibility to examine the 'interplay between fixed and flexible' (Schlanger 1994) and thus to elaborate on some issues such as intentionality and the transmission of knowledge.

It has been seen in the present research that in the case of the working of clay, the *chaîne opératoire* approach indicated some elements of transmission of knowledge as seen through the organisation of the technological activities into a constellation of knowledge. Some conclusions about the social contexts concerned can also be drawn and will be looked at in this part of the chapter. Firstly, this will be done through looking at the connections between the use of clay and the places it is found in. This leads to considering elements of intra- and inter-site variability in the two studied regions. Secondly, some links visible between sites of aggregation and the use of clay can be drawn, again in both the studied regions, as *chaîne opératoire* studies also make it possible to look at techniques used in the gatherings and social occasions in which actors are engaged in the performance of society (Gamble 1999: 82).

The use of clay in the Upper Palaeolithic and places

It has been seen that in all the sites in which clay was used in the Magdalenian of the French Pyrénées or in the Pavlovian in Moravia, a striking point is the local origin of the raw material used as if the use of clay in the Upper Palaeolithic was closely associated to certain places and to the specific forms of symbolic expression taking place there. Clay was used near the source where it was found. There is no evidence for it being transported over great distance or exchanged as is seen later in some Neolithic contexts in which the clay substance was sometimes given symbolic meaning in itself, as a symbol for flesh for example (Chris Fowler, personal communication March 2004).

Looking at the variability visible inside the sites themselves (intra-site) and between the sites of the region (inter-site) considered with regard to the use of clay will be the mean to looks at the relation between the use of clay and places seen in the Upper Palaeolithic, first in the French Pyrénées, then in Central Europe.

Regional variability in the French Pyrénées

As seen previously, all the sites considered in the French Pyrénées are caves. Gamble (1999: 75) reminds us that the cave is an example of place in the Upper Palaeolithic that although not symbolically created as in the case of architecture will nonetheless structure behaviour by the constraints of space it imposes. He adds that it allowed privacy away from constant communal monitoring or sharing as well as defined an arena for social performance in the case of cave art.

In the French Pyrénées, the art on clay appears consistently made in the close vicinity of a clay source naturally available in the cave sites. The clay is found as a plastic material, used for its plastic qualities, and abandoned after fashioning still as a plastic material: this is seen in the majority of the studied sites. There is no transformation of the nature of the material through human action in these cases. Through this quality of plasticity, clay is useful in looking at parietal art as a process rather than as cultural output, as advocated by Conkey (1987: 67), as the material keeps easily the marks indicating the steps of its working.

The Magdalenian is traditionally seen as a period of cultural stability, with much uniformity and consistency especially in the art, but with important regional variations (Dobres 2000: 189; Sieveking 1979: 160). As Dobres points out (2000: 189), research has mainly concentrated on looking at variability between the main regions in which the Magdalenian culture is found, but very little on examining intra-regional variability of the Magdalenian. Clay studies provide here an opportunity to look at variability within the region, notably through the repartition of the clay techniques found in the caves.

It has been seen earlier in this chapter that the technological variations between sites and even within sites was great in the French Pyrénées. The techniques for working clay have been described as situated applications of knowledge as they differ even when the same technique is considered across the sites of the region.

A common knowledge seems to be shared, however, among the Magdalenian groups in the French Pyrénées about the working of clay; the knowledge that in cave sites, clay could be found and used in the making of art and a notion of what could be done with it. The Magdalenian people of the

French Pyrénées had integrated the knowledge that clay was suitable and acceptable to use in the making of art. This was not the case in other regions in the Magdalenian period.

The organisation of the Magdalenian groups considered is the likely context in which some of the knowledge and know-how related to clay was transmitted. In most of the cave sites studied, it seems likely the art was made by a small group of individuals as the restricted space would not allow the presence of a number of people. Footprints, when they have been preserved seem to comfort this view (Clottes 1995). The fact that these small groups could be familial units can be considered, again because children and teenager footprints have been found, indicating their presence even in the deep caves. The idea of possible initiation rites of adolescents has often been considered regarding this presence (Bégouen and Breuil 1952). Whether this explanation is accepted or not, there is certainly a form of 'initiation' the children present in the caves were exposed to: learning, in passive or active manner, about the techniques used in the making of the art and about the suitable conditions and contexts for this making.

It has not been possible to determine up to date the age and gender of the people making Upper Palaeolithic cave art; it is normally assumed it is made by adults. In cases in which the art made shows high technical skills combined with talent in the representation depicted, it is indeed likely that it would have been made by an individual with years of practice, most likely an adult. But it is also true that a consequent part of Upper Palaeolithic cave art can be described as clumsy or badly drawn; could it be assumed it was done by the least experienced, the younger members of the groups in their learning process?

As an illustration for the working of clay, the group of four bison found in the terminal gallery of Tuc d'Audoubert can be mentioned again (Bégouen *et al.* 1977) (figure 4.5). It has been seen that the two larger bison of the group showed great skill and talent in their making, and that the small bison sculpture and the partly engraved/partly modelled bison on the floor nearby showed lower levels of skills and were very different from each other. It could be assumed the large bison were the work of (a) skilled adult(s), and the two others were made by two other individuals, possibly children or adolescents, while copying the adults and experimenting on the same theme. The fact that children or young teenagers' footprints were found on the clay floor of the Chamber of the Heels only a few metres away from the group of bison (Bégouen and Breuil 1958) makes this hypothesis plausible.

The possibility of an informal transmission of knowledge, possibly through observation, related to the working of clay in the context of each cave site can be raised here. This would have occurred among a small group of individuals and would explain the great variability in the technical processes seen between the sites in the French Pyrénées; each small group, and possibly the individuals inside the group, would apply the knowledge and know-how of the working of clay they possessed. If these were acquired through learning by observation and personal experience of interpreting this observation, this would explain the technological variability visible.

But it is also true that the tradition of using clay is a regional phenomenon in the French Pyrénées. It seems that at least the notion that clay was a suitable material to use in art and the overall manner of using it in the making of this art were common knowledge shared by the Magdalenian people in the region. This statement implies that some form of contacts and exchanges existed between the small groups making most of the cave art, which allowed sharing of this knowledge. It is likely the existing system of contacts and exchanges was a complex one that allowed the strong unity seen today in the Pyrenean Magdalenian as a whole (Clottes 1989: 89). One context in which gatherings are known to have existed in the region is the context of aggregation. The aggregation site is an example of place in which knowledge could be shared and transmitted. The links between clay and aggregation sites is looked at below.

Inter and intra-site variability in Central Europe

It has been seen that in Central Europe, there appear to be a great uniformity of the technical processes behind the making of the fired clay figurines in the Czech Republic. However, some elements do indicate variability, marking the differences between the sites studied. They are mostly the sizes of the body extremities and themes of representation visible as well as the location of the ceramics within the sites of Dolni Vestonice I and Pavlov I.

The study of the recognised body extremities from the sites of Dolni Vestonice I and Pavlov I was carried out in Chapter 3. It allowed underlining striking differences between the two neighbouring sites in the size range of the pieces considered.

In Dolni Vestonice I, the body extremities considered are overall larger than in Pavlov I (figure 3.42). It was assessed, by comparison with the size range of whole figurines normally seen in the art of the region (Table 3.30), that it was likely most of these body extremities were fashioned as separate pieces with no intention of attaching them to an animal body, as the resulting whole animal would have much larger than what is seen in the art context.

In Pavlov I, the majority of the record is under 2.5 cm long (87%, see figure 3.42). This is much smaller than in Dolni Vestonice I.

There seem to be more variability in size in the record of body extremities in Pavlov I than in Dolni Vestonice I. The body extremities inventoried are also much more numerous in Pavlov I (117) than in Dolni Vestonice I (73 pieces).

It has been seen that in the themes of representation in the clay record in Moravia, the most commonly found are the female figurine, the feline heads and the mammoth figures.

Again, different situations were found between Dolni Vestonice I and II and Pavlov I when looking at the female representations. The Black Venus (DVI1) was apparently used as a model or following itself a model for making female figurines in Dolni Vestonice I and II, with a degree of resemblance that would indicates a close connection between the two sites. This model is absent from Pavlov I altogether. In Dolni Vestonice I, only two types of female figurines are found. In Pavlov I, five different types of female figurines are described. Again this seems to point towards more variability in Pavlov I.

The representations of feline heads in ceramic are found in fairly similar quantities in Dolni Vestonice I and Pavlov I, but differences exist when looking at the types of muzzle encountered (table 4.6); two different muzzle types are found in Dolni Vestonice I, three in Pavlov I. However, it can also be pointed out that two heads, stylistically very similar, have been found, one in each site, indicating some form of contact between them. Altogether, the feline representations are possibly the theme of representation showing the most visual uniformity across the Dolni Vestonice-Pavlov locale. This might be indicative of a special symbolic meaning attached to these representations.

When looking at the mammoth representations, much variability is again noticed between the sites and within them. Only one mammoth figurine is found in Dolni Vestonice I but there are five in Pavlov I. In Pavlov I, there is no uniformity in style visible among the five figurines. This seems to indicate different makers for each of them. It has also been seen that part of the record of body extremities could possibly be interpreted as mammoth legs; if this is accepted, 27 more mammoth pieces could be added to the Pavlov I record, thus making the mammoth representations the most commonly found by far at this site. This would point out to a very special role of this animal in the symbolic world associated to art in Pavlov I, something that is not found in Dolni Vestonice I or II.

Variability is one of the elements underlined by Conkey (1980) which are expected from aggregation sites: could there be a difference in nature or in the type of aggregation between the two sites of Dolni Vestonice I and Pavlov I visible through the analysis of the clay record? The greater variability visible in the detail of the fired clay figurines in Pavlov I certainly underlines a degree of difference with Dolni Vestonice I.

The repartition of the ceramics within the sites of Dolni Vestonice I and Pavlov I also underlines some elements of variability. The possibilities of a study of the spatial repartition of the Moravian ceramics are unfortunately limited as the objects at the sites were only recorded by 2 by 2 metre squares and only for the most recent excavations. This record can also be completed by some notes or sketches done for the 'best' pieces (Svoboda 1994; Jarošova 1997: 169; Oliva 2000). The ceramics coming from the areas of Pavlov I excavated in 1957 were plotted as a tri-dimensional map in spite of these imprecision (Figure 4.15) This allowed seeing that both unshaped pellets and shaped pieces concentrated in a striking manner in square 9, which correspond to the dwelling XII. Little bone, antler and ivory industry were found in this same square, but there is a clear concentration of art objects (Jarošova 1997). 1,280 ceramic fragments out of 1,772 from the 1957 excavation come from square 9. A high concentration of stone tools, stone debris and colouring material were found in the same square (Klima 1997a: 42). In this dwelling, like in most cases at the Pavlovské Hills sites, the ceramics were found directly associated with a hearth.

Conclusion

In the French Pyrénées, a wide technological variation is visible: each site seems to show a situated application of knowledge that is probably connected to the individuals carrying the action themselves. But there is some uniformity at the scale of the site and the region through shared style and conventions, even if much variations is also seen in the representation of details for example.

In Central Europe, the manufacture process is much standardised but some individualisation is recognisable in variations in style or in the themes of representations between the sites at the locale, and also within them. The wider regional art context shows some common points with the art on clay, notably through the themes of representation, but they are few: some representations are unique to the art on clay in the context of the Pavlovian culture.

Figure 4.15: diagram of spatial distribution of the ceramics around square 9 from Pavlov I, 1957 excavation, indicating their concentration (from Jarošova 1997: figure 12).

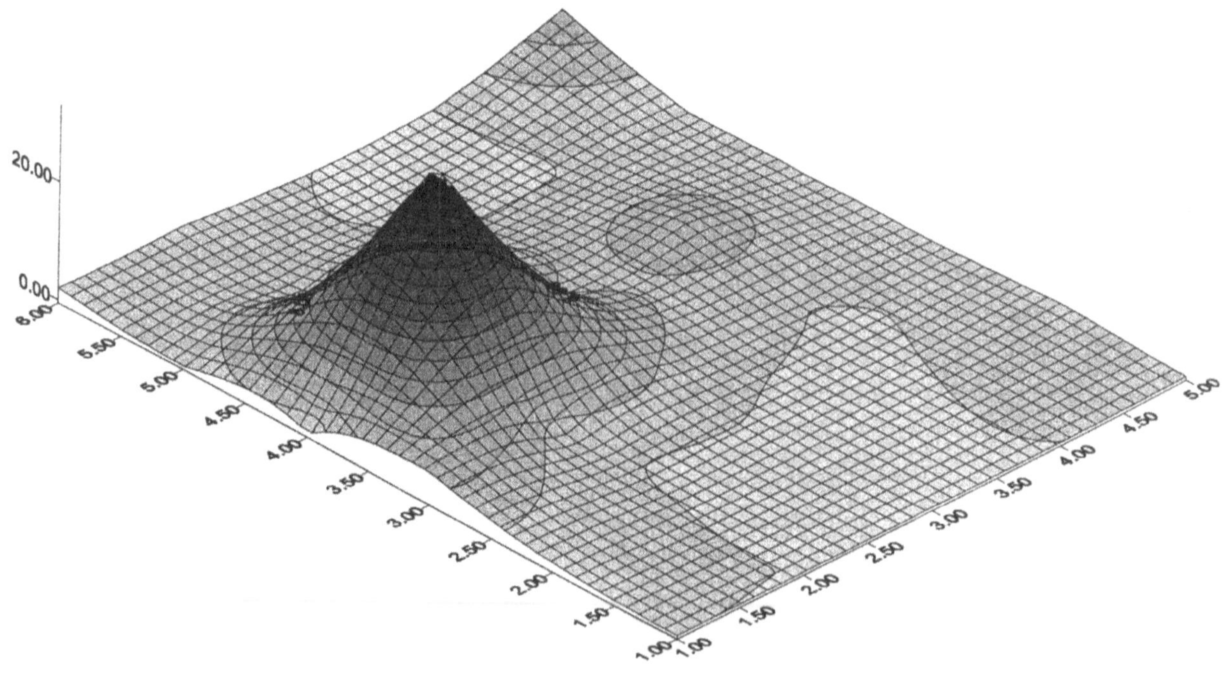

The use of clay and Aggregation

In both the studied region it appears that many of the sites considered could be described as aggregation sites. The large sites of Moravia are usually thought to be such places (Soffer 2000: 59; Gamble 1999; Svoboda *et al.* 1996). In the French Pyrénées, Mas d'Azil has long been thought to be a super-site where seasonal gatherings took place. The enormity of the site for the archaeological material it contains, for the size of the cave, the ease to find it in the landscape, to access it and the numerous 'exotic' objects found there all go in the sense of an aggregation site (Sieveking 1976; Clottes 1989: 46). For other clay sites, such as Enlène or Bédeilhac, it has been argued they could have been aggregation sites, but not all researchers agree on this point (Clottes 1989: 46; Bahn 1982: 263). A number of the other sites in which clay uses are found are in close geographical association with a possible aggregations site: Tuc d'Audoubert is very near Enlène (Bégouen and Clottes 1984 a, 1984b); Niaux is thought to have had close links with La Vache (Bahn 1982; Clottes and Delporte 2003); Oxocelhaya-Hariztoya and Erbérua are part of the same karst network as Isturitz (Laplace and Larribau 1984: 283; Larribau and Prudhomme 1989: 480).

After reminding the elements that are expected to be found in an aggregation site, elements related to the use of clay which can contribute to recognising sites as aggregation will be presented first for the French Pyrénées and second for Central Europe. These are notably the use of clay in rituals connected with the aggregation, possibly with the breakage and taking away the objects during the detaching rituals at the end of the aggregation; and the idea that technical innovation can often be related to the context of aggregation and could be visible in the use of clay.

What makes an aggregation site

Conkey (1980: 612) reminds us that 'an aggregation site among hunter-gatherers is a place in which affiliated groups and individuals come together'. This corresponds to ethnographic observations of modern hunter-gatherers, which are often seen to follow an annual cycle characterised by periods of concentration of the human groups during aggregation, and of dispersion. This model is not universal, but it is thought it existed among the hunter-gatherers of Upper Palaeolithic Europe (Conkey 1980: 612).

Aggregation can be motivated by a combination of various factors which can be ecological, as well as social, economic and ritual. This can result in a great variability between the different types of aggregation sites (Conkey 1980: 609).

Conkey (1980: 612) determined the following indicators as the elements to investigate in archaeological excavation in order to recognise aggregation sites:

- Indication of the gathering of a large group with a large spatial extent of the occupation.
- Elements indicating a seasonal occupation, possibly repeated.
- The structuring of activities carried out in the context of the aggregation.
- The maintenance of relevant site features.
- A greater total range of activities that is seen in other sites, with much diversity (from ethnographic reference such as the !Kung San study by Yellen 1977).
- At least some activities are different from those that took place at smaller (dispersion) sites and possibly unique to the aggregation site.
- Traces of ecological factors that may have prompted or contributed to the aggregation.
- A mixture of regional personnel.

To this, it can be added that rituals are normally an important part of the aggregation/dispersion pattern as they help stabilising the large group of people normally gathered (Conkey 1980: 610). As art is often seen as the strongest of all archaeological evidence of increasing ritual behaviour, it is to be expected that a quantity of art is found in aggregations sites (Conkey 1985: 300). As human groups with different identities gather and are the makers of this art, it is also to be expected that a wide stylistic variety of the art will be found, as well as some form of art unique to the context of the aggregation. These systems of visual representation could be taken as ritual communication which can have an important part in constituting the social order of the aggregation context. (Conkey 1985: 315). Gamble also argues for the importance of leaving quantities of quality residues from the gathering behind when leaving it in order to mark the place of aggregation itself as an enduring locale and to leave traces of the interactions at the gathering (Gamble 1999: 71).

It has also been pointed out that it is in the context of aggregation that some innovations could have taken place, as innovations are thought to happen in order to meet some new socially determined need. The context of the aggregation itself would be a creator of a new social stress, thus creating problems that could be solved by innovation (Soffer 2000: 64, 65). The notion that borrowing from a neighbouring group is actually very similar to inventing (Leroi-Gourhan 1943: 334) can also be emphasised in the context of aggregation since it calls for recombining already existing elements and creating new associations of technical elements (Audouze 2002: 285). Thus innovation can also be drawn from existing elements.

Aggregation sites in the French Pyrénées

Among the sites with clay uses in this region, Mas d'Azil, Enlène and Bédeilhac have been recognised as potential aggregation sites.

There is little doubt that Mas d'Azil was an important aggregation site in the region as seen by the huge variety and quantity of archaeological material found at the site as well by its potential for holding very large human groups (Sieveking 1976; Clottes 1989: 46). It was possible to estimate through faunal analysis that the cave was occupied nearly permanently in the Magdalenian, but more intensely in autumn and winter (Clottes 1989: 51). The range of activities possibly related to ritual behaviour seen at the site is found in the abundance and variety of the art techniques used and the art works produced. Parietal art was found in some deeper galleries. Alteirac and Vialou (1984: 394) note that 'imagination seems to play an important part in the organisation of the parietal art in Mas d'Azil', as some rare themes (fish, phallus) are found beside the traditional animal themes of representation such as bison, horses and cervid the art (Alteirac and Vialou 1984: 394). The portable art is very rich and varied and it had been seen in Chapter 2 that it included portable objects made in clay and fired clay (Péquart 1960). Such objects are very rare in the region and the making of fired clay objects can even be described as an innovation. It could be argued this goes in the sense of the site being an aggregation site as it shows an increased degree of technological diversity. The objects concerned are art objects, which are possibly made in the ritual context of the aggregation. It must also be noted that the other clay and fired clay objects that have been found in Mas d'Azil (see Appendix 1) are ornaments for three of them, or could have been pigment blocs. They also carry a strong potential symbolic charge as social elements characterising individuals (Taborin 2004).

Enlène is sometimes also considered as a super-site by the abundance of its portable art and its connections with two major cave art sanctuaries, Trois-Frères and Tuc d'Audoubert (Clottes 1989: 53). The habitation site seems to have been reoccupied over a long period. The common points between the Middle Magdalenian layers of Enlène and of Mas d'Azil are so many that they go beyond simple convergences due to a common civilisation; exactly similar objects are found, for example fossil shark teeth used as ornament, similar lignite beads, thin bone discs with numerous holes on their edge and reindeer antler sculptures with incrustations for the eyes and much more (Clottes 1989: 53). It is true that only a 13 km distance as the crow flies separates the two sites.

Enlène is also the only other site with clay uses studied in the French Pyrénées in which portable objects made of clay and fired clay have been found. The clay plaquettes from Bédeilhac are not included here as they are clearly part of a different phenomenon unique to the site they are found in. The use of clay in Enlène can be seen to show an increased technological complexity, which can relate to the context of an aggregation site. The proximity of the site with Mas d'Azil could explain the similarities in the archaeological record; however, many similarities are also found between Enlène, Mas d'Azil and the super-site of Isturitz in the western Pyrénées, such as a similar intentionally broken bison sculpture made of sandstone, perforated sandstone elements, identical contours découpés, osseous artefacts and ornaments (Bégouen *et al.* 1987; Bégouen 1921). Isturitz is the other large site, together with Mas d'Azil, in the Pyrénées region which is widely accepted as being an important aggregation site at all periods of the Upper Palaeolithic (Clottes 1989: 53). These numerous similarities could indicate that these three sites had roles in the network of exchanges visible in the region notably through the visual uniformity of the Magdalenian art.

As for Bédeilhac, this site is not usually found mentioned as a recognised aggregation site in the region although some of its characteristics, notably through the use of clay found there, will allow arguing that it could have been one.

It is true that the size of the cave would be very suitable for large gatherings as it is enormous, on a similar scale than Mas d'Azil. A specificity of the art of Bédeilhac is that, except for sculpture, all the techniques used in the Magdalenian Franco-Cantabrian cave art are found there, including many clay techniques: *modelage*, engraving on silt, engraving on walls, monochrome and polychrome painting, drawing with clay on the walls as well as combinations of these techniques. The site also yielded very important portable art objects, such as an example of a spear thrower with 'foal and birds', a theme of representation found copied in a number of sites in the Pyrénées (Gailli et al, 1984: 375).

Bédeilhac was unfortunately excavated very early, by many different people, mostly without proper recording of the finds and without publications. The collections obtained from the excavations have also been scattered in various museums on both sides of the Atlantic (Bahn 1984). As a result of these problems, the site is quite badly known, and most of the objects kept in various collections have not yet been published and are unknown. Sauvet (2006) mentions as many as about 900 portable art objects existing for this site. His recent research has highlighted the potential still born by the study of these objects. The collections of clay plaquettes found in Bédeilhac are part of this new potential.

The clay plaquettes are absolutely unique to the site of Bédeilhac up to date. They are a clear example of a technical innovation at this site. Other plaquettes are also found there, made of at least eight different types of sandstone and limestone (Sauvet 2006), most of which were brought from outside the cave. The decoration of plaquettes by engravings and/or painting and their intentional destruction has been indicated as a possible ritual associated with this type of objects. It has also been assumed some of these plaquettes sites could be aggregation sites, the ritual of destruction being attached to the nature of the site (Davidson 1997: 130). Enlène, for example is the most important plaquettes site found in the whole of the French Pyrénées (Clottes 1989): this is another element indicating its potential role as aggregation site. The presence of stone plaquettes at the site indicates that this type of object was desired. The idea of transposing the making of plaquettes to the clay material can be seen as an adaptation of the knowledge people had to a new material in a form of technological innovation.

The fact that other sites with clay uses are found in fairly close geographical association with sites recognised as potential aggregation sites, such as Tuc d'Audoubert with Enlène (Bégouen and Clottes 1984a, 1984b), or Oxocelhaya-Hariztoya and Erbérua with Isturitz (Laplace and Larribau 1984: 283; Larribau and Prudhomme 1989: 480) might simply be acknowledged as indicating that ritual activity was an important part of the aggregation, and that making cave art was one form of this ritual.

Aggregation sites in central Europe

All the sites of the Pavlovské Hills in Moravia present many characteristic which have encouraged recognising them as sites of aggregation. Many of the ritual activity recognised in the Pavlovian are associated with ritual to be performed in aggregation such as burials and the ceramic production (Svoboda 1996: 197). Svoboda argues these rituals would have had an important social function in supporting the efficiency of the Pavlovian adaptation (Svoboda 1996: 295). Indeed it appears that most of the Pavlovian sites identified in the region present characteristics of aggregation sites (Soffer 2000: 66). Soffer suggests there was an increase in seasonal mobility of small groups in the Pavlovian which would have put a strain on mating, information exchanges and risk sharing networks, and that cold weather aggregations were their social solution for this strain (Soffer 2000: 66). The context of aggregation however, would have created other problems such as the intensification of the food procuring efforts; solving these problems would have been done by technological innovations which are numerous in the Moravian sites. They are for example the common use of microlithic tools or net hunting (Soffer 2000: 67).

As for clay in the sites of Dolni Vestonice and Pavlov, the ritual use of this material would have taken the form of a performance associated with fire in which the fashioned figurines were made to explode. Soffer argues that this use of clay in a pyrotechnology is a characteristic of the aggregation at these sites because of the uniqueness of this technology (Soffer 2000: 67). The use of clay at the site of the Pavlovské Hills could be seen as much standardised as it could correspond to a ritualised behaviour with an important social effect. Gamble (1999: 413) connects the ritual of exploding ceramic to the attaching and detaching rituals that would have taken place at the beginning and at the end of the aggregation period; this event would help create the place of the gathering. In the creation of this ritual of exploding ceramics, it is clear a technological innovation took place. Soffer (2000: 65) argues that the reason for the number of innovations seen at the Moravian sites could be a need to solve problems created to a social stress possibly due to the changing group size.

It must be pointed out that this mode of using clay was possibly also a characteristic of the other Pavlovian sites recognised in the region as it has been seen in Chapter 3 that fragments of fired clay were found in all of them, although in much smaller numbers than in Dolni Vestonice and Pavlov.

Conclusion

The context of aggregation recognised for a number of the inventoried sites with clay uses can also be considered as a context for learning as there is no doubt the exchanges taking place between individuals present would involve knowledge. This is indeed possible to see when one considers the visible technologies unique to some sites, as these technologies would have been transmitted during the gathering of the aggregation. The fact that these are not found in other sites would seem to indicate that they were connected to the gathering itself and possibly the rituals happening there.

At other levels, the aggregation site can also be seen as an increased opportunity to learn, possibly through the observation of individuals from other groups in action. In the French Pyrénées, it could be argued that the great visual uniformity seen in the art of the Magdalenian period could result from knowledge acquired during this type of gathering.

Conclusion: the use of clay in Palaeolithic societies

The way clay was used in the Upper Palaeolithic has a great potential for informing us about the individuals using the material thanks to its very special properties. Dobres (2000: 210) reminds us that the chaîne opératoire besides being an analytic methodology is also 'a conceptual framework for linking the static material traces of ancient techniques, gesture, knowledge, values, judgment and skills to the socially constituted technicians which practice them.' It can inform about social relations of production, meaningful interactions, and embodied being-in-the-world form of existence (Dobres 2000: 209). And indeed, through the use of clay, it has been possible to underline two very different Upper Palaeolithic social contexts in which this material was used in completely different manners.

In the French Pyrénées, it can be argued that the transmission of knowledge seen of the technical processes of working clay was not subjected to an organised form of learning. The transmission possibly happened by observation, which would imply the clay was used among a group of people. Because we are mostly considering cave art here, this also has implications for indicating that cave art was not done in secrecy away from the group, but possibly by allowing the presence of all individuals during its making. This correlates indications such as children footprints found in the caves (Clottes 1995). The elaboration of the techniques of working the clay would then be left to each individual through the development of personal know-how as is visible from the many different ways the detail of working the material was carried out. As Dobres (2000: 209) points out, this possibly allowed for subtle social transformations through the individual skills acquired and showed although through the themes of representation chosen the belonging to the same cultural groups was expressed and reinforced. The conclusion obtained for the working of clay in the French Pyrénées can apparently be widened to the Magdalenian cultural context in the region for the wider material culture.

In Central Europe, the situation is very different; there the study of the use of clay allows underlining a much standardised way of doing things, one for which a process of formal learning could be implied. The process of the working of clay seen is a complex one which uses one after the other the unique qualities of the material. The fact that this standardisation is connected to the making and exploding of the clay figurines allows putting emphasis on the importance of this process at the Pavlovian sites: an importance possibly related to the ritual role of this use of clay in the context of aggregation. There appear to be less standardisation in the working of osseous material in general at the same sites for example although some very specific types of production are found which could imply standardised *chaîne opératoires* for artefacts intended for a specific purpose. Thus in the Pavlovian of Moravia, it seems that the way of making things was more set than what is seen in the French Pyrénées, with possibly less emphasis on the individual person: Could this also indicate a society with a more organised hierarchy?

In conclusion, it is hoped the present research will play a role in putting emphasis on the potential of clay study for increasing our understanding of Upper Palaeolithic complexity. Clay is a topic that has been rather neglected up to date for this period. It is hoped the present research has demonstrated its interest. A number of further routes to follow in research on this topic have been underlined as a way to indicate that the present synthesis is only a first step in a study that will need to be pursued.

BIBLIOGRAPHY

Abramova, Z.A. (1995), *L'art Paléolithique d'Europe Orientale et de Sibérie,* L'homme des origines, Grenoble: Jérome Millon.

Absolon, K. (1949), 'Moravia in Palaeolithic Times', *American Journal of Archaeology* 53, pp 19-28.

Adovasio, J.M., Klima, B. and Soffer, O. (1996), 'Upper Palaeolithic fibre technology: interlaced woven finds from Pavlov I, Czech Republic, circa 26,000 years ago', *Antiquity* 70 (269), September, pp 526-533.

Adovasio, J.M., Hyland, D.C. and Soffer, O. (1997), 'Textiles and cordages: a preliminary assessment', in Svoboda, J. (Dir.) (1997), *Pavlov I-Northwest, the Upper Palaeolithic burial and its settlement context,* The Dolni Vestonice Studies 4, Brno: The Academy of Sciences of the Czech Republic.

Alteirac, A. and Bahn, P. (1982), 'Premières datations radiocarbones du Magdalénien moyen de la grotte du Mas d'Azil (Ariège)', *Préhistoire ariégeoise - Bulletin de la Société Préhistorique d'Ariège* 37, pp 107-110.

Alteirac, A. and Vialou, D. (1984), 'Grotte du Mas d'Azil', in *L'Art des Cavernes, Atlas des grottes ornées paléolithiques françaises* (1984), Paris: Ministère de la Culture, pp 389-394.

Arnold, D.E. (1989), *Ceramic Theory and Cultural Process*, Cambridge: Cambridge University Press.

Audouze, F. (2002), 'Leroi-Gourhan, a Philosopher of Technique and Evolution', *Journal of Archaeological Research* 10 (4), December, pp 277-306.

Aujoulat, N. (1984a), 'Grotte de La Calévie', in *L'Art des Cavernes, Atlas des grottes ornées paléolithiques françaises* (1984), Paris: Ministère de la Culture.

Aujoulat, N. (1984b), 'La Grotte de La Forêt', in *L'Art des Cavernes, Atlas des grottes ornées paléolithiques françaises* (1984), Paris: Ministère de la Culture.

Aujoulat, N. (2005), *The Splendour of Lascaux. Rediscovering the greatest treasure of prehistoric art,* London: Thames and Hudson.

Aujoulat, N. and Dauriac, N. (1984) 'Grotte de Bara-Bahau', in *L'Art des Cavernes, Atlas des grottes ornées paléolithiques françaises* (1984), Paris: Ministère de la Culture.

Baffier, D. and Girard, M. (1998), *Les Cavernes d'Arcy Sur Cure*, Paris: La Maison des Roches.

Bahn, P. (1978), 'Palaeolithic Pottery, the History of an Anomaly', *Anthropos* 50, pp 98-108.

Bahn, P. (1980), 'Histoire d'eau: l'art pariétal préhistorique des Pyrénées', *Travaux de l'Institut d'Art Préhistorique de Toulouse* 22, pp 129-135.

Bahn, P. (1984), *Pyrenean Prehistory: a Palaeoeconomic survey of the French Sites*, Warminster: Aris and Phillips.

Bahn, P. (1986), 'Palaeolithic Pottery in England', in Colcutt, S.N. (ed.) (1986), *The Palaeolithic of Britain and its nearest neighbours: recent trends*, Sheffield: University of Sheffield, Department of Archaeology and Prehistory.

Bahn, P. (1998), *The Cambridge Illustrated History of Prehistoric Art*, Cambridge University Press.

Bahn, P. and Otte, M. (1985), 'La poterie 'paléolitique' de Belgique: analyses récentes', in *Helinium* XXV, Universa, Wetteren, pp 238-241.

Bahn, P. and Renfrew, C. (1996), *Archaeology: Theories, Methods and Practice*, Thames and Hudson, London.

Bahn, P. and Vertut, J. (1997), *Journey through the Ice Age*, Berkeley and Los Angeles: University of California Press.

Banesz, L. (1961), 'Zlomek hlinenej platiky z paleolitickej stanice v Kasove', in *Archeologicke Rozhledy* 13, pp 774-780.

Banesz L. (1996), 'Entdeckungen Der Kunstobjekte Im Palaolithikum Der Ostslowakei', in Svoboda J. (ed.) (1996), *Paleolithic in the Middle Danube region,* Spisy Archeologického Ústavu AV ČR v Brně Svazek 5, Archeologický Ústavu AV ČR, Brno, pp 279-281.

Banesz, L., Hromada, J., Desbrosse, R., Margerand, J., Kozlowski, J.K., Sobczyk, K. and Pawlikowski, M. (1992), 'Le site de plein air du Paléolithique Supérieur de Kasov I en Slovaquie Orientale', in *Slovenska Archeologia* 40 (1), pp 5-28.

Barnett, W. and Hoopes, J. (eds.) (1995), *The Emergence of Pottery: Technology and Innovation in Ancient Societies*, Washington: Smithsonian Institution.

Barrière, C. (1973), 'Les gravures en tracé digital de Gargas', *Bulletin de la Société Préhistorique de l'Ariège* 28, pp 79-102.

Barrière, C. (1976a), *Palaeolithic Art in the Grotte de Gargas*, Oxford: British Archaeological Reports Supplementary Series 14 (i & ii).

Barrière, C. (1984a), 'Grotte de Rouffignac', in *L'Art des Cavernes, Atlas des grottes ornées paléolithiques françaises* (1984), Paris: Ministère de la Culture.

Barrière, C. (1984b), 'Grotte de Gargas', in *L'Art des Cavernes, Atlas des grottes ornées paléolithiques françaises* (1984), Paris: Ministère de la Culture.

Beasley, B. (1986), 'Les bisons d'argile de la grotte du Tuc d'audoubert', *Bulletin de la Société Préhistorique Ariège-Pyrénées* 41, pp 23-30.

Bednarik. R. (2001), 'Cupules: the oldest surviving rock art', International *Newsletter on Rock Art* 30: pp 18-23.

Bégouen, H. (1912), 'Les statues d'argile de la caverne du Tuc d'Audoubert (Ariège)', *L'Anthropologie* 23, pp 655-665.

Bégouen, H. (1921), 'Découverte d'argile cuite dans un gisement Magdalénien', *L'Anthropologie* 31, pp 524-530.

Bégouen H. and Casteret N. (1923), 'La caverne de Montespan (Haute-Garonne)', *Revue Anthropologique* 33 (11-12), pp 533-550.

Bégouen, H. and Bégouen, L. (1936), 'Deux petites statuettes magdaléniennes de la caverne des Trois-Frères (Ariège)', in *12e Congrès préhistorique Français*, Toulouse-Foix, pp 297-9.

Bégouen, H. and Breuil, H. (1958), *Les cavernes du Volp: Trois-Frères- Tuc D'Audoubert*, Paris: Arts et Métiers Graphique.

Bégouen, R., Clottes, J. and Delporte, H. (1977), 'Le retour du petit bison au Tuc d'Audoubert', *Bulletin de la Societé Préhistorique Française* 74, pp 112-120.

Bégouen, R. and Clottes, J. (1981a), Nouvelles fouilles dans la Salle des Morts de la Caverne d'Enlène à Montesquieu-Avantès (Ariège), Extrait du Congrès Préhistorique de France XXIe session, Quercy, 1979 (1981), 33-70, in Clottes, J. (1999), *La vie et l'art des Magdaléniens en Ariège*, Paris: La Maison des Roches.

Bégouen R. and Clottes J. (1981b), Apports mobiliers dans les cavernes du Volp (Enlène, les Trois-Frères, le Tuc-d'Audoubert), Altamira Symposium, pp 157-188 published in Clottes, J. (1999), *La vie et l'art des Magdaléniens en Ariège*, Paris: La Maison des Roches, pp 218-247.

Bégouen R. and Clottes J. (1982), 'Des ex-votos magdaléniens?', *La Recherche* 132, Avril 1982, pp 518-520.

Bégouen, R. and Clottes, J. (1983), 'A propos d'une datation radiocarbone de l'habitat magdalénien du Tuc D'Audoubert', *Bulletin de la Société Préhistorique de l'Ariège* 38.

Bégouen R. and Clottes J. (1984a), Grotte des Trois-Frères. in *L'Art des Cavernes, Atlas des grottes ornées paléolithiques françaises* (1984), Paris: Ministère de la Culture, pp 400-409.

Bégouen, R. and Clottes, J. (1984b), Grotte du Tuc d'Audoubert, in *L'Art des Cavernes, Atlas des grottes ornées paléolithiques françaises* (1984), Paris: Ministère de la Culture, pp 410-415.

Bégouen R. and Clottes J. (1987), 'Les Trois-Frères after Breuil', *Antiquity* 61, pp 180-187.

Bégouen, R., and Clottes, J. (1988), 'Matériaux pour servir à l'histoire de Montespan', *Bulletin de la Société Préhistorique Ariège-Pyrénées* 43, pp 13-33.

Bégouen R., Briois F., Clottes J. and Servelle C. (1984-1985), Art mobilier sur support lithique d'Enlène (Montesquieu-Avantès, Ariège), collection Bégouen du musée de l'Homme. Ars Prehistorica T. III-IV, p. 25-80. published in Clottes, J. (1999), *La vie et l'art des Magdaléniens en Ariège*, La Maison des Roches, Paris, pp 321- 388.

Bégouen R., Clottes J., Giraud J.P. and Rouzaud F. (1987), 'Un bison sculpté en grès a Enlène', *Bulletin de la Société méridionale de spéléologie et de préhistoire*. 27, pp. 23-27 published in Clottes, J. (1999), La vie et l'art des Magdaléniens en Ariège, La Maison des Roches, Paris,

Bégouen, R., Clottes, J., Giraud, J.P. and Rouzaud, F. (1989), 'Les foyers de la caverne d'Enlène (Montesquieu-Avantès, Ariège)', in *Actes du colloque de Nemours 1987*, Mémoires du Musée de Préhistoire d'Ile-de-France 2, pp 165-179, published in Clottes, J. (1999), La vie et l'art des Magdaléniens en Ariège, La Maison des Roches, Paris,

Beltran, A., Robert, R. And Gailli, R. (1967), *La Cueva de Bédeilhac*, Monografias Arqueologicas II, Zaragoza.

Beltran, A. and Saura Ramos, P.A. (1999), *The cave of Altamira*, New York: Harry N. Abrams.

Bernaldo de Quiros, F.H. (1991), 'Reflections on the Art of the Cave of Altamira', *Proceedings of the Prehistoric Society* 57 (I), pp 81-90.

Bisson, M.S. and Bolduc, P. (1994), 'Previously undescribed figurines from the Grimaldi caves', *Current Anthropology* 35 (4), pp 458-468.

Bordes F. (1961), *Typologie du Paléolithique ancien et moyen*. Delmas, Publications de l'Institut de Préhistoire de l'Université de Bordeaux, Mémoire n° 1, réédition CNRS 1988.

Bordes, F. (1968), *Le Paléolithique dans le monde*, Paris: Hachette.

Bosinski, G. (1990), *Les Civilisations de la préhistoire: les chasseurs du Paléolithique Supérieur (40,000-10,000 av JC)*, Paris: Errance.

Bougard, E. (2003), 'Ceramic in the Upper Palaeolithic', in Gibson, A. (ed.), (2003) *Prehistoric Pottery: People, Pattern and Purpose,* Prehistoric Ceramic Research Group: Occasional Publication N°.4, BAR International Series 1156, pp 29-34.

Bourdu, R. (1994), 'Des empreintes pour témoigner', in *Argiles*, Laboratoires Beaufour, Paris, pp 11-18.

Boureux, C.-J. (2004), 'L'Homme et l'argile au Paléolithique supérieur dans l'espace Franco-cantabrique', *Préhistoire, Art et Sociétés*, Revue éditée par la Société Préhistorique Ariège-Pyrénées 59, pp 93-108.

Bravard, JP., Cammas, C., Nehlig, P., Poupet, P., Salvador, PG. and Wattez, J. (1999), *La Géologie, les Sciences de la Terre*, Collections Archéologiques, Errance.

Breuil, H. (1903). 'Rapport sur les fouilles dans la grotte du Mas d'Azil (Ariège)', *Bulletin Archéologique du Comité des Travaux Historiques et Scientifiques*, pp 421-436.

Breuil, H., (1952), *Four Hundred Centuries of Cave Art*, Centre d'Etudes et de Documentation Préhistorique, Montignac, Dordogne: F. Windels.

Breuil, H., Alcalde del Rio, H. and Sierra, L. (1911), *Les Cavernes de la Région Cantabrique (Espagne)*, Monaco: A.Chêne,.

Cahen D. and Caspar J.P. (1984), 'Les traces d'utilisation des outils préhistoriques', *L'Anthropologie* 88 (3), pp. 277-308.

Clot, A. (1984), 'Grotte du Bois du Cantet', in *L'Art des Cavernes, Atlas des grottes ornées paléolithiques françaises* (1984), Paris: Ministère de la Culture.

Clottes J. (1977), 'Information archéologiques, circonscription de Midi-Pyrénées', *Gallia Préhistoire* 20 (2), pp. 517-559.

Clottes, J. (1984), 'Grotte de Niaux', in *L'Art des Cavernes, Atlas des grottes ornées paléolithiques françaises* (1984), Paris: Ministère de la Culture.

Clottes, J. (1989), 'Le Magdalénien des Pyrénées', in *Le Magdalénien en Europe, Actes du colloque de Mayence 1987*, ERAUL 38, pp. 281-357, published in Clottes, J. (1999), La vie et l'art des Magdaléniens en Ariège, La Maison des Roches, Paris, pp 30-89.

Clottes, J. (1992), 'Les Civilisations du Paléolithique Supérieur dans les Pyrénées françaises', in *Actas Congresos Internacional Historia de los Pirineos*, Cervera, Novembre 1988, published in Clottes, J. (1999), La vie et l'art des Magdaléniens en Ariège, La Maison des Roches, Paris, p.15-29.

Clottes J. (1993), 'Analyses des peintures magdaléniennes de grottes ornées ariégeoises', in *Actes du congrès international des sciences préhistoriques*, Bratislava, 1-7 septembre 1991, tome 4, pp 347-357 published in Clottes, J. (1999), La vie et l'art des Magdaléniens en Ariège, La Maison des Roches, Paris, pp 163-175.

Clottes, J. (1995), *Les cavernes de Niaux, Art Préhistorique en Ariège*, Paris: Seuil.

Clottes, J. (1996a), *The Cave Beneath the Sea: Paleolithic Images at Cosquer*, New York: Harry N. Abrams.

Clottes, J. (1996b), 'Thematic Changes in Upper Palaeolithic Art: a View from the Grotte Chauvet', *Antiquity* 70 (268), June, pp 278-287.

Clottes J. (1999), *La vie et l'art des Magdaléniens en Ariège*, Paris: La Maison des Roches.

Clottes, J. (ed.) (2003), *Return to Chauvet Cave: Excavating the Birthplace of Art; The first Full Report*, London: Thames and Hudson.

Clottes, J. and Simonnet, R. (1972), 'Le réseau René Clastres de Niaux (Ariège)', *Bulletin de la Société Préhistorique Française* 69 (1), pp 293-323 published in Clottes, J. (1999), *La vie et l'art des Magdaléniens en Ariège*, Paris: La Maison des Roches, pp 499 530.

Clottes, J. and Simonet, R. (1984), 'Le réseau René Clastres', in *L'Art des Cavernes, Atlas des grottes ornées paléolithiques françaises* (1984), Paris: Ministère de la Culture.

Clottes, J., Rouzaud, F. and Wahl, L. (1984), 'Grotte de Fontanet', in *L'Art des Cavernes, Atlas des grottes ornées paléolithiques françaises* (1984), Paris: Ministère de la Culture, pp 433-437.

Clottes, J. and Gailli, R. (1984), 'Grotte de Massat', in *L'Art des Cavernes, Atlas des grottes ornées paléolithiques françaises* (1984), Paris: Ministère de la Culture.

Clottes, J., Menu, M. and Walter, Ph. (1990), 'La préparation des peintures magdaléniennes des cavernes ariègeoises', *Bulletin de la Société Préhistorique Francaise* 87 (6), pp 170-192.

Clottes J., Valladas H., Cachier H. and Arnold M. (1992), 'Des dates pour Niaux et Gargas', *Bulletin de la Société préhistorique francaise* 89 (9), pp. 270-274 published in Clottes, J. (1999), La vie et l'art des Magdaléniens en Ariège, La Maison des Roches, Paris, pp 176-181.

Clottes J., Garner M. and Maury G. (1994), 'Bisons Magdaléniens des cavernes ariégeoises', *Bulletin de la Société Préhistorique Ariège-Pyrénées* XLIX, pp 15-49, published in Clottes , J. (1999), 'La vie et l'art des Magdaléniens en Ariège', Paris: La Maison des Roches, pp. 106-137.

Clottes, J. and Lewis-Williams, D. (1996), *The Shamans of Prehistory: Trance and Magic in the painted caves*, New York: Harry N. Abrams.

Clottes J. and Delporte H. (2003), *Grotte de la Vache en Ariège. Comité des travaux Historiques et Scientifiques,* Document préhistorique No 16. (Two volumes), Paris: Editions du Comité des travaux historiques et scientifiques, Editions de la Réunion des musées nationaux.

Clottes J. and Delporte H. (2003a) 'L'Art Mobilier Figuratif', in Clottes J. and Delporte H. (2003), *Grotte de la Vache en Ariège. Comité des travaux Historiques et Scientifiques,* Document préhistorique No 16. (Two volumes), Paris: Editions du Comité des travaux historiques et scientifiques, Editions de la Réunion des musées nationaux, Second Volume, pp 7-26.

Clottes J., Courtin J. and Vanrell L. (2005), *Cosquer redécouvert*, Paris: Seuil.

Combier, J. and Roudil, J.L. (1984), 'Grotte de la Vacheresse', in *L'Art des Cavernes, Atlas des grottes ornées paléolithiques françaises* (1984), Paris: Ministère de la Culture.

Combier J., Porte J.L., Ayroles P. and Gély B. (1984), 'Abri du Colombier', in *L'Art des Cavernes, Atlas des grottes ornées paléolithiques françaises* (1984), Paris: Ministère de la Culture, pp 621-625.

Conkey M.W. (1980), 'The identification of Prehistoric hunter-gatherer aggregation sites: the case of Altamira', *Current Anthropology* 21 (5), October 1980, pp 609-630.

Conkey, M.W. (1985), 'Ritual Communication, Social Elaboration, and the variable trajectories of Paleolithic Material Culture', in Price, T. D and Brown, J. A. (1985), *Prehistoric Hunter-Gatherer: the emergence of cultural complexity*, London: Academic Press, pp 299-323.

Conkey, M. (1987), 'Interpretive problems in Hunter-Gatherer Regional Studies : Some thoughts on the European Upper Palaeolithic', in Soffer, O. (ed.) (1987), *The Pleistocene Old World, Regional Perspectives*, New York, London: Plenum.

Conkey, M.W., Soffer, O., Stratmann, D. and Jablonski N.G. (eds) (1997), *Beyond art: Pleistocene Image and Symbol*, California: University of California Press.

Couraud, C. (1983), 'Pour une étude méthodologique des colorants préhistoriques', *Bulletin de la Société Préhistorique Française* 80 (4), pp 104-110.

Couraud, C. and Laming-Emperaire, A. (1979), 'Les colorants', in Leroi-Gourhan, Arl. and Allain, J. (dir.) (1979), *Lascaux Inconnu*, Gallia Préhistoire supplément 12, Paris: Editions du Centre National de la Recherche Scientifique, pp 153-169.

Damblon, F. (1997), 'Anthracology and Past vegetation reconstruction', in Svoboda, J. (Ed.) (1997), *Pavlov I-Northwest, the Upper Palaeolithic burial and its settlement context*, The Dolni Vestonice Studies, Vol. 4, Brno: Academy of Sciences of the Czech Republic, pp 437-442.

Davidson, I. (1997), 'The Power of Pictures', in Conkey, M.W., Soffer, O., Stratmann, D. and Jablonski, N.G. (eds) (1997), *Beyond art: Pleistocene Image and Symbol*, California: University of California Press, pp 125-159.

Delluc, B. and Delluc, G. (1984a), 'Grotte de la Croze à Gontran', in *L'Art des Cavernes, Atlas des grottes ornées paléolithiques françaises* (1984), Paris: Ministère de la Culture.

Delluc, B. and Delluc, G. (1984b), 'Grotte de La Martine', in *L'Art des Cavernes, Atlas des grottes ornées paléolithiques françaises* (1984), Paris: Ministère de la Culture.

Delluc, B. and Delluc, G. (1984c), 'Grotte de Sous-Grand-Lac', in *L'Art des Cavernes, Atlas des grottes ornées paléolithiques françaises* (1984), Paris: Ministère de la Culture.

Delluc, B. and Delluc, G. (1984d), 'Grotte de Saint-Cirq', in *L'Art des Cavernes, Atlas des grottes ornées paléolithiques françaises* (1984), Paris: Ministère de la Culture.

Delluc, B. and Delluc, G. (1997), 'Les gravures de la Grotte ornée de Bara-Bahau (Le Bugue, Dordogne)', *Gallia Préhistoire* 39, pp 109-150.

Delluc, B. and Delluc, G. (2003), *La vie des Hommes de la Préhistoire*, Rennes: Ouest-France.

Delporte, H. (1984), *Archéologie et réalité: essai d'approche épistémologique*, Paris: Picard.

Derev'anko, A. (1998), 'The Paleolithic of the Russian Far East', in Derev'anko, A., Shimkin, D. and Roger Powers, W. (eds) (1998), *The Paleolithic of Siberia, New Discoveries and Interpretations*, Urbana and Chicago: University of Illinois Press, pp 274-285.

Dextreit, R. (1997), *L'Argile qui guérit, Mémento de Médecine Naturelle*, Editions Vivre en Harmonie.

Djindjian, F., Koslowski, J. and Otte, M. (1999), *Le Paléolithique Supérieur en Europe*, Paris: Armand Colin.

Dobres M.A. (1996), 'Variabilité des activités magdaléniennes en Ariège et en Haute-Garonne, d'après les chaînes opératoires dans l'outillage osseux', *Bulletin de la Société Préhistorique Ariège-Pyrénées* 51, pp 149-194.

Dobres M.A. (1999), 'Technology's links and chains: the processual unfolding of technique and technician', in Dobres M.A. and Hoffman C. (1999), *The Social Dynamics of Technology, Practice, Politics and world views*, Smithsonian Institution Press, pp 124-146.

Dobres, M.A. (2000), *Technology and Social Agency*, Oxford: Blackwell Publishers.

Dobres, M.A. and Hoffman, C. (1994), 'Social Agency and the Dynamics of Prehistoric Technology', *Journal of Archaeological Method and Theory* 1 (3).

Dougherty, J. and Keller, C.M. (1982), 'Taskonomy: a practical approach to knowledge structures', *American Ethnologist* 5, pp 763-774.

Dowson, T.A. and Porr, M. (2001), 'Special Objects – Special Creatures: Shamanistic Imagery and the Aurignacian art of southwest Germany', in Price, N., (2001), *The Archaeology of Shamanism*, London, New York: Routledge, pp 165-177.

Drouot, E. (1984), 'Grotte de La Baume-Latrone', in *L'Art des Cavernes, Atlas des grottes ornées paléolithiques françaises* (1984), Paris: Ministère de la Culture.

Duday, H. and Garcia, M.A. (1983), 'Les empreintes de l'Homme préhistorique. La grotte du Pech Merle à Cabrerets (lot) : une relecture significative des traces de pieds humains', *Bulletin de la Société Préhistorique Française*, 80, pp. 208-215.

Duday, H. and Garcia, M.A. (1993), 'Les empreintes de mains dans l'argile des grottes ornées', *Dossier d'Archéologie* 178, Janvier, pp 56-59.

Encyclopædia Britannica. 2007. Encyclopædia Britannica Online. 17 May 2007 <http://www.britannica.com/eb/article-9053140>.

Eluère, C. (1980), 'Fabriquer des liens et des cordages préhistoriques', *Dossier d' Archéologie* 46.

Estrade, J. (1994), 'Argiles, Thérapeuthiques d'Antan', in *Argiles*, Laboratoires Beaufour, Paris, pp 126-133.

Eygun, G. (2001), 'Clays and Neolithic potters', in Druc, I.C. (ed.) (2001), *Archaeology and clays*, British Archaeological Report International Series 942, Oxford.

Fladerer, F.A. (2001), 'The Krems-Wachtberg camp-site: mammoth carcass utilization along the Danube 27,000 years ago', *the world of elephants- international congress*, Rome 2001, accessed at www.cq.rm.cnr.it/elephants2001/pdf/432_438.pdf

Fosse, G. (1984), 'Grotte du Renard', in *L'Art des Cavernes, Atlas des grottes ornées paléolithiques françaises* (1984), Paris: Ministère de la Culture.

Fritz, C. (1999), *La Gravure dans l'Art Mobilier Magdalénien, du Geste à la Représentation: Contribution de l'Analyse Microscopique*, Dossier d'Archéologie Française, Paris: Maison des Sciences de l'Homme.

Gabori, M. (1968), Archaeology and Prehistory of the Hungarian loess Palaeolithic, in La Préhistoire, problèmes et tendances, CNRS editions, pp. 183-189.

Gailli, R., Paihaugue, N. and Rouzaud, F. (1984), 'Grotte de Bédeilhac', in *L'Art des Cavernes, Atlas des grottes ornées paléolithiques françaises* (1984), Paris: Ministère de la Culture.

Gailli, R. (2004), *La Grotte Préhistorique du Ker à Massat (Ariège),* Nîmes: Editions Lacour.

Gamble, C. (1982), 'Interaction and Alliance in Palaeolithic Society', *Man*, New Series 17 (1) (March 1982), pp 92-107.

Gamble, C. (1991), 'The Social Context for European Palaeolithic Art', *Proceedings of the Prehistoric Society* 57, part I, pp 3-15.

Gamble, C. (1993), 'The Center at the Edge', in Soffer, O. and Praslov, N. D. (eds), (1993), *From Kostenki to Clovis – Upper Paleolithic Paleo-Indian adaptations*, Interdisciplinary contributions to Archaeology, New York: Plenum Press, pp 313-321.

Gamble, C. (1999), *The Palaeolithic Societies of Europe*, Cambridge: Cambridge World Archaeology.

Garanger, J, (Dir.) (1992), *La Préhistoire dans le monde*, Presses Universitaires de France.

Gibson, A. and Woods, A. (1990), *Prehistoric Pottery for the Archaeologist*, Leicester, London and New York: Leicester University Press.

Gonysevova, M. (1999), 'Fabrication expérimentale d'artéfacts gravettiens en terre cuite et essai de reconstitution de la variante d'un 'four' (Moravie, République Tchèque)', *L'Anthropologie* 103 (4), pp 519-29.

Goutas, N. (2004), 'Fiche exploitation des matières dures d'origine animales au Gravettien', in Ramseyer, D. (2004), *Industrie de l'os préhistorique. Matière et techniques*, cahier XI, Editions de la Société Préhistorique Francaise, pp 53-66.

Hachi, S. (1985), *Figurines en terre cuite du gisement Ibéromaurusien d'Afalou Bou Rhummel*, Aix-en-Provence: Travaux du LAPEMO, étude 9.

Hachi, S., Frohlich, F., Gendron-Badou, A., Lumley, H., Roubet, C. and Abdessadock, S. (2002), 'Figurines du Paléolithique Supérieur en matière minérale plastique cuite d'Afalou Bou Rhummel (Babors, Algérie). Premières analyses par spectroscopie d'absorption Infrarouge', *L'Anthropologie* 106, pp 57-97.

Hahn, J. (1993), 'Aurignacian Art in Central Europe', in Knecht, H., Pike-Tay, A. and White R. (Eds) (1993), *Before Lascaux: the complex record of the Early Upper Palaeolithic*, Florida, London: Boca Raton CRC Press.

Henderson, J. (2000), *The Science and Archaeology of Materials, an investigation of inorganic materials*, London and New York: Routledge.

Hodges, H. (1964), *Artefacts: an introduction to early materials and technology*, New Jersey: Humanity Press, London: John Baker.

Hyland, D.C., Zhushchikhovskaya, I.S., Medvedev, V.E., Derevianko, A.P. and Tabarev, A.V. (2001), *Pleistocene textiles in the Russian Far East: Impressions From Some of the world's oldest pottery*. Report available from http://mai.mercyhurst.edu/PDFs/Hylans_Amur.pdf consulted on 10th July 2004.

Imamura, K. (1996), *Prehistoric Japan: New Perspectives on Insular East Asia*, London: UCL Press.

Ingold, T. (1990), 'Society, nature and the concept of technology', *Archaeological Review from Cambridge* 9:1, pp 5-17.

Jarošova, L., Cílek, V., Oches, E. and Snieszko, Z. (1996), 'Petřkovice, Excavations 1994-1995', in Svoboda J. (ed.) (1996), *Paleolithic in the Middle Danube region,* Spisy Archeologického Ústavu AV ČR v Brně Svazek 5, Archeologický Ústavu AV ČR, Brno, pp 191-206.

Jarošova, L. (1997), 'Spatial distribution of artefacts', in Svoboda, J. (Dir.) (1997), *Pavlov I-Northwest, the Upper Palaeolithic burial and its settlement context*, The Dolni Vestonice Studies, Vol. 4, Brno: The Academy of Sciences of the Czech Republic.

Jeunesse, C. (2008), 'Une théorie de l'invention préhistorique', *La Recherche* 418 (April), pp 46- 50.

Kaner, S. (2003), 'The oldest Pottery in the World', *Current World Archaeology* 1, pp 44-49.

Karlin, C. and Julien, M. (1994), 'Prehistoric Technology: a Cognitive Science ?', in Renfrew, C. and Zubrow, E.B.W. (Eds) (1994), *The Ancient Mind, Elements of Cognitive Archaeology*, Cambridge: Cambridge University Press, pp 152-164.

Keally, C. T. (2004), 'Bad science and the distortion of history: radiocarbon dating in Japanese Archaeology', *Sophia International Review* 26 (February).

Keally, C. and Kuzmin, Y. V. (2001), 'Radiocarbon chronology of the earliest Neolithic sites in East Asia', *Radiocarbon* 43 (2B), pp 1121-1128.

Keally, C., Taniguchi, Y. and Kuzmin, Y. (2003), 'Understanding the beginnings of Pottery technology in Japan and Neighbouring East Asia', *The review of Archaeology* 24 (2), pp 3-14.

Kenrick, D.M. (1995), *Jomon of Japan: the world's oldest pottery*, London and New York: Kegan Paul International.

Klima, B. (1955), 'Přínos nové paleolitické stanice v Pavlově k problematice nejstarších zemědělských nástrojů', *Pamatky Archeologické* 46, pp 7-29.

Klima, B. (1994), Die Knochenindustrie, zier-und Kunstgegenstände, in Svoboda, J. (ed.) (1994), *Pavlov I: Excavations 1952-53*, ERAUL 66. The Dolni Vestonice Studies Vol. 2, pp 97- 159.

Klima, B. (1997a), 'Grabungsgeschichte, Stratigraphie und Fundumstande- Excavation History, Stratigraphy and context', in Svoboda, J. (Ed.) (1997) *Pavlov I-Northwest, the Upper Palaeolithic burial and its settlement context*, The Dolni Vestonice Studies, Vol. 4, Brno: The Academy of Sciences of the Czech Republic, pp 13-51

Klima, B. (1997b), 'Die Knochenindustrie, Zier-und Kuntstgegenstande- Bone industry, Decorative objects and Art', in Svoboda, J. (Ed.) (1997), *Pavlov I-Northwest, the Upper Palaeolithic burial and its settlement context*, The Dolni Vestonice Studies, Vol. 4, Brno: The Academy of Sciences of the Czech Republic, pp 227-286.

Klima, B. (1997c), 'Zur Frage der Mikroindustrie aus Pavlov- The wuestion of Microindustry from Pavlov', in Svoboda, J. (Ed.) (1997), *Pavlov I-Northwest, the Upper Palaeolithic burial and its settlement context*, The Dolni Vestonice Studies, Vol. 4, Brno: The Academy of Sciences of the Czech Republic, pp 289-312.

Knecht, H. (Ed.) (1997a), *Projectile technology*, New York and London: Plenum Press.

Knecht, H. (1997b), 'Projectile points of bone, antler and stone', in Knecht, H. (1997), *Projectile Technology*, New York and London: Plenum Press.

Kozlowski, J.K., Koumouzelis, M., Escutenaire, C., Sitlivy, V., Sobczyk, K., Valladas, H., Tisnerat-Laborde, N., Wojtal, P., Ginter, B., Kaczanowscka, M., Kazior, B. and Zieba, A. (2001a), 'La fin du Paléolithique Moyen et le début du Paléolithique supérieur en Grèce: la séquence de la grotte 1 de Klissoura', *L'Anthropologie 105*, pp 469-504.

Kozlowski, J.K., Bar-Yosef, O., Koumouzelis, M., Pawlikowski, M., Ginter, B., Maria Albert, R., Litynska-Zajac, M., Stworzewicz, E., Wojtal, P., Lipecki, G., Tomek, T., Bochenski, Z.M. and Pazdur, A. (2001b), 'The early Upper Palaeolithic in Greece: the excavations in Klissoura Cave', *Journal of Archaeological Science* 28, pp 515-539.

Kralik, M. et Novotný, V. (2005) 'Dermatoplyphics of ancient ceramics', in Svoboda, J. (ed.) (2005), *Pavlov I Southeast: a window into the Gravettian lifestyles*, The Dolni Vestonice Studies, Vol. 14, Academy of Sciences of the Czech Republic, Institute of archaeology at Brno, pp 449-497.

Lacorre, F. (1960), *La Gravette, le Gravettien et le Bayacien*, Laval: Imprimerie Barnéaud.

Laplace, G., Valicourt, E., Boucher, P. and Lauga, M. (1984), 'Grotte Etxeberri', in *L'Art des Cavernes, Atlas des grottes ornées paléolithiques françaises* (1984), Paris: Ministère de la Culture.

Larribau, J.D. and Laplace, G. (1984), 'Grotte Oxocelhaya-Hariztoya', in *L'Art des Cavernes, Atlas des grottes ornées paléolithiques françaises* (1984), Paris: Ministère de la Culture.

Larribau, J.D. and Prudhomme, S. (1984), 'Grotte d'Erbérua', in *L'Art des Cavernes, Atlas des grottes ornées paléolithiques françaises* (1984), Paris: Ministère de la Culture.

Larribau, JD, and Prudhomme, S. (1989), 'Etude préliminaire de la Grotte d'Erbérua (Pyrénées-Atlantiques)', *L'Anthopologie* 93 (2), pp 475-493.

Le Feu Apprivoisé: Le feu dans la vie quotidienne des hommes préhistoriques, catalogue d'exposition, 11 May-31 December 1987, Musée de Préhistoire d'Ile-de-France, A.P.R.A.I.F.

Lemonnier, P. (1990), 'Topsy Turvy Techniques: Remarks on the Social Representation of Techniques', *Archaeological Review from Cambridge* 9 (1), pp 27-37.

Lemozi, A. (1929), *La grotte-temple du Pech-Merle, un nouveau sanctuaire préhistorique*, Paris: Auguste Picard.

Leroi-Gourhan, A. (1943), *Evolution et techniques I: L'Homme et la Matière*, Paris: Albin Michel.

Leroi-Gourhan, A. (1965), *Préhistoire de l'Art Occidental*, Paris: Mazenod.

Leroi-Gourhan, A. (1984), 'Grotte du Cheval', in *L'Art des Cavernes, Atlas des grottes ornées paléolithiques françaises* (1984), Paris: Ministère de la Culture.

Leroi-Gourhan, Arl. (Dir.) (1979), Lascaux Inconnu, Supplément à Gallia Préhistoire 12, Paris: Editions du CNRS.

Lewis-Williams, D. (2002), *The Mind in the Cave*, London: Thames and Hudson.

Lorblanchet, M. (1977), 'From naturalism to abstraction in European rock art', in Ucko P. (ed.) (1977), *Form in indigenous art*, Australian Institute of Aboriginal studies, Canberra.

Lorblanchet, M. (1984), 'Grotte du Pech-Merle', in *L'Art des Cavernes, Atlas des grottes ornées paléolithiques françaises* (1984), Paris: Ministère de la Culture.

Lorblanchet, M. (1995), *Les grottes ornées de la préhistoire. Nouveaux regards,* Paris: Errance.

Lorblanchet, M. (1999), *La Naissance de l'Art. Genèse de l'Art Préhistorique*, Paris: Errance.

Lumley, H. (Dir.) (1984), *Art et Civilisations des chasseurs de la Préhistoire 34 000-8000 ans av. J.C.*, (Catalogue de l'exposition), Laboratoire de Préhistoire du Musée de l'Homme, Paris et Musée des Antiquités Nationales de Saint-Germain-En-Laye.

Malvesin-Fabre, G., Nougier, R. and Robert, R. (1953), 'Les bisons à cupules sur argile des grottes Pyrénéennes', *Bulletin Archéologique du Comité des Travaux Historiques et Scientifiques*, Presses Universitaires de France.

Martin, H. (1921), 'Deux fragments de céramique paléolithique trouvés près de Nemours (Seine et Marne)', *L'Anthropologie* 31, pp 524-526.

McDermott, L. (1996), 'Self-representation in Upper Paleolithic female figurines', *Current Anthropology* 37 (2), pp 227-275.

Méroc, L. (1959), 'Prémoustériens, Magdaléniens et Gallo-romains dans la caverne de Labouiche (Ariège)', *Gallia Préhistoire* 2, pp. 1-37.

Mons, L. (1974), 'Notes de technologie et de morphologie de l'art paléolithique mobilier: IV. Cheval modelé "au doigt" sur une plaquette de limon de la grotte de Bédeilhac (Ariège)', *Bulletin publié par le Musée des Antiquités Nationales* 6.

Mons, L. and Delporte, H. (1973), 'Notes de technologie et de morphologie de l'art paléolithique mobilier: III bison gravé sur plaquette de limon de Bédeilhac (Ariège)', *Antiquités Nationales* 5, pp 20-32.

Mons, L. and Delporte, L. (1980), 'Gravure sur os et argile', *Dossier d' Archéologie* 46 (Sept-Oct).

Musil, R. (1994), 'Hunting game from the culture layer at Pavlov', in Svoboda, J. (ed.) (1994), *Pavlov I: Excavations 1952-53*, ERAUL 66/ Dolni Vestonice studies 2, pp 170-196

Musil, R. (1997), 'Hunting Game Analysis', in Svoboda, J. (Ed.) (1997) *Pavlov I-Northwest, the Upper Palaeolithic burial and its settlement context,* The Dolni Vestonice Studies 4, Brno: the Academy of Sciences of the Czech Republic, pp 443-468.

Nature et Fonction des Foyers Préhistoriques: Actes du Colloque International de Nemours 12-13-14 Mai 1987, published in 1989 by Association pour la promotion de la recherche archéologique en Ile-de-France.

Neugebauer-Maresch, C. (1999), *Le Paléolithique en Autriche,* Série Préhistoire d'Europe N°8, Collection l'Homme des Origines Grenoble: Jérôme Millon,.

Octobon, E. (1939), 'Peintures et gravures sur paroi ou sol de la grotte de Bédeilhac (Ariège)', *Revue Anthropologique*, pp 222-235.

Oliva, M. (1993), 'The Aurignacian in Moravia', in Knecht H., Pike-Tay A., White R. (1993), *Before Lascaux: the complex record of the EUP*, London, Florida: Boca Raton, CRC Press.

Oliva, M. (2000a.), 'Some thoughts on Pavlovian adaptations in Moravia', in Roebroeks, W., Mussi, M., Svoboda, J. and Fennema, K. (2000), *Hunters of the Golden Age, the mid Upper Palaeolithic of Eurasia 30,000 - 20,000 BP*, University of Leiden, pp 219-229.

Oliva, M. (2000b), 'Dolni Vestonice I- Une révision de la stratigraphie culturelle', *Anthropologie*, 38 (3), pp 283-290.

Oliva, M. (2000c), 'The Brno II Upper Palaeolithic burial', in Roebroeks W., Mussi M., Svoboda J. and Fennema K. (2000), *Hunters of the Golden Age, the mid Upper Palaeolithic of Eurasia 30,000 - 20,000 BP*, University of Leiden, , pp 143-153.

Opravil, E. (1994), 'The vegetation', in Svoboda, J. (ed.) (1994), *Pavlov I: excavation 1952-53*, Liège: ERAUL 66.

Otte, M. (1981), *Le gravettien en Europe Centrale*, Dissertationes Archaeologicales Gandenses 20, Bruges: De Tempel.

Otte, M. (2003), *Recherches sur le Paléolithique Supérieur*, British Archaeological Reports International Series 1107, Oxford.

Oxford English Dictionnary, Second Edition J.A.Simpson and E.S.C..Weiner (eds) Carendon Press, Oxford 1989.

Paillet, P. (1989), 'La Galerie aux peintures d'Etcheberri'ko Karbia', *L'Anthropologie* 93 (2), pp 493-512.

Pales, L. (1972), *Les ci-devant Venus stéatopyges aurignaciennes*, in Santander Symposium, Actas del symposium internacional de arte prehistorico.

Pales, L. and Vialou, D. (1984), 'Grotte de l'Aldène', in *L'Art des Cavernes, Atlas des grottes ornées paléolithiques françaises* (1984), Paris: Ministère de la Culture.

Pawlikowski, M., Koumouzelis, M., Ginter, B. and J.K. Kozlowski (2000), 'Emerging ceramic technology in structured Aurignacian Hearths at Klissoura Cave 1 in Greece', *Archaeology, ethnology and anthropology of Eurasia* 4 (4), pp 19-29.

Pearson, R., emeritus professor of the University of British Columbia: invitation to Jomon Archaeology. From website www.jomon.or.jp/e2.html accessed on 10th July 2004.

Pélégrin, J. (1990), 'Prehistoric lithic technology: some aspects of research', *Archaeological Review from Cambridge* 9:1, pp 116-125.

Pelegrin, J. (mars 1991) « Les savoir-faire : une très longue histoire », *Terrain*, numéro 16 - *Savoir-faire*, put online on 21st July 2005, URL : http://terrain.revues.org/document3001.html. Accessed 16 April 2007.

Péquart, M. and S-J. (1960), 'Grotte du Mas d'Azil (Ariège), une nouvelle Galerie Magdalénienne', *Annales de Paléontologie* 46, Paris: Masson.

Perlès, C. (1977), *La Préhistoire du Feu*, Paris: Masson.

Peyrony, D. (1934), 'La Ferrassie', *Préhistoire* 3, pp 1-92.

Piette, E. (1875), 'Nouvelles fouilles dans la grotte de Gourdan', *Bulletin de la Société d'Anthropologie de Paris*, séance du 15 avril 1875, pp 279-296.

Pigeaud, R. (2002), 'La Grotte Ornée Mayennes-Sciences (Thorigné-en-Charnie, Mayenne): grotte-limite aux marges du monde anté-Magdalénien', *L'Anthropologie* 106, pp 445-489.

Pigeot, N. (1987), *Magdaléniens d'Etiolles: Economie de débitage et organisation sociale*, XXV supplément à Gallia Préhistoire, Centre National de la Recherche Scientifique, Paris.

Pigeot, N. (1990), 'Technical and Social actors: flintknapping specialists and apprentices at Magdalenian Etiolles', *Archaeological Review from Cambridge* 9 (1), pp 126-141.

Plassard, J. (1999), *Rouffignac, le sanctuaire des mammouths*, Paris: Seuil.

Plassard, F. and Plassard, J. (2000), 'Figures inédites de la Grotte de Rouffignac', *Gallia Préhistoire* 42, pp 85-106.

Poignant, J. (1984), 'Grotte du Croc-Marin', in *L'Art des Cavernes, Atlas des grottes ornées paléolithiques françaises* (1984), Paris: Ministère de la Culture.

Pollard, A.M. and Heron, C. (1996), *Archaeological Chemistry*, Cambridge: The Royal Society of Chemistry.

Porr, M. (2004), 'Individual reflections: Gender and the Aurignacian art of Southwest Germany', EAZ, Ethnogr.-Archäol. 45, pp 257-269.

Praslov, N.D. (1985), 'L'art du Paléolithique Supérieur à l'est de l'Europe', *L'Anthropologie* 89 (2), pp 181-192.

Price, T. D. and Brown, J.A. (1985a), 'Aspects of hunter-gatherer complexity', in Price, T. D and Brown, J.A. (1985), *Prehistoric Hunter-Gatherer: the emergence of cultural complexity*, London: Academic Press, pp 3-20.

Price, T.D. and Brown, J.A. (1985b), 'Complex hunter-gatherers: retrospect and prospect', in Price, T. D and Brown, J.A. (1985), *Prehistoric Hunter-Gatherer: the emergence of cultural complexity*, London: Academic Press, pp 435-442.

Ramseyer, D. (2004), *Industrie de l'os préhistorique. Matière et techniques*, cahier XI, Editions de la Société Préhistorique Francaise.

Rapp, G. and Hill, C. (1998), *Geoarchaeology, the Earth-Science Approach to Archaological Interpretation*, New Haven and London: Yale University Press.

Rice, P.C. and Patterson, A.L. (1986), 'Validating the cave art archaeofaunal relationship in Cantabrian Spain', *American Anthropologist* 88, pp 658-667.

Rivenq, C. (1984), 'Grotte de Gantiès-Montespan', in *L'Art des Cavernes, Atlas des grottes ornées paléolithiques françaises* (1984), Paris: Ministère de la Culture.

Roebroeks, W., Mussi, M., Svoboda, J. and Fennema, K. (2000), *Hunters of the Golden Age, the mid Upper Palaeolithic of Eurasia 30,000 - 20,000 BP*, Leiden: University of Leiden.

Roudil, J.L. (1984), 'L'Art Pariétal Languedocien: Problèmes de Conservation', *L'Anthropologie* 88 (4), pp 499-502.

Roussot, A. (1984), 'Grotte de Cournazac', in *L'Art des Cavernes, Atlas des grottes ornées paléolithiques françaises* (1984), Paris: Ministère de la Culture.

Roussot, A. (1997), *L'Art Préhistorique*, Editions Sud Ouest.

Roussot, A. (1998), *Petit glossaire de l'art préhistorique au Paléolithique*, Toulouse: Editions Confluences.

Sacco, F. and Sauvet, G. (Dir.) (1998), *Le Propre de l'Homme, Psychanalyse et Préhistoire*, Collection Champs Psychanalytiques, Paris: Delachaux and Niestlé.

Sauvet, G. and Tosello, G. (1998), 'Le mythe paléolithique de la caverne', in Sacco, F., and Sauvet, G. (dir.) (1998), *Le Propre de l'Homme, Psychanalyse et Préhistoire*, CollectionChamps Psychanalytiques, Paris: Delachaux and Niestlé.

Sauvet, G. (2006), *La Grotte Ornée Paléolithique de Bédeilhac (Ariège)*, First accessed October 2006 at http://w3.univ-tlse2.fr/utah/bedeilhac/index.html

Saxon, E.C. (1976), 'Pre-Neolithic Pottery: New Evidence from North Africa', *Proceedings of the Prehistoric Society* 42, December.

Saxon, E.C., Close, A., Cluzel, C. and Shackleton N.J. (1974), 'Results of recent investigations at Tamar Hat', *Lybica* XXII, pp 49-91.

Sayre, E.V., Vandiver, P.B., Druzik, J. and Stevenson, C. (1988), *Materials issues in arts and archaeology*, Pittsburg, Pensylvania: Material Research Society symposium proceedings 123.

Ščelinskij, V.E. and Širokov, V.N. (1999), *Höhlenmalerei im Ural. Kapova und Ignatievka die altsteinzeitlichen Bilderhöhlen im südlichen Ural*, Thorbecke Spelao.

Shchekinsky, V.E. (1989), 'Some results of new investigations at the Kapova Cave in the Southern Urals', *Proceedings of the Prehistoric Society* 55, pp 181-191.

Schlanger, N. (1994), 'Mindful technology: unleashing the *chaîne opératoire* for an archaeology of the mind', in Renfrew C. and Zubrow E.B. (eds) (1994), *The Ancient Mind: Elements of Cognitive Archaeology*, New Directions in Archaeology, Cambridge: Cambridge University Press, pp 143-151.

Schlanger, N. (2004), 'Suivre les Gestes, éclat par éclat –la chaine opératoire d'André Leroi-Gourhan', in Audouze F. and Schlanger, N. (Dir.) (2004) *Autour de l'Homme: contexte et actualité d'André Leroi-Gourhan*, Publication du CRA/CEPAM Nice.

Sieveking, A. (1979), *The Cave Artists*, Thames and Hudson, London.

Simonnet, G., Simonnet, L. and Simonnet R. (1984), 'Grotte de Labastide', in *L'Art des Cavernes, Atlas des grottes ornées paléolithiques françaises* (1984), Paris: Ministère de la Culture.

Sinclair, A. (1995), 'The technique as a symbol in late glacial Europe', in Shennan, S. (1995), 'Symbolic aspects of early technologies', *World Archaeology* 27, June 1995, pp 50-62.

Sinclair, A. (2000), 'Constellations of knowledge. Human agency and material affordance in lithic technology', in Dobres, M.A. and Robb, J.E. (2000), *Agency in Archaeology*, London and new York: Routledge, pp 196-212.

Škrdla, P., Cílek, V. and Přichystal, A. (1996), 'Dolni Vestonice III, excavations 1993-1995', in Svoboda J. (ed.) (1996), *Paleolithic in the Middle Danube region*, Spisy Archeologického Ústavu AV ČR v Brně Svazek 5, Archeologický Ústavu AV ČR, Brno.

Skrdla, P. (1997), 'The Pavlovian lithic technologies', in Svoboda, J. (Dir.), 1997, *Pavlov I-Northwest, the Upper Palaeolithic burial and its settlement context*, The Dolni Vestonice Studies 4, the Academy of Sciences of the Czech Republic, Brno.

Soffer, O. (1985), 'Patterns of intensification as seen from the Upper Paleolithic of the Central Russian Plain', in Price, T. D and. Brown, J. A (1985), *Prehistoric Hunter-Gatherer: the emergence of cultural complexity*, London: Academic Press, pp 235-270.

Soffer, O. (1989), 'Storage, Sedentism and the Eurasian Palaeolithic record', *Antiquity* 63, pp 719-32.

Soffer, O. (1993), 'Upper Paleolithic Adaptations in Central and Eastern Europe and Man-Mammoth interactions', in Soffer, O. and Praslov, N.D. (1993), *From Kostenki to Clovis. Upper Paleolithic Paleo-indian Adaptations*, New York and London: Plenum Press, pp 31-49.

Soffer, O. (2000), 'Gravettian technologies in social contexts', in Roebroeks, W. *et al.* (eds) (2000), *Hunters of the Golden Age*, University of Leiden, pp 60-75.

Soffer, O., Vandiver, P., Klima, B. and Svoboda, J. (1990), 'Les plus anciennes céramiques du monde', *La Recherche* 226, Novembre 1990, Volume 21.

Soffer, O., Vandiver, P., Klima, B. and Svoboda, J. (1993), 'The pyrotechnology of performance art: moravian venuses and wolverines', in Knecht, H., Pike-Tay, A. and White, R., (ed.) (1993), *Before Lascaux: the Complex Record of Early Upper Palaeolithic*, Florida, London: Boca Raton C.R.C. Press, pp 259-275.

Soffer, O. and Vandiver, P. (1994), 'The ceramics', in Svoboda, J. (Ed.) (1994), *Pavlov I: Excavations 1952-53*, ERAUL 66, Liege. The Dolni Vestonice Studies Vol. 2.

Soffer, O. and Vandiver, P. (1997), 'The ceramics from Pavlov I- 1957 excavation', in Svoboda, J. (dir.) (1997), *Pavlov I- Northwest, the Upper Paleolithic burial and its settlement context*, The Dolni Vestonice Studies, Vol. 4, the Academy of Sciences of the Czech Republic, Brno.

Soffer, O., Adovasio, J.M., Illingworth, J.S., Amirkhanov, H.A., Praslov, N.D. and Street, M. (2000), 'Palaeolithic perishables made permanent', *Antiquity* 74, pp 812-21.

Svoboda, J. (ed.) (1991), *Dolni Vestonice II- Western Slope*, ERAUL 54.

Svoboda, J. (ed.) (1994), *Pavlov I: Excavations 1952-53*, ERAUL 66. The Dolni Vestonice Studies Vol. 2.

Svoboda, J. (1995), 'L'Art Gravettien en Moravie: contextes, dates et styles', *L'Anthropologie* 99 (2/3), pp 258-272.

Svoboda, J. (1996), 'The Pavlovian: typology and behaviour', in Svoboda J. (ed.) (1996), *Paleolithic in the Middle Danube region*, Spisy Archeologického Ústavu AV ČR v Brně Svazek 5, Archeologický Ústavu AV ČR, Brno, Pp 283-301

Svoboda, J. (ed.) (1996), *Paleolithic in the Middle Danube region*. Spisy Archeologického Ústavu AV ČR v Brně Svazek 5, Archeologický Ústavu AV ČR, Brno.

Svoboda, J. (ed.) (1997a), *Pavlov I-Northwest, the Upper Palaeolithic burial and its settlement context*, The Dolni Vestonice Studies (4), Brno: the Academy of Sciences of the Czech Republic.

Svoboda, J. (1997b), 'Lithic industries of the 1957 area', in Svoboda, J. (Dir.) (1997), *Pavlov I-Northwest, the Upper Palaeolithic burial and its settlement context*, The Dolni Vestonice Studies (4), Brno: the Academy of Sciences of the Czech Republic.

Svoboda, J. (1999), *Čas Lovců Dějiny paleolitu*, Zvláště na Moravě, Akademie Věd České Republiky- Archeologický ústav Brno.

Svoboda, J. (2003), 'The Gravettian of Moravia: Landscape, settlement and dwellings', in Vasilev, S. *et al.* (Eds) (2003), *Perceived landscapes and built environments*, BAR 1122, pp 121-129.

Svoboda, J. (ed.) (2005), *Pavlov I Southeast: a window into the Gravettian lifestyles*, The Dolni Vestonice Studies, Vol. 14, Academy of Sciences of the Czech Republic, Institute of archaeology at Brno.

Svoboda, J., Klima, B. and Skrdla, P. (1995), 'The Gravettian project: Activities during the 1991-1994 period', *Archeologicke Rozhledy* 47 (2).

Svoboda, J., Lozek, H. and Vleck, E. (1996), *Hunters between East and West: the Paleolithic of Moravia*, Plenum Pub Corp.

Taborin, Y. (2004), *Langage sans parole: la parure aux temps préhistoriques,* Paris: La Maison des Roches.

Tixier, J., Inizan, M.-L. and Roche, H. (1990), *Préhistoire de la pierre taillée 1 : terminologie et technologie*, Paris: CREP.

Trésor de la langue francaise. Dictionnaire de la langue du XIXeme et du XXeme siècle (1789-1960) Gallimard 1985, Nancy: CNRS Institut National de la langue française.

Trombe, F. and Dubuc, G., (1947), *Le Centre Préhistorique de Gantiès-Montespan (Haute-Garonne),* Archives de l'Institut de Paléontologie Humaine Mémoire 22, Paris: Masson.

Valoch, K. (1980), 'Comments', in Conkey M.W. (1980), 'The identification of Prehistoric hunter-gatherer aggregation sites: the case of Altamira', *Current Anthropology* 21 (5), October 1980, pp 609-630.

Valoch, K. (1996), *Le Paléolithique en Tchéquie et en Slovaquie*, Série Préhistoire d'Europe N° 3, Grenoble: Jérôme Millon.

Van Der Leeuw, S.E. (1994), 'Cognitive Aspects of 'Technique', in Renfrew, C. and Zubrow, E.B.W. (Eds) (1994), *The Ancient Mind, Elements of Cognitive Archaeology,* Cambridge University Press, pp 35-142.

Van Der Leeuw, S.E., and Pritchard, A.C. (eds.) (1984), *The Many Dimensions of Pottery, Ceramics in archaeology and anthropology,* Universiteit van Amsterdam, Netherlands.

Vandiver, P.B. (1988), 'Reconstructing and interpreting the technologies of ancient ceramics', in Sayre, E.V., Vandiver, P.B., Druzik, J. and Stevenson, C. (1988) *Materials issues in arts and archaeology*, Material Research Society symposium proceedings 123, Pittsburg, Pensylvania, pp 89-102.

Vandiver, P.B. (1997), 'Pavlov I pigments and their processing', in Svoboda, J. (Dir.) (1997), *Pavlov I-Northwest, the Upper Palaeolithic burial and its settlement context, the Dolni Vestonice Studies* 4, Brno: The Academy of Sciences of the Czech Republic.

Vandiver, P.B., Soffer, O, Klima, B. and Svoboda, J. (1989), 'The Origins of Ceramic Technology at Dolni Vestonice, Czechoslovakia', *Science* 246, pp 1000-8.

Vandiver, P.B., Soffer, O., Klima, B. and Svoboda, J. (1990), 'Venuses and Wolverines: the origins of ceramic technology, c.a. 26,000 B.P', in Kingery, W.D. (ed.) (1990), *The changing roles of ceramics in society: 26,000 to the present*, Ceramics and Civilization volume V, The American Ceramic Society, U.S.A., pp 13-81.

Vasile'v, S.A. (1985), 'Une statuette d'argile paléolithique de Sibérie du Sud', *L'Anthropologie* 89 (2), pp 193-196.

Verpoorte, A. (1997), 'Along the peripheries of a radiolarite concentration : the lithic industry of 1956/ABC and 1958', in Svoboda, J. (Ed.) (1997), *Pavlov I-Northwest, the Upper Palaeolithic burial and its settlement context*, The Dolni Vestonice Studies 4, Brno: The Academy of Sciences of the Czech Republic, pp 211-226.

Vialou, D. (1986), *L'Art des Grottes en Ariège Magdalénienne*, XXIIe Supplément à Gallia Préhistoire, Paris: Editions du C.N.R.S.

Vialou, D. (1998), *Our Prehistoric Past: Art and civilization*, New Horizions, London: Thames and Hudson.

Vlcek, E. (1997), 'Human remains from Pavlov and the biological anthropology of the Gravettian human population of South Moravia', in Svoboda, J. (Dir.) (1997), *Pavlov I-Northwest, the Upper Palaeolithic burial and its settlement context*, the Dolni Vestonice Studies. 4, Brno: The Academy of Sciences of the Czech Republic.

Whitbread, I.K. (2001), 'Ceramic Petrology, Clay Geochemistry and Ceramic Production – from Technology to the Mind of the Potter', in Brothwell, D.R. and Pollard, A.M. (Eds) (2001), *Handbook of Archaeological Sciences*, Chichester: John Wiley and Sons Ltd.

White, R. (1993), 'Technological and social dimensions of Aurignacian-Age' body ornaments across Europe, in Knecht, H., Pike-Tay, A. and White, R. (ed.) *Before Lascaux: the complex record of the E.U.P.*, Boca Raton, CRC Press, Florida, London.

White, R. (1997), 'Substantial Acts: From Materials to Meaning in Upper Palaeolithic Representation', in Soffer, O. *et al.* (1997), *Beyond Art: Pleistocene Image and Symbol*, University of California Press.

White, R. (2003), *Prehistoric Art, the Symbolic Journey of Humankind*, New York: Harry N. Abrams.

Whittaker, J.C. (1994), *Flintknapping, Making and Understanding Stone Tools*, Austin: University of Texas Press.

Wilson, L. and Pollard, A.M. (2001), 'The Provenance Hypothesis', in Brotthwell, D.R. and Pollard, A.M. (Eds) (2001), *Handbook of Archaeological Sciences*, Chichester: John Wiley and Sons.

Yellen, J. (1977), *Archaeological Approaches to the Present: Models for Reconstructing the Past*, New York: Academic Press.

RÉSUMÉ

Ce travail de recherche présente une étude de synthèse sur l'utilisation d'un matériau, l'argile, au Paléolithique Supérieur (environ 30 000 à 10 000 BP) en Europe, matériau qui a été très peu étudié dans ce contexte jusqu'à aujourd'hui.

Cette utilisation apparaît uniquement liée à l'expression symbolique pour cette période, en particulier pour la fabrication de modelages et figurines, à la différence des périodes plus récentes comme le Néolithique pour lesquelles l'argile fut un matériau utilitaire important (poterie).

Il est apparu également qu'en plus d'une série de cas isolés d'utilisation d'argile qu'on retrouve tout au long du Paléolithique Supérieur, mais qui ne seront pas détaillés ici (voir Table 1.2), on trouve en Europe deux contextes paléolithique différents dans lesquels il a existé une véritable tradition liée à l'utilisation de l'argile : la culture Pavlovienne (au Gravettien) en Moravie, République Tchèque, et le Magdalénien dans les Pyrénées françaises (Table 1.1). Les bases de données issues de l'analyse de ces corpus régionaux forment la base sur laquelle cette étude a été menée par une approche technologique et stylistique.

Le terme 'argile' utilisé ici englobe une définition large du mot qui dépasse sa définition strictement chimique. Ce sont en effet les qualités de la matière reconnues et utilisée par les Paléolithiques qui nous intéressent ici. Ainsi seront parfois appelés 'argile' des matériaux aux propriétés similaires, à savoir la plasticité et la capacité à durcir par séchage ou au feu principalement, et qui ont été utilisés de la même manière. C'est le cas de certains lœss et limons ou encore parfois du mondmilch pour sa plasticité.

L'utilisation par les Paléolithiques des qualités uniques au matériau argileux (plasticité, capacité à durcir par séchage, par passage au feu, à éclater au feu, réversibilité avant cuisson) ou partagées avec d'autres matériaux (dureté de pierre tendre, de matière osseuse, de pierre dure après cuisson, propriétés colorantes) permet d'aborder l'étude technologique en constatant les choix effectués au cours du travail. La reconstitution de chaines opératoires et schémas techniques associés est alors possible.

L'état actuel de la recherche sur le thème de l'utilisation de l'argile au Paléolithique Supérieur est assez limité. Le dogme selon lequel la céramique est apparue uniquement avec le Néolithique continue à avoir la vie dure malgré les récentes découvertes qui font remonter cette technique de plus en plus loin dans le temps, avec de la poterie en contexte pré-mésolithique datée entre 16,000 et 14,000 BP en Asie de l'est par exemple (Kaner 2003). Certaines utilisations du matériau en contexte Paléolithique sont connues de longue date, telles les fameux bisons d'argile du Tuc d'Audoubert (découverts en 1912) ou les figurines en céramique de Dolni Vestonice (connues depuis 1924), sans pour autant qu'on ait cherché à mieux comprendre leur contexte technologique global. Pour les Pyrénées françaises, le seul travail de synthèse existant résulte d'un travail de Master publié en 2005 (Boureux), le reste étant inclus dans des monographies ou articles sur des sites précis. Pour les sites de Moravie, les publications relatives aux céramiques ont été très partielles jusqu'à la fin des années 1980. Après cette date, l'étude systématique des résultats des fouilles anciennes a été reprise et peu à peu publiée sous la direction de Jiri Svoboda, processus qui est toujours en cours de nos jours. C'est dans ce cadre que les céramiques ont été étudiées par O. Soffer et P. Vandiver et leur équipe, avec un accent mis sur leur aspect technologique.

L'étude de l'utilisation de l'argile pour la période du Paléolithique Supérieur est un sujet peu étudié au potentiel bien plus vaste qu'on ne le pense habituellement. Pour cette raison, la présente recherche espère surtout montrer le potentiel important du sujet et dégager des voies à poursuivre mais ne prétend surtout pas à l'exhaustivité. Le sujet s'avère en effet trop complexe pour en faire le tour en une seule recherche doctorale.

Une approche technologique a été choisie pour aborder l'étude car c'est un facteur qui permet de lier lieux et périodes dans cette synthèse. Une étude formelle a aussi été menée car c'est d'expression symbolique qu'il s'agit ici, pour laquelle les forme et styles sont nécessairement importants. L'approche théorique choisie passe par la reconstitution des chaines opératoires associées (Schlanger 1994 ; Ingold 1990 ; Audouze 2002 ; Pélégrin 1991, White 1993, 1997 entre autres) et des schémas techniques correspondants (Karlin et Julien 1994 ; Knecht 1997) et par le regroupement de ces éléments en constellations de connaissances techniques (Dougherty and Keller 1982 ; Sinclair 1995, 2000). Ces éléments permettent de considérer la transmission de connaissance et de savoir-faire dans les différents contextes d'apprentissage des techniques du travail de l'argile.

Le Chapitre 2 présente l'analyse des données liées au travail de l'argile issues des sites Pyrénéens qui sont au nombre de quatorze : Tuc d'Audoubert, Enlène, Labouiche, Niaux, Mas d'Azil, Fontanet, Bédeilhac, Massat, Montespan, Labastide, Bois du Cantet, Erbérua, Oxocelhaya-Hariztoya et Etcheberri (Table 2.1) et qui regroupe un total d'au moins 289 unités graphiques reconnues (Table 2.2). Tous ces sites correspondent à des grottes, pour la plupart également ornées avec d'autres techniques que celles de l'argile.

L'inventaire des utilisations d'argile est détaillé pour chaque site avec un rappel sur le contexte artistique correspondant, une présentation des techniques de travail reconnues (modelage, gravure, tracés digitaux, impression, sculpture en 3 D) pour chaque unité graphique, ainsi que les relations reconnues ou pas avec le reste de l'art de la grotte ou avec la topographie de cette grotte.

Les thèmes de représentation qu'on retrouve dans l'art sur argile (Table 2.32) suivent ceux retrouvés habituellement dans le reste de l'art Paléolithique : animaux, humains et signes. On remarque dans les Pyrénées magdaléniennes des spécificités régionales telles qu'une dominance des représentations de bisons/aurochs dans la partie est des Pyrénées, alors que plus on avance vers l'ouest plus le cheval domine. En considérant uniquement l'art sur argile on remarque cependant que les deux thèmes, bison (ou aurochs) et cheval, ne se trouvent associés que dans deux sites (Niaux et Bédeilhac) sur les 14 inventoriés. Il est vrai que d'autres techniques artistiques se trouvent généralement aussi dans ces sites, qui peuvent compléter les associations animalières. La troisième espèce la plus représentée dans l'art sur argile est le bouquetin, ce qui correspond aux statistiques trouvées dans les autres techniques artistiques. Quand on regarde le détail des sites on constate cependant qu'on ne trouve cet animal dans les représentations sur argile que dans deux sites (Niaux and Massat). D'autres animaux représentés sur argile sont par ordre d'importance poissons, cervidés, ours et félins, qu'on ne trouve à chaque fois que dans deux sites. Un mammouth est aussi connu à Montespan et quatre indéterminés sont inventoriés.

Poisson, ours, félins et mammouth sont des thèmes rares dans l'art des cavernes magdaléniennes. A part pour Niaux et Bédeilhac, la variété des thèmes animaliers représentés sur argile est plutôt limitée dans les sites avec entre trois et un thème.

En recherchant s'il existe une correspondance entre le thème représenté et la technique employée pour la représentation (Table 2.33), on remarque que seules des représentations de bisons et de chevaux, avec une seule exception, l'ours de Montespan, ont été représentées par la technique du modelage, c'est-à-dire en trois dimensions. Les autres animaux sont représentés en deux dimensions par gravure ou tracés digitaux. Cette recherche de modelé naturaliste peut être vue comme une manière des renforcer l'importance symbolique de ces animaux dans les croyances magdaléniennes de l'époque.

Il est aussi intéressant de remarquer que la technique d'impression, à l'opposé, n'a servi que pour faire des représentations non figuratives.

Aucune représentation humaine complète n'a été trouvée sur argile. Deux vulves sont cependant inventoriées, une à Bédeilhac, une à Montespan.

Les signes quant à eux sont nombreux et constituent un peu plus de la moitié des unités graphiques reconnues sur argile (Table 2.34). Il y a une grande variété dans les types de signes mais deux types sont nettement plus récurrents : les signes linéaires formés d'associations de cupules (10 exemples) et les signe formés de lignes parallèles (15 exemples). Cela correspond aussi aux types de signes que l'on trouve dans le reste de l'art pariétal, si l'on compare l'effet rendu par les signes à cupules à ceux des signes faits de points juxtaposés. Il est vrai que le geste à effectuer est similaire entre appliquer un doigt recouvert de peinture sur une paroi ou appuyer sur une surface argileuse pour la marquer du bout du doigt. Il existe cependant un nombre de cas où les cupules formant le signe n'ont pas été faites simplement en imprimant du bout du doigt, mais avec un outil ou par modelage au doigt. Elles sont en général plus larges. L'intention dans ces cas semble bien être d'obtenir un effet visuel similaire, achevé cependant avec des moyens différents.

Les signes avec cupules ou formés de lignes parallèles ainsi que deux signes barbelés et un claviforme nous rappellent sans difficulté le plus large contexte artistique connu au Magdalénien dans les Pyrénées.

Quant aux techniques utilisées pour faire ces signes (table 2.35), ils sont en grande majorité (60.2%) gravés, puis imprimés (17.3%), tracés au doigt (14.3%) et enfin modelés pour quelques cas bien précis au Tuc d'Audoubert et à Fontanet uniquement (8.2% du total).

On peut donc conclure que la technique du modelage a été principalement utilisée pour représenter du figuratif (animaux et vulves), ce qui semble correspondre à un désir de donner du volume aux figures pour accentuer leur réalisme. Les signes quant à eux restent principalement en deux dimensions, n'exploitant pas les possibilités de rendre le volume offertes par le matériau argileux : cela peut être vu comme une confirmation de leur nature abstraite et symbolique. La technique de l'impression est appliquée uniquement pour représenter des signes.

Au niveau formel, les représentations réalisées sur argile par les techniques du modelage, de la gravure, du tracé digital, et de l'impression s'inscrivent clairement dans le plus large contexte de l'art pariétal magdalénien avec lequel ils partagent des conventions et ils sont souvent associés d'ailleurs. Elles doivent donc être étudiées dans le cadre de ce contexte artistique. Le choix de techniques différentes appliquées aux formes de l'art magdalénien des Pyrénées, celles de l'argile, est cependant un élément régional particulier.

Pour chaque technique du travail de l'argile reconnue, il est intéressant de déterminer si les propriétés propres à cette matière ont été reconnues et utilisée ou si des techniques connues et appliquées habituellement à d'autres matériaux ont été transférées. Cela permet de reconnaître la présence ou l'absence de transmission de connaissances particulières.

Le modelage est une technique artistique caractéristique de l'art des cavernes du Magdalénien des Pyrénées. En plus des exemples inventoriés, un nombre assez important d'accumulations artificielles d'argile notamment au Tuc D'Audoubert, à Labouiche et à Montespan sont interprétées comme des restes de modelages, aujourd'hui informes. Les modelages peuvent être en trois-dimensions, en haut-relief et en bas relief. Le schéma technique de la fabrication des modelages d'argile a été reconstitué (Table 3.37). On constate que les étapes obligatoires dans le travail sont peu nombreuses, le reste étant très flexible. La technique est souvent combinée à d'autres. Le modelage repose sur la propriété de plasticité de l'argile, donc à partir d'une matière malléable. Le fait que dans la majorité des cas reconnus de modelage (22 sur 35) (Table 2.38) de l'argile ait été transportée pour obtenir le résultat final voulu indique bien que cette technique n'est pas seulement appliquée de manière opportuniste, mais souligne l'intentionnalité de cette pratique.

La technique de gravure à l'aide d'un outil n'utilise pas les propriétés particulières de l'argile. Elle semble être une transposition directe des gravures sur parois de grottes, sans tenir compte de la nature particulière des surfaces. En ce sens un choix technique particulier est effectué, le choix de travailler avec un outil une surface qu'on pourrait simplement marquer au doigt. On remarque cependant une dominance des gravures sur sols argileux, ce qui semble être particulier aux Pyrénées.

Les tracés digitaux quant à eux utilisent la qualité de plasticité propre à l'argile, qui permet de la travailler simplement avec la main sans l'intermédiaire d'un outil. C'est la seule technique concernant un matériau argileux qu'on retrouve sur une aire géographique qui dépasse les Pyrénées françaises (Table 2.1). Les tracés digitaux se trouvent surtout sur des parois de grottes, à la différence des gravures. Le schéma technique suivi dans leur travail est le plus simple possible. On remarque que le plus souvent les surfaces marquées sont naturellement disponibles et n'ont pas été préparées par apport d'argile. Cette technique semble plus opportuniste.

La technique d'impression de l'argile n'a été utilisée que pour des images abstraites dans les Pyrénées françaises, formées d'accumulations de cupules, faites avec les doigts ou avec un outil. Les impressions de retrouvent surtout sur les sols. L'impression est une technique qui utilise pleinement la qualité de plasticité de l'argile.

Des sculptures mobilières se trouvent aussi sur argile dans les Pyrénées, avec quelques cas de céramique reconnus. Seuls trois sites ont livrés ce type d'objets qui semblent donc à part. A Enlène et Mas d'Azil ce sont quelques objets, dont certains cuits, qui ont été trouvés. Le processus technique est complexe, mais leur rareté ne permet pas de certifier l'intentionnalité de tout ce processus. Avec le cas de Bédeilhac, on trouve quelque chose de tout à fait unique au site : une série importante (au moins 122 objets, probablement beaucoup plus) de plaquettes de limon travaillé. Le matériau a été travaillé de manière rationnelle à différents stades de sa plasticité pour obtenir les résultats voulu, mais il ne semble pas y avoir de passage au feu de ces objets. Les propriétés de plasticité, et de capacité à durcir par séchage et passage au feu ont été utilisées dans ces cas.

L'argile a aussi parfois été utilisée pour ses qualités colorantes comme pigment mais on se heurte là au problème de la définition exacte des termes 'ocre' et 'argile' qui varie selon les sources et se confond parfois rendant une distinction mal aisée.

En conclusion, on peut dire qu'on remarque une combinaison des différentes techniques applicables au travail de l'argile dans de nombreux cas, ce qui montre une pleine appréciation des propriétés du matériau sous sa forme non cuite, et aussi une extrême versatilité dans son travail. Le transport et l'accumulation d'argile, parfois même quand la technique utilisée ne le demandait pas (Table 2.49), renforce l'impression que ce matériau a été utilisé avec intention dans un but d'expression symbolique par les Magdaléniens des Pyrénées.

Le Chapitre 3 concerne l'utilisation de l'argile en Europe Centrale, c'est à dire dans le Pavlovien (sous-culture régionale du Gravettien) de Moravie en République Tchèque. C'est là qu'on trouve les plus anciennes céramiques connues mondialement, sous la forme de figurines, entre 27 000 et 25 000 années BP. Les sites de Dolni Vestonice et Pavlov sont les plus connus mais ce phénomène de production de céramique s'avère de plus en plus comme un phénomène régional avec un total de 16 sites reconnus jusqu'à nos jours (Table 3.1). Les artefacts retrouvés pour ce contexte sont tous des figurines ou fragments de figurines, ou des fragments d'argile cuite sans traces claires de façonnage.

Les sites ayant fourni les plus grand nombre d'artéfacts en céramique sont Dolni Vestonice I et II et Pavlov I, avec un total de plus de 11 000 objets. Parmi ces objets, la grande majorité de l'inventaire figuratif a été analysée dans le cadre de notre étude, soit un total de 316 figurines ou fragments de figurines qui forment notre base de données.

Les thèmes de représentation trouvés dans cet art céramique se répartissent comme suit (Table 3.7) : des représentations humaines avec une domination des figurines féminines qui correspond bien au contexte Gravettien (22 figurines) ; des animaux (62) ; ce qu'on a appelé des extrémités corporelles (190 pièces) soit des pieds, jambes, pattes, possibles museaux, cornes ou trompes ; d'autres pièces qui ne rentrent pas dans les catégories ci-dessus : boules, fragments travaillés indéterminables (42).

Les figurations humaines se répartissent en figurines féminines, probablement féminines, masculines et anthropomorphes. Les figurines féminines et probablement féminines (identifiées par comparaison car fragmentaires) (Table 3.12) semblent suivre des modèles en particulier celui de la fameuse 'Venus Noire' de Dolni Vestonice I, la plus connue des figurines en céramique. Six figurines semblent se rattacher à de modèle. Elles proviennent toutes de Dolni Vestonice I et II. La présence de ceintures est un autre critère commun à ces figurines. Les représentations féminines trouvées à Dolni Vestonice I et II semblent plus homogènes que celles de Pavlov I, aux styles plus variés. Deux figurines sont identifiées comme masculines, une à Dolni Vestonice I qui est fragmentaire mais très réaliste (figure 3.27), et une autre plus stylisée à Pavlov I (figure 3.28). Six anthropomorphes ont aussi été identifiés (Table 3.13) qui sont trop imprécis, bien qu'étant souvent entiers, pour être sexués.

Sur les 62 représentations animales, la moitié est constituée d'indéterminés car les pièces sont trop fragmentaires ou imprécises pour être identifiées. Les 31 autres pièces sont dominées par des représentations d'espèces dites dangereuses, à savoir les félins et les mammouths (Table 3.14).

Les félins sont au nombre de 11. Ce sont tous des têtes isolées, très finement façonnées, pour lesquelles on a pu déterminer quatre types selon les détails du façonnage des museaux (Table 3.17). On y décèle à nouveau une plus grande uniformité stylistique à Dolni Vestonice I, mais cette fois des liens existent clairement avec Pavlov I car certaines pièces extrêmement similaires ont été trouvées dans les deux sites.

Les mammouths sont au nombre de 5, avec une seule pièce à Dolni Vestonice I et 4 à Pavlov I. Cette fois ce sont tous des corps de mammouth. Les pièces de Pavlov I sont toutes extrêmement différentes en style, ce qui accentue à nouveau la plus grande variété stylistique visible dans ce site.

Quatre représentations de chevaux sont également trouvées à Pavlov I et Dolni Vestonice II. Trois figurines d'oiseaux (toutes de Dolni Vestonice I), deux têtes de rhinocéros (une de Dolni Vestonice I et une de Pavlov I), deux rennes, un ours, un capriné, un jeune bovidé, et une petite tête de mustélidé complètent l'inventaire animalier.

On remarquera (Table 3.24) que les espèces représentés dans l'art en céramique ne correspondent pas à la faune retrouvée en fouille sur les mêmes sites, comme souvent en art paléolithique. Une importance symbolique spéciale semble donnée aux félins et mammouths, qui sont par ailleurs ce qu'on considère comme des espèces dangereuses. Il est aussi intéressant de voir des associations entre une espèce particulière et le fait qu'elle ne soit représentée que par des corps entiers ou par des têtes isolées, ce qui semble indiquer un phénomène intentionnel et non pas un résultat du hasard de la fragmentation.

Les extrémités corporelles forment de loin la catégorie de représentations en céramique la plus nombreuse dans le corpus étudié avec 190 objets, soit 60,3 % du total de l'inventaire. Ce thème inhabituel semble indiquer un système de représentations basées sur la synecdoque (soit une partie utilisée pour représenter un tout) largement appliqué sur les sites Pavloviens. Un nombre d'éléments appuient cette interprétation. Le fait que des types bien précis soient reconnaissables parmi ce corpus indique que les formes données suivent des modèles précis. Sept types différents, avec des sous-types ont pu être déterminés (Table 3.26 et

3.27). Pour la majorité leur attribution à une espèce précise reste sujette à caution. Pour quelques cas, une interprétation est proposée. Ainsi une série de pattes avec pieds est attribuée à des mammouths et une autre à des pieds humains, nus ou possiblement chaussés de bottes. On remarque également qu'une partie de ces extrémités corporelles sont raisonnablement trop grandes pour avoir été rattachées au corps d'une figurine entière par comparaison avec les tailles habituelles de l'art mobilier de cette époque (Figure 3.42). L'étude du degré de fragmentation des pièces permet aussi de voir qu'un certain nombre d'entre elles, même parmi les plus petites, sont en fait entièrement façonnées et n'ont donc jamais été rattachées à un corps.

Ces éléments renforcent l'hypothèse d'une représentation par synecdoque comme étant appliquée de manière systématique dans l'art céramique de Dolni Vestonice I et Pavlov I. On peut voir dans les représentations de têtes isolées, de félins notamment, le même phénomène. L'analyse stylistique du corpus céramique permet donc de déterminer des tendances notamment entre les sites, mais aussi au cœur du même site, avec une variabilité bien plus grande à Pavlov I qu'à Dolni Vestonice I.

L'analyse technique du matériel céramique a permis de déterminer précisément les chaines opératoires et schémas techniques liés à la fabrication (Table 3.33) tant pour la procuration du matériel, que le façonnage et la cuisson. On peut en conclure que la manière de fabriquer les céramiques moraves est tout à fait la même pour toute la durée de fabrication de ces objets, dans un processus quasiment standardisé. Les étapes complexes du travail sont parfaitement maitrisées. On voit une compréhension du matériau utilisé et de ses propriétés. Cette constation implique qu'une transmission précise des connaissances liées à ce travail de l'argile ait existé pendant toute la durée de fabrication des objets, soit une période de peut-être 2000 ans, d'une manière très stable, sans changement visible dans le processus appliqué. On remarque cependant associée à cette stabilité des techniques, une grande variabilité des styles des figurines, comme si le nombre de participants au façonnage était large.

Les contextes de connaissance reconnus au travers de l'analyse du travail de l'argile au Paléolithique Supérieur en Europe forment l'objet du chapitre 4. Ils y sont abordés en détaillant les quatre éléments de bases présents dans toute constellation de connaissance : les connaissances liées aux matériaux utilisés, aux outils du travail, aux processus techniques et aux finalités obtenues par ce travail (Table 4.1), avant d'aborder les aspects du contexte social visible.

Dans les Pyrénées, le contexte de l'utilisation de l'argile est celui de la grotte qui est aussi le lieu où l'on trouve naturellement cette matière. Il faut savoir la reconnaître et tester son état de plasticité. Sa préparation est réduite à son minimum, avec parfois un transport et une accumulation à un endroit voulu. Il n'y a pas d'action active sur la nature même de la matière dans le contexte Magdalénien : on la trouve dans des conditions de plasticité convenables pour toutes les utilisations liées à l'art pariétal. Les grottes fournissent ainsi in stock naturel d'argile plastique dont il fallait juste savoir qu'on pouvait l'utiliser.

En Moravie, le lœss des sols sur lesquels les habitats sont implantés est la source de matière première, lœss qui a pu être extrait depuis l'intérieur des habitats, avec un contrôle plus ou moins grand sur les personnes autorisées à voir ce processus, ou à leur alentours immédiats et donc probablement en vue des membres du groupe. Savoir se procurer la matière première n'est donc pas un processus compliqué. Par contre, la connaissance de ce qu'on pouvait faire avec cette matière est plus complexe : on ne peut pas deviner à partir de l'extraction simple du lœss brut que celui-ci devient plastique quand on le malaxe avec la bonne quantité d'eau ou qu'on peut le durcir de manière permanente ou encore le faire éclater intentionnellement sous l'action du feu. Ces savoirs ont besoin d'être transmis d'une manière plus organisée. Le lœss était systématiquement malaxé après ajustement du contenu en eau pour obtenir la plasticité souhaitée. Il est possible de préparer cette matière à l'avance si on peut la stocker d'une manière appropriée pour conserver son contenu en eau.

Un élément commun est en tous cas discernable entre les deux régions étudiées : l'origine locale de la matière première utilisée comme si dans ces contextes paléolithiques l'utilisation de cette matière appliquée uniquement à l'expression symbolique était aussi associée étroitement aux lieux mêmes où on l'utilisait, d'une manière peut-être plus visible que pour d'autres techniques artistiques. Cette association pourrait aider à expliquer qu'on ne soit pas passé à des applications utilitaires de l'argile au Paléolithique Supérieur.

Aucun outil spécifiquement dédié au travail de l'argile n'a été déterminé en fouilles jusqu'à présent, mais l'aspect extrêmement versatile de cette matière première implique que la plupart des outils existants pouvaient être appliqués à son travail. Sous forme malléable, l'argile est d'ailleurs un des seuls matériaux qui peut se travailler sans outil, simplement avec la main. Sous une forme partiellement séchée, le travail à la main peut encore être possible, mais devient plus difficile. Tout outil en matière végétale (bois) ou animale peut convenir, mais plus les détails du façonnage sont fins, plus l'outil lui-même doit être fin. Une matière argileuse bien séchée quant à elle acquiert une dureté équivalente à celle d'une pierre tendre, ce qui implique des outils d'une dureté supérieure. Après ce processus de façonnage, la construction d'une structure de combustion est nécessaire si on veut cuire l'argile. Ce peut être un simple foyer, mais à Dolni Vestonice, les fouilles ont fourni deux structures de fours primitifs sous forme de fosses creusées surmontées au moins partiellement par une couverture en lœss mouillé. Construire et utiliser des structures de combustion adaptées demande de comprendre et maitriser les étapes du processus de cuisson, ce que l'on peut considérer comme un savoir de spécialiste, nécessitant connaissance et savoir-faire. Le travail de l'argile sous ses autres états est plus simple.

Dans l'étude du concept de chaine opératoire, il est largement reconnu que les processus de l'action technique impliquent deux types de savoirs pour obtenir un résultat réussi : la connaissance et le savoir-faire. La connaissance regroupe les concepts de formes idéales et de matières premières et les gestes associées à l'action, dans une connaissance abstraite qui peut être acquise par simple observation des autres en action ou transmise pas enseignement même hors contexte. Le savoir-faire est un savoir physique, qui est à la fois une forme de connaissance et une forme de pratique. Il implique la capacité d'appliquer des concepts abstraits et d'évaluer le résultat d'opérations techniques, impliquant dextérité et jugement. Le savoir-faire est fortement lié à l'expérience pratique de l'individu et n'est pas vraiment transmissible, même s'il est nécessaire d'avoir du savoir-faire pour pouvoir transmettre des connaissances (Pélégrin 1991 ; Ingold 1990 ; Schlanger 1994, 2004).

On a pu voir que certaines propriétés de la matière sont prises en compte et comprises lors de son travail dans les contextes Paléolithiques concernés. D'autres techniques sont cependant appliquées sans raison techniques valables, comme dans le cas de la gravure à l'outil sur surface argileuse alors qu'un travail au doigt suffirait. Dans ce cas les raisons du choix de la technique appliquée doivent être considérées comme autres que techniques.

Dans les Pyrénées, la propriété de plasticité de l'argile est la plus utilisée et semblait recherchée ; la capacité à durcir par séchage et passage au feu a aussi été utilisée mais pas de manière aussi systématique, et seulement dans le contexte spécifique de certains sites. De grandes variations sont cependant visibles, quasiment au cas par cas. Mais on peut affirmer que dans l'art des cavernes des Pyrénées Magdaléniennes, c'est la propriété de plasticité de la matière argileuse qui a été principalement associée à son travail : on trouvait l'argile sous forme plastique, on l'utilisait sous cette forme et on abandonnait l'œuvre finie toujours sous forme plastique. La nature de la matière elle-même est très peu affectée par l'application des processus techniques.

En Europe Centrale l'utilisation principale d'argile visible par l'archéologie est la fabrication de figurines en céramique, ce qui implique une véritable transformation de la nature du matériau lui-même. Les propriétés recherchées de la matière sont celles permettant d'obtenir le résultat voulu : la capacité à durcir au feu de manière permanente et la capacité à éclater au feu. On pense en effet que les figurines ont été intentionnellement soumises à ce traitement, probablement pour des raisons rituelles. Le lœss utilisé pour les figurines est en effet naturellement très résistant aux chocs thermiques provoquant l'éclatement au point que certains chercheurs pensent qu'un processus particulier est nécessaire pour obtenir l'éclatement voulu (Vandiver *et al.* 1993a ; Soffer *et al.* 1993). La propriété d'éclatement au feu n'est pas normalement retenue en association avec l'argile, simplement car elle n'a pas d'application utilitaire. Elle est normalement vue comme un problème potentiel qu'on essaie d'éviter lors de la fabrication de poterie par l'ajout de dégraissant. L'ensemble du processus technique pour l'obtention des figurines céramiques utilise tour à tour les différentes propriétés du matériau qui permettent de passer de l'état de plasticité, à une dureté réversible par séchage puis à une dureté permanente sous l'action du feu, qu'il y ait ou non éclatement. Les propriétés uniques de la matière sont donc toutes comprises et prises en compte dans un processus technique complexe qui résulte en la transformation complète de la nature d'un matériau par une activité humaine. Il n'est pas étonnant de retrouver cette matière associée à une forme de symbolisme. Ce processus nécessite certainement un degré important tant de connaissance que de savoir-faire.

Les schémas techniques liés aux différentes techniques appliquées au travail de l'argile ont permettent de voir si des éléments récurrents sont présents et de déduire les connaissances et savoir-faire nécessaires et donc d'apercevoir le degré de transmission des connaissances visible ou inféré.

Dans les Pyrénées on ne peut pas proposer un seul schéma technique lié au travail de l'argile. Il semble que la connaissance de ce qu'on pouvait faire avec l'argile plastique et des contextes appropriés où le faire ait été un élément partagé et donc transmis au sein des populations magdaléniennes car cela donne sa base même à une tradition régionale particulière. Cette connaissance est simple et peut s'acquérir par simple observation. Ce n'est pas un savoir-faire. Les détails du travail de l'argile quant à eux semblent varier pratiquement au cas par cas, comme s'ils étaient laissés au libre choix de chacun.

Une comparaison avec le travail de la matière osseuse dans le même contexte du Magdalénien des Pyrénées a été faite à partir d'une étude de Dobres (1996, 1999, 2000), qui enregistre une importante variabilité dans les traitements technologiques appliqués à l'échelle d'un site comme à l'échelle régionale. Elle conclut qu'il n'y a pas de modèle Magdalénien de quoi faire, comment et où le faire mais qu'on voit plutôt une application au cas par cas de connaissance et niveau d'expertise. Dans ce contexte on entrevoit que les règles techniques de procédures et de comportement liées au contexte culturel Magdaléniens étaient simplement des grandes lignes partagées par les populations, le détail des applications dépendant des individus concernés. L'étude de la variabilité du travail de l'argile arrive à des conclusions similaires: on constate que la connaissance liée aux possibilités offertes par la plasticité de l'argile, notamment sa capacité à donner du volume aux représentations pariétales, étaient partagée par les Magdaléniens de cette région. Le détail exact du savoir-faire appliqué à ce travail ne semble pas quant à lui avoir fait l'objet d'une transmission.

A coté de cette variabilité technique, un élément remarquable dans le Magdalénien des Pyrénées est une grande uniformité stylistique dans l'art. Le respect des formes à partager semble avoir été un élément bien plus important que l'application de techniques particulières. On peut aller plus loin en impliquant qu'alors seules les connaissances étaient transmises, par observation passive ou par message oral, mais que les savoir-faire, notamment ceux liés à l'art et au travail des matières osseuses, ne faisaient pas l'objet d'une transmission précise, notamment par apprentissage. Il sera intéressant d'élargir ces observations à l'ensemble de la culture magdalénienne des Pyrénées pour faire des comparaisons.

En Europe Centrale, bien qu'il s'agisse également d'un contexte d'expression symbolique, le travail de l'argile se présente bien différemment. Le schéma technique de la fabrication des céramiques est très linéaire et homogène et montre un processus complexe avec une maitrise parfaite des diverses propriétés de la matière, et cela pendant toute la durée d'occupation des sites. Cela semble indiquer des connaissances et un savoir-faire partagés et donc transmis précisément au cours des générations par une forme d'apprentissage organisé. Il est vrai que la versatilité du matériau argileux fait qu'un apprentissage ne requière pas nécessairement d'être long, à la différence de la taille du silex par exemple, pour maitriser les savoir-faire. On remarque cependant des différences importantes de savoir-faire dans l'aspect formel du façonnage des figurines : certaines montrent un talent affirmé, d'autres ont pu être faite par un jeune enfant. On est tenté de penser que le processus de fabrication des figurines n'était pas nécessairement effectué par une seule personne d'un bout à l'autre, mais qu'il pouvait impliquer la participation d'un grand nombre de personnes, notamment pour les étapes de façonnage. Il y avait peut-être une séparation des savoir-faire liés a certaines étapes, comme le passage au feu, contrôlé par seulement quelques personnes, qui en quelque sorte ont permis la pérennité du processus.

Dans ce même contexte Gravettien de Moravie, le travail des matières osseuses montre également une adaptation totale et une grande compréhension des matériaux utilisés et de leur nature. Les techniques appliquées sont nombreuses et variées, mais semblent dédiées jusque dans leurs détails à l'obtention d'un artéfact précis, comme si la manière précise de fabriquer cet artéfact avait été transmise par un apprentissage des connaissances aussi bien que des savoir-faire. Un certain degré de standardisation technologique semble donc exister, même s'il faut rester prudent car les recherches dans ce domaine sont encore peu nombreuses.

D'un point de vue stylistique il y a peu en commun entre les objets faits en ivoire par exemple et l'art céramique. La matière travaillée semble donc aussi liée au type de représentation réalisée. L'art céramique est principalement naturaliste et réaliste, ce qui contraste avec un art sur ivoire très stylisé et abstrait. Le fait que certains objets très similaires se retrouvent en plusieurs exemplaires au sein des sites à céramique ou dans plusieurs d'entre eux, comme certaines têtes de félin, indique également que des critères très précis dans certaines représentations ont été partagés et suivis.

Par l'étude des utilisations d'argile, on remarque donc deux contextes culturels aux caractéristiques très différentes. Dans les Pyrénées françaises, les connaissances liées au travail de l'argile semblent partagées par les Magdaléniens mais pas les savoir-faire précis liées à ce travail. Une uniformité importante est par contre visible dans les formes données aux représentations. En Europe Centrale, le travail de l'argile apparaît comme presque standardisé, ce qui implique une transmission organisée des connaissances et savoir-faire associés. Certaines formes artistiques élaborées suivent également des critères très proches. Ceci est combiné avec une grande variété de styles pour d'autres représentations.

On trouve donc dans les Pyrénées une association du modelage au naturalisme pour donner un volume plus réaliste aux œuvres d'art. On remarque que l'art sur argile est principalement rencontré sur des sols et dans des galeries terminales ou des parties terminales de galeries secondaires. Il est vrai qu'au vu de la fragilité des œuvres en argile cette constatation peut être influencée par des facteurs de conservation. Cet art est aussi habituellement associé à d'autres techniques artistiques. La série de plaquettes sur argile trouvées à Bédeilhac est unique par la nature du matériau, mais s'inscrit dans un ensemble de découvertes d'autres sites avec plaquettes de pierre décorées dans la région, comme Enlène, Labastide ou Mas d'Azil entre autres. Il faut aussi noter qu'à Bédeilhac des plaquettes en pierre ont aussi été retrouvées. La tendance vers une sculpture monumentale sur argile qu'on voit au Tuc d'Audoubert et à Montespan semble être une caractéristique propre à l'art sur argile dans la région et diffère du plus large contexte artistique.

En Europe Centrale l'utilisation de l'argile est associée aux représentations réalistes en trois dimensions. Les représentations d'animaux dits dangereux (félins, ours, rhinocéros, mammouths) s'inscrivent dans le plus large contexte des thèmes de représentation artistique du début du Paléolithique Supérieur (Aurignacien et Gravettien ancien). Les représentations féminines participent bien du plus large contexte artistique Gravettien. La question de la destruction intentionnelle d'objet d'art mobilier lors d'activités rituelles est posée par l'étude des céramiques de Moravie.

Dans les Pyrénées, les utilisations d'argile fournissent une opportunité d'étudier la variabilité intra-régionale de l'art ; la notion que l'argile était un matériau convenable pour l'expression artistique et l'idée générale de l'utilisation de sa plasticité semblent avoir été des connaissances partagées parmi les groupes magdaléniens, ce qui implique l'existence d'échanges et de contacts.

En Europe Centrale, les représentations de félins semblent être le thème de représentation qui montre le plus d'uniformité formelle dans tous les sites à céramique, ce qui peut impliquer une signification symbolique très particulière attachée à cet animal. On remarque des variations dans le détail de l'inventaire des pièces avec par exemple des extrémités corporelles plus grandes dans leur ensemble à Dolni Vestonice I qu'a Pavlov I, où il semble y avoir plus de variabilité, ce qui se retrouve aussi dans les différents styles rencontrés. Les représentations de mammouth, si on y associe les pattes de mammouth, sont aussi largement dominantes à Pavlov I, ce qui devient une caractéristique de cette localité.

La nature de ces variations peut être aussi regardée avec l'idée que dans les deux régions étudiées, de nombreux sites où l'argile a été utilisée sont considérés comme des sites possibles d'agrégation (voir Conkey 1980, 1985), où des rassemblements réguliers de chasseur-cueilleurs prenaient place. Il s'agit de Dolni Vestonice I, Pavlov I, Mas d'Azil et de peut-être Bédeilhac et Enlène. On y trouve normalement des activités artistiques en quantité et de grande variété stylistique, résultant d'activités rituelles, et c'est souvent un contexte propice au développement d'innovations. Les utilisations d'argile ont pu parfois se développer dans ce type de contexte. Des techniques résultant en la fabrication d'artéfacts particuliers et uniques à certains sites peuvent en effet s'expliquer ainsi : le fait que les figurines en céramiques soient uniques, dans l'état actuel de la recherche, aux sites des collines de Pavlov en Moravie par exemple ; ou l'existence de plaquettes de limon travaillé à Bédeilhac qui correspondent à une innovation propre au site. Dans ces cas, les utilisations d'argile semblent correspondre à des innovations dans des contextes sociaux très particuliers. La versatilité de cette matière lui donne un potentiel important pour être utilisée dans ces applications particulières et aussi pour nous informer grâce à l'étude des utilisations faites au Paléolithique Supérieur sur ces contextes sociaux particuliers. Dans ce sens, l'étude de ce matériau peut encore apporter beaucoup et a besoin d'être poursuivie.

APPENDIX 1: CATALOGUE OF USES OF CLAY IN THE FRENCH PYRENEES

The following inventory gathers the recognised individual graphic units found in the cave sites of the Magdalenian in the French Pyrénées; the graphic units can be figurative in the case of an animal or vulvae representations for example, or can be described as signs, often made of separate elements used in relation with each other (i.e. a series of dots forming a pattern). These elements were determined from clear mentions in the literature or were seen during fieldtrips. This inventory forms the database from which the present study of the use of clay in the Upper Palaeolithic of the French Pyrénées was carried out.

The graphic units are presented for each of the studied sites, in the same order of presentation as in the main part of the research (Chapter 2 in volume 1); they are designated by the inventory (reference) number given to them for the study, which only corresponds to the present catalogue. When an original inventory number previously existed, it is also mentioned in bracket. Then for each graphic unit, measurements (always given in their maximal dimensions) and a short description are given together with an illustration, when it was possible to obtain one, and the corresponding sources of reference.

Beside recognisable graphic units, the presence of macaronis finger-tracings was indicated separately in the sites where they are found; by their nature, they are difficult to quantify but their presence remains important as evidence of the technique used.

Any traces found on the clay of the cave that reflects possible human activity in the site (clay removals, clay balls, foot, hand and finger prints, various structures, small mounds) are also listed at the end of each section. They are part of a poorly studied and poorly known body of evidence that would seem to indicate some forms of human activities in the cave, possibly connected with the rituals involved in the making of cave art. They have been added here simply as an indicator for possible further research.

Tuc D'Audoubert, Ariège

Reference number	Graphic unit	Size	Description	Illustration (when available)	References
TA 1	Female bison *modelage*	61 cm by 11 cm	Three-dimensional sculpture representing a female bison, clearly sexed, situated in front the second bison sculpture at the site. The whole body is shaped and modelled by hand from a large clay plaque, and traces smoothing by fingers and hands are still visible on the whole surface. Details are added by engraving with tools and finger-tracing to indicate the mane, some of the hair, details of the head with the eye indicated by a hole. Tail and horns are modelled in relief. The bulk of the female bison was put into its final position by supporting it with lumps of clay set between it and the rock against which it is lying. The whole figure shows a great mastery of the technique used and much talent in the representation. It is found in the terminal gallery in Tuc d'Audoubert.	 (From Bégouen and Breuil 1958, plate 26)	Clottes *et al.* 1994:126; Bégouen and Clottes 1984b; Bégouen *et al.* 1977; Beasley 1986.
TA 2	Male bison *modelage*	63 cm by 13 cm	The second large three-dimensional bison representation in Tuc d'Audoubert is very similar to the first one but variations in details individualise it. It is slightly larger and more massive than the other, and thus has been interpreted as a male bison although the sex is not indicated. Its eye is indicated in relief. Stones have been used to support it against the large rock support. It can also be described as a masterpiece among Upper Palaeolithic art works. It is found in the terminal gallery in Tuc d'Audoubert.	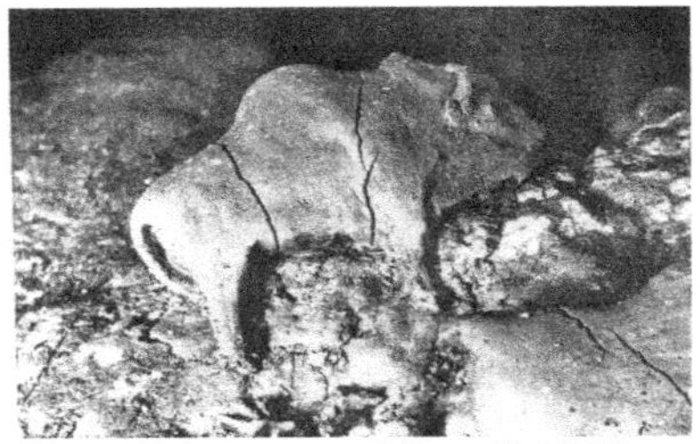 (From Bégouen and Breuil 1958, plate 26)	Clottes *et al.* 1994:126; Bégouen and Clottes 1984b; Bégouen *et al.* 1977; Beasley 1986.

Tuc D'Audoubert, Ariège

TA 3 (MAN 58030)	Small bison sculpture	11 cm by 11.5 cm	Small portable sculpture of a bison found on the floor next to the two large bison sculptures. The material is dried only. It is brownish-greyish. Inclusions are visible in the deep groove making the belly line and the back legs. The hump on its back is typical of the bison. The head is missing. The back and the rear of the animal are very thin in comparison to the base and the head. The edge forming the back is much smoothed and worked; it nearly seems polished. It was worked with a tool and shows a continuation of small flat surfaces, which combine to form the curve of the back, as if they were made one after the other. The tail is marked by a groove. The line under the belly is quite large and was roughly smoothed, but it does not show the same type of patina as the edge of the back. On the other side, some rough scraping marks are visible, and the surface is littered with little agglomerates of clay, probably from the floor on which it was posed. It was found in the Terminal Gallery, near the block with the bison *modelages.*	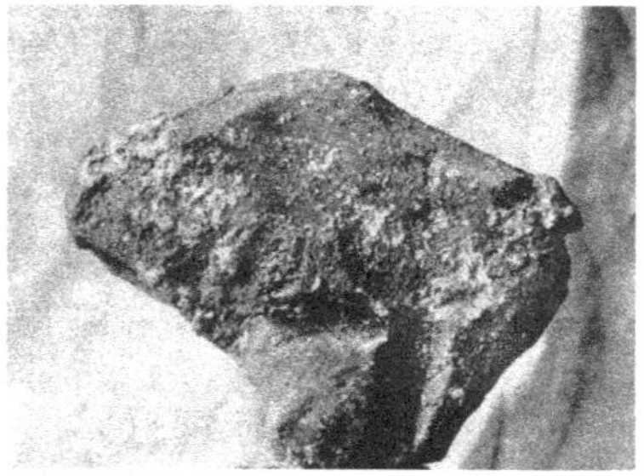 (From Bégouen and Breuil 1958, plate 29)	Bégouen *et al.* 1977.
TA 4	Bison	51 cm long	A partly engraved, partly modelled bison is found directly on the clay floor in the vicinity of the large bison sculptures in the Terminal Gallery. The line of the back is deeply marked and modelled in relief; the tail and the rear leg well marked. The belly, the head and the front legs are hardly indicated with quick finger strokes; one horn at least was indicated in relief by a modelled piece of clay that broke off but that is still visible. Remains of the second horn are mentioned by Bégouen (1912) as having been crushed by a fallen stone. For Bégouen *et al.* (1977: 117), this bison was finished and intentionally left like it is found today: they disagree with all the previous publications considering it as the preparation for extracting it from the floor to make another three-dimensional bison *modelage*.	 (after Bégouen *et al.* 1977, fig. 1)  (from Bégouen *et al.* 1977, fig. 5)	Bégouen *et al.* 1977: 117; Bégouen 1912.

Tuc D'Audoubert, Ariège

TA 5 a, b, c, d, e	Modelled clay coils		On the floor of the chamber of the Heels, five modelled clay coils were found. They form coil-like fragments that were originally interpreted as phalluses (Bégouen and Breuil 1958), but that are now rather possibly thought to be remains left while testing the plasticity of the clay (Beasley 1986).	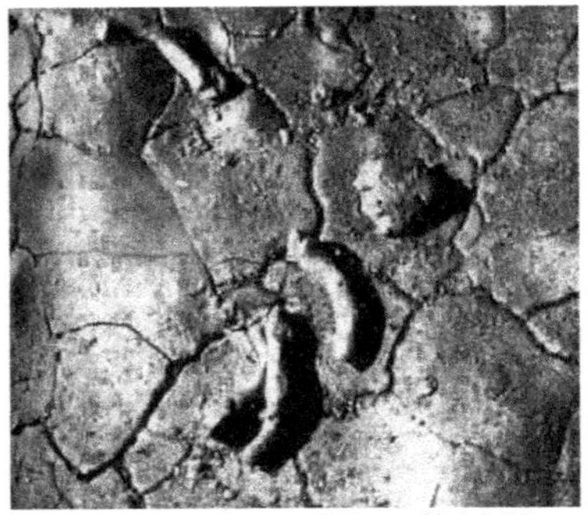 (From Bégouen and Breuil 1958, plate 28)	Bégouen and Clottes 1984b: 412; Bégouen and Breuil 1958: 98; Bahn and Vertut 1997; Beasley 1986.
TA 6	Engraved oval sign		One large oval sign is mentioned as engraved on the floor of the Chamber of the Heels in Tuc d'Audoubert, without more description or illustration of it existing.		Bégouen and Breuil 1958: 101; Bégouen and Clottes 1984b: 413
TA 7	Elongated 'pseudo-barbed' engraved sign		A thin layer of clayey stalagmite naturally covered all the floor surfaces in the Chamber of the Heels. It forms a fragile layer that was possible to detach with precaution without damaging the prints underneath, on the edges of the accessible area (Bégouen and Breuil 1958). Thus natural moulding of surface traces indicating the slightest details were obtained such as the 'pseudo-barbed' sign now kept in the Musée de l'Homme in Paris. It forms a line with sorts of commas arranged symmetrically on each side of it.	 (From Bégouen and Breuil 1958, plate 29)	Bégouen and Breuil 1958: 101; Bégouen and Clottes 1984b: 413
TA 8	Impressed circular sign		A dozen of holes made by finger tip impressions are arranged in a circular shape in the Chamber of the Heels and might be in connection with other linear figures.		Bégouen and Breuil 1958: 101

Tuc D'Audoubert, Ariège

TA 9	Impressed linear sign 1		9 impressed fingertip holes forming a line on the floor of the Chamber of the Heels.		Bégouen and Breuil 1958: 101
TA 10	Impressed linear sign 2		12 impressed fingertip holes forming a line on the floor of the Chamber of the Heels.		Bégouen and Breuil 1958: 101
TA 11	Impressed linear sign 3		15 impressed fingertip holes forming a line on the floor of the Chamber of the Heels.		Bégouen and Breuil 1958: 101

Finger-tracings	Panel 1: a panel of macaroni tracings is found in the side gallery above the 'Salle Nuptiale' in Tuc d'Audoubert together with engravings, all this in the middle level of the Tuc karst, lower than the terminal gallery where the bison are found (Bégouen and Clottes 1984b: 413). Panel 2: a few dozen metres away from the 'cat flap' marking the beginning of the upper level of Tuc d'Audoubert, a panel of engraved macaronis are found on the ceiling. There is no trace of art between this panel and the Chamber of the Heels near the very end of the gallery, where the art found is also on clay (Bégouen and Clottes 1984b: 413; Bégouen and Clottes 1981b).

Traces of human activity on clay	Some masses of clay are described as smoothed and pierced, possibly in some ritual activity (Clottes 1989: 80; Bégouen and Clottes 1984b: 414). Footprints can be counted 'by hundreds' in Tuc d'Audoubert (Bégouen and Breuil 1958: 96) and finger prints are also very numerous (1958: 101). Traces of clay removal are found near the Chamber of the Heels (Bégouen and Breuil 1958: 97). Four lumps of worked clay, rather shapeless, are seen next to the large bison sculptures: two are left between the two bison and still show fingerprints on them, the two others are near the head of the female bison (Bégouen et al 1977:116). In the Chamber of the Heels, there are traces of the removal of a large bloc of clay, with a broken stalactite abandoned on its edge. It is thought one of the clay plaques to make the large bison was extracted from there (Bégouen and Breuil 1958: 97). Among the traces in the chamber, about 50 heel prints are found, which have been interpreted as children participating in some sort of ritual, or simply playing. Some footprints are also visible in this Chamber and all seem to belong to children by their size (Clottes 1989: 80; Bégouen and Clottes 1981b; Bégouen and Clottes 1984b: 414).

Enlène, Ariège

Reference number	Graphic unit	Size	Description	Illustration (when available)	References
E1	Animal figurine	2.7 cm by 1.9 cm	Headless animal figurine made of fired clay, from the 'Salle des Morts' in Enlène. The head is missing. The front and back limbs are indicated by two roughly fashioned stumps. A rounded protuberance is marking the tail. The body is twisted in a position indicative of movement. Some traces of fashioning under the front legs and the line of the belly form a sort of hook, and could represent the front legs folded as would be seen in the position of a jumping animal. This, as well as the rounded tail, would fit well an interpretation as a mountain ibex. Two striations made with a flint tool are also visible at the neck.	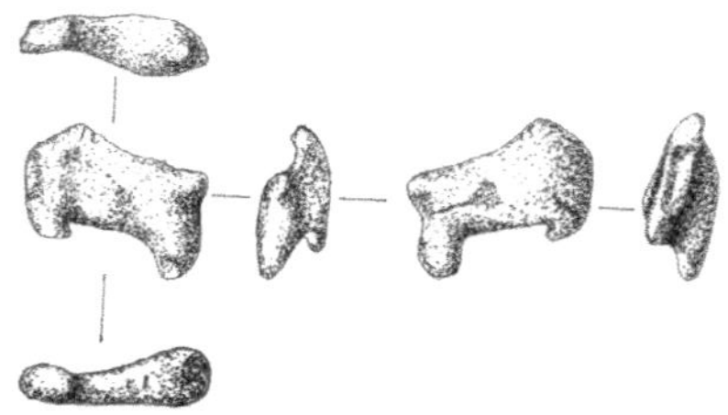 (Begouen private collection)	Bégouen and Bégouen 1936; Bougard 2003
E2 (ESF70 in Bégouen private collection)	Possible animal head	5.5 cm by 4.7 cm by 3.5 cm	This object comes from the 'Salle des Morts' area. It seems made of heavy fired clay with a fragile powdery surface. It is much damaged with a whole side apparently destroyed, as no trace of working of any sort is visible on it. A few elements on one side allow putting forward an interpretation as an animal head. Various traces of removal of material are visible on the better-preserved side of the object that mark the chin, the mouth and the forehead. A small flat surface forming a sort of oval shape of 2.1 cm by 1.7 cm in maximal dimension is still well defined on one side. Its surface is rather irregular, and many traces of working with a very small pointed tool are still visible, as if it was used to level the surface. A face forming a generally oval surface, about 5 cm by 3.5 cm, is also very roughly flattened. If we admit this is an animal head, this surface would be the neck.	  (Begouen private collection)	Bougard 2003

Enlène, Ariège

E3 (ESF70 in Bégouen private collection)	Facetted block with geometrical engravings	4 cm by 3.8 cm by 3.6 cm	This object is from the 'Salle des Morts' in Enlène. It appears made of unfired clay, or possibly of clay fired at such a low temperature that it did not undergo a ceramic change (under 500oC). This is a small block with flattened and engraved faces. It has a base, a top face not exactly parallel to the base, with four flat faces and two small irregular faces (probably breakages) connecting them. Three of the faces bear engravings of lines that combine to form a chevron motif on one of the faces (see tracing opposite). The whole object is not very regular and the surfaces are very worn and fragile. It is a sort of yellow ochre colour, with slight traces of dark red colorant in places. The broken base shows a material much harder inside the object that on the surfaces.	 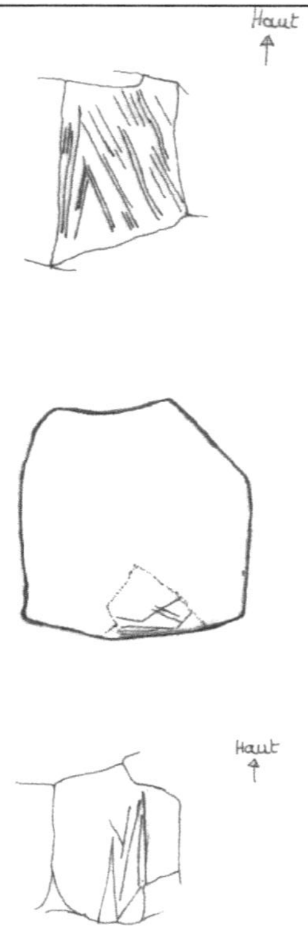 (Begouen private collection)

Traces of human activity on clay	In Enlène, some other traces of clay uses have also been found. They are notably some brick-like fragments of fired clay (Bégouen 1921:524)), and what looks like the levelling of the floor of a living area with clay (Clottes 1995).

Labouiche, Ariège

Reference number	Graphic unit	Size	Description	Illustration (when available)	References
	Labouiche bison sculpture	12 cm by 8 cm by 3 cm	A statuette of a headless bison made in dried clay was found at the end of the Dunac Gallery in Labouiche cave. It is described as being similar to a low relief as one side is better fashioned than the other as if the object was possibly intended to be seen only from one side. The animal is represented headless. The whole front leg and the bottom half of the rear leg is missing. These mutilations are ancient, as the section of the breaks is irregular and does not show the same careful smoothing than the rest of the edge. It looks rough, not very finished, but realistic at the same time. The front carefully is fashioned, with the enormous back hump that does not allow any doubt about the represented species, a bison. The strongly marked depression of the side of the hump was done by a finger, which print is clearly visible.	 (From Méroc 1959, fig. 11)	Méroc 1959; Pascal Allard, personal communication August 2004.

Traces of human activity on clay	A structure, possibly a hearth, is dug out in the clay floor. It is circular, 1.1 m in diameter, 30 cm deep, with a 10 high rim of clay around it. It is found at the entrance of this part of the Dunac Gallery. Numerous small holes between 20 and 30 cm in diameter are dug out in the clay floor next to each other, in a zone with low ceiling, with apparently accumulations of the excess material in between the holes and a few with charcoal pieces in them. Méroc (1959) seems quite puzzled by them, noting that he is not even sure they are artificial. They are further towards the end of this part of the gallery. Clay removals are seen in at least two locations in the vicinity of the hearth where engraved plaquettes were found, one being very close to the bison modelling. Clay balls (number not précised) and lumps (at least four) are also found next to the hearth number 3 at the foot of the walls. There are also prints of fingers in the clay banks in at least two places. (Méroc 1959)

Niaux, Ariège

Reference number	Graphic unit	Size	Description	Illustration (when available)	References
N1 N2	Fish N° 47, Fish N° 48	30 cm, by 11 cm; 13 cm	The two fish engravings are found a few dozen metres before the Salon Noir, on the floor on the right of the gallery when heading towards it. The largest one is still well visible today: it shows details such as eye, ears, fins, a horizontal line splitting the body, the tail. It is thought to possibly represent a trout. The second fish, the smaller one, has nearly disappeared. At the time of discovery, it was turned towards the right in front and slightly below the other fish, with the contour, the fins and the tail well drawn. Clottes (1995) does not indicate which fish corresponds to which reference number.	(From Clottes 1995: figure 106) 	Clottes 1995: 100
N3	Sign		Between the two fish engravings N1 and N2 were a dozen of short parallel lines, which have been destroyed today but are still visible on some old photographs.		Clottes 1995: 100
N4 N5	Fish N°60 Fish N°61	Both about 40 cm long	Placed underneath the deer figure 55 of panel 3 in the Salon Noir. They represent basic fish shapes engraved on the clay floor. They face towards the right.	(From Clottes 1995: figure 115) 	Clottes 1995: 103
N6	Group of Signs		Next to the fish N4 and N5, Clottes mentions half a dozen lines of various length, one T shaped sign, one oblique line with a barb on the right that reach the head of the second fish, another line that intersects with its tail and a series of six parallel lines intersected by one other line.		Clottes 1995: 103
N7	Group of Signs N° 85		Under the large horse 74 in panel 4 in the Salon Noir, signs are visible engraved on the floor and described as one barbed line and one curved line.		Clottes 1995: 109

Niaux, Ariège

N8	Horse N° 87	70 cm long	It is the most readable figure on the floor composition: a horse, turned toward the right, quite simple but very recognizable, with a well drawn head with oval eye and a beard. The ear is indicated, the mane is doubled. The line of the back is interrupted on both sides. The rear leg with its hoof and the tail are well detailed but the belly and the front leg are unfinished. Two lines on the shoulder are followed on the left by a large angular sign and a series of 7 bars that possibly represent the hair.	N8 and N9 (From Clottes 1995: figure 125) 	Clottes 1995: 110
N9	Undetermined animal N° 88	25 cm long	This figure is superimposed to the horse 87. Breuil interpreted it as a rhinoceros turned towards the left, but Clottes (1995: 111) points out that it could be as well a badly drawn bison, or many other animals and says it is best left undetermined.		Clottes 1995: 111
N10	Bison N° 89		Only the front of the animal turned to the left was represented. The horns are forward. The line of one horn attaches to the top of the forehead, the line of the other to the mane. The eye and the ear are represented as well as the hump, the beard, the sketch of the front leg and the beginning of the line of the belly. It has been interpreted as possibly a young male.	(From Clottes 1995: figure 126) 	Clottes 1995: 111
N11	Bison N° 90		This figure is found perpendicular to the bison 89. It represents the rear side of a bison, also turned to the left. It can be identified thanks to its hooves. The tail is raised vertically.		Clottes 1995: 111

Niaux, Ariège

N12:	Bison No 91		The front part of a bison, roughly engraved, is turned to the right, with the head, the two horns, the short mane and the beard indicated. The eye is represented by two lines. The sketch of the front leg is stretched forward and the head is lowered as if it was sniffing.	(From Clottes 1995: figure 130) 	Clottes 1995: 111
N13	Ibex No 92		Two parallel lines could be all that remain (after M. Garcia who traced the engravings) of a small ibex turned to the right traced by Breuil and nowadays disappeared. It could be the remains of its horns.	(From Clottes 1995: figure 127) 	Clottes 1995: 111
N14	Series of signs		Together with the preceding animals, one can add half a dozen series of lines, more or less parallel, of 3, 8, 7, 8, 15 lines that are found on the edges of the engravings.		Clottes 1995: 111
N15	Bison 141 and 142		Drawn by Breuil, but nowadays mostly destroyed: with only the tail of the lower bison remaining. Breuil indicated and represented two whole bison turned in opposite directions, one above the other. The one above turned toward the right is interpreted as an adult female in a wailing position. The one below, head lowered and the tail slightly raised, is thought to be an adult male, ready to charge, nervous. There is a barbed sign just above its horns.	(From Clottes 1995: fig. 147, after Breuil) 	Clottes 1995: 111, 117, 120

Niaux, Ariège

N16	Bison N°119	70 cm	This figure is a large bison head, turned to the left, with a circular eye, the contour of which is indicated by three large lines, and thin horns. The muzzle and the forehead are done in one line. Just below, oblique parallel lines represent the beard.	N16 (bison 119), N17 (Bison 120), N18 (Ibex N121), N19 (Horse 122), N20 (Horse Drawing of the group (From Clottes 1995: figure 141) 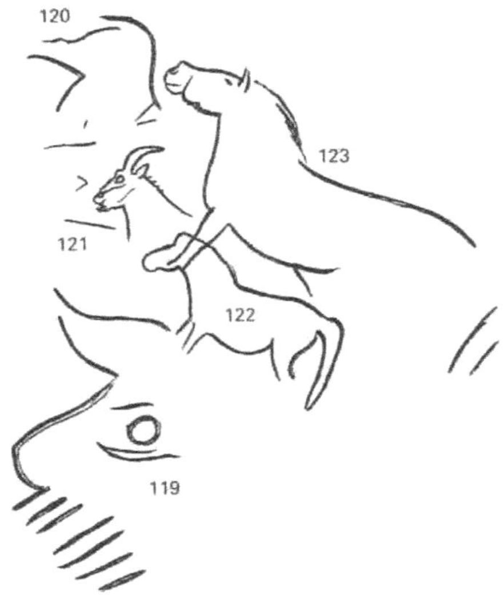	Clottes 1995: 117
N17	Bison N°120		This figure probably represents the hind side of a bison turned to the right.		Clottes 1995: 117
N18	Ibex N° 121	Height 20 cm	Above the large bison head (N° 119), an ibex head (N° 121) is turned to the left. Two short horizontal lines are found in front of the head. An angular sign points towards its forehead. It has one horn and one ear. The eye is oval, the forehead and the muzzle are well marked as well as the chest and the beginning of the line of the back.		Clottes 1995: 117
N19	Horse N° 122	47 cm	Just above and slightly behind the horns of the bison a whole horse is found, with a long thick tail, turned to the left. The silhouette is simple without details for the head, the mane or the hooves, but it is well proportioned with only one rear leg and the two front legs, the one on the right being sketched with a short line. This is the engraving that is the most visible in this group today because it is very deeply engraved and better preserved.		Clottes 1995: 117
N20	Horse N° 123	80 cm	A larger horse was partly superimposed to the previous one. Breuil represented it facing left, with the eye, ear, muzzle, mane, beard, chest, unique front leg, the lines of the belly and the back. The rear end is missing. It has been destroyed.	Photograph of Bison 119 and horse 122 (From Clottes 1995: figure 166) 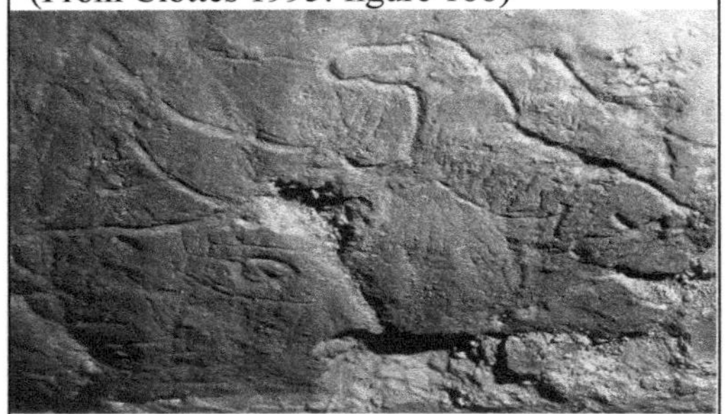	Clottes 1995: 117

Niaux, Ariège

N21	Ibex N° 136	30 cm	It is part of a group of animals (together with the 'bear paw' sign 137 and the aurochs 138) engraved behind a chaos of large blocs in what form a small chamber at the entrance of the Salon Noir. The small ibex faces left. It has one long and wide horn and might represent an adult male. A line marks the separation between the belly and the side. The body is massive, without much detail. The line of the back is intersected y a long line.	(From Clottes 1995: figure 151) 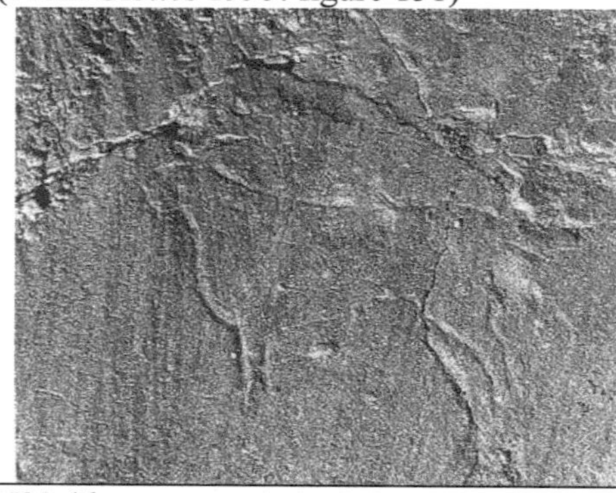	Clottes 1995: 119, 120
N22	'Bear paw' sign N° 137	about 18 cm in diame-ter	This strange figure is found a few centimetres below the ibex 136. It is rounded in shape with five globular shapes on the top, underlined by a curved line. Breuil thought it was a closed fist, Beltran an undetermined animal print, Clottes thinks it is the reproduction of a bear footprint likeable to some real ones that are found in some parts of Niaux.	N22 ('bear paw print' 137), N23 (aurochs 138) and N21 (ibex 136) (From Clottes 1995: figure 146) 	Clottes 1995: 119, 120

Niaux, Ariège

N23	Aurochs N° 138	40 cm	Below the 'bear print', an aurochs figure faces left, recognizable because of its horns pointed forward. It seems rather thin, possibly representing a young in spite of the length of the horns. The legs are roughly sketched. The eye is a natural cupule that existed before the figure was drawn, exactly like for the bison with cupules N27. A long double-barbed sign reached the chest, a shorter one the top of the side.	(see above)	Clottes 1995: 119, 120
N24	Horse N° 139		This animal is part of the second group or floor engravings found in the small side chamber near the entrance of the Salon Noir, which is on a higher part of the floor, horizontal this time. The horse faces right. It is well preserved and recognisable by it straight mane and two lines on the shoulder. The muzzle is square; the four legs are indicated but sketchy. The body bears three signs, one being angular. The animal seems to be in movement.	(From Clottes 1995: figure 148) 	Clottes 1995: 120
N25	Bison N° 140	45 cm	In an opposite position compared to the previous horse, the bison is facing left. It is quite roughly made with a badly drawn head and its hump is too backward on the back. It seems to be a female in movement that is calling its young. Breuil did not locate it rightly in comparison with the horse. In fact the line of the back of this bison follows the natural contour of the floor in which the hump fits well and its head is orientated toward the hind side of the horse.	(From Clottes 1995: figure 149) 	Clottes 1995: 120

Niaux, Ariège

N26	Aurochs figure with circles	110 cm	This figure is the only work of art found on the floor of the 'Galerie de l'éboulis' together with some red dots. It is turned toward the right. It is recognisable by its parallel horns directed forward and by the line of the back being different from that of bison. The drawing is basic even if the legs are indicated and their perspective respected. Details are absent. Three circles of different sizes have been added to the top of the body. They remind of the red dot on the bison with cupules and of the barbed signs found on many of the animals of the Salon Noir. A few other lines are visible on the animal: a line that intersects with the circle on the neck, two curved lines on the rear and three converging signs forming a barbed sign that reach the animal in the chest. This engraving was walked on on the rear part of the animal.	(From Clottes 1995: figure 85 and figure 39)   Clottes 1995: 86
N27	The bison with cupules		This figure of bison has been drawn around existing natural cupules on the clay floor: one has been used to make the eye, and the others on the body have arrow-shaped signs reaching them. This can be seen as a form of use of natural relief or shapes in the art on the clay floor, in the same manner that natural relief is commonly integrated into the representations drawn on the walls of decorated caves.	(From Clotte 1995: figure 40)   Clottes 1995: 120, Malvesin-Fabre et al. 1953

The figures listed above are the ones for which it was possible to find a description and/or illustration. A number of 41 signs is also mentioned in Vialou 1986 and Clottes 1995 without more description. These have been taken into account in the inventory tables but are not detailed here.

Traces of human activity on clay	Numerous footprints and some finger traces were found in Niaux and in the Réseau Clastres that continues the cave (Clottes and Simonnet 1972: 522-523, Clottes 1984, 1995).

Mas D'azil, Ariège

Reference number	Graphic unit	Size	Description	Illustration (when available)	References
MA 1	Possible feline head	3.2 cm long	It is a possible feline head described by the Péquart (1960) as a small 'lump of silt mixed with ochre and some fat that gives it a beautiful red brick colour'. For them, it was clearly modelled, with a massive muzzle and a wide-opened mouth, severed at the neck intentionally. From examination, this object appears rather made of fired clay It is much worn and this has unfortunately erased most of the details of the working of the surface. Only the general shape remains, and its interpretation as a feline head is rather debatable as only the general shape vaguely reminds us of one. It was found in 'Galerie des Silex' in the cave.	One side:  Reverse side:  (Mas d'Azil museum, Péquart Collection)	Péquart 1960: 278
MA 2	A small bison figurine	6 cm by 1 cm	This figure was recognised by the Péquart as a headless and limbless bison because of the typical hump on the back. It was modelled out of clay and unfired. It has unfortunately been destroyed after its discovery. It was found in 'Galerie des Silex' in the cave.		Péquart 1960: 249

Mas D'azil, Ariège

Traces of human activity on clay

A number of other portable objects made of clay come from Mas D'azil. They are indicated here as traces of human activities as they do not represent symbolic expression although it is always possible to argue that a form of symbolism is indeed visible in the ornaments found. It has been decided here to present them in this section. They were all found by the Péquart in their excavations in 'Galerie des Silex'.

MA 3	Large bead	2.8 cm	This object is a large flattened ball of clay pierced by a bi-conical hole, made of clay. The Péquarts (1960) thought the clay was mixed with some fatty material and interpreted it as a possible loom weight. This was because a large fragment seems to have flaked off the edge of the hole, a possible result of its use. They say it could show that weaving was done in the Magdalenian period. However, it is true that by looking at the object one can see that the perforation is not centred and this could have been a problem for its use in weaving. It might simply have been an ornament. The material, although partly covered with a white deposit, seems to be made of dried clay or silt: it is yellowish in colour, of a dull sort of yellow. This object is kept in the Mas d'Azil Museum.	 (Mas d'Azil museum, Péquart Collection)	Péquart 1960
MA 4	Small bead	0.5 cm	This object is a small clay ball pierced when still soft and left to dry, probably a bead. It was seen in the Mas d'Azil museum. One side of the object is reddish in colour, the other is dark brown, and with a sort of colour variation that suggests the effect of a bonfire firing on clay. Also, its surface is fairly irregular, much more than for ivory beads for example.	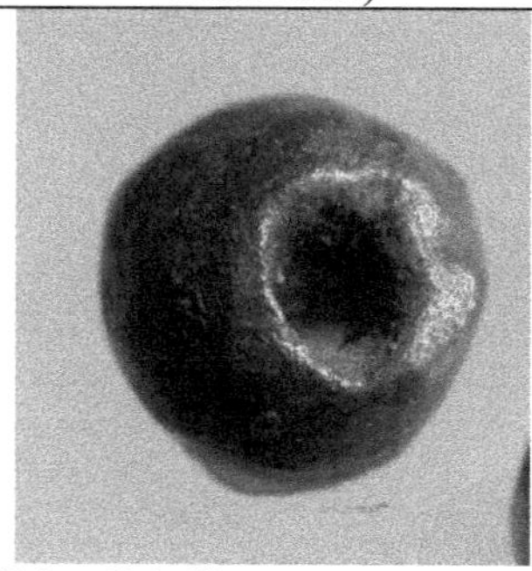 (Mas d'Azil museum, Péquart Collection)	Péquart 1960
MA 5	Dried clay ball		This object was illustrated in the Péquart s' 1960 publication, but is not described in their text.		Péquart 1960

Mas D'azil, Ariège

Traces of human activity on clay

MA 6	The 'tattooing' plaque	5 cm by 1.5 cm thick for the larger fragment; 3.5 cm by 0.8 cm thick for the other.	This object has been described by its discoverers (Péquart 1960) as being part of tattooing equipment. It is in two fragments today. It forms a flat plaque described as made of clay and ochre bound by some fatty material, then kneaded and hardened; it is not fired. The whole surface of the larger fragment is covered with deep holes, which do not all go through to the other side. The smaller fragment is only pierced with one hole that goes through the object. It was recovered originally in three fragments that could not be refitted together, but only two of these are today kept in the collection. The colour of this object differs from the other clay objects from Mas d'Azil examined; it is certainly due to the addition of ground ochre pigment to the clay as the object is purple red in colour. Traces on the material are very clear, and the marks left by fingers that smoothed and thinned the edge of the smaller fragment can be seen, as well as the slight rim left around the hole that indicates that it was made on a wet material before being left to dry. The use of this object remains enigmatic.	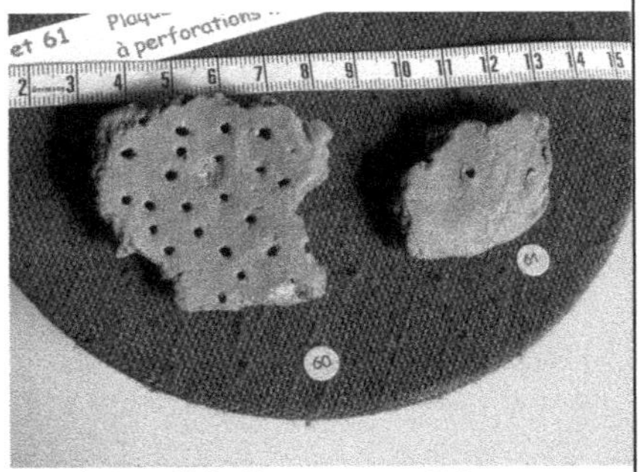 (Mas d'Azil museum, Péquart Collection)	Péquart 1960
MA 7	Shapeless worked lump		This object is described by Péquart as made of the pure and compact silt that forms the primitive soil in the gallery. They say it has been worked by the human hand and looks roughly like a disc with some man-made removals on its edges and two small lines on it, but it is very worn out and fragile. This object could not be found again in the Péquart collection in the Mas d'Azil Museum.		Péquart 1960: 249
MA 8	Shapeless worked lump		This object has also been described by Péquart as made of pure silt. It showed a deep groove in the middle, and was thought to possibly have been used to polish bone as the material has some mild abrasive properties. This object could not be found again in the Péquart collection in the Mas d'Azil Museum.		Péquart 1960:249

Mas D'azil, Ariège

Traces of human activity on clay

MA 9	Fashioned lump of silt		Péquart describes an object made of clay mixed with ochre and possibly some fat, similar to the previous one. It was much worn, but still reminded them of an animal muzzle. This object could not be found again in the Péquart collection in the Mas d'Azil Museum.		Péquart 1960: 249
MA 10	Large fired clay bead	3.2 cm	This object is not mentioned in Péquart's publication, but was seen as part of their finds in the Mas d'Azil Museum: it is a bead, oval and flattened, pierced with a circular hole 0.8 cm in diameter. It is the object that looks most like well-fired clay out of all the ones seen from Mas d'Azil, with a red-orange shade, and irregularities on the surface that look like inclusions in the material.	 (Mas d'Azil museum, Péquart Collection)	Péquart 1960
	Breuil's bead		Breuil mentions that he found a small bead of dried earth (clay), 7mm in diameter in the middle, 5 mm at the 'poles', with a well calibrated hole 2.5 mm in diameter and 4 mm long. He describes it as yellow and very compact, adhering strongly to the clay matrix. He found it himself in a hole between large stones, together with an engraved bone in his 1902 excavations (Breuil 1903: 427). Further in the same publication, he also mentions finding another dried clay bead the previous year but in the waste of the excavation, so he did not talk about it. He refers to Piette (1875) who found some similar objects in Gourdan in the Upper Part of the layer with engravings, and also to the cave of Bruniquel in the Tarn in which some absolutely similar beads were found. He concludes that as these beads are not made of fired clay but simply of dried clay there should not be any reason why they could not be attributed to the Magdalenian period (Breuil 1903: 430-431). Currently, however, it appears that all these objects have been lost.		

Fontanet, Ariège

The numbers given to the sign correspond to the number used in Vialou's publication (1986)

Reference number	Graphic unit	Size	Description	Illustration (when available)	References
F1	Sign N° 163		This sign is a concentration of disorganised punctuations impressed with the finger on the floor on a narrow strip at the foot of a wall. It is found in front of two vertical engraved bison, at the limit of the decorated zone. .	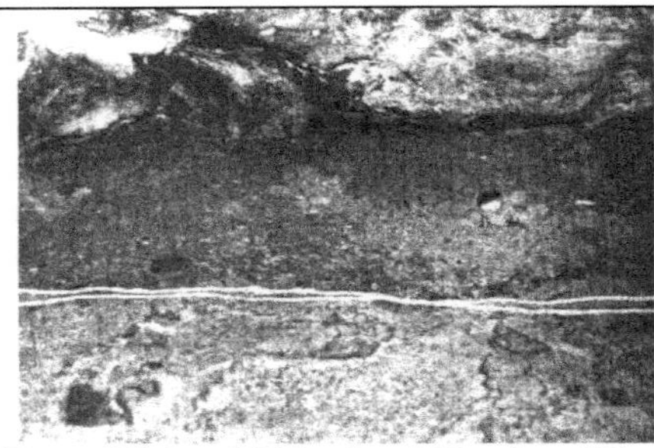 (From Vialou 1986: sign 163, plate 12)	Vialou 1986: 71
F2	Sign N° 167	20 to 30 cm long	This sign is made by a dozen cupules forming a 20 to 30 cm long line. These cupules are circular, flat and wide (about 2 cm in diameter), clearly different from finger prints. They appear modelled. They are made on the sloping side of a clay accumulation that forms the floor.	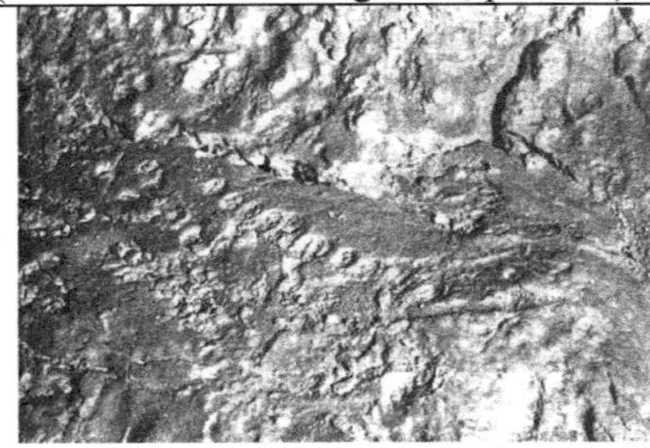 From Vialou 1986: sign 167, plate 13, 2)	Vialou 1986: 71
F3	Sign N° 168		This sign is an alignment of four large and deep cupules, near to the limestone wall of the cave. The two middle cupules are separated by a space twice the size of the two other intermediate spaces.	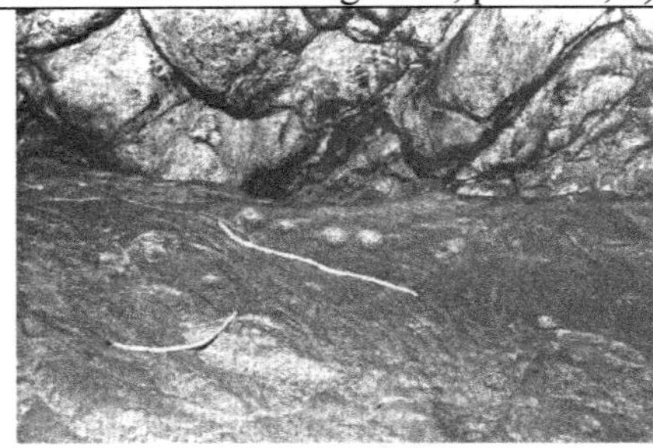 (From Vialou 1986: sign 168, plate 13, 3)	Vialou 1986

Fontanet, Ariège

F4	Sign N°169	Diameter about 12.5 cm	This sign is a circular motive made on a calcited surface on which fifteen holes, irregularly placed, draw a circle, with six more holes inside it roughly making a second circle, and a large impresses surface in its middle. The surface was hardened before the motive was made, and each hole cracked the calcited surface. Next to it, four large (5 cm wide) cupules are also found in an alignment.	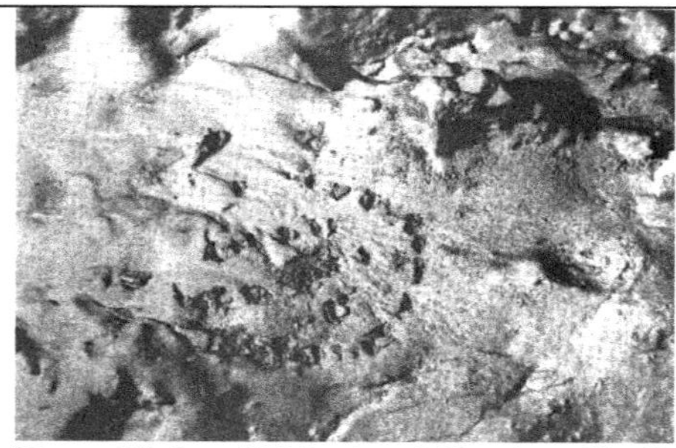 (From Vialou 1986: sign 169, plate 13, 5)	Vialou 1986: 71
F5	Sign N° 170		This sign is made by five finger prints forming a curve, with smaller cupules and clay removal nearby.	 (From Vialou 1986: sign 170, plate 13, 4)	Vialou 1986: 71
F6	Sign N° 171	10 cm long	This sign is a motive of seven cupules (2 to 3 cm in diameter), forming two parallel lines of three cupules reaching the final cupules at their extremity, forming a geometrical shape. All the cupules are slightly different in size, which shows they have been modelled individually with the finger or a rounded instrument. This sign is placed at the very edge of the dangerous cliff one needs to climb down to access the rest of the cave, hidden out of the normal passage route.	 (From Vialou 1986: sign 171, plate 13, 6)	Vialou 1986: 71

Fontanet, Ariège

F7	Sign N° 172	24 cm by 12 cm	This figure is constituted of over twenty holes, covering a roughly rectangular surface of 24 cm by 12 cm. They were made by impressing on a calcite-hardened clay surface that cracked under pressure of finger or tool for each hole.	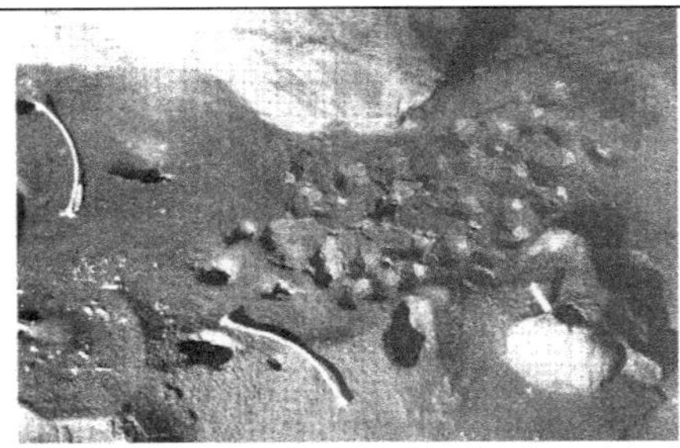 (From Vialou 1986: sign 172, plate 14, 1)	Vialou 1986
F8	Sign N° 173	17.5 cm by 10.5 cm	This figure is a triangular motive formed by fifteen elongated and asymmetrical punctuations for the contour, and with twelve more cupules filling the inside. It was made on soft clay. The motif is clearly associated with the footprint of a foot wearing a shoe. The punctuations are fairly close to each other to the point that they overlap at times. They do not all appear made in the same manner as they are variations in depth, width, length and irregularities of the contour.	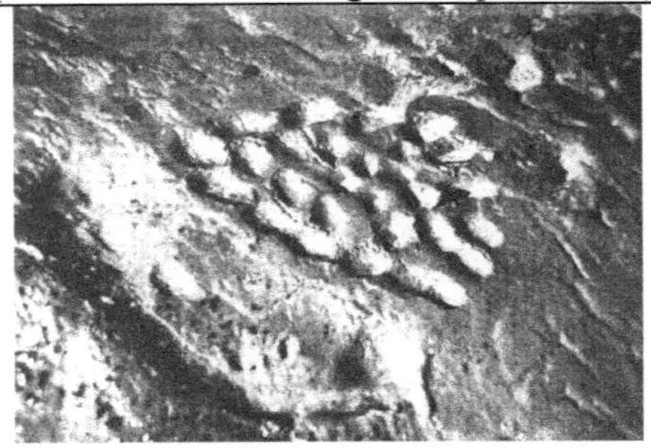 (From Vialou 1986: sign 173, plate 14, 2)	Vialou 1986: 72
F9	Sign N° 174	8 cm long	This sign is made of an alignment of five holes that are quite flat, about 1.5 cm in diameter, probably made with the finger. They are all identical.	 (From Vialou 1986: sign 174, plate 14, 3)	Vialou 1986: 72

Fontanet, Ariège

F10	Sign N° 175	60 cm long	This figure is a long snake-like composition found in the terminal part of the galleries. It is constituted by about 50 impressions that are very different from the digital impressions or the modelled cupules. They are mostly small and deep. They are not all identical, and might have been made with different tools. Some have a quadrangular section for example that exclude the use of a finger. They are found on a sloping surface near the wall in the terminal part of the area visited by prehistoric people. The graphic unit is complex with an axis and other punctuations organised around it.	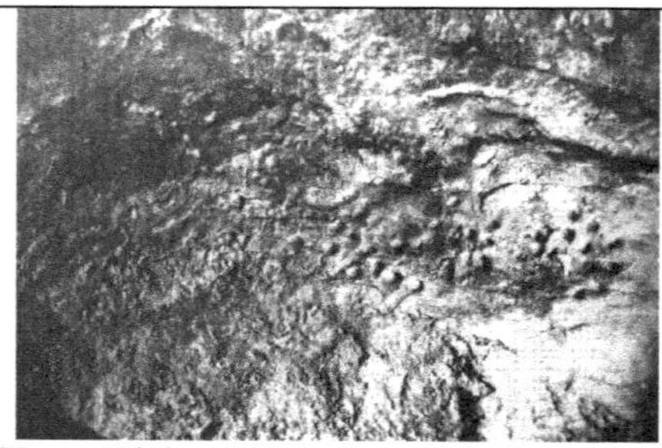 (From Vialou 1986: sign 175, plate 14, 4)	Vialou 1986: 72
F11	Group of 11 parallel vertical incisions		This group of vertical incisions have been found nested above a stalagmitic shelf about 2.5 metre above the normal floor level in the cave of Fontanet, near the inhabited part of the site. A reindeer antler was found planted in the clay nearby on the wall.	 (From Boureux 2004: figure 12)	Vialou 1986; Clottes et al. 1984: 434; Clottes 1989: 74; Boureux 2004

Traces of human activity on clay	A horizontal hole modelled in clay has been found nearby hearth number 2 (further than the main art panels): it is also located high on the wall, on a sort of small platform, 2.2 m higher than the floor level, and 1.8 m higher than a nearby rock on which it was possible to climb to make the hole (Clottes et al. 1984: 434). Just before the well, traces of clay removal are visible as well as two clay balls. There are more traces of clay removal and more clay balls near the extremity of the gallery (Clottes et al. 1984: 434). Also, in Fontanet, many prints are found on the floors: prints of feet, hands, fingers, most apparently belonging to children and teenagers (Clottes et al. 1984: 435).

Bédeilhac, Ariège

Out of the 123 clay plaquettes existing in Bédeilhac, 57 have been studied for the present study and are described here (reference B17 to B74). The others are known to exist but have not been published or studied for this research. The rest of the clay inventory for Bédeilhac is also presented (reference B1 to B16).

Reference number	Graphic unit	Size	Description	Illustration (when available)	References
B1	Modelled bison 1	29 cm by 18 cm	This bison modelage in low-relief is the best preserved of a group of four modelled bison that originally existed next to each other. It faces towards the left. Its body is marked with a cupule. It is found in the Jauze-Mandement gallery.	 (From Boureux 2004: Figure 16)	Clottes et al. 1994; Octobon 1939; Roussot 1998; Gailli et al. 1984: 373
B2	Modelled bison 2	19 cm long	The second bison *modelage* seems to follow the first one. It seems surrounded by a groove made with the finger. It is found in the Jauze-Mandement gallery.		Octobon 1939; Gailli et al. 1984: 373
B3	Modelled bison 3	6 cm long	The third small bison of the group is badly damaged and hardy visible today. It is found in the Jauze-Mandement gallery.		Octobon 1939; Gailli et al. 1984: 373
B4	Modelled bison 4	19 cm long	The fourth bison appears more massive than the others; only the line of its back remains. It is found in the Jauze-Mandement gallery.		Octobon 1939; Gailli et al. 1984: 373

Bédeilhac, Ariège

B5	Modelled vulvae		This figure is found in the continuation of the frieze formed by the four bison, It represents a realistic vulva; a small fragment of stalagtite was even added in the place of the clitoris. It is made in low-relief. It is found in the Jauze-Mandement gallery.	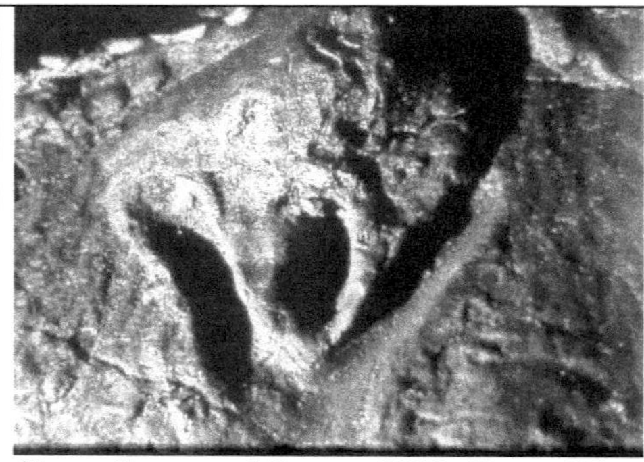 (From Sacco and Sauvet 1998: figure 32)	Gailli et al. 1984: 373
B6	Double bracket sign		This sign is described as a deeply engraved double bracket sign found on the same panel as the group of bison. It is found in the Jauze-Mandement gallery.		Gailli et al. 1984: 373
B7	Horse head		This figure is the front part of a horse. It is made of clay that seems worked as a *contour découpé*, which was then stuck on the wall in the vicinity of the panel with the *modelages*. It is found in the Jauze-Mandement gallery.		Octobon 1939; Gailli et al. 1984: 373; Bégouen et al. 1977; Vialou 1986
B8	A diamond-shaped sign	12 cm by 8 cm	This sign was found on the floor and forms a diamond-shape figure which has also been found on some portable art from the middle Magdalenina layers in Bédeilhac. This sign has been damaged and cannot be seen anymore. It was found in the Jauze-Mandement gallery.		Octobon 1939; Vialou 1986
B9	Animal covered with arrows		M. Mandement describes a bloc covered with clay with tracings on it, in which he saw the figure of an animal covered with arrows. This work has disappeared nowadays. It was found in the Jauze-Mandement gallery.		Octobon 1939

Bédeilhac, Ariège

B10	Engraved bison	23 cm long	This figure was drawn on the clay floor in the terminal chamber in Bédeilhac and is part of a group of floor engravings or drawings found there. It was probably drawn with a finger and has three cupules made on it with finger tips. A few figures of bison are known from the French Pyrénées that bear these cupule marks, like the modelled bison from the 'Galerie des Modelages' in Bédeilhac or the floor engraving N27 from Niaux.	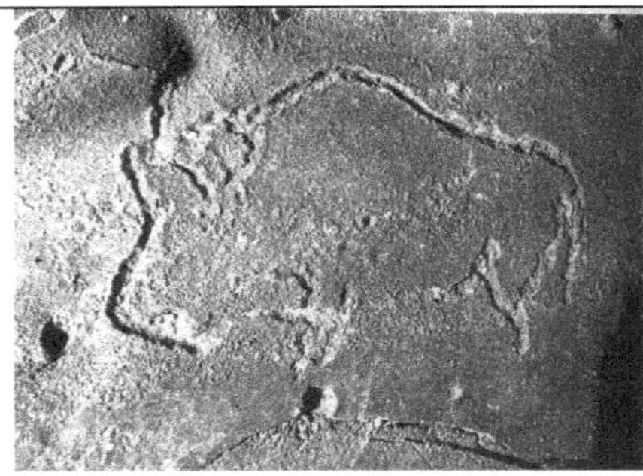 (From Gailli *et al.* 1984: figure 7)	Octobon 1939; Gailli et al. 1984: 374
B11	Small horse head	11 cm long	A small horse head is also found on the floor of the Main Gallery. It seems engraved.	 (From Breuil 1952: 216 figure 195)	Breuil 1952: 216; Octobon 1939; Gailli et al.1984: 373
B12	Horse head		This figure was found in the Terminal Chamber. It was published by Breuil (1952) with a photograph but has disappeared nowadays.	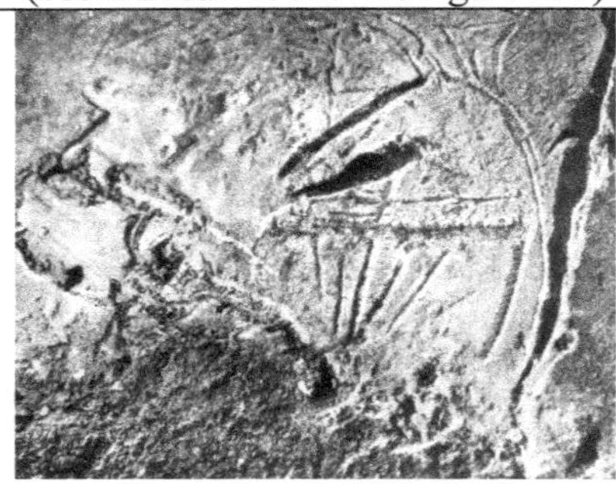 (From Breuil 1952: 217)	Breuil 1952; Gailli et al.1984: 373

Bédeilhac, Ariège

B13	Radiating sign		This figure has been described as an enigmatic radiating sign and was found in the Terminal Chamber. It cannot be found nowadays in the cave.		Octobon 1939
B14	Cervid	82 cm long	This finely drawn cervid leg is engraved on the clay floor in a part of the terminal chamber in which the floor appears to have shifted since the Magdalenian. It is just on the edge of an area where the floor seems to have collapsed, possibly destroying the rest of the figure.	 (From Breuil 1952: 217)	Octobon 1939; Gailli et al 1984: 373
B15	Horse	42 cm long	This figure is partly engraved, and partly in low relief on the floor of the Terminal Chamber. It shows a much stylised animal, very angular, which has been traditionally accepted as a horse representation.	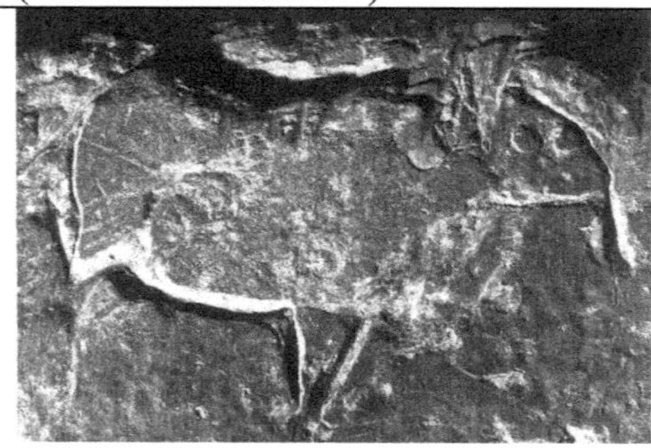 (From Gailli *et al.* 1984: figure 6)	Gailli et al. 1984: 374; Octobon 1939
B16	Possible bison head		This figure could have been an attempt at modelling a bison head in clay. It was found in the Terminal Chamber but has been destroyed since.		Octobon 1939

Bédeilhac, Ariège

B 17: (MAN 76.253)	Plaquette with horse	14 cm by 20 cm by 2.5 cm	The front part of a horse is recognisable on this plaquette, with the two front legs extended as if in movement, and the lower part of the neck indicated. The rest is broken off. It is very finely modelled using relief as well as engraving with a fine tool. Some traces of modelling on a wet material are visible, a well as some engravings that were made on a dry support. It is a very dynamic. (Mons 1974; Mons and Delporte 1973)	 (Cliché Musée des Antiquités Nationales, Saint-Germain-en-l'Haye)
B18: (MAN 76.967 C)	Plaquette	16.5 cm by 12.5 cm by 2 cm	This object is a flat plaquette with well smoothed surfaces. It is also a *contour découpé* as the edges are fashioned, but it is not figurative. It bears various traces of fashioning such as incisions and impressions. (Mons 1974; Mons and Delporte 1973)	 (Cliché Musée des Antiquités Nationales, Saint-Germain-en-l'Haye)
B19: (MAN 76.254)	Plaquette	18 cm by 30 cm by 2.3 cm	This large plaquette shows a finely engraved bison body. Another engraving of an animal, partly erased, is visible under the bison figure. (Mons 1974; Mons and Delporte 1973)	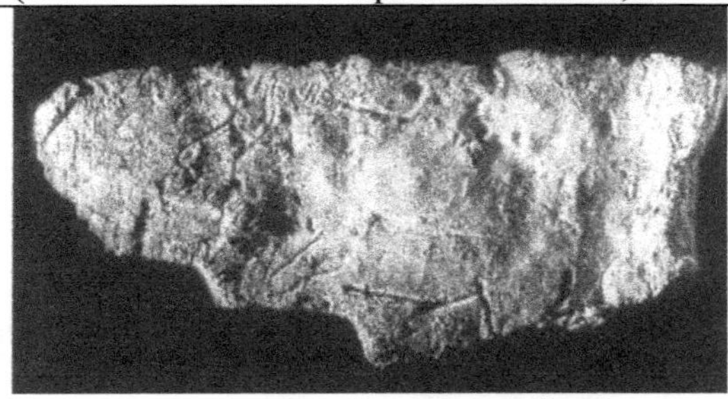 (From Boureux 2004: figure 14; collections M.A.N.)

Bédeilhac, Ariège

B20: (MAN 76.959)	Plaquette	9.8 long	This object shows a bison finely engraved on a plaquette. The front part of the animal is shown with the mane indicated by a series of short lines. The horns are done exactly in the same way than on the object B19. The eye is well marked. (Mons 1974; Mons and Delporte 1973)	 (From Boureux 2004: figure 5; collections M.A.N.)
B21: (MAN 76.250)	Plaquette	12.3 cm long	Fashioned as *Contour découpé* and slightly modelled as well, like a flattened sculpture. It is engraved on the two sides and has been interpreted as a possible bear or bison, unless it is a composite animal; it has horns, but its profile evokes a bear. (Mons 1974; Mons and Delporte 1973)	 (From Boureux 2004: figure 8; collections M.A.N.)
B22: (MAN 76.952)	Plaquette		This object shows the front part of an aurochs engraved on a clay plaquette. It is very precisely traced. (Vialou 1986)	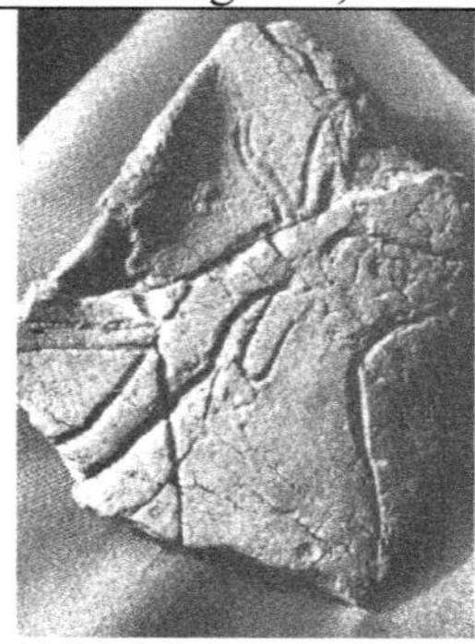 (From Vialou 1986: plate 20; collections M.A.N.)

Bédeilhac, Ariège

B23: (298 and 298.2)	Plaquette	12.6 cm by 8.7 cm by 1.4 cm	In two fragments with no clear trace of modelling.
B24: (292)	Plaquette	5 cm by 2 cm by 0.7 cm	Small plaquette with at least two lines criss-crossing each other incised on one side.
B25: (281)	Plaquette	7.6 cm by 5.2 cm by 1.5 cm	Plaquette shaped as the *contour découpé* of the hind side of an animal, a bison by the bulk of the body, with the hump clearly indicated.
B26: (290)	Plaquette	8 cm by 5.2 cm by 1.3 cm	Small plaquette with 3 deeply engraved parallel lines on it. It is fragmentary.
B27: (295)	Plaquette	6.5 cm by 4.8 cm by 1.4 cm	This plaquette appears to be a *contour découpé.* Its edges are fashioned, but it is nothing clearly recognisable.
B28: (297)	Plaquette	4.3 cm by 4.5 cm by 0.8 cm.	This piece is a carefully fashioned *contour découpé*, fashioned smoothly on all surfaces even the edges of the profile. The general shape is rounded and could represent an animal head although it seems fashioned in champleve manner with relief on one side which are not obvious to interpret.
B29: (294)	Plaquette	5.7 cm by 4.7 cm by 1 cm	This is a plaquette worked in *contour découpé* with care, but representing nothing recognisable.
B30: (296.2)	Plaquette	4.2 cm by 8.3 cm by 1.2 cm	Deeply engraved with parallel lines that could result from scraping with a flint tool. It appears fragmentary.
B31: (293)	Plaquette	4.2 by 3.6 cm by 1 cm	Small plaquette covered with very light striations on one side, possibly from scraping.
B32: (296)	Plaquette	7.3 cm by 4.3 cm by 0.9 cm	Plaquette is worked in *contour découpé* but is not figurative
B33: (289)	Plaquette	8.3 cm by 6.7 cm by 1.5 cm	Engraved with very deep lines forming an abstract zigzag motif on one side.
B34: (286)	Plaquette	8 cm by 7.2 cm by 2.5 cm	This object is very thick in comparison to most of the other plaquettes. Some rough traces of fashioning are visible on one side

Bédeilhac, Ariège

B35: (287)	Plaquette	13.7 cm by 11 cm by 1 cm	Large plaquette with a few engraved parallel and intersecting lines on it and traces of fashioning on its edges.
B36: (288)	Plaquette	14 cm by 15.2 cm by 1 cm	Large plaquette broken in four fragments. Some engraved straight intersecting lines are visible on it, but they remain abstract. At least one of the edges appears worked
B37: (285)	Plaquette	12 cm by 7.6 cm by 3 cm	This is a very thick and rough plaquette with little traces of fashioning.
B38: (284)	Plaquette	9.5 cm by 11.8 cm by 2 cm	This plaquette does not show any trace of fashioning, but it appears that a shape was imprinted on it when it was still soft.
B39: (367)	Plaquette	6 cm by 3.5 cm by 0.5 cm	Smoothed on one side and very rough on the other side, which is broken. The edge is possibly worked to make a notch.
B40: (362)	Plaquette	8.2 cm by 4.6 cm by 0.9 cm	Contour of what could be the hind side of an animal, which is well fashioned and rounded. The other edges are broken.
B41: (361)	Plaquette	5.7 cm by 7.9 cm by 1.4 cm	Most of the edges have been rounded and smoothed to form a *contour découpé* in this object. It resembles a triangle with straight edges and rounded corners. It is broken on one side.
B42: (360)	Plaquette	5.8 cm by 5.2 cm by 1 cm	Plaquette with well smoothed faces. There is also some trace of working on one of the edges (possible notch).
B43: (311)	Plaquette	6.6 cm by 4 cm by 0.9 cm	Part of this plaquette appears to have been worked in 'champlevé' by lowering the surrounding surfaces in order to make a shape stand out on the object. Again, it seems to be abstract representations. It is worked as a *contour découpé*.
B44: (312)	Plaquette	6.8 cm by 4.4 cm by 0.8 cm	Also worked in champlevé with a smoothing of the surfaces. One notch is visible on one edge.
B45: (313)	Plaquette	4.8 cm by 7.7 cm by 1.3 cm	This object appears to be the *contour découpé* of the profile of the hind side of an animal, but no other detail are given. One notch is visible on one edge. The other rounded contour is fashioned.
B46: (310)	Plaquette	5.6 cm by 5 cm by 2 cm	Very thick small plaquette that is much damaged. It is hard to see any fashioning traces. There are possible notches on the edges
B47: (309)	Plaquette	8 cm by 4.3 cm by 0.6 cm	Very thin *contour découpé* that possibly represents an animal body as it has an appendix which could be interpreted as a leg. The contour is nearly entirely fashioned, smoothed and prepared. The surface on one side separated into two parts by wavy lines which are not just engraved but for which edges are also worked in champleve. What could also be interpreted as notches are visible on two of the edges.

Bédeilhac, Ariège

B48: (308)	Plaquette	10.5 cm by 7 cm by 1.2 cm	This is a plaquette with one surface left in relief by champleve work on one side, the other side is smooth and flat. There is a large semi-circular notch-like area on one edge.
B49: (307)	Plaquette	8.7 cm by 6 cm by 0.7 cm	Very thin plaquette with an area worked in champlevé as a depression in the middle of one face. Again one face is worked into a sort of notch. Surfaces are smoothed.
B50: (306)	Plaquette	6 cm by 5.4 cm by 0.5 cm	This plaquette forms a *contour découpé* that has also been worked in champlevé. The contour was fashioned in order to shape two rounded shapes on two angles of the piece. There is a notch on one side.
B51: (304)	Plaquette	5.6 cm by 5.7 cm by 1.1 cm	Small plaquette with smooth surfaces, a fashioned contour, some incisions visible on one side and notch-like working on the edges. .
B52: (305)	Plaquette	8.6 cm by 5 cm by 1.2 cm	This object forms an elaborate sculpture. The surfaces are worked at different levels of relief to fashion complex shapes. A very marked notch is visible on one side.
B53: (302)	Plaquette	9.7 cm by 8.9 cm by 2.3 cm	Thick plaquette with well flattened surfaces. On one side deep incisions are repeated to form two abstract patterns (unless they are interpreted as vegetation or branches) and possibly the beginning of a third one. The pattern can be described as three lines branching off one stem. .
B54: (299)	Plaquette	5.7 cm by 4.5 cm by 1.2 cm	This plaquette seems much damaged. Not much is visible on it. At least one notch visible on one side.
B55: (303)	Plaquette	5.8 cm by 3.7 cm by 1.9 cm	This object was made by combining *contour découpé* and champlevé techniques. It shows a well-rounded contour shape, oval, separated from the rest of the object by a groove obtained by champleve fashioning. It seems broken off a larger body on one side. The shape is neat and well delimited but interpretation is not obvious. It can also be noted that the fashioning makes two marked notches on opposite faces.
B56: (300)	Plaquette	7.3 cm by 6.6 cm by 1 cm	This object is rather damaged with uneven surfaces. It appears worked at least on one edge where a marked notch is visible.
B57: (301)	Plaquette	13 cm by 8.7 cm by 2 cm	This object shows elaborate working on one side with the juxtaposition of different surfaces obtained by champlevé, among which one small circle.
B58: (454)	Plaquette	4.7 cm by 6.1 cm by 1.5 cm	This object has rough and damaged surfaces. Two deep lines are intersecting. Marked grooves on two opposite sides forming notches.

Bédeilhac, Ariège

B59: (456)	Plaquette	7.5 cm by 8.8 by 1.8 cm	Roughly smoothed on one side with some incisions visible. Small notches are seen on two edges.
B60: (458)	Plaquette	7.3 cm by 5.8 cm by 2.7 cm	Very thick plaquette. One side appears smoothed and two intersecting incisions are visible on it. It appears damaged.
B61: (460)	Plaquette	6.2 cm by 7.7 cm by 1.8 cm	This object is much damaged. There could be one notch on one side.
B62: (185)	Plaquette	9.7 cm by 5.4 cm by 1.8 cm	This plaquette has been literally soaked in a consolidation agent, which covers it completely and prevents the reading of the original surfaces. A notch could be visible on one edge.
B63: (190)	Plaquette	8.2 cm by 9.3 cm by 1 cm	This piece appears flattened but the surfaces are very uneven. Small notches could be visible on two or three sides.
B64: (448)	Plaquette	10.7 cm by 4.6 cm by 2.4 cm	This plaquette is worked in *contour découpé* with a rounded edge in a shape that evokes the hind side of an animal. The surfaces are damaged. Notches are visible on the edges.
B65: (112)	Plaquette	2.1 cm by 2 cm by 1 cm	This small object appears to be a first rough out for the carving of a semi-circular disc shape that is starting to stick out of the lump of material. In doing so one notch appear on one side.
B66: (111)	Plaquette	2.7 cm by 1.6 cm by 2 cm	Broken fragment of a plaquette rather than as an object in itself. Possibly not worked.
B67: (198)	Plaquette	8 cm by 5.7 cm by 0.8 cm	This is a very flat object, broken in two parts, with a much damaged surface on one side. Two small notches are visible on opposite sides.
B68: (496)	Plaquette	22.4 cm by 20.3 cm by 1.8 cm	Large plaquette, with well smoothed surfaces worked by scraping. It is engraved with the hind side of a feline body recognisable by the very long thin tail and the general bulk of this part of the body. Straight lines are visible on the surface also, one of them overlapping with the feline. What is very interesting is that the extremities of a few of these lines clearly connect with notches on the edges, at least four that are well visible: could they be interpreted as the imprints of strings used to tie or suspend the object?
B69: (449)	Plaquette	12 cm by 7 cm by 2.1 cm	This object is falling into pieces and needs to be kept in its plastic bag. Not possible to see if worked.
B70: (424)	Plaquette	6.7 cm by 5.3 cm by1 cm	This is a plaquette with rough surfaces on which no traces of modelling are visible. One notch however is clearly marked on one edge.

Bédeilhac, Ariège

B71: (455)	Plaquette	10 cm by 4 cm by 2.1 cm	This object is also in poor state of conservation and is crumbling into pieces. Not possible to see traces of working
B72: (452)	Plaquette	9.5 cm by 5 cm	This object is not really a plaquette as it is not flat. It bears some fashioning traces but is also much damaged. Again some notch-like surfaces could be used for suspending the object.
B73: (450)	Plaquette	13.5 cm by 12 cm by 2 cm	This piece seems flattened but is much damaged. Notch-like areas visible on two edges.
B74: (451)	Plaquette	18 cm by 14.5 cm by 3.5 cm thick	This object is much damaged. One surface seems to have been flattened and worked. One notch is visible.

Use as pigment	Clay is mentioned to have been used as pigment in Bédeilhac (Clottes 1989: 77). Octobon (1939) adds that the outline of a horse is drawn with clay on the wall in the Jauze-Mandement gallery, its mane being represented by juxtaposed dots.

Traces of human activity on clay	In the Jauze-Manadement gallery, Octobon (1939) mentions a series of slight relief that would suggest that other *modelages* were done and destroyed in Bédeilhac. He also mentions traces of clay removal by fingers, prints of knees, traces of fingers on the ceiling, very deep small heel prints in the clay of the floor, mostly visible in front of the black horse, the print of a naked foot (Octobon 1939) In the Terminal gallery, he points out traces of many removals of clay from the floor, possibly to make the plaquettes.

Massat, Ariège

Reference number	Graphic unit	Size	Description	Illustration (when available)	References
M1	Large grid pattern	160 cm by 150 cm	This large grid pattern is partly engraved, partly completed with finger tracing on clay surface. It is found in the Louis Méroc Chamber. The squares forming the grid are irregular in sizes. It is located on a reclining wall, covered by a 2 to 3 cm layer of clay and partly destroyed by vandalism today. In the destroyed parts the layer of clay flaked off and it is possible to see traces of the tool that engraved the grid faintly on the rock beneath	 (From Gailli 2004: figure 30)	Gailli 2004: 61-62
M2	Snake-like sign	About 80 cm long	At the base of a destroyed finger-tracing panel on the right of the grid sign in the Louis Méroc gallery, this large sign is drawn with the finger on the reclining wall: the line is quite horizontal for about 40 cm then heads downward while meandering making several continuous S shapes on a 40 cm long distance. It stops 45 cm above the floor level.	 (From Gailli 2004: figure 32)	Gailli 2004: 64

Massat, Ariège

M3	Group of 9 horizontal parallel lines	23 cm by 20 cm	This group of lines is found 25 cm below the snake-like sign on the left of the Louis Méroc chamber. They are slightly arched and are placed inside a small concave relief on the wall. The 4th, 5th, and 6th line descending are longer and end up curved toward the bottom. The 7th is the longest of all and goes back upward meeting the 3rd line. All these lines are spaced between 2 and 3 cm away. They have suffered from modern damage.	 (From Gailli 2004: figure 33)	Gailli 2004: 65
M4	Ibex head 1	57 cm by 17 cm	This ibex figure is placed 1.5 m above ground level, facing towards the entrance of the Louis Méroc gallery. It represents a specific species of ibex: a mountain ibex called 'Isard' with very different horns from the usually represented ibex known in parietal art. A large flake that seems to represent an eye is in fact a modern damage to the figure. The eyes were not originally represented. It must have been engraved with a burin as a fine V-shaped line is visible at the bottom of the incision. The ear is indicated, abnormally long.	 (From Gailli 2004: figure 34)	Gailli 2004: 66
M5	Ibex head 2		Just below the previous ibex head, it seems a similar figure existed that is now nearly destroyed: only one ear and two horns similar to the other ibex are still visible.		Gailli 2004: 67
M6	Ibex head 3	24 cm by 8 cm	This figure is found just above the ibex head M4 in a nearly horizontal position. It is facing towards the end of the gallery. The line of the neck continues towards the horns, then divides into two and curves upwards elegantly. A small dot anciently made by removing clay could indicate the eye.	 (From Gailli 2004: figure 36)	Gailli 2004: 68

Massat, Aiège

M7	Bovine head 1	24 cm long	This figure is found above the ibex heads group, about two metres above floor level in the Louis Méroc Chamber. The heads faces towards the entrance of the gallery. The very rough, shallow line forms a vague animal profile with two lines for the horns, another for the ear.	 (From Gailli 2004: figure 37)	Gailli 2004: 69; Clottes and Gailli1984: 397
M8	Bovine head 2	45 cm by 45 cm	In the same style than the previous one, this figure is traced with a shallower engraving. The head faces towards the end of the gallery and is situated about two metres above floor level. The profile of the head that is very elongated, the ear and the beginning of a horn are visible. A triangular removal (ancient) figures the eye.	 (From Gailli 2004: figure 38)	Gailli 2004: 71
M9	Impressed pattern	20 cm by 11.5 cm.	In front of the ibex, in the Louis Méroc Chamber, a sign was made in a recess on the wall by impressing about 30 holes with the finger tips. It is situated about 80 cm above floor level. Some of the holes were made by a perpendicular impression but others are at an angle. There does not seem to be any sequence or pattern in their making. They have suffered from modern damage.	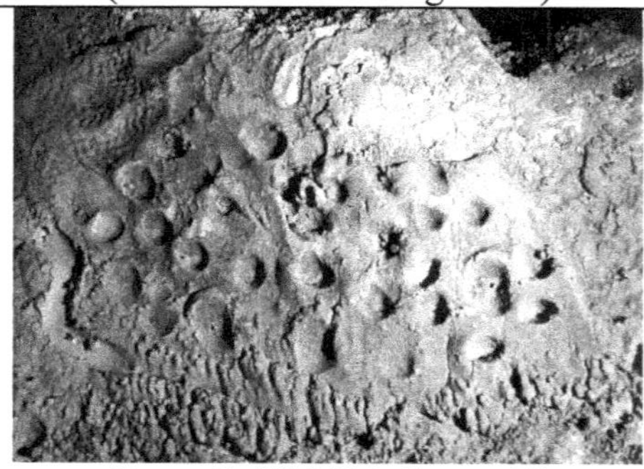 (From Gailli 2004: figure 40)	Gailli 2004: 73; Clottes and Gailli 1984; Vialou 1986

Massat, Ariège

M10	Fish-like sign	38 cm by 14 cm	This sign is found in a deep recess on the same wall as most of the art in the Louis Méroc gallery. It was drawn with the fingers. It is situated about 1.7 m above floor level. The head is rounded, there are no details. Five punctuations are visible on the body: two near the head and one of them could indicate the eye; the three other are found at the centre of the body and are slightly elongated. The tail is not traced but the natural relief of the wall suggests it perfectly.	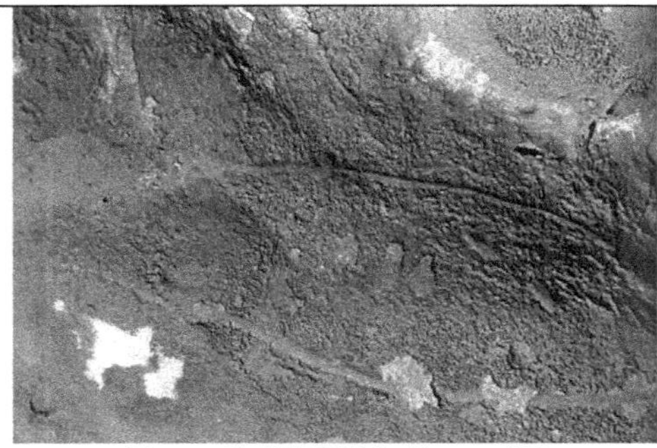 (From Gailli 2004: figure 39)	Gailli 2004: 72; Clottes and Gailli 1984: 398
M11	Complex sign or of a group of signs	48 cm long	This group of signs is found in the deep Gallery, five metres away from the Louis Méroc gallery where the engravings on clay are found. It looks like a large line with a club-like' extremity. It is partly drawn with a finger on the right wall.		Gailli 2004: 75; Clottes and Gailli 1984: 398
M12	Claviform-like sign	18 cm long	Another line is vertical with the 'club' end completely closed resembling a claviform sign. It was also drawn with the fingers.	 (From Gailli 2004: figure 41)	Gailli 2004: 76; Clottes and Gailli 1984: 398

Massat, Ariège

M13:	Sign with 14 parallel lines	18 cm long	This sign is made of 14 vertical parallel engraved lines. They are located about 1 m after the claviform in the Louis Méroc chamber, on the same wall and are all about the same length. They seem engraved with a burin because of V-shaped section of the line.	 (From Gailli 2004: figure 42)	Gailli 2004: 77

Finger-tracings	On the ceiling of the Jacques-Paloumé chamber, towards the end, near to the passage to the Louis Méroc chamber, a series of macaronis tracings are done with the finger on the clay (Gailli 2004: 60). There are also some at the beginning of the Louis Méroc chamber on the right of the large quadrangular sign. Among these tracing were some zigzags, triangular shapes, radiating shapes, on about one square metre. They have been destroyed since their discovery (Gailli 2004: 63).

Traces of human activity on clay	Clottes and Gailli (1984: 398) mention many traces of fingers visible in the clay in Massat. Clottes (1995) also mentions some engravings on floors in Massat, but without more detail.

Montespan, Haute-Garonne

Reference number	Graphic unit	Size	Description	Illustration (when available)	References
MO1	Series of arches	25 to 30 cm high	This sign is made of a series of arches, some of which are intersecting, drawn with the fingers on the clay surface. It is found n the Casteret – Godin gallery.	 (From Trombe and Dubuc 1947: figure 106)	Trombe and Dubuc 1947: 95
MO2	Alignment of holes on filled fissure 1		In the Casteret –Godin gallery, a natural fissure in the limestone wall was filled with clay, which was then perforated with many holes about 1 cm in diameter and 1 to 2 cm deep.		Trombe and Dubuc 1947: 95
MO3	Horse head *contour découpé*	8 cm	The *contour découpé* in clay of the front part of a horse was found in Montespan In the Casteret –Godin gallery. It is made of a small clay plaque which was cut out and then stuck on the wall of the gallery. It was detached from the wall after its discovery and removed from the cave for preservation.	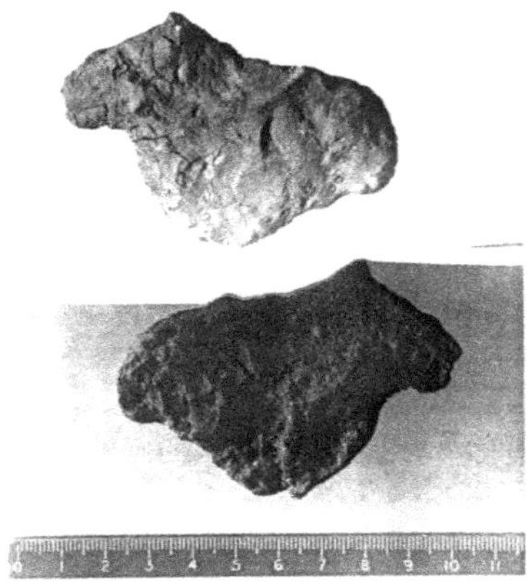 (From Méroc 1959: figure 12)	Trombe and Dubuc 1947: 96; Méroc 1959
MO4	Alignment of holes on filled fissure 2		In the Casteret –Godin gallery, a horizontal natural fissure in the limestone was filled with clay and pierced with round holes placed close to each other. It is close to the horse head in *contour découpé*.		Trombe and Dubuc 1947: 96

Montespan, Haute-Garonne

MO5	Possible modelled vulva		Three clay balls were found in the Casteret –Godin gallery on a clay bank artificially carved by removal of material; one of them was interpreted as a possible modelled vulva as it is made of a small ball laying on a wider flattened oval piece of modelled clay inside a circular depression dug out in it. An engraved line completes it.	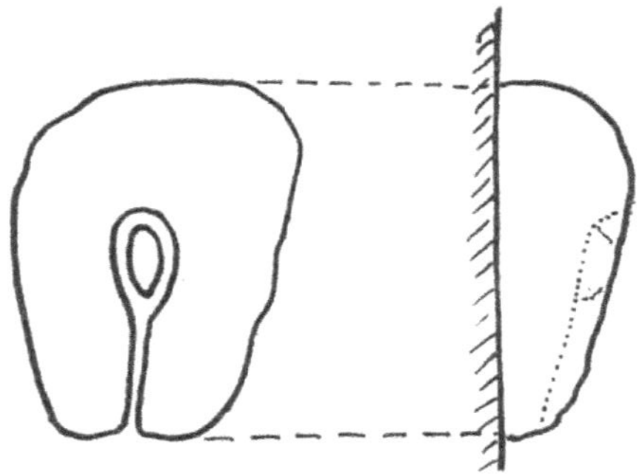 (From Trombe and Dubuc 1947: figure 109)	Trombe and Dubuc 1947: 96; Bégouen and Clottes 1988: 29
MO6	Mammoth	80 cm by 36 cm	This figure is found near the clay balls on the same side of the Casteret –Godin Gallery, in a panel of macaroni finger-tracings. It is the only recognisable figure in it.	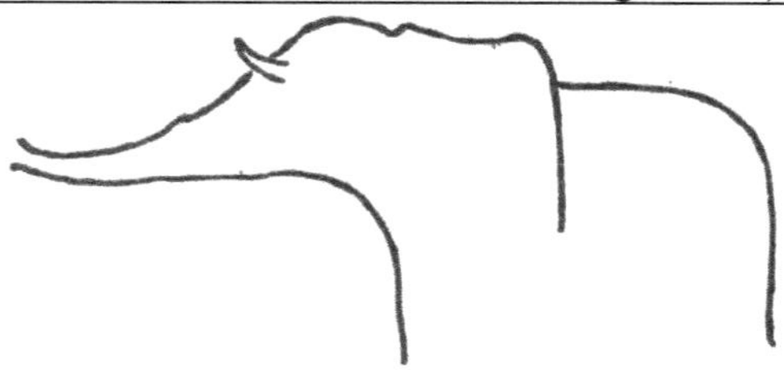 (From Trombe and Dubuc 1947: figure 110)	Trombe and Dubuc 1947: 96, 98
MO7	Diamond-shape sign		The sign is finger traced on the clay wall in the Casteret –Godin gallery.	 (From Trombe and Dubuc 1947: figure 111)	Trombe and Dubuc 1947: 96

Montespan, Haute-Garonne

MO8	Horse or feline modelled high relief 1	1.6 m by 0.7 m	Three very large animal *modelages* in high relief existed in Montespan, in the Casteret –Godin gallery. Tey are today very fragmentary. They have been interpreted as horses or felines. MO8 is the most complete; it still has the neck and the front of the body remaining, as well as part of the tail and the back leg. It is full of holes. Lumps fallen from the modelage litter the floor nearby. The rest of the body can be guessed from the marks left on the bank against which it was lying.	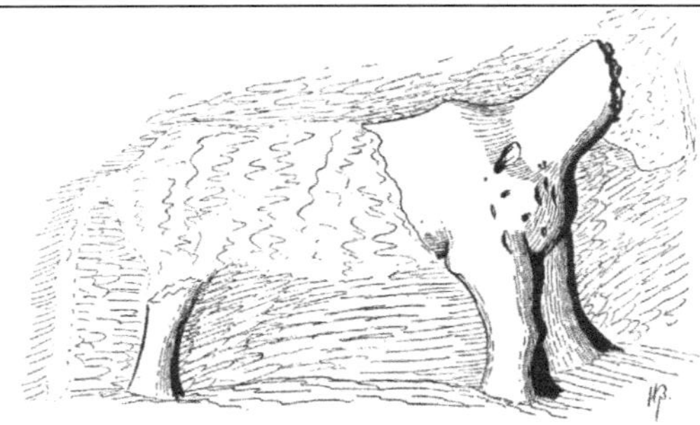 (From Trombe and Dubuc 1947: figure 59)	Trombe and Dubuc 1947: 72; Rivenq 1984: 441
MO9	Horse or feline modelled high relief 2	1.5 m	This animal *modelage* has completely collapsed and is visible by the traces it left on the wall it was lying against. Some remains of it can be seen on the floor nearby.		Trombe and Dubuc 1947: 73; Rivenq 1984: 441
MO10	Horse or feline modelled high relief 3	1.5 m	This animal *modelage* has completely collapsed and is visible by the traces it left on the wall it was lying against. Some remains of it can be seen on the floor nearby.	(From Trombe and Dubuc 1947: figure 58)	Trombe and Dubuc 1947: 72; Rivenq 1984: 441
MO11	Horse head	12 cm by 12 cm	The head of a horse was found drawn and slightly modelled on a horizontal clay bank about 50 cm above and 1 m behind the first feline modelage, about 80 cm above floor level. The head looks towards the centre of the gallery. It was engraved with a sharp point, with a line 2 to 3 mm deep; it is calcited; two prints of fingers are visible on the neck; the forehead of the animal is slightly in relief.	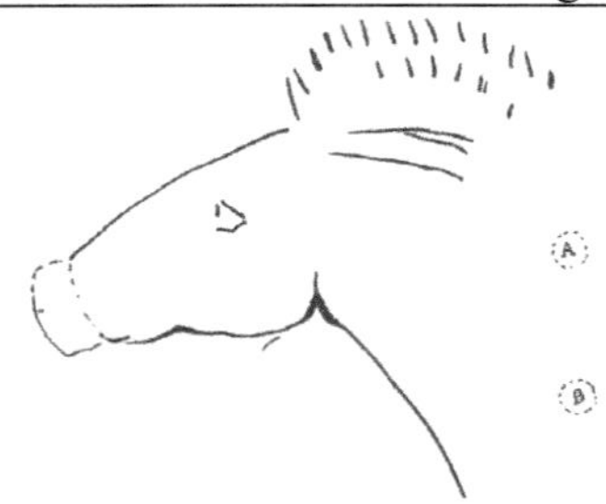 (From Trombe and Dubuc 1947: figure 60)	Trombe and Dubuc 1947: 72; Rivenq 1984: 441

Montespan, Haute-Garonne

MO 12	Alignment of holes on filled fissure 3		This sign was found opposite the felines on the other side of the Casteret –Godin gallery. A large fissure in the limestone was filled with clay and perforated with fingerprints.	 (From Trombe and Dubuc 1947: figure 113)	Trombe and Dubuc 1947: 96
MO13	Bear	1.1 m by 0.6 m	The most famous work in clay from Montespan is a large statue of a headless bear, the 'bear dummy'. It is seen lying in a 'sphinx-like' position. It is fully rounded, with the area of the haunches and the stomach being carefully shaped. The paws are represented with the claws incised. It is estimated that a mass of 700 kilograms of clay was necessary to fashion the statue: this represents a considerable amount of work in order to accumulate the material. Much speculation has been raised about the missing head of the bear. It was noticed that the body is very polished, and this has been interpreted by suggesting that a real bear hide may have been wrapped over the body during some ritual ceremonies. It would thus be a 'bear dummy'. A hole with a triangular section is seen on the neck, and the skull of a bear cub was found between the front paws of the bear. It was therefore added that the skull of the animal was possibly attached to the body, and then covered with a bear hide. This would have served in a hunting magic ritual as the bear is covered with many holes, interpreted as spears piercing it.	 Reconstructive drawing by N. Casteret or L. Capitan (From Bégouen and Clottes 1988: figure 3)  (From Trombe and Dubuc 1947: figure 61) 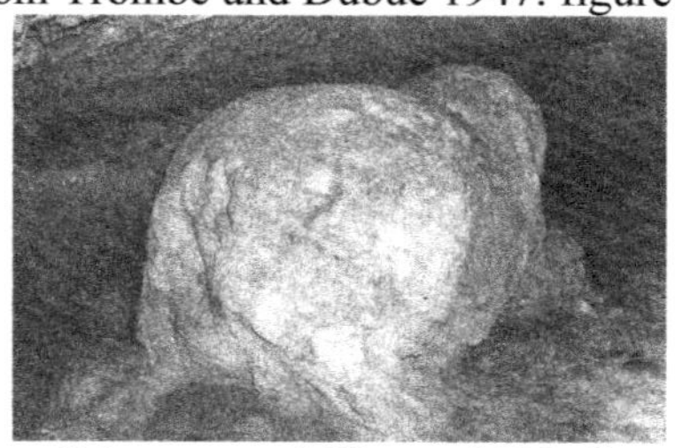 (From Bégouen and Clottes 1988: figure 2)	Trombe and Dubuc 1947: 75, 76, 96; Rivenq 1984: 441; Clottes 1995; Bahn and Vertut 1997

Montespan, Haute-Garonne

MO14	Modelled horse	68 cm	This figure is found next to the bear, on the floor. It is partly modelled-partly drawn and is much damaged. The neck, the tail and part of the head of the horse have been represented in low-relief with the lowering of the level of the clay outside the figure notably near the back. The head is slightly raised, the back is straight, and the tail is thick and lifted. It is thought to represent a horse in movement, notably because of the series of parallel lines drawn at a straight angle from the chest of the animal that would represent the movement. The horse also bears lines engraved on its body: one on the belly, three vertical lines on the neck.	 (From Trombe and Dubuc 1947: figure 63)	Trombe and Dubuc 1947: 77; Rivenq 1984: 441
MO15	Headless horse	70 cm	The headless body of a horse is modelled with a relief between 5 and 10 cm high in the Casteret –Godin gallery. The head, the neck, the legs and tail were missing at the time of the discovery. Only the bulk of the body remains with 4 deep incisions done, two parallel in the lower belly after the front leg, two parallel in a similar position but on the back.	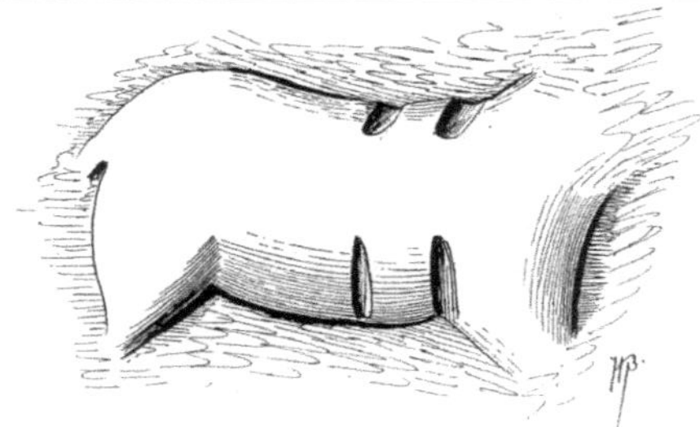 (Drawing by H. Breuil, From Bégouen and Clottes 1988: figure 6)	Bégouen and Clottes 1988: 27
MO16	The 'pregnant mare'	50 cm	This figure has been modelled on the floor on a clay bank in the middle of the Casteret –Godin gallery. It is interpreted as a pregnant mare and was originally found with three large clay balls placed above the back. They have disappeared nowadays. It has an enormous belly, a raised tail and is headless. The feet and the end of the tail have also disappeared. The low-relief is 9 cm high in the middle of the figure.	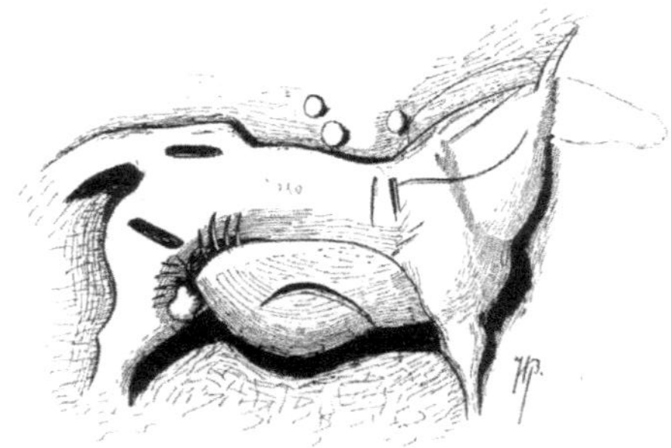 (Drawing by H. Breuil, From Bégouen and Clottes 1988: figure 5)	Bégouen and Clottes 1988: 27; Trombe and Dubuc 1947: 78; Rivenq 1984: 442

Montespan, Haute-Garonne

MO17	Group of vertical lines		Before the hunting frieze, on the left of the Hunting Frieze gallery, deep vertical lines are done with fingers on the clay wall.	 (From Lorblanchet 1995: 111)	Trombe and Dubuc 1947: 48 Clottes 1995
MO18	Vertical horse head		In the Hunting Frieze Gallery, a long frieze engraved and impressed on clay is found on a wall associating 3 horses with many vertical lines and deep holes made with a conical implement: it is the famous 'hunting scene' ('scène de la Chasse'), that played an important role for the believers of hunting magic as an explanation for cave art. They claimed that the horses had been ritually killed with spears in order to try and influence the results of subsequent hunts (Trombe and Dubuc 1947). This has been strongly questioned, but it does seem to represent a scene, and this is rare in Upper Palaeolithic art. The whole scene is 3 metres long. The engraving of a horse head is found placed vertically at the beginning of the Hunting Frieze. It was finely traced with a pointy tool and it seems that the artist took care to smooth the contours and erase the smudges that resulted from the process of tracing in the soft clay. One especially notices the decreasing depth of the incisions tracing the mane. The contour of the head and the ear seem to have been softened by a slight modelling after the engraving. The supporting surface around this drawing shows scraped areas and some small swellings resulting from applying pressure with the fingers; it is an artificially obtained surface. This figure was not pierced in any way and thus contrasts with the nearby horses which are nearly destroyed by holes.	 (From Lorblanchet 1995: 101)	Rivenq 1984: 444; Trombe and Dubuc 1947: 118

Montespan, Haute-Garonne

MO 19	Vertical lines		Many vertical lines are found on the Hunting Frieze, sometimes interpreted as marking some sort of fencing used in the hunting.	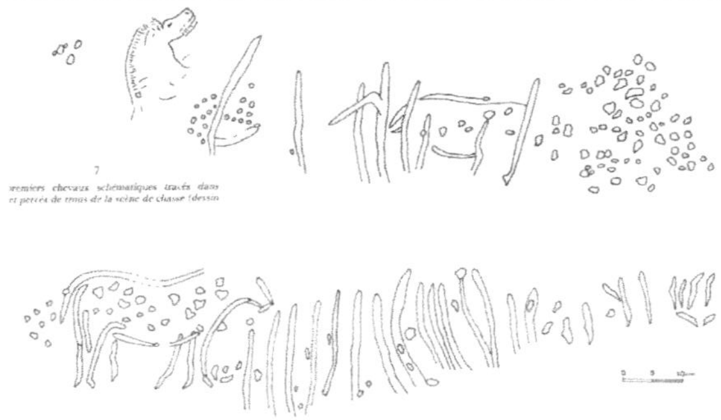 (the whole Hunting scene From Riveng 1984: fig. 7)	Riveng 1984: 444
MO 20	Many holes		In the Hunting Frieze Gallery This scene is also covered with deep holes of different sizes and inclination that indicate they were probably not made by the same persons.		Riveng 1984: 444
MO 21	horse	30 cm	Two finger-traced horses are still visible in the Hunting Frieze underneath the lines and holes covering them. The whole scene is situated towards the end of a gallery, after a very narrow passage, on the right wall. It is three metres long. The surface on which it was made was prepared by adding clay to it. The base of the frieze is slightly higher than the floor of the gallery, which is itself only 1.2 m high in this part. The horses were interpreted as being pushed through a fence towards a trap (hole) while the spears of the hunters reach them.	 (From Lorblanchet 1995: 111	Riveng 1984: 444; Trombe and Dubuc 1947: 51, 119
MO 22	Horse				

Finger – tracings	On the right wall of the Casteret-Godin gallery, numerous macaroni tracings are visible near the entrance. There is also another panel in the middle of the gallery, where the mammoth representation if found (Trombe and Dubuc 1947: 96, 98).

Montespan, Haute-Garonne

Traces of human activity on clay	They are extremely numerous in Montespan. A 'font-like' structure is described as a sort of basin, oval, three quarter vertical on the wall, 50 to 60 cm long, 30 cm deep. It was dug out in the clay of the wall, with a sort of flattened rim all over it. The bottom is rounded, the walls are thin, all the surface of the inside is covered with calcited finger prints. It is found near the entrance of the Casteret-Godin gallery (Bégouen and Clottes 1988: 29; Trombe and Dubuc 1947: 96). Two clay balls are described near the possible modelled vulva MO5 found horizontally on an artificial clay bank (Trombe and Dubuc 1947: 96). Some torn off pieces of clay are found in all surfaces and corners, without it being very clear what is prehistoric what is modern (Rivenq 1984: 441). Small mounds of accumulated clay are found at the foot of the right wall of the Casteret-Godin gallery, over a 2m long surface; they are mentioned as possible *modelages* damaged by dampness by H. Breuil (Trombe and Dubuc 1947: 84). Traces of removal of a large quantity of clay that forms a depression dug in the floor in the bear chamber near the opposite wall: it was possibly where the clay to make the bear statue came from. Other possible melted *modelages* are found in the bear chamber. They resemble molehills, 30 to 70 cm in diameter, 10 to 15 cm high. (Rivenq 1984: 442; Trombe and Dubuc1947: 78). A sort of cup was fashioned in clay and attached to the wall of the Castert-Godin gallery by the Magdalenian (Trombe and Dubuc 1947: 86). In other parts of the cave, in the Trombe-Dubuc gallery numerous footprints, prints of fingers and hands are found (Trombe and Dubuc 1947: 44).

Labastide, Hautes-Pyrénées

Reference number	Graphic unit	Size	Description	Illustration (when available)	References
L1	Modelled bison		The only work in clay in the cave is a deeply engraved back and rear of a bison on a boulder covered with clay in the terminal part of the cave. It is actually the last representation found in the cave, the furthest from the entrance.	 (From Clottes 1977)	Simonnet *et al*. 1984: 532; Clottes 1977

Use as pigment	Clottes (1989: 77) mentions clay used as pigment in the site of Labastide without more detail.

Traces of human activity on clay	Clottes (1989: 87) also mentions groups of cupules found in the clay in this site. Prints of fingers on a clay bank are also mentioned but it is difficult to be sure of their age (Simmonet *et al*. 1984: 529).

Bois du Cantet, Hautes-Pyrénées

Reference number	Graphic unit	Size	Description	References
BO1	Impressed sign		Two masses of clay bear seems ancient with surface patina in the higher part of the 'large chamber'; one is pierced by 16 holes that seem to go upward, that are cylindrical at their base and conical at the extremity, evoking traces left by spears. They are organised into two slightly wavy lines of 6 holes with two more holes added in between at on e extremity, and one hole between the two lines towards their middle, and then one hole completely isolated in a corner.	Clot 1984

Traces of human activity	Some finger traces are found on clay surfaces (Clot 1984). Groups of cupules are also described as being found in the cave (Clottes 1989: 87).

Erberua, Pyrénées-Atlantiques

Reference number	Graphic unit	Size	Description	References
ER1	Horse *modelage* 1	20 cm	This horse and the three other in the group have been sculpted in clay specially brought to this location and then stuck on the wall. Part of the clay has flaked off the wall surface, but it has left the imprint of the figure on it. For this horse, only the head and the four limbs remain. It is found in the 'chamber with the clay sculptures'.	Larribau and Prudhomme 1989: 480, 481
ER2	Horse *modelage* 2	14 cm	Only the clay of the tail survives for this horse *modelage* but the imprint of the animal on the wall is quite clear, except at the level of the leg where it stops. It is found in the 'chamber with the clay sculptures'.	Larribau and Prudhomme 1989: 480, 481
ER 3	Horse *modelage* 3		For this horse *modelage*, only two legs remain, one at the front and one at the back. The bent knees seem to indicate that the horse was represented in movement. It is found in the 'chamber with the clay sculptures'.	Larribau and Prudhomme 1989: 480, 481
ER4	Horse *modelage* 4	28 cm	This is the better preserved and the largest of the group of four horses. Its legs are well detailed.	Larribau and Prudhomme 1989: 480, 482
ER5	Cross-shaped sign		Two lines are traced with the fingers in a narrow area before the narrow passage exit: they form a cross-shape.	Larribau and Prudhomme 1989: 482

Erberua, Pyrénées-Atlantiques

ER6	Series of short lines		A series of short lines, calcited in places, are drawn on a clay wall.	Larribau and Prudhomme 1989: 482
ER7	Head of a doe	17 cm	It is traced with the fingers on the ceiling of one chamber in the IIc group next to four engraved animal figures in the vicinity of the exit passage for this area. It represents a doe with one very long ear.	Larribau and Prudhomme 1989: 485
ER8	Radiating sign 1		This is a sign radiating from a centre, with 7 branches. It was drawn with the fingers. It is found on the ceiling of chamber in the IIc group.	Larribau and Prudhomme 1989: 485
ER9	Radiating sign 2		This is a sign radiating from a centre, with 6 branches. It was drawn with the fingers on the ceiling of chamber in the IIc group.	Larribau and Prudhomme 1989: 485
ER10	Sign with parallel lines		This sign shows two parallel lines with a small circle juxtaposed on each side. It is drawn with the fingers on the ceiling of chamber in the IIc group.	Larribau and Prudhomme 1989: 485
ER11	Rectangle with handles sign		This sign can be described as a rectangle with a double semi-circle attached on its two opposite longer sides. It is finger-traced on the ceiling of chamber in the IIc group.	Larribau and Prudhomme 1989: 485
ER12	Horse	60 cm	The line of the belly and the lower limbs of the horse figure are missing. Some signs seem superimposed on it. It is found on the ceiling of chamber in the IIc group.	Larribau and Prudhomme 1989: 485

Finger-tracings	One panel of macaronis tracings is described in this cave as being done in a semi-concentric shape, slightly oblique, with wide lines, shallow, concentrated in one small niche.	Larribau and Prudhomme 1984:277; 1989: 480

Oxocelhaya-Hariztoya, Pyrénées-Atlantiques

Reference number	Graphic unit	Size	Description	References
O1	Triple line sign	60 cm by 30 cm	This sign is finger-engraved on a clay surface on a bloc and forms a triple digital line. It is found before the entrance of Gallery Larribau.	Laplace and Larribau 1984: 283
O2	Horse 1		Two small horses are found on a slightly calcited clay surface. They seem to follow each other on a 1.2 m long surface. Their heads face towards the left. They are done by finger-tracings. They are found before the entrance of Gallery Larribau.	Laplace and Larribau 1984: 285
O3	Horse 2		Two small horses are found on a slightly calcited clay surface. They seem to follow each other on a 1.2 m long surface. Their heads face towards the left. They are done by by finger-tracings and for horse 2, the mane in done by engraving. They are found before the entrance of Gallery Larribau.	Laplace and Larribau 1984: 285

Finger-tracings	In the final part of the cave, a panel measuring 1.5 m by 1.3 m is formed by macaronis finger-tracings with triple or quadruple digital lines .	Laplace and Larribau 1984: 285

Traces of human activity on clay	Some prints of fingers have been noted at the site.	Laplace and Larribau 1984: 285

Etcheberri, Pyrénées-Atlantiques

Reference number	Graphic unit	Size	Description	Illustration (when available)	References
ET1	Horse	30 cm	Two horses have been made by engraving and finger-tracing on a mass of clay specially brought there and previously smoothed. They are facing towards the right and they are superimposed. They are located on a 'cornice' above a fissure at the very end of the cave, above the mouth of a 7 m deep abyss (unexplored by Prehistoric people) in a dangerous location. Three holes are found with the horses: two on the neck of the largest horse, and one in front of it.	 From Boureux 2004: figure 4)	Laplace et al. 1984: 269, 271
ET2	Horse	20 cm	This horse is made by the same technique as the previous one to which it is superimposed. The only major difference with it is that it is smaller.	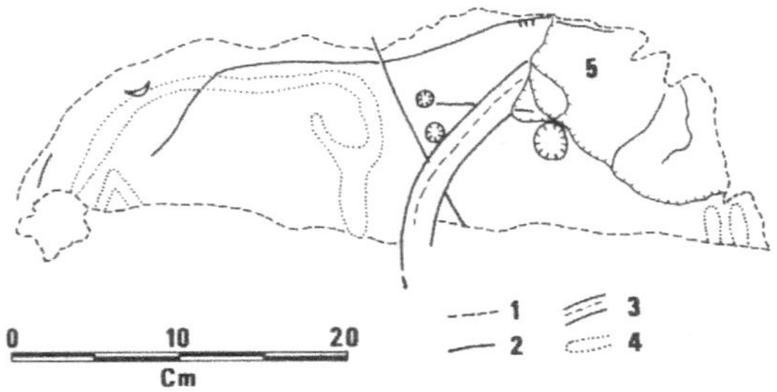 (From Laplace *et al*. 1984: figure 6)	Laplace et al. 1984: 269, 271

Use as pigment	Clottes (1989: 77) mentions that clay was used as pigment in the cave of Etcheberri to draw a horse.

APPENDIX 2: CATALOGUE OF PORTABLE FIRED CLAY OBJECTS FROM DOLNI VESTONICE I AND II AND PAVLOV I

The present catalogue is based on the ceramic objects seen during visits to the Anthropos Institute in Brno, Czech Republic, for the material from Dolni Vestonice I and to the Archaeological Institute, Academy of Sciences in Dolni Vestonice for the material from Dolni Vestonice II and Pavlov I. For the purpose of the present study the objects have been given an inventory number, which does not normally correspond to their reference numbers in the respective museum collections. The original reference number is however mentioned in this catalogue in bracket after my own reference. In the presentation of each object, the two reference numbers are given, then the name of the object –usually being descriptive-, its measurements, in most cases in their maximal dimensions, a full description and illustrations, for most pieces a view from each side (front and reverse). All objects from collections from **the Anthropos Institute in Brno** for the Dolni Vestonice I artefacts and from **the Archaeological Institute, Academy of Sciences in Dolni Vestonice** for the Dolni Vestonice II and Pavlov I artefacts.

Dolni Vestonice I

Reference number	Graphic unit name	Size	Description	Illustrations
DVI 1: (No reference)	Female figurine: 'the Black Venus'	Height: 11 cm; maximal width: 4 cm at the hips	This is the best known of the Moravian ceramics. It represents a female figure, naked, with a deep incision at the hips, which seems to indicate a belt. There is an emphasis on the heavy breasts and the belly, which are very realistically modelled. A deep bellybutton is indicated by an oval-shape hole. The extremities of the body, as often in the Gravettian figurines, are schematic: the legs are represented by a conical extremity, broken off at the end, with a deep straight incision. The feet are not indicated, and do not seem to have ever been there even if the extremity is slightly broken. In the same manner, the top part of the body is schematic. The shoulders and the collarbone are much stylised, in a rather angular manner, but which is aesthetically very successful. The upper part of the arms only is shaped, but the viewer sees them continuing under the breast, lying on the belly, even though this is only visually suggested. The head is a sort of rounded stump, with no clear separation between the neck and the face. The only features are two symmetrical slits where the eyes would have been. On the front of the face, running vertically from the point between the 'eyes' to the collar bone, one can see a series of parallel lines which seem to be the result of a scraping with possibly a micro denticulate tool. On the top of the head, there are four holes, oval in section, arranged in a quadrangular shape, made by imprinting with some sort of tool. On the back of the figurine, the spine is indicated by a deep depression with an incision at the bottom of it, incision that continues the one indicating the separation of the legs. This incision actually breaks the one made to indicate the belt. On this side the upper part of the arms are indicated by deep incisions. In the middle of the back there are two parallel nearly horizontal incisions on each side of the spine, which seem to indicate some sort of body decoration. The figurine is broken: all its right leg and part of the right hip separated from the rest of the body at some point, but can be neatly refitted. The surface is altogether very smooth with some irregularities, brown-black colour. Inside the break the material is uniformly charcoal grey and not at all shiny like the rest of the outside surface. This is due to the conservation product use, which also changes the colour, giving it a brownish tint.	 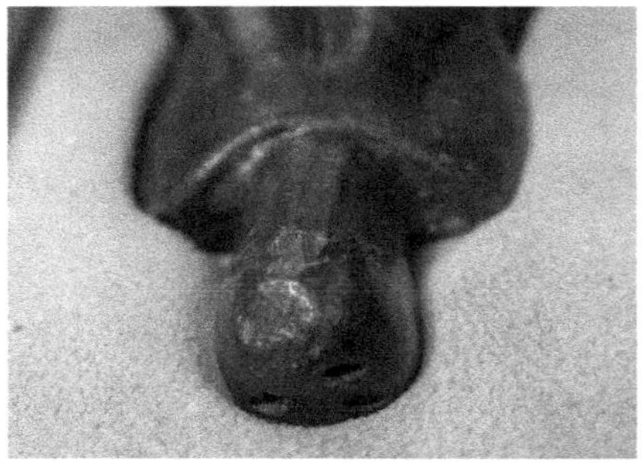

Dolni Vestonice I

DVI 2: (DV13)	Fragmentary female figurine	Height: 4 cm; width: 3 cm; thickness: 2 cm	Fragmentary female figurine with only the right part of the waist area (belly and side) surviving. It is recognisable by comparison with the other known female figurine: part of the incised belt is visible, as well as the bellybutton indicated by a hole. There are also two deep incisions coming from the top part of the figure. It must have been a rather large figurine when complete. The surface of the broken section is very smooth, as if it had been intentionally flattened. The figure also bears some reddish traces that remind of ochre.	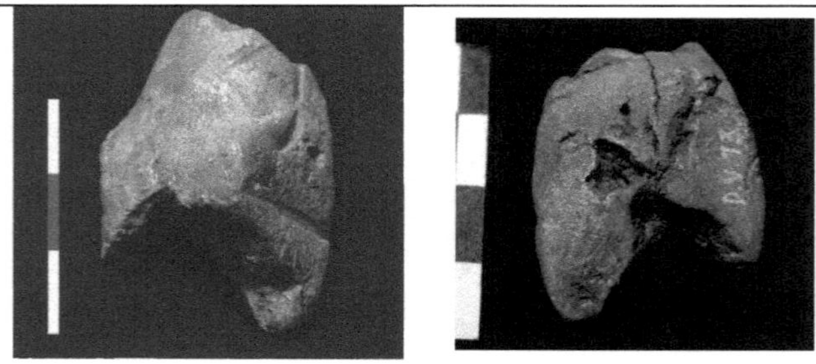
DVI 3 and DVI 4: (DV8 and 9)	Female figurine in two fragments	Height: 5.4 cm; width: 2.8 cm; thickness: 1.6 cm	Fragmentary female figurine, broken in two at the belt, that seems built on the same model than the Black Venus: it has a deep incision marking the belt at the hips; another incision at the front and one at the back of a conical stump to mark the legs; the feet are not indicated; a depression indicates the spine on the back, with two incisions on each side of it; the arms are indicated by incisions; the buttocks are large. It is broken at the neck and at the belt. The front of the bust is flat with no indication of the breasts or of a prominent belly. This part of the figure is damaged. We can see traces of two rectilinear incisions on the belly that would cross at a perpendicular angle if that part was not broken; the rest of the surface is very irregular and was possibly restored as some fragments are missing. It is brown-black.	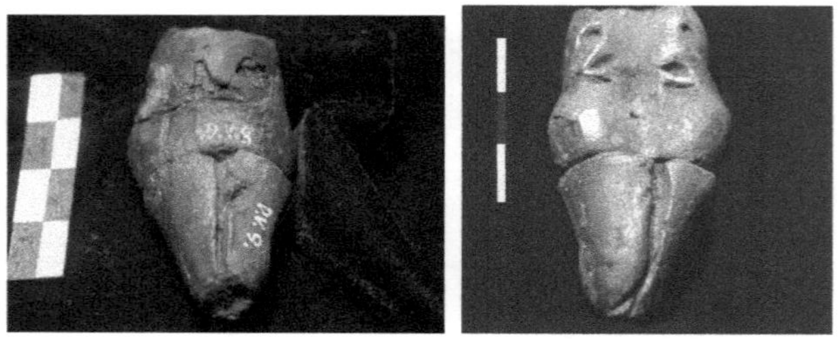
DVI 5: (DV 10)	Bust of a female figurine	Height: 3.2 cm; maximal width: 2.4 cm; maximal thickness: 1.5 cm	Fragmentary female figurine with only part of the bust surviving. It is recognisable mainly by comparison with the Black Venus thanks to the typical parallel twin incisions on each side of the spine on its back. The top of the back is missing, and the back is broken just above the buttocks. We can guess the beginning of two legs and of a rounded shoulder on the left side and then it is broken. The front of the figure seems very flat, but its irregular surface might mean that part of it is missing.	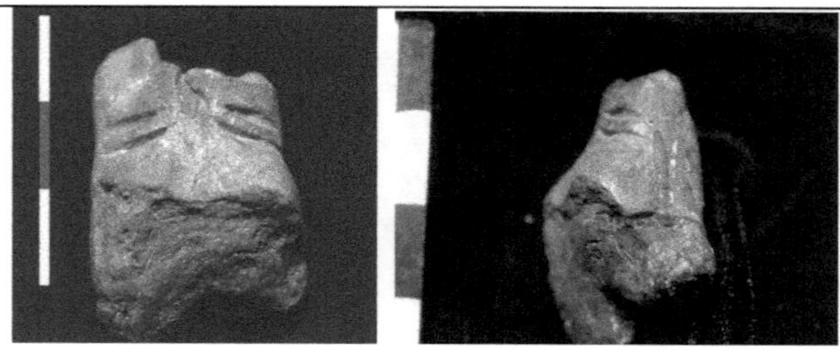
DVI 6: (DV 12)	Anthropo-morph with crossed legs	Height: 4.1 cm; maximal width: 2.3 cm; maximal thickness: 2.2 cm	This piece is a fragmentary human figurine: the top of the legs (from just above the knees) and the lower part of the belly survived. One leg seems to cross over the other indicating movement. The bellybutton is indicated; the belly is prominent. The back part is much damaged. The material is irregular, not smooth, with different colours.	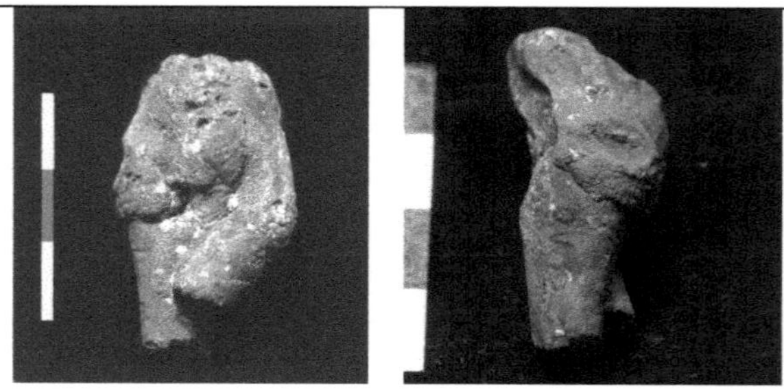

Dolni Vestonice I

DVI 7: (DV 7)	Female figurine 'the 'Tattooed' Venus'	Height: 3 cm high; width: 1.7 cm; thickness: 1.5 cm	Small fragmentary female figurine with patterns drawn on the surface as if to indicate some form of body decoration. The left side of the body remains. The head is broken, as well as a large part of the right side. On the left side the leg is indicated by a short conical stump with no foot. It is whole. The arm is indicated by an incision, and it vanishes at the waist without indication of a hand; the waist is marked by an incision at the front and the back of the body, indicating a belt. The female genitals are marked by an incision between the legs. The breasts are not indicated. There is a large rounded belly with a marked bellybutton. On the back, an incision seems to mark the spine and continues downwards to indicate the buttocks, breaking the line of the belt. The body decoration is made of impressed small dots forming patterns: a circle of dots around the bellybutton; a few dots in a line at the neck seem to indicate a necklace, but it is broken off; and the remaining arm is impressed with a line of these dots from one end to the other.	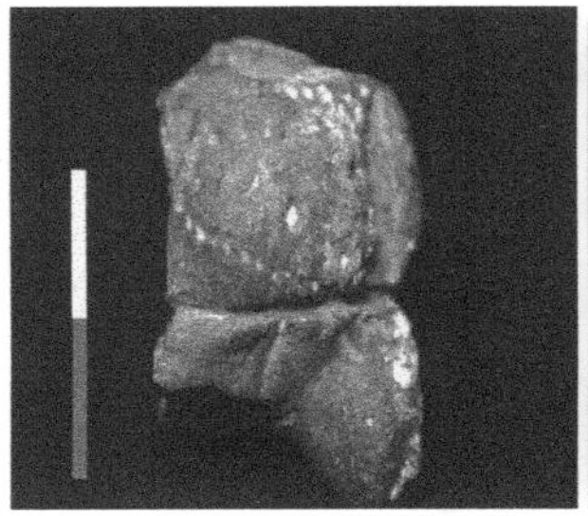 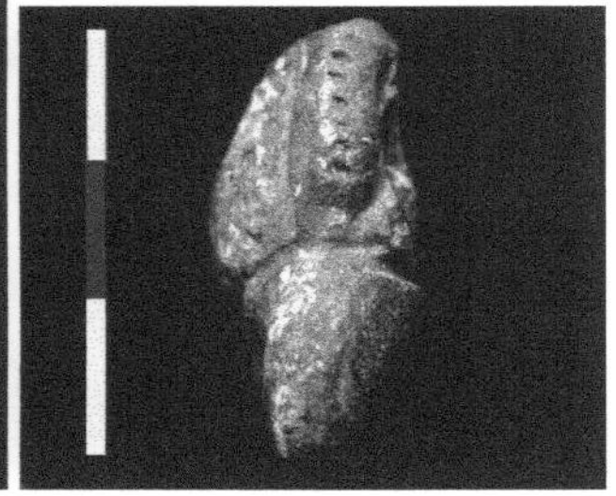
DVI 8: (DV 14)	Breasts of a female figurine	Height: 3.4 cm; width: 2 cm	Two breasts, separated from the rest of the body, are similar to the black Venus' ones. They are probably broken off a statuette. Some traces of a necklace are visible at the neck. The breasts are asymmetrical.	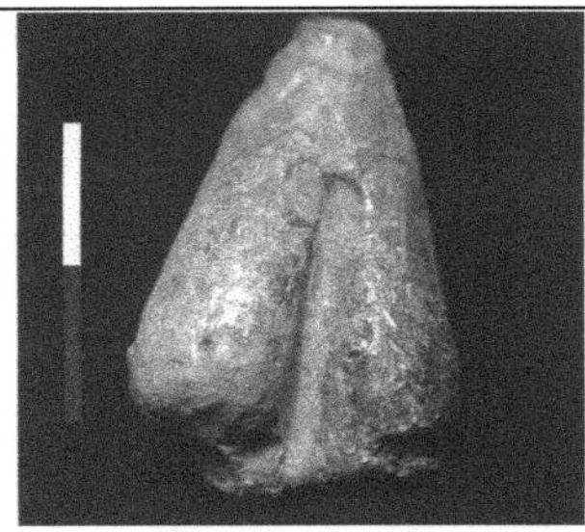 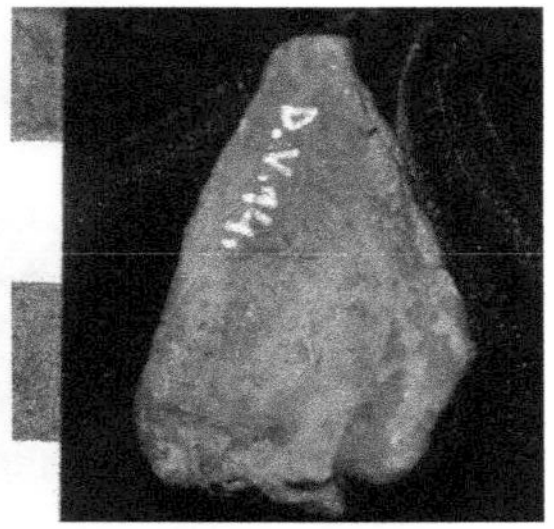
DVI 9: (DV 15)	Anthropo-morph	Height: 4 cm; width: 1.7 cm; thickness: 2 cm	This object is a small human figurine. One side of the body is clearly recognisable with a little stump indicating the arm, a deep incision marking the belt at the waist and going all round the body, the top of the leg clearly fashioned with incisions to mark the groin. The head is indicated by a very small protuberance, which seem to have been made this way intentionally. A whole side of the body is flattened as if it had been thrown onto a flat surface after fashioning when still wet. The belt is a factor in indicating this could be a female figurine but as there are no other determining elements this piece is considered as an anthopomorph without more precision. The object is light brown with black spots. The surface is not very smooth or regular.	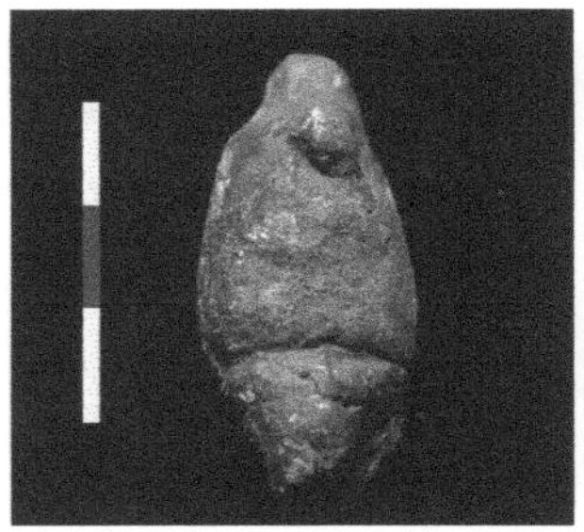 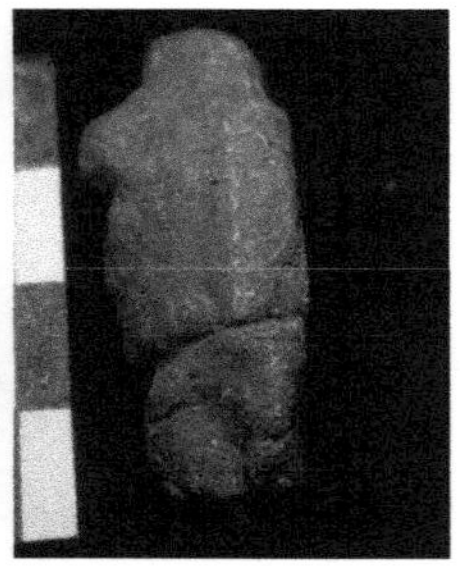

Dolni Vestonice I

DVI 10: (DV 17)	Male figurine	Height: 2.7 cm high; width: 2 cm; thickness: 1.6 cm	This piece is a fragmentary male human figurine: it shows the pelvis area and the top part of the legs, which are thin. The testicles are clearly indicated in a realistic manner, but the penis is broken off. The figure is black in colour, even inside the broken section of the body. This suggests a firing in reducing conditions (without oxygen).	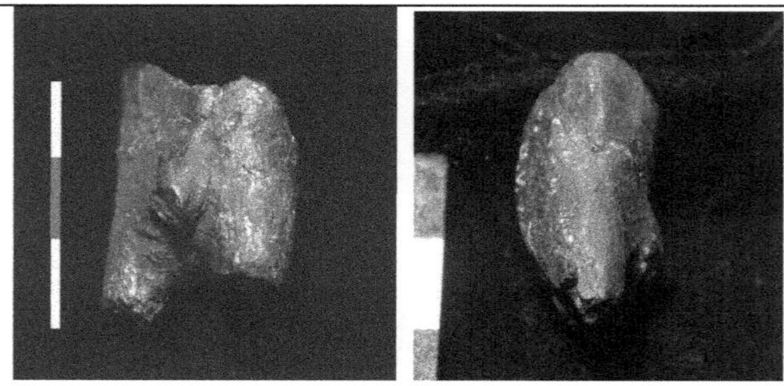
DVI 11: (DV 181)	Body extremity	Height: 2.5 cm	This piece is the lower part of a leg with a foot. The heel, which indicates a human foot, is well modelled. The foot is at an angle with the leg. The material appears porous with some inclusions. It is mostly red in colour with darker areas.	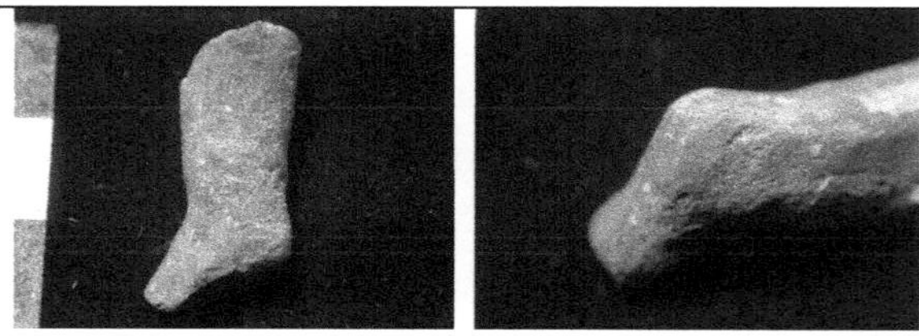
DVI 12: (DV 187)	Body extremity	Height: 2.5 cm; width: 2.1 cm; thickness: 1.5 cm	This body extremity forms a massive 'stump' with a foot at the end. The heel is indicated, and it can be seen as a booted human foot. It is much stylised.	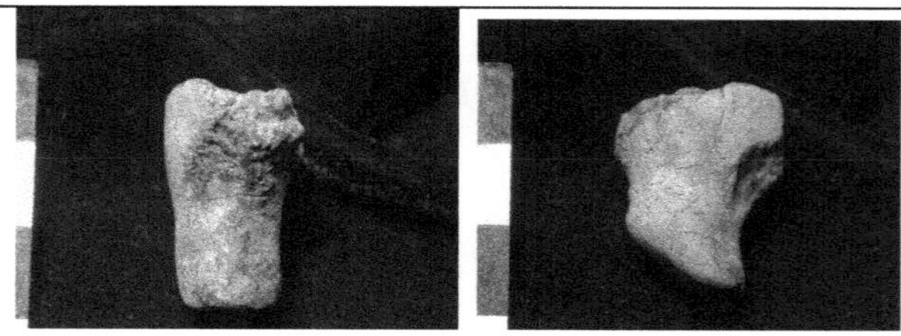
DVI 13: (DV 32)	Anthropo-morph	Height: 3.5 cm; width: 1.2 cm; thickness: 1.5 cm	This piece forms a sinuous elongated shape that appears to be a human figure. This is supported by the fact that the top of one of the legs is clearly visible, with an incision to separate it from the second leg, which completely broken off. The rounded end at the top would be the head. A small protuberance on it could be the nose. The arms are not indicated. This would be a stylised human figure in movement, as the sinuous line of the body seems to indicate. The surfaces are very smooth, except in damaged areas that seem to have undergone restoration. This object has been compared to an 'owl' (DVI 30) figure from the same site on the basis of a similar overall shape, but I do not find in it any element to support this interpretation	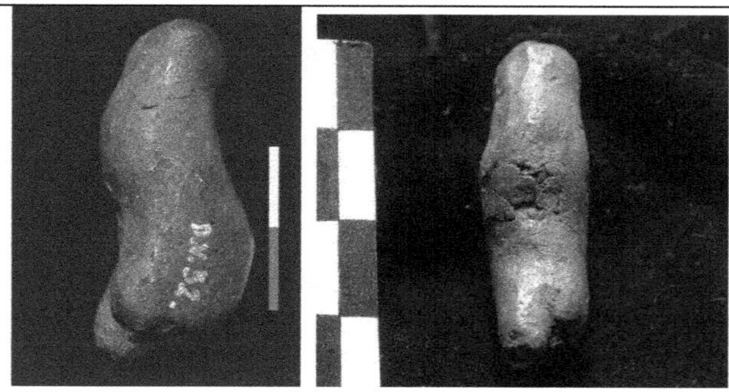

Dolni Vestonice I

<table>
<tr><td>DVI 14: (No reference)</td><td>Rhinoceros head</td><td>Height: 2.8.cm high; length: 3.8 cm; maximal width: 1.4 cm</td><td>This piece represents the head of a rhinoceros, recognisable by the horn and the bump on top of the skull. One eye is indicated by a hole on one side of the head. On the other side no eye is marked. Two circular and asymmetrical holes indicate the nostrils. The broken surface of the neck is very irregular. The surface of the object is altogether shiny and dark brown, but also irregular with many asperities, and some incised lines, or possibly traces left during scraping.
M. Oliva (Pers. Com. May 2004) said the horn has been probably entirely restored by Absolon.
Reference: published in Absolon 1949, Plate V</td><td>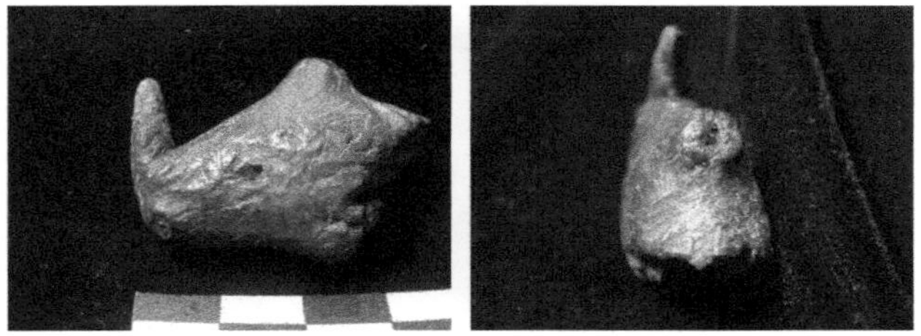</td></tr>
<tr><td>DVI 15: (DV 47)</td><td>Feline head</td><td>Height: 2.1 cm; length: 3.1 cm; width: 2 cm</td><td>This is a fragmentary piece and is made of two parts that have been glued back together. It can be interpreted as a feline head by comparison with the other feline heads of the site, as the same convention is used in making the eyes: they are indicated by vertical slits, 0.7 cm long, on each side of the forehead. Only one eye is visible here as the other side of the head is missing.</td><td>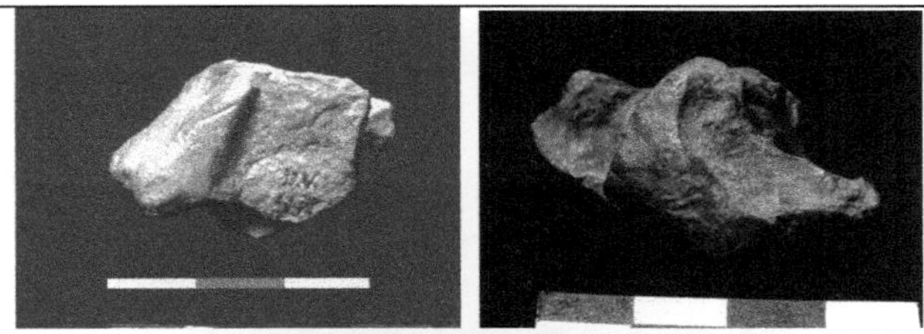</td></tr>
<tr><td>DVI 16: (DV 16)</td><td>Bear figurine</td><td>Height: 3.2 cm; length: 7.4 cm; width: 2.2 cm</td><td>This figurine represents a nearly whole bear. It is one of the most complete ceramic figurines of the site. The face is roughly shaped with a pointed muzzle, and possibly a slight swelling indicating the mouth, but without any other detail. The emphasis seems to be on the line of the back, which is carefully modelled to give the silhouette typical of the animal, probably the most important detail to recognise the species.
The front legs seem neatly broken at the same level. They are well shaped at their base, as well as the line of the belly. The two back legs seem represented by a large stump that is damaged, and in which we can guess the beginning of a separation into two legs. The break is less neat than for the front legs, and it affects also the rest of the rear of the animal, where we can only guess the top of the tail.
The surface is shiny, dark brown with asperities, holes and missing parts. It is not very regular. M. Oliva (pers. com. May 2004) mentioned that the statuette was in fact broken in two at the discovery, and that it was restored by Absolon, although we are not sure to what extent, as no written traces were left. The crack of the breakage can still be seen.
Reference: Absolon 1949, plate V</td><td>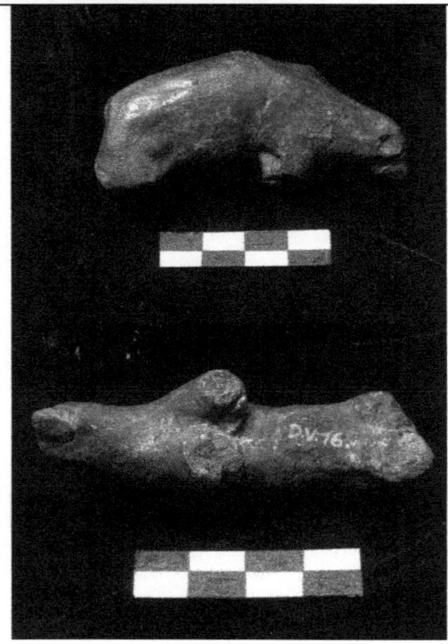</td></tr>
</table>

Dolni Vestonice I

DVI 17: (DV 17)	Fantastic animal head	Maximal height: 2.1 cm; length: 6.6 cm; maximal width: 2.3	K. Absolon (1949) points out that this animal head was first described as a crocodile at the time of its discovery as it looks 'somewhat exotic' (p.25), but that is was later identified 'beyond doubt as the head of reindeer' (p. 25). This interpretation has remained since, but a close examination of the object reveals some features that are not fitting with it. These are notably a sort of crest running from the extremity of the muzzle to the beginning of the mane, and two small protuberances under the chin that remind of long fangs or teeth. The surface of the object is entirely covered with incisions forming parallel lines, which is unique in the ceramic record of Dolni Vestonice and Pavlov. We can see a muzzle and a small protuberance that could be the ear. Then, above the very elongated neck, there is a 1.5 cm wide flattened surface that apparently forms a mane. It is also covered with parallel incisions running across the width, in a different direction than for the rest of the surfaces. There is a sort of crest on the muzzle, ending up in a small flattened nose broken at one end and stopped by the flat surface of the mane at the other end. It is quite surprising to see two small protuberances neatly marked on each side of the chin; they evoke two teeth. On each side of the nose, there is also a swollen area that reminds of the lips of a feline or a canine. The ears are also small and rounded like the ones of the felines. Now, it is true that the general shape of the head, the presence of a mane stretched on a long neck would resemble more a reindeer or a horse. The deep incisions covering the surface could evoke the fur of an animal, but they are rather deep and widely spaced, and have no known equivalent in the other objects of the site. This animal head evokes a fantastic animal, composite and unique. The area of the neck is quite shapeless, but the flattened and incised surface of the top of the head seems to continue up to the very edge of the break, as if this end was given a rather conical shape. The part where the neck would have attached to a body has been worked and smoothed; this suggests that it was not attached to a body. Reference: Absolon 1949, Plate V	

Dolni Vestonice I

DVI 18: (DV 18)	Feline head	Height: 2 cm; length: 4.4 cm	This is an elongated feline head with a long fine muzzle. Long vertical slits, 0.8 cm and 1.2 cm long, are used to indicate the eyes. Small ears are indicated on the sides of the head; they are not exactly symmetrical. The back of the head forms a sort of conical stump. It could indicate that the head was modelled alone without any intention to do the body. It could also form the head of a second animal using the same ears. The object is light brow in colour. Most of the surface is well smoothed. The nose of the feline is very fine and pointed, its details shaped by incisions. The head is damaged underneath. An irregular hollow shape is visible underneath, possibly the print of a finger.	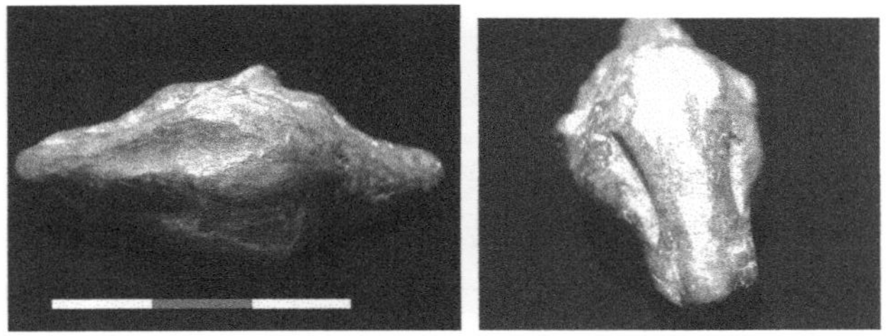
DVI 19: (DV 19)	Reindeer head	Height: 1.7 cm; length: 3.8 cm; width: 1.6 cm	This object is an elongated animal head. The mouth and the nostrils are indicated by incisions and the eyes by protuberances, which are not symmetrical, on each side of the head. Between the ears on top of the head, a damaged surface is visible with two incisions at the bottom of it, which indicate it could have been made intentionally. The surface is rather smooth, with some accidents in the material. The head is broken at the neck. It is dark brown Reference: Absolon 1949 Plate V, published as reindeer head	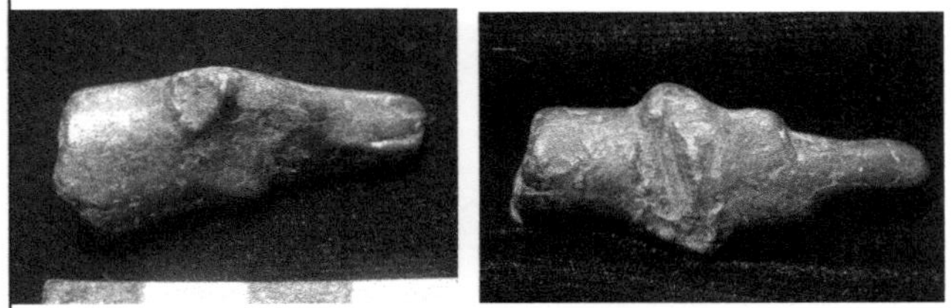
DVI 20: (DV 20)	Feline head	Measures 3.6 cm long, 2.4 cm large, and 2.3 cm maximal height	This piece is a feline head with the ears in a low position on the side. The ears are well modelled and in relief; they are quite large. The eyes are indicated by vertical incisions, 0.6 and 0.7 cm long, that run parallel to the line of the forehead. The nose is well modelled with the lips indicated by incisions. Four parallel incisions on top of the head are quite deep and large (3 to 4 mm), and about 5 to 6 mm long. They seem to indicate the mane. The neck is broken at an angle. The surface of the break is irregular. Reference; Absolon 1949, Plate V	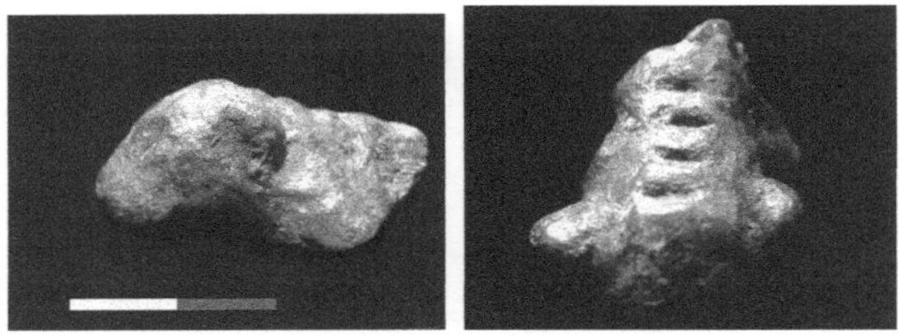
DVI 21: (DV 37)	Feline muzzle	Height: 1.5 cm	This object is the end of the muzzle of a feline, broken off a head. It is recognisable as it follows the same conventions as the other feline heads in representing the muzzles: the nose, the nostrils and the lips are delimited by incisions.	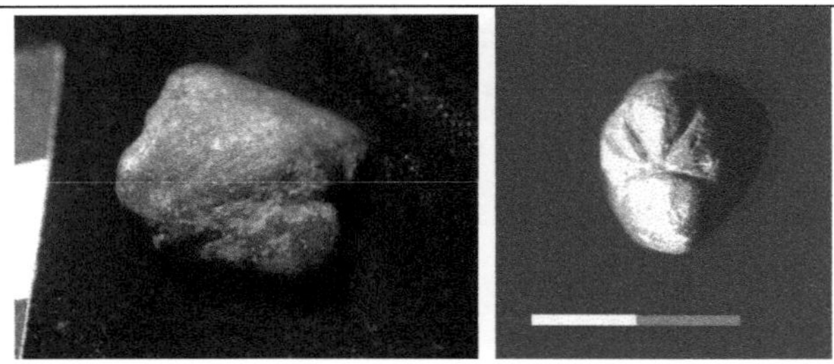

Dolni Vestonice I

DVI 22: (DV 31)	Animal body	Height: 1.4 cm; Length: 2.8 cm	This object represents the fragmentary body of an animal, with the hind side nearly intact, and the front part (head and front legs) broken off.	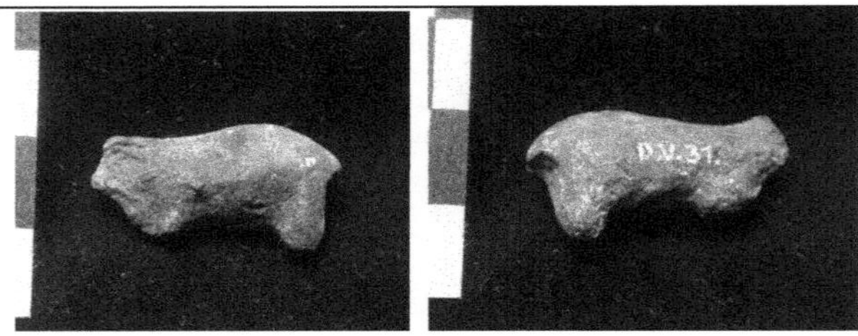
DVI 23: (DV 23)	Feline head	Maximal height: 2.4 cm; length: 3.8 cm; width: 2.6 cm	This is another 'cut-off' feline head. It is very detailed and finely modelled. The chin is very prominent and forms a protuberance. The mouth, nostrils and lips are well delimited by incisions, and are also modelled. The two ears are indicated and form protuberances on the sides of the head, in a low position. One of them is broken. The forehead and the top of the head are well modelled as well. The object is light brown in colour. The eyes were originally indicated by vertical slits, 0.6 cm long for the left one, like on the other feline heads. On the right side, the slit is no longer visible as a 0.5 cm large circular hole was pierced later just on it. It is very deep, but does not pierce completely the object. Just above it, next to the ear, another circular hole, 0.3 cm in diameter, was made and this time it does pierce the whole object to emerge on the broken surface of the neck. This feline head might have been transformed into an ornament by the addition of these holes. Reference; Absolon 1949, Plate V	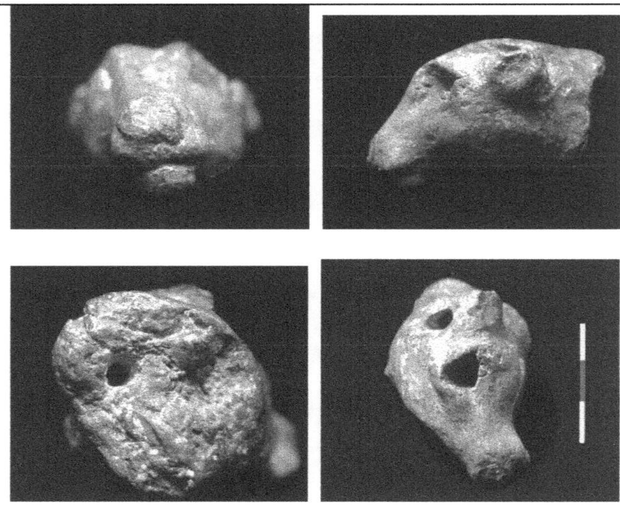
DVI 24: (DV24)	Feline head	Length: 3.6 cm; width: 1.7 cm	This piece is a feline head broken at the neck. It shows the same stylistic elements as the other feline heads; the eyes are made by vertical slits; a vertical sunken depression runs along the forehead; the muzzle is modelled by incisions to mark the lips and the nose (although the mouth itself is not indicated here); the ears are indicated by small protuberances on the side of the head.	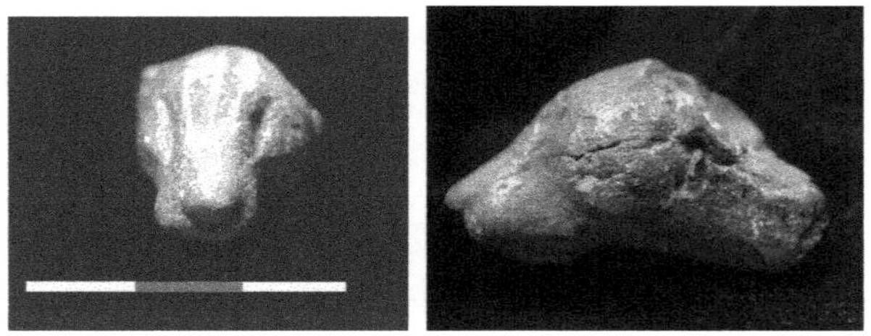

Dolni Vestonice I

DVI 25: (DV 25)	Animal head	Height: 3.6 cm; length: 4.2 cm	This piece represents an animal head, possibly of an herbivore because of its overall shape. It is much stylised, without much detail. The surface is well smoothed. One ear indicated on the left, a line is impressed in the neck on the right.	 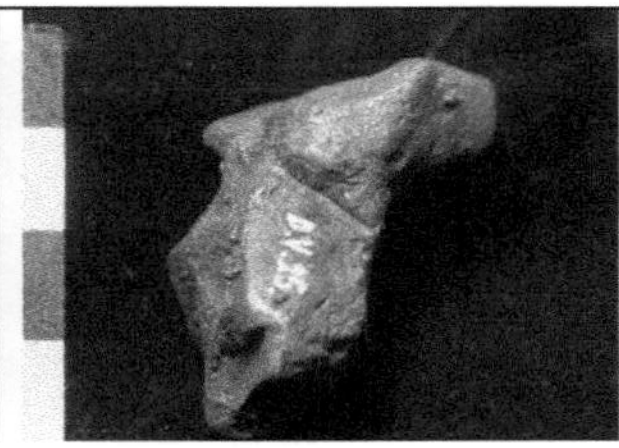
DVI 26: (DV 26)	Animal head	Length: 2.8 cm; width: 1.2 cm	This object appears to be a small animal head, roughly shaped without details, with an elongated pointed muzzle and a fine neck. The top of the head is flattened and the extremity of the muzzle is broken off. There are three holes in the material: two that seem made by impressing, and another larger and less regular.	 
DVI 27: (DV 33)	Bovine figurine	Height: 2.5 cm; length: 4.8 cm	This object represents an animal body. The head and the front legs are well detailed; the rear legs are missing. It reminds of a bovine, possibly a calf. The eyes are prominent surfaces pierced by very small holes. Two holes indicate the nostrils. The right ear is broken off, the left one is present. The two front legs are indicated as superimposed, with an incision separating them. It is different in style from the rest of the ceramic objects. M. Oliva (pers. com. May 2004) mentioned that it is very similar to the numerous Neolithic clay figurines found in the same region, and that he would not be very surprised if it had been mixed up during the ancient excavations.	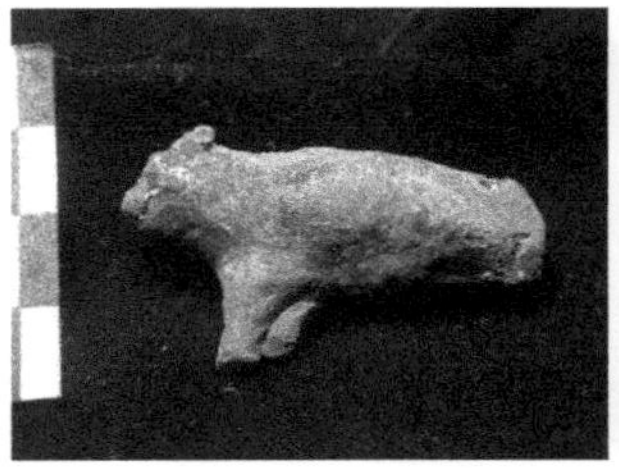 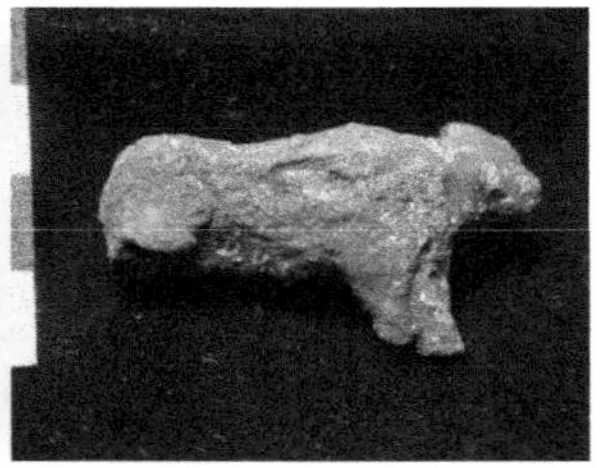
DVI 28: (DV 28)	Possible bird	Length: 11.8 cm; maximal width: 1.6 cm	This piece is the largest of the ceramic objects from Dolni Vestonice. It has a cylindrical narrow elongated main part that ends up in a conical shape at one end, and in two pointed 'ears' at the other end for the part that still exists. It can be interpreted as a stylised bird by comparison with DVI 29, probably a type of owl because of the 'ears'. A groove also appears to run all around the object in the proximity of the 'horns'. Just at the base of the 'horns', traces remain of two holes drilled at the ends of what would have made a rectilinear axis. The object broke off along this axis, and one half of the holes can still be seen. They were at least 0.5 cm deep, but they did not completely pierce through the object. The object is much damaged, but the original surfaces were carfefully smoothed.	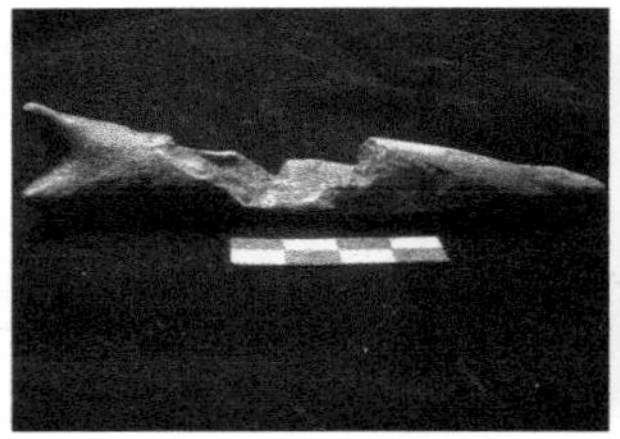 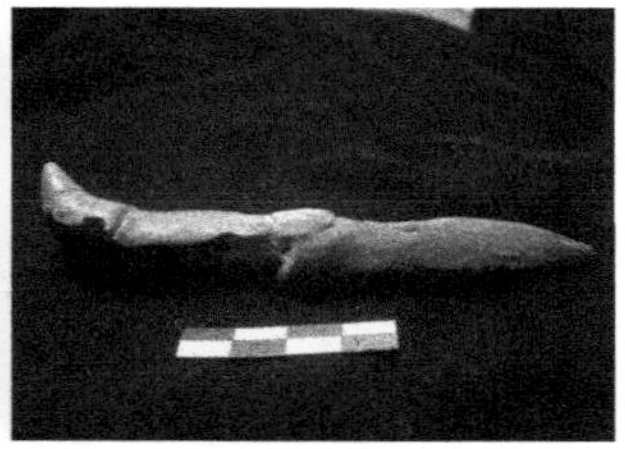

Dolni Vestonice I

DVI 29: (DV 29)	Bird	Height: 3.8 cm; diameter: 0.6 cm	This object represents the head of a bird of prey, possibly an owl, because of the shape of the beak and of the presence of two 'ears' on the head. This head is at the top of an elongated body that ends in a conical extremity. It is extremely similar to DVI 28 above.	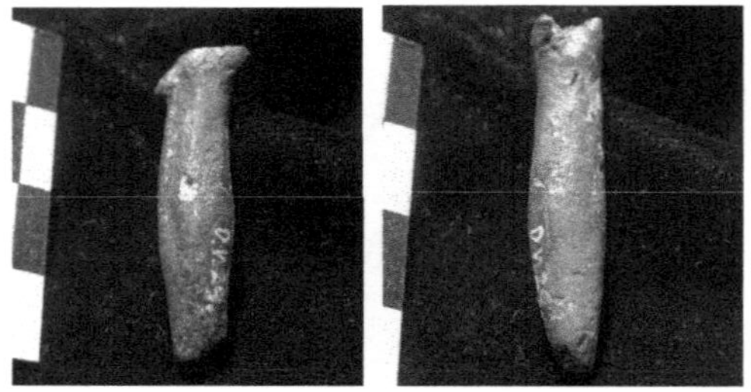
DVI 30: (DV 30)	'Owl'	Height: 4.5 cm; width: 1.5 cm	Figurine with an elongated sinuous body. The head is formed by a rounded end in which two holes are made for the eyes, and with incisions to mark the limit with the neck. The lower part forms a sharp angle, and then is broken off just after. The surface is very smooth. The general shape of the object reminds of an owl with the typical twist of the head in relation to the body.	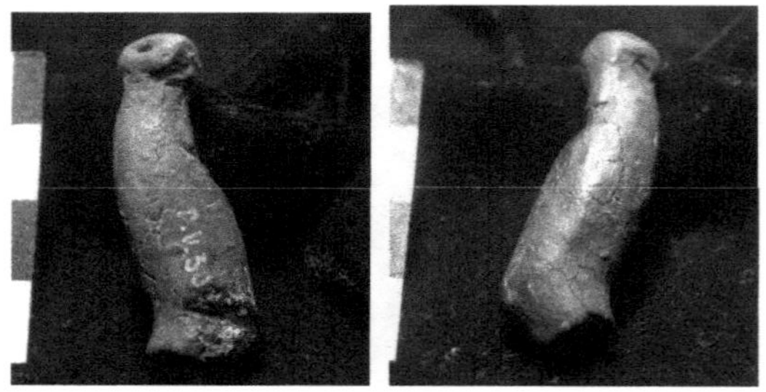
DVI 31: (DV 34)	Animal head	Height: 1.6 cm; length: 2.8 cm	Roughly fashioned animal head, broken on one side, with a rather elongated muzzle. The nostrils are represented by short angular incisions and the mouth is indicated by a continuous line. There are no details to identify the species.	
DVI 32: (DV 35)	Animal body	Height: 1.3 cm; length: 2.8 cm	This object represents a small animal body with the head missing. The top of the legs is indicated, but they are broken in the lower part. The tail is well marked by a large protuberance.	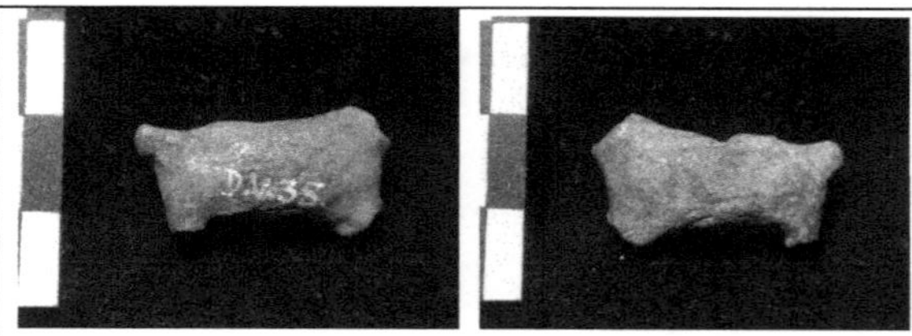

Dolni Vestonice I

DVI 33: (DV 36)	Fragmentary animal	Height: 2.3 cm; Length: 3 cm	This piece represents the rear side of an animal recognisable at its curved shape, and an incision that marks the top of the separation for the legs, which are broken off. There is a series of 7 incisions on the narrow edge they are short and parallel. There are 5 others, parallel and at an angle with the orientation of the previous, just on the side of them.	
DVI 34: (DV 38)	Possible animal head	Height: 1.2 cm; length: 2.1 cm	This object is very schematic and forms a rounded stump, curved and pointy where the muzzle would be. The break of the neck is a smooth surface, as if it had been cut off neatly. Two small holes indicate eyes.	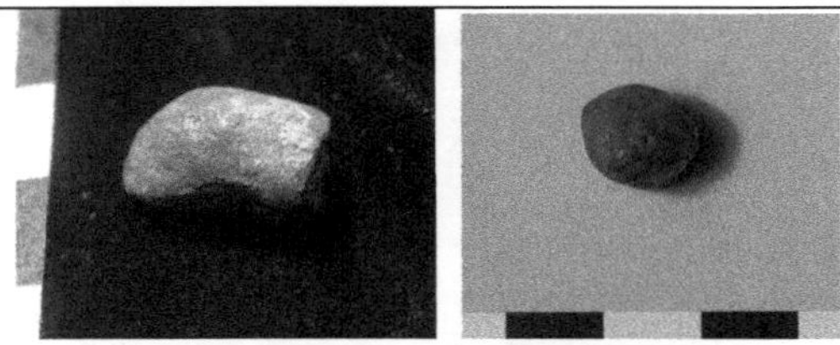
DVI 35: (DV 39)	Body extremity	Height: 1.2 cm; length: 2.5 cm	This forms a sort of stump broken and wide at one end and with a much pointed shape at the other end. It could possibly be a muzzle.	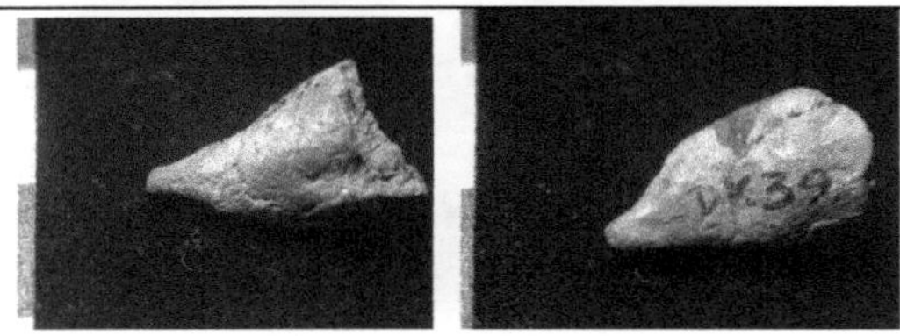
DVI 36: (DV 40)	Fragmentary animal?	Height: 3 m; length: 1.8 cm; width: 1.2 cm	This object forms a fragment shaped and rounded all around, and broken at each end. Series of short incisions parallel are placed around it forming 6 regularly spaced columns. One side is rather curved.	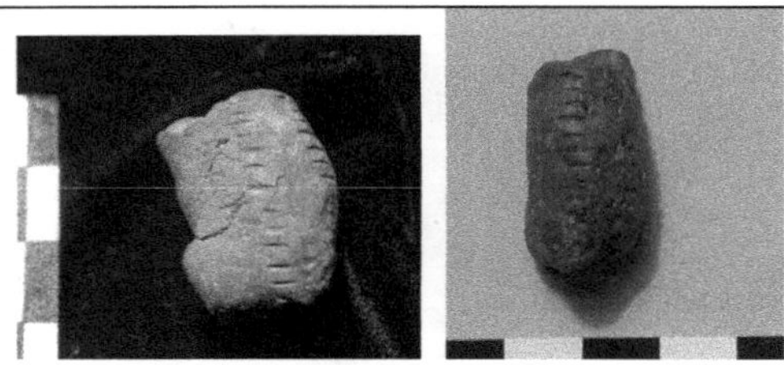
DVI 37: (DV 41)	Fragmentary animal	Height: 2.4 cm; length: 2.9 cm; width: 2 cm	Two broken stumps indicate the front legs. The base of the neck is shaped; the rest of the head is missing. All the back of the body is broken off just after the base of the front legs. The original surface is well smoothed. The animal represented was quite massive.	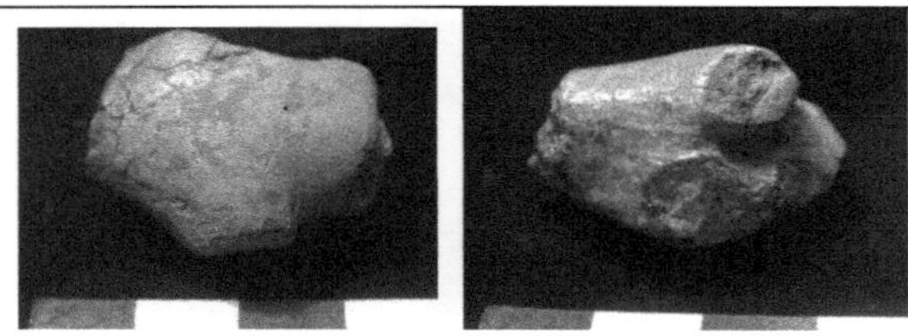

Dolni Vestonice I

DVI 38: (DV 42)	Animal head?	Height: 2.1 cm; length: 3cm; width: 3 cm	This piece is a block without a definite shape but with traces of fashioning. It is smooth and shiny on one side, porous on the other. Many inclusions are visible. Some red ochre traces are visible on it. The general shape reminds of an animal head.	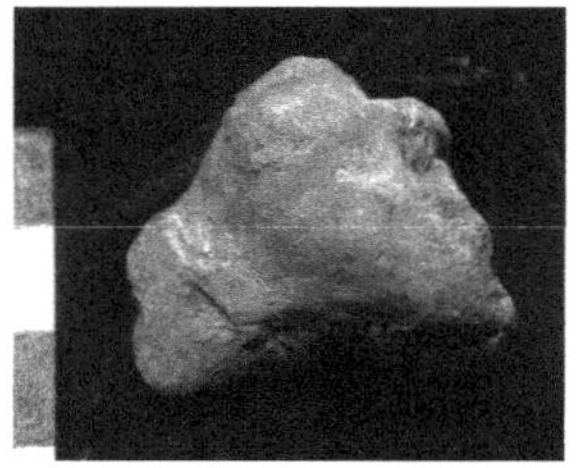
DVI 39: (DV 43)	Fragmentary animal	Height: 1.8 cm; length: 2.4 cm	This object represents the rear side of an animal, with the tail and the separation of the legs indicated by incisions, and the top of the back legs visible. All the front of the body is missing.	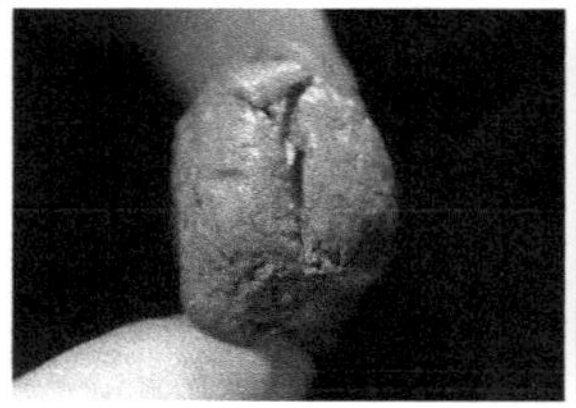 
DVI 40: (DV 44)	Undetermi-ned animal	Height: 3.2 cm; length: 4.7 cm	This object is roughly shaped with a pointy part, which could be the muzzle and two small protuberances for the ears. The front legs are indicated by a stump. The back part is broken. One side is better preserved than the other. It is rather smooth, light brown and shiny. The other side is very irregular and presents a mixture of different colours: black, yellow and brown. It is slightly flattened.	 
DVI 41: (DV 45)	Fragmentary animal	Height: 1.9 cm; length: 3.5 cm	This is the rear side of an animal that was broken in many pieces and reconstructed. The beginning of the rear right leg, the belly, part of the back and the area around the tail are visible. The original surface of the animal is very smoothed and shiny.	
DVI 42: (DV 46)	Body extremity	Height: 1.6 cm; length: 2.4 cm	This object forms a rounded and slightly curved stump. The material is different from the other figurines, being much more yellow on one side, with black stains on the other. There is a rather elongated shape. The yellow side is very soft and smooth; it is not shiny at all on the contrary to most of the other objects. It seems porous, with small holes and asperities. It could be interpreted as a muzzle.	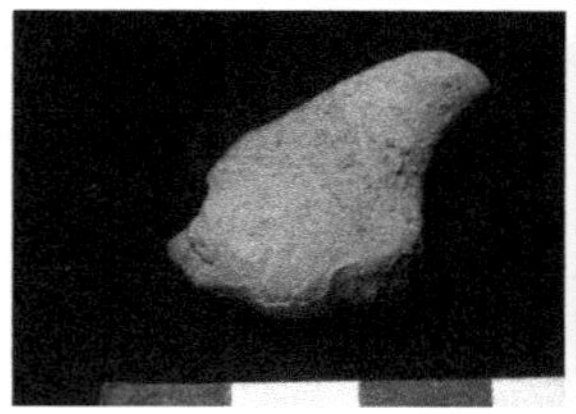 

Dolni Vestonice I

DVI 43: (DV 72)	Body extremity	Height: 3.3; width: 2.5 cm	This piece forms a flattened shape, rounded at one end and broken at the other. The material is very porous. It can be interpreted as an animal leg.	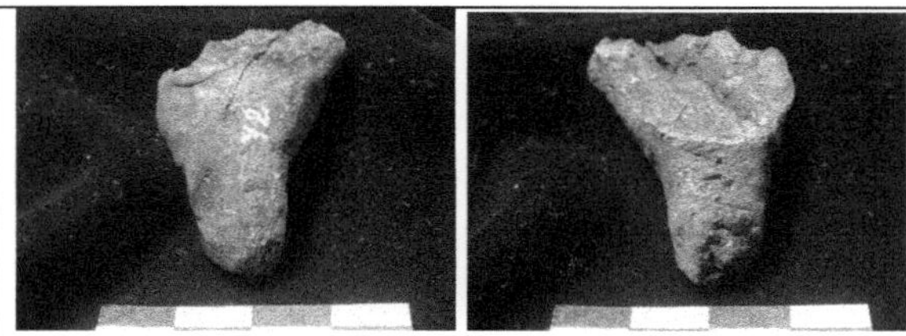
DVI 44: (DV 73)	Body extremity	Length: 4.5 cm; width: 4.5 cm	This object is a body extremity rounded at one end and broken at the other. Its overall shape is conical. The surface is smooth but irregularly coloured. It seems to be a massive leg.	
DVI 45: (DV 74)	Fashioned fragment	Height: 3 cm; length: 3 cm; width: 2 cm	This is a thick fashioned fragment of ceramic that is not readily interpretable. It evokes the rear side of an animal.	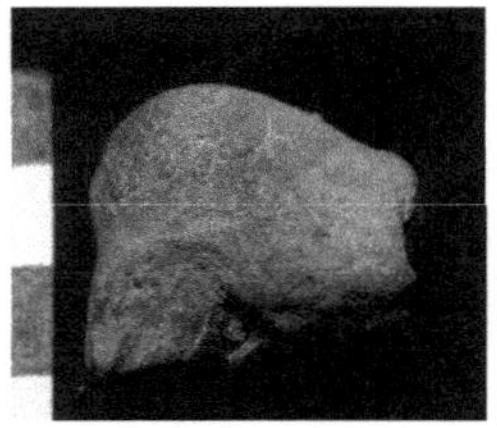
DVI 46: (DV 75)	Body extremity	Height: 3 cm; length: 4.5 cm; width: 2.8 cm	This object has been previously interpreted as the head of a wolverine on the ground of its overall shape, but there is no detail indicated. It seems it was attached to a much larger body, as remaining fragments indicate. One end is much thinner than the other and is curved and slightly flattened. It would represent the muzzle of the animal.	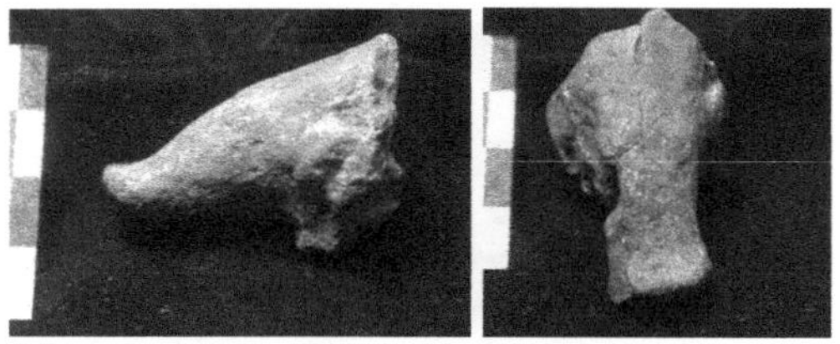
DVI 47: (DV 76)	Body extremity	Height: 3 cm; maximal width: 2.2 cm	This object forms a massive conical stump, rounded at the thinner end, broken at the top and on one of the sides. The other side is very smooth. The colour is irregular: brown on one side, very dark on the other.	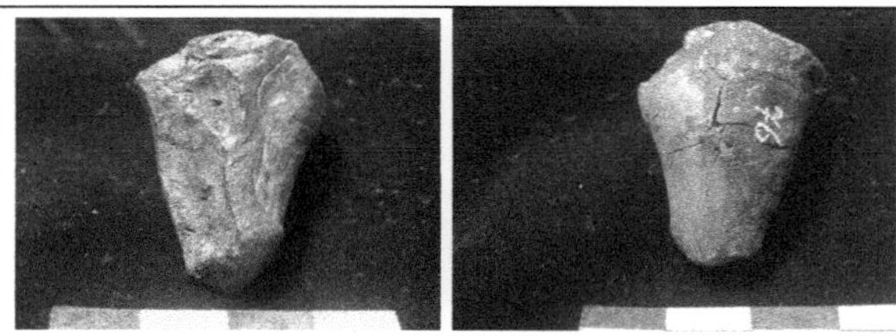

Dolni Vestonice I

DVI 48: (DV 77)	Body extremity	Height: 2.5 cm; Width: 0.9 cm	This object represents a small leg with an animal foot at the end. The foot sole is flat underneath. There are some visible traces of scraping on it.	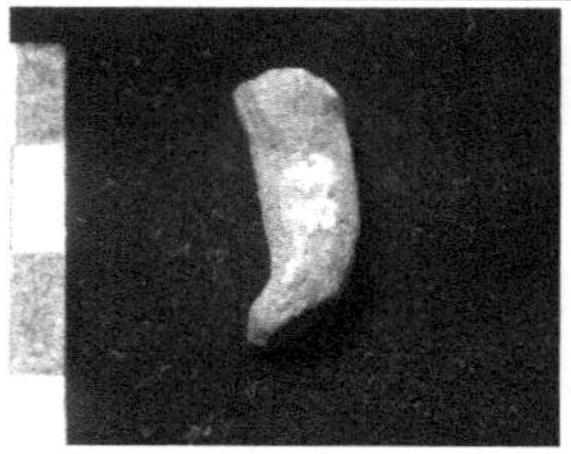 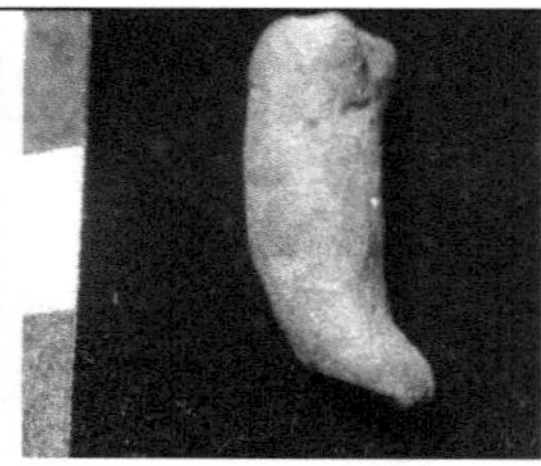
DVI 49: (DV 78)	Body extremity	Height: 3.5 cm; width: 2.5 cm; thickness: 0.8 cm	This piece is a flattened body extremity, very thin in comparison to its width. It is larger at one end, which is completely broken. It narrows towards the extremity before widening again slightly at the other end. The material is porous, not very smoothed, and there are traces of red ochre on the surfaces.	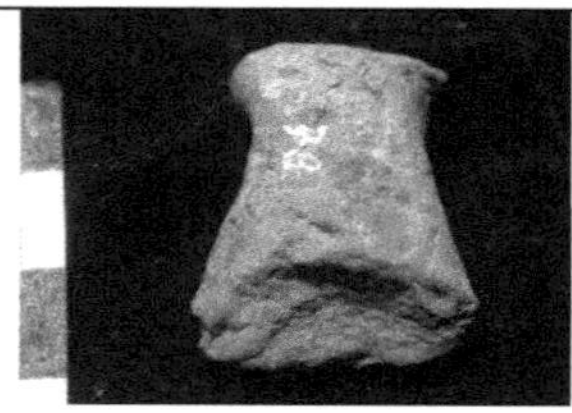 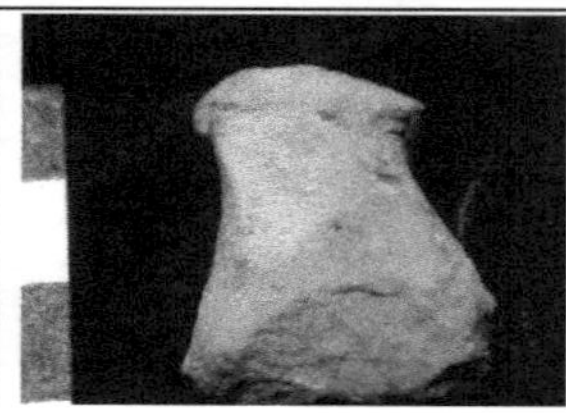
DVI 50: 1 (DV 79)	Flattened ball	Diameter: about 3 cm; thickness: 0.7 cm	This object is a flattened circular piece of fired clay with regularly spaced deep incisions around it; four of them are visible only as the other side is broken. It is dark grey to black in colour with some red stains on it, possibly of red ochre.	 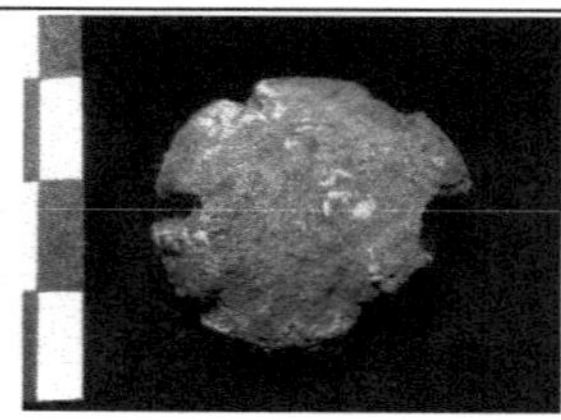
DVI 51: (DV 80)	Animal body	Maximal height: 1.1 cm; Length: 2.8 cm	This is a small animal body, headless. The legs are formed by stumps. The colour of the surface is brown with darker black areas that indicate a bonfire firing.	 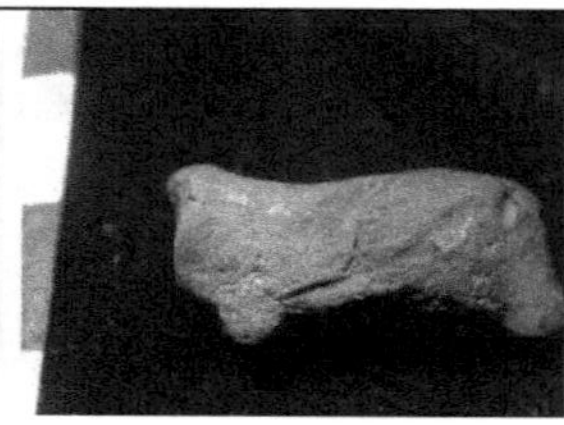
DVI 52: (DV 182)	Body extremity	Height: 3.4 cm; Width: 1.6 cm; maximal thickness: 1.4 cm	This piece forms a small leg with a foot at the end, possibly an animal foot. The sole is flattened and is at an angle with the vertical. The colour is irregular and the material porous.	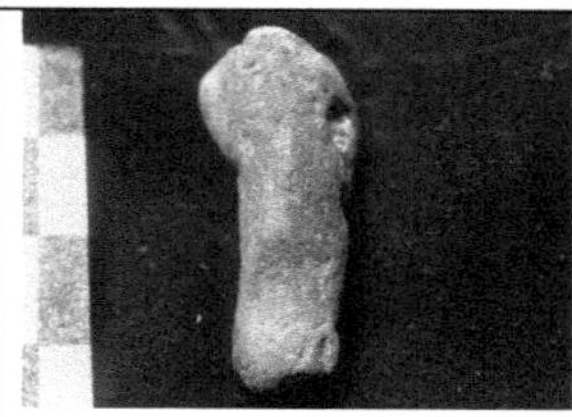 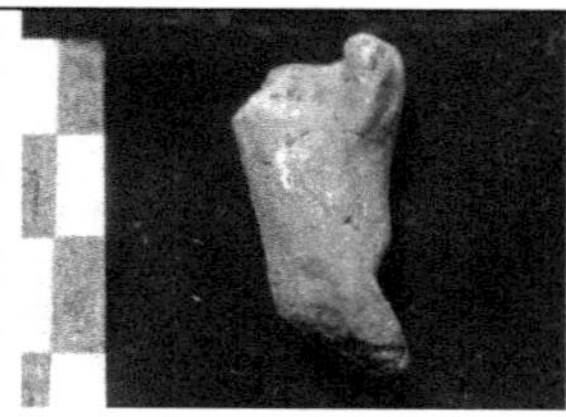

Dolni Vestonice I

DVI 53: (DV 183)	Body extremity	Height: 2.8 cm; maximal width 2.4 cm	This body extremity forms a massive conical stump, with the larger end broken and the thinner end rounded and smoothed. The surfaces are smooth and porous. The object feels lighter than most of the other ceramic objects from the site, as if it was less dense. It is light brown in colour.	
DVI 54: (DV 184)	Ovoid fragment	Height: 2 cm; length: 3 cm	This is a roughly ovoid fragment of fired clay, irregular in shape, but fashioned, with a deep incision at one end of it.	
DVI 55: (DV 185)	Body extremity?	Height: 2.9 cm; maximal width: 2.8 cm	This is possibly a fragmentary body extremity. One side of the object has been well fashioned and smoothed; all the other side is broken. It is wider at one end than at the other. It is a light brown colour on the smoothed side and black on the broken side. The material feels compact and dense.	 
DVI 56: (DV 186)	Body extremity	Height: 2.6 cm; width: 2.4 to 0.9 cm	This object is a flattened stump of fired clay, much wider at one end than at the other. One side of it is broken, with an irregular surface. The other side is well smoothed. It is yellow-light brown in colour.	 
DVI 57: (DV 188)	Body extremity	Height: 2.7 cm; width 1.4 cm to 0.9 cm	This represents a fragment of limb, whose foot has broken off. It is broken at each end, cylindrical with a wider diameter at one end than at the other.	 

Dolni Vestonice I

DVI 58: (DV 189)	Angular fragment	Height: 1.6 cm; length: 2.1 cm; width: 1.2 cm	This piece is a fragment showing a perfectly straight and smooth corner, delimited by a deep incision on one edge. The rest seems made of broken surfaces. The object is black.	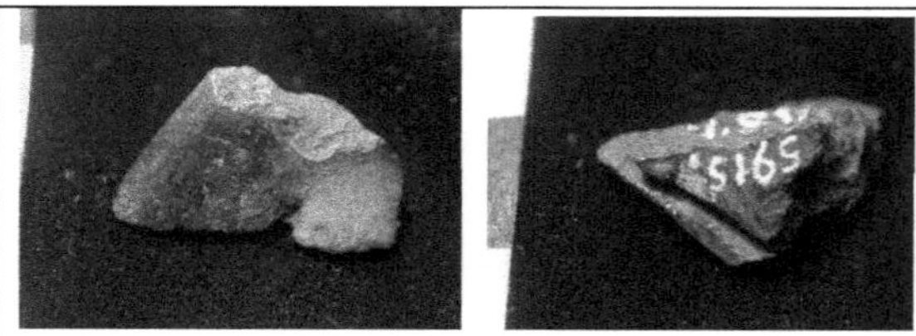
DVI 59: (DV 190)	Fragment with parallel incisions	Length: 2.7 cm; width: 1.2 cm	This object is a black flat fragment with 6 parallel incisions visible on one side. The other side is completely broken. It is also completely broken on the edges.	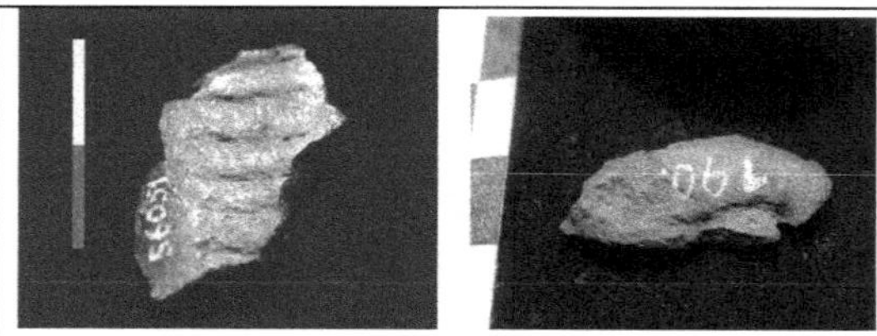
DVI 60: (DV 191)	Fragment with incision	Length: 1.8 cm; width: 1.4 cm	This object is a small fragment bearing one straight deep incision on a black smooth surface. The rest shows broken surfaces.	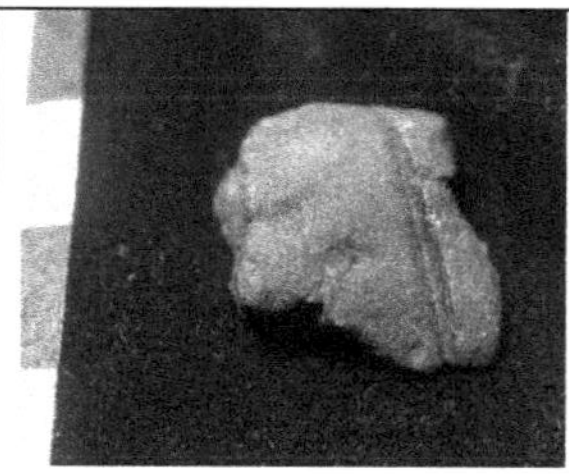
DVI 61: (DV 193)	Body extremity	Length: 3.2 cm; diameter: varies between 0.7 and 0.9 cm	This is a slightly curved thin cylinder that ends in three small points. It is well smoothed and detailed. It is only broken at one extremity of the cylinder. It is interpreted here as a mammoth trunk.	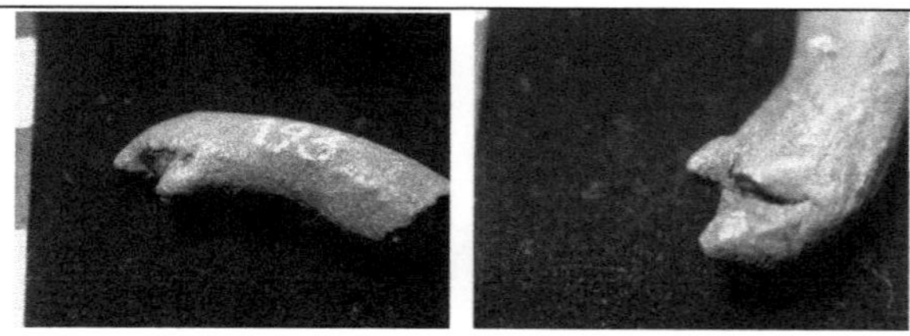
DVI 62:	Mammoth figurine	Height: 2.6 cm; length: 3.6 cm	This figurine represents a mammoth, possibly a young because of the rather rounded silhouette. It is schematic, with roughly fashioned legs separated by a marked arch at the belly, a rounded profile of the back lacking the usual bumps marking the head and the top of the back and a trunk that detaches enough from the body to leave a gap and that is then stuck between the front legs. It is nearly whole. This is the only figurine of a mammoth body found in Dolni Vestonice I. (Absolon 1949 plate IV).	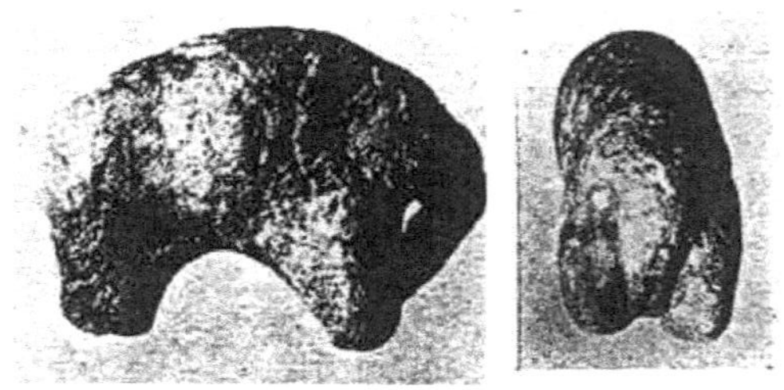

Dolni Vestonice I

DVI 63:	Weasel head	Length 0.9 cm; width 0.6 cm wide	This figurine has a very pointed and elongated muzzle and a slightly snub-nosed. There are two elongated ears, the eyes are indicated by very small slits under the ears, the neck is clearly formed, and the surfaces are quite rough. These elements evoke clearly the head of a weasel.	 
DVI 64:	Body extremity	Length 3 cm; height from heel to top 1.5 cm; width at the ankle: 1 cm	The foot is long and has a flat sole 2.4 cm long, and the heel well marked. This indicates a human foot, possibly a naked foot here with a small relief indicating the big toe. It would then be a left foot.. It is yellow ochre in colour.	 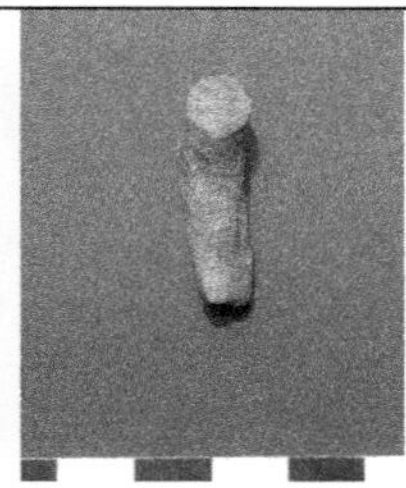
DVI 65:	Body extremity	Length 3.3 cm; width 0.8 cm	This object is a thin and wavy body extremity. It is rounded at one end, the break is neat at the other end as if it was cut off.	 
DVI 66: (4651?)	Body extremity	2.5 cm by 0.8 cm by 0.7 cm	This body extremity is straight and thin, broken at one end.The other end seems fashioned (rounded); traces of scraping on dry material visible on its side.	 
DVI 67:	Body extremity	3.3 cm by 0.9 cm by 0.7 cm	This body extremity is altogether straight, slightly curved at the extremity, black, rounded at both ends, thin. The surfaces are irregular in places.	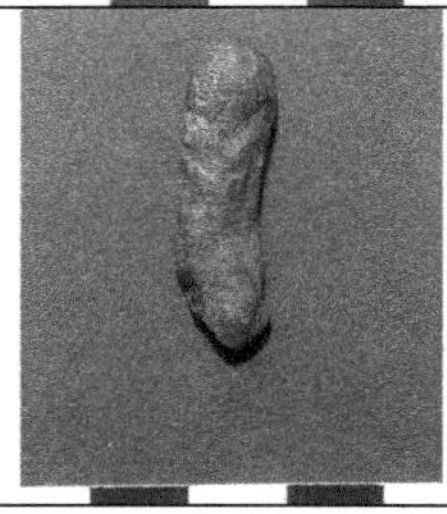

Dolni Vestonice I

DVI 68:	Body extremity	2.6 cm by 1 by 1 at to top and 0.6 cm the thin end	This object is yellow in colour, with one rounded extremity and a swollen end:; the other end thinner. It is broken at the larger end. There are some remains of two holes pierced just at the limit of the breakage: half of the section is still visible for one, the other can only just be guessed.	
DVI 69:	Body extremity	2.8 cm by 0.8 at top and 0.6 at bottom	This is a body extremity with a foot. It is black and narrows to form the foot at one end. The heel is not marked. The other extremity seems cut off with a very neat section.	
DVI 70:	Body extremity	3.3 cm by 1.1 by 1 at the foot and 1.3 at the top	This body extremity is black, the foot is well-marked and the two extremities are smoothed. It is not broken. The foot is rounded underneath, the knee is indicated by a swelling.	
DVI 71:	Body extremity	2.1 cm by 0.9 at top and 0.6 at bottom	The top of this body extremity is much larger than the other end, indicating the thigh, with smoothed and rounded surfaces. The foot end is much narrower. There are some traces of scraping to shape the foot end.	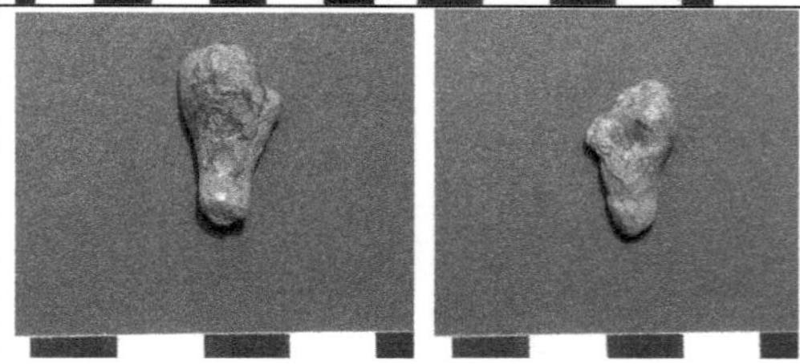
DVI 72:	Body extremity	2 cm by 1.2 cm by 0.8 cm	This body extremity is broken at both ends. It has straight sides converging slightly toward the foot end. Deep grooves are incised on both sides, apparently done when the material was dry. The piece curves very slightly. It is black and brown.	

Dolni Vestonice I

DVI 73:	Body extremity	3.6 cm by 1.6 at top and 0.9 at bottom by 1 cm thick	The piece is broken vertically on all the side. The foot end was rounded and wider than ankle part, the leg widens towards the top to form the thigh. It is black and brown.	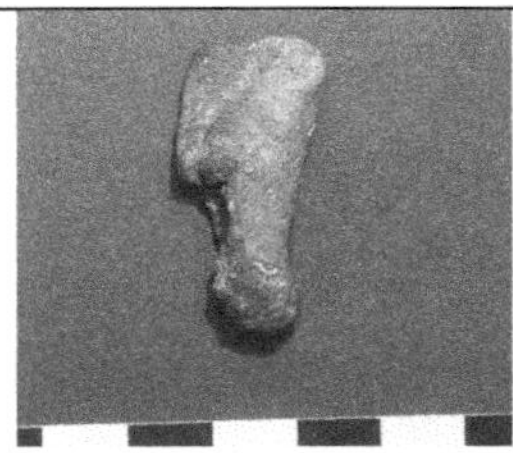 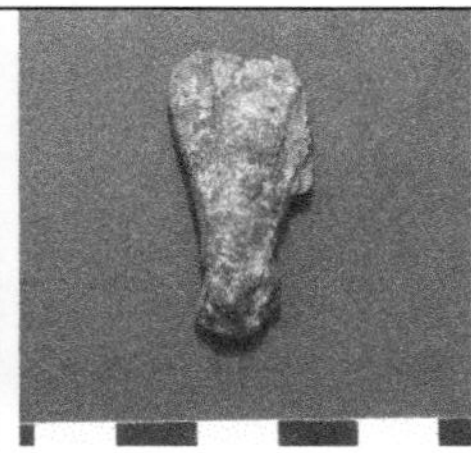
DVI 74:	Body extremity	Length 3 cm; width 1.3 cm	This object is a leg with a flattened foot. It is circular at the top.	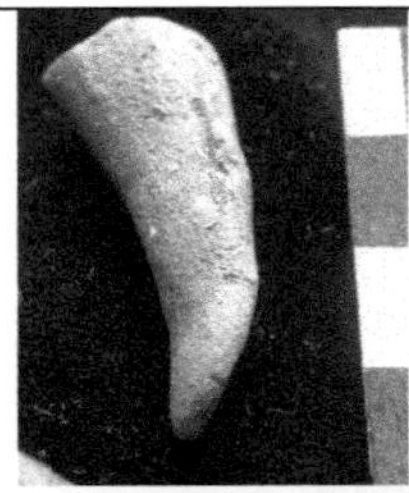
DVI 75:	Body extremity	Length 2.5 cm; maximal width 2 cm	This body extremity is shaped as a conical pointy stump. It has straight sides and is quite even.	
DVI 76	Fragment with palm print	4.1 cm; maximal thickness 1 cm, minimal	This is a thin fragment, quite shapeless, with digital and palm prints on both sides. It was broken into two and glued back. Looks like a fragment of clay pressed in one hand.	 
DVI 77:	Clay ball	2.5 cm by 2.2 cm	This ball is not fully round. It is slightly flattened with one groove on each side and a hole.	 

Dolni Vestonice I

DVI 78:	Clay ball	2.5 cm by 2 by 1.8 cm	Flattened clay ball, broken in three pieces and glued back. The entire surface is smoothed.	
DVI 79:	Clay ball	2.3 cm by 2 by 2.1	Round clay ball. It is not broken but the surfaces are uneven.	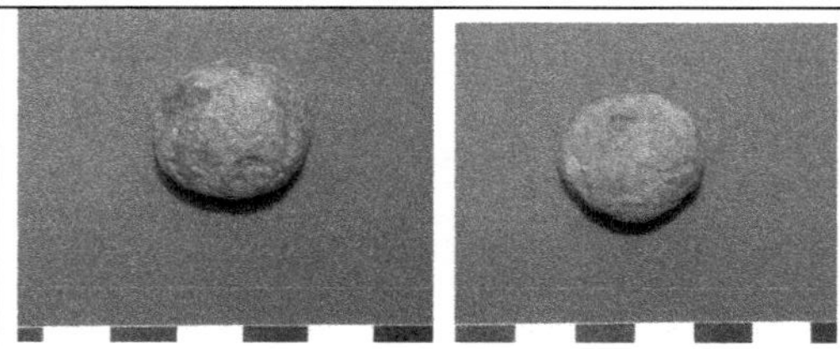
DVI 80:	Clay ball	2 cm by 1..8 by 1.4	This clay ball is half broken. One very small hole is visible, possibly made with a needle.	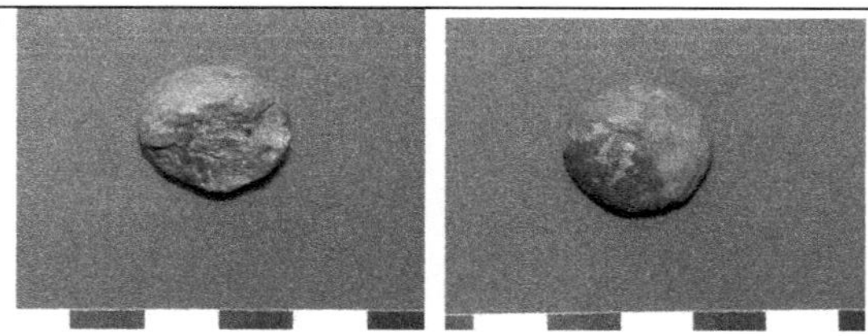
DVI 81:	Clay ball	1.9 cm by 1.4 by 1.2	This ball is black, with an uneven surface. It is the less well modelled of the group of balls.	
DVI 82:	Clay ball	1.5 cm by 1.5 by 1.2.	It is rounded and well smoothed on all sides. It is very similar to DVI 83 and DVI 84.	
DVI 83:	Clay ball	1.5 cm by 1.5 by 1.2	It is rounded and well smoothed on all sides. It is very similar to DVI 82 and DVI 84.	
DVI 84:	Clay ball	1.5 cm by 1.5 by 1.3.	It is rounded and well smoothed on all sides. It is very similar to DVI 83 and DVI 82.	

Dolni Vestonice I

DVI 85: (54151)	Body extremity	0.9 cm by 1 cm by 0.4 cm	This body extremity is the middle part of a leg broken at both ends.	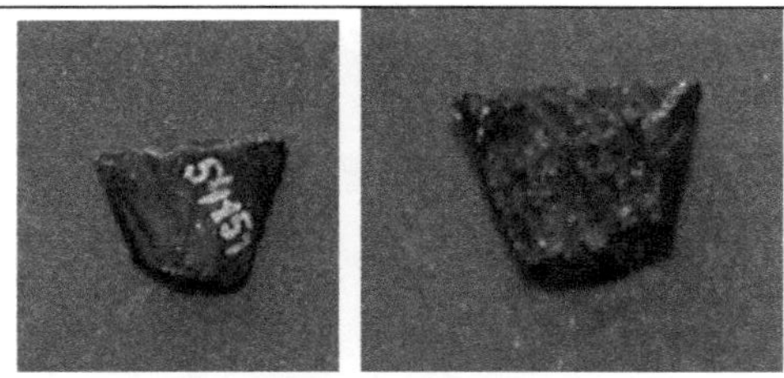
DVI 86: (52454)	Body Extremity	1.2 cm; width 0.6 cm; thickness 0.4 to 0.3 cm.	This is a leg with a larked foot, ball-shaped. It is broken in the upper part.	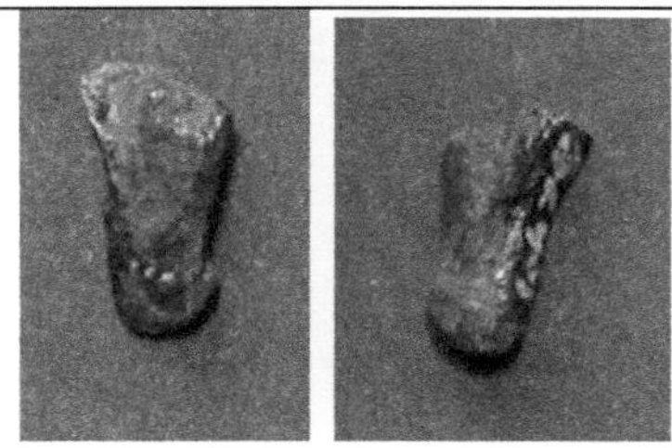
DVI 87: (5?)	Flattened piece	1.5 cm by 1.5 cm by 1 cm.	This is a flattened piece of fired clay with finger prints on it.	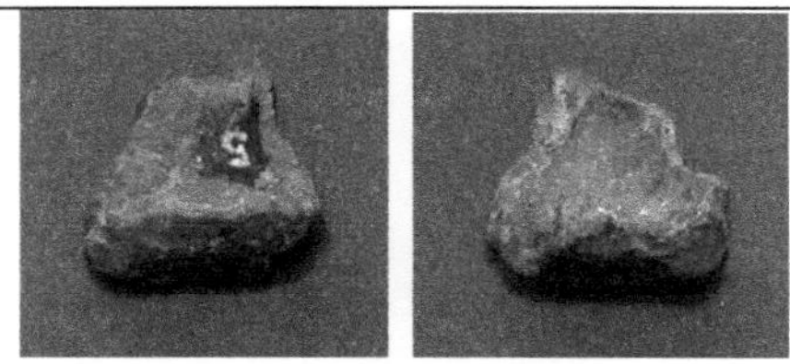
DVI 88: (54051)	Body extremity	1.3 cm by 0.6 cm by 0.5 cm.	This body extremity is a fragment of a limb, slightly flattened, broken at both end.	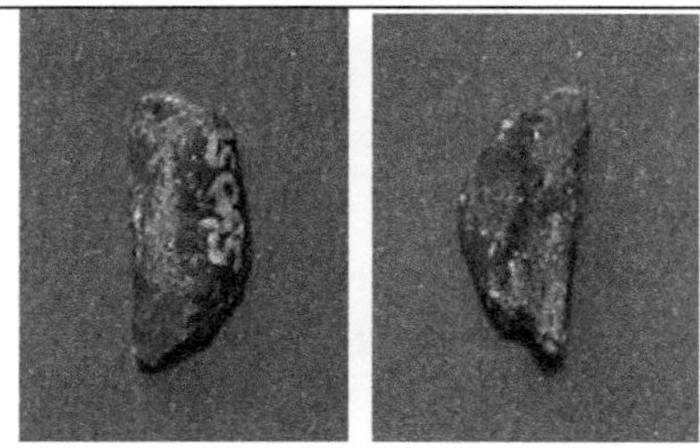
DVI 89: (500?)	Body extremity	0.8 cm by 0.5 cm	The edges of this body extremity are straight. It is very slightly conical. It belongs to the mammoth type, with a circular foot and a marked rim indicating the foot. It is broken at one end.	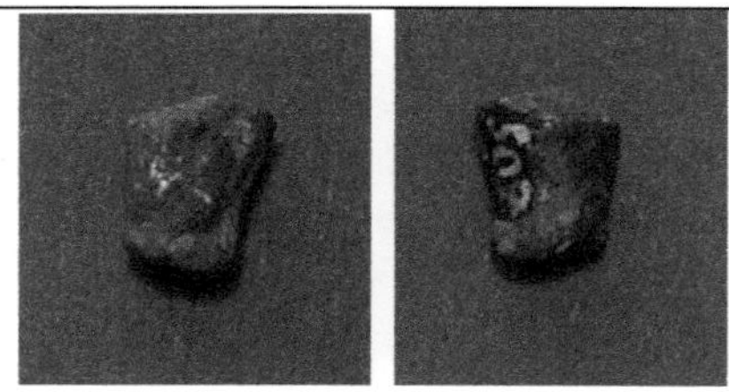

Dolni Vestonice I

DVI 90: (501?)	Body extremity	1.2 cm by 0.6 cm	It is similar to DVI 89 but larger, with a circular foot.	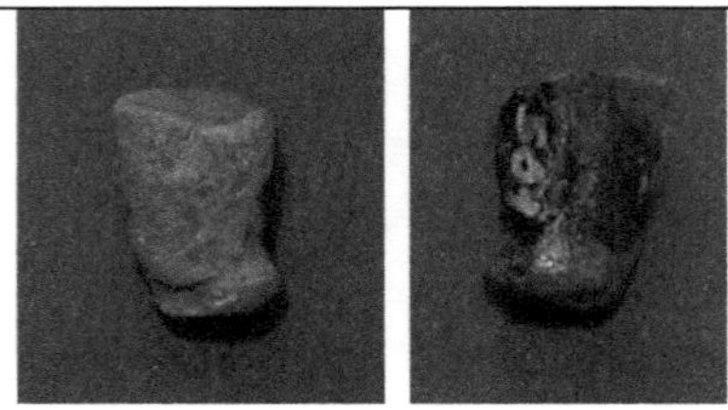
DVI 91: (56751)	Body extremity	4 cm by 2.5 at top and 1.3 at bottom; 0.8 thick	This piece is cut into two on the length of the object. It widens at the top for the thigh. It is black in colour.	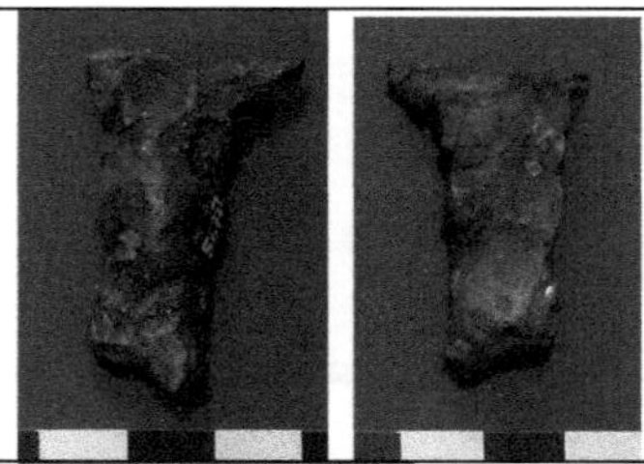
DVI 92: (581)	Body extremity?	1.3 cm by 1.2 cm by 0.8 cm	This object forms like a small volcano on the completely flat base that seems cut with a tool. It could be a short rounded body extremity.	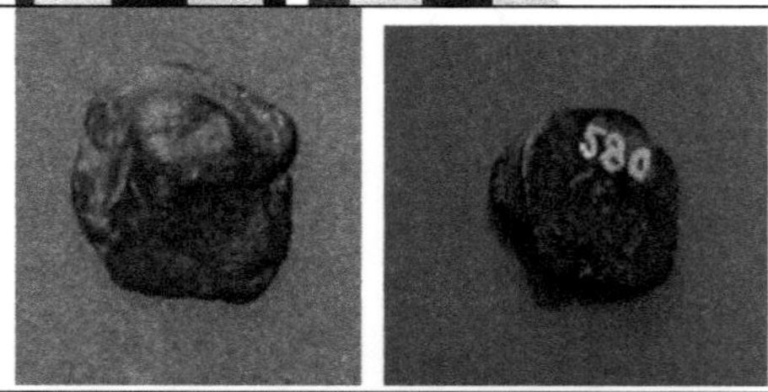
DVI 93: (542)	Body extremity?	1.1 cm by 0.7 cm by 0.3cm	This is a possible flattened limb fragment, broken at both ends.	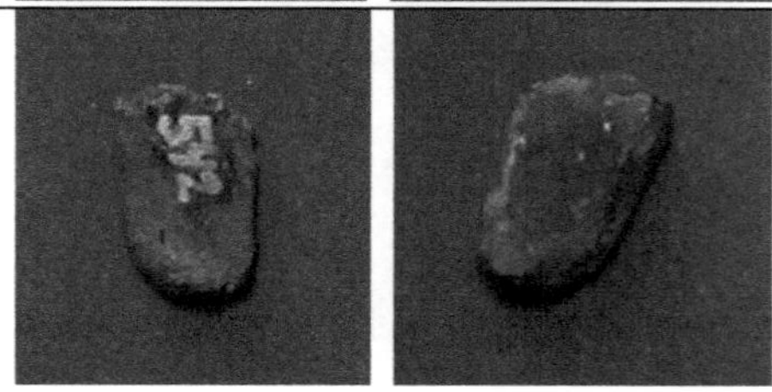
DVI 94:	Possible bead?	1.3 cm by 1 cm by 0.3 cm	This is a flattened piece, pierced with small hole, with traces of the piercing well visible from one side.	

Dolni Vestonice I

DVI 95: (56651)	Body extremity	3.3 cm by 2.6 cm by 1.5 cm	This object shows a conical shape with a wide base, broken on one side, well smoothed on the other. It is large and yellow in colour.	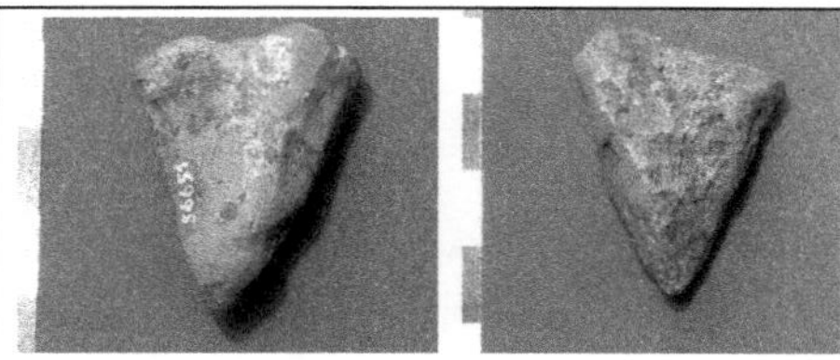
DVI 96:	Fashioned fragment	1.6 cm by 1.5 cm by 0.8 cm	This fragment looks like the muzzle of an animal from one side and like the rear part of a body from the other.	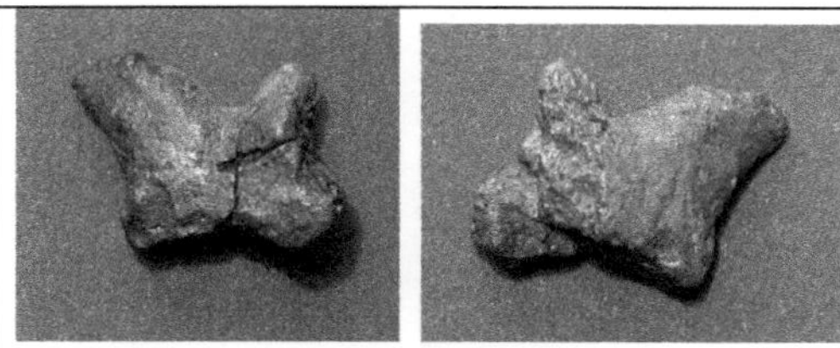
DVI 97: (49751)	Body extremity	2.4 cm by 1.6 cm by 1.1 at top	This body extremity is a leg of the mammoth type with a circular foot slightly hollow underneath. It widens at top then it is broken.	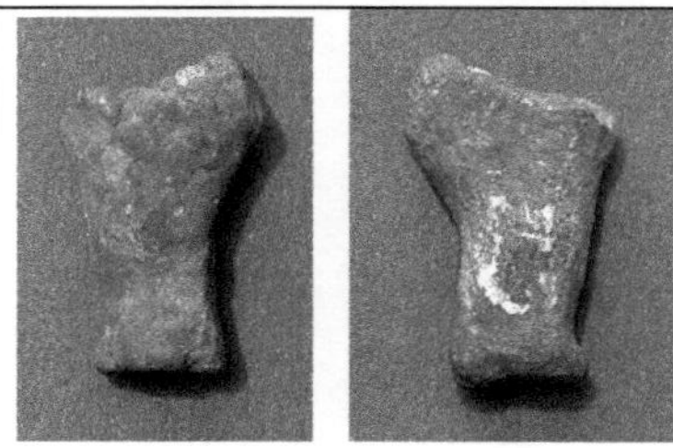
DVI 98: (49851)	Body extremity	2 cm by 1.1 cm by 1 cm	This is a leg fragment shaped like a cylinder that narrows then widens again for foot. It is broken at both ends. It is black.	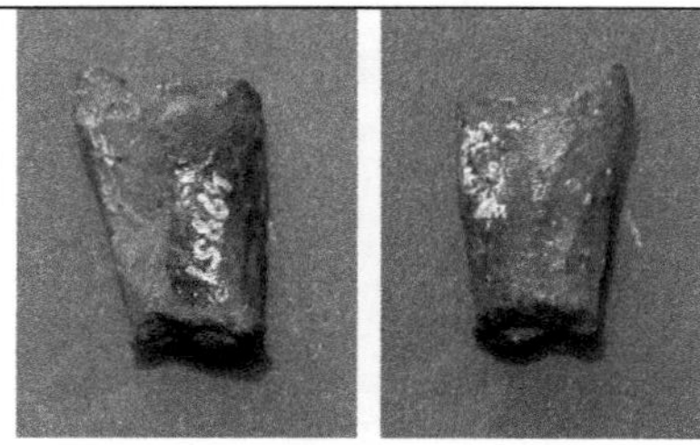
DVI 99: (15015)	Body extremity	2.3 cm by 1.1 cm by 0.7 cm	This is a leg with foot, broken vertically on all the length. Part of the break is completely flat and smooth as if it had been cut. The whole surface is not very even (some holes). The foot is also uneven.	

Dolni Vestonice I

DVI 100: (51251)	Body extremity	2.2 cm by 1 cm by 0.6 cm	Body extremity with a well marked heel. The sole is smoothed and at a sharp angle with the leg. One extremity is broken, the sides are damaged, the end of the foot is pointed and smoothed.	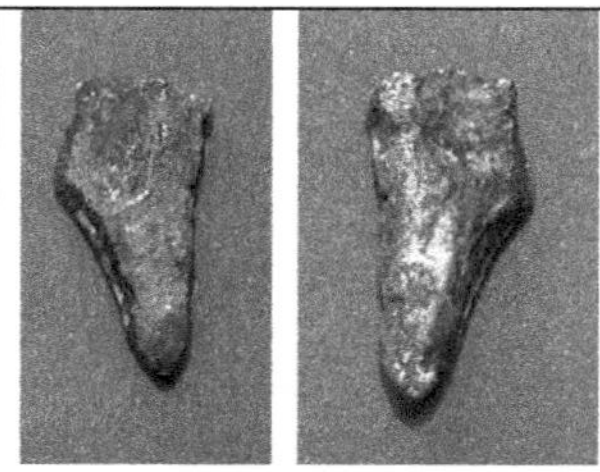
DVI 101: (49651)	Body extremity	1.8 cm by 1.3 cm by 1.2 at top	It has a wide base, slightly swollen. The extremity is very thin, the larger end is broken. Underneath and on one side the foot is flattened by scraping, then it is broken towards the extremity.	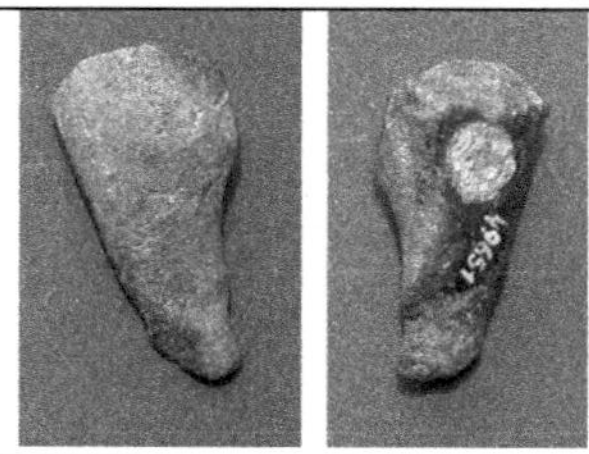
DVI 102: (54851)	Body extremity	3.5 cm by 0.8 cm by 0.6 cm at top	This piece is an elongated black piece. It is smoothed and finished by a point at one end. At the other end, an incision crosses the middle of the extremity which seemed shaped.	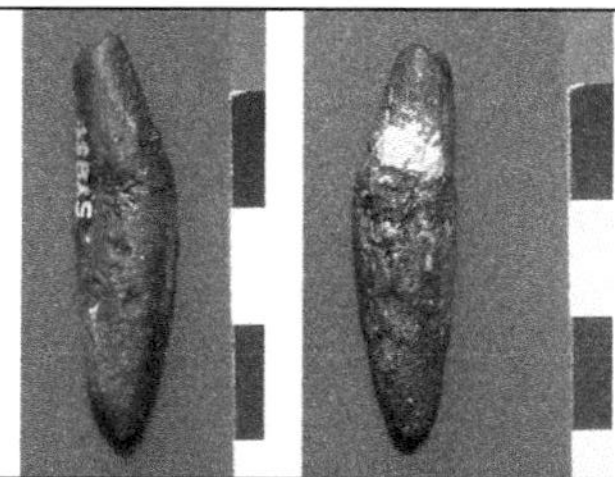
DVI 103: (50551)	Body extremity	1.6 cm by 1.1 cm by 0.8 cm	This is a leg with foot. The surface is flattened to mark the top of the foot. It looks like a boot. One side is damaged, the foot is rounded underneath to make the junction with the leg.	
DVI 104: (522)	Body extremity	1.8 cm by 0.8 cm by 0.5 cm	This piece is a segment of a limb, flattened and slightly larger at one end. It has with straight sides. It is broken at both ends.	

Dolni Vestonice I

DVI 105: (554)	Body extremity	1.4 cm by 0.6 cm by 0.4 cm	This body extremity is exactly like DVI 104 but smaller.	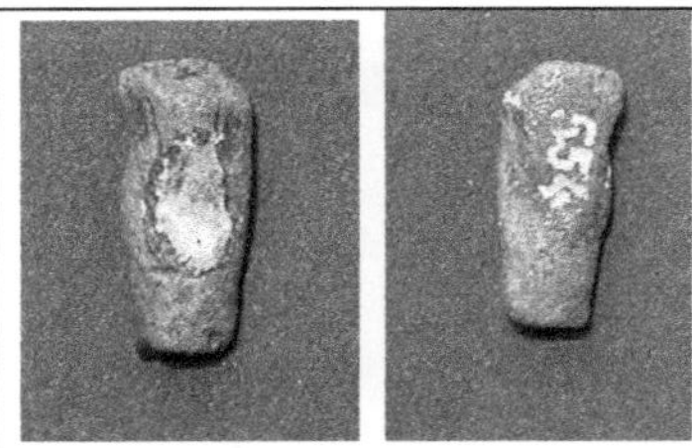
DVI 106: (53951)	Body extremity	1.2 cm by 0.7 cm by 0.6 cm	Body extremity with foot, rounded at one end, broken on the side.	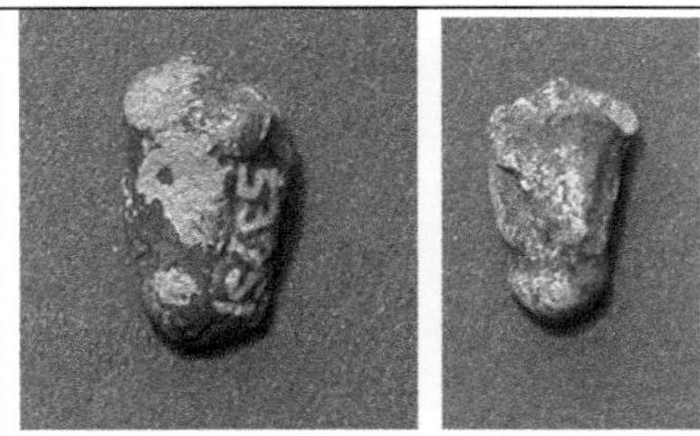
DVI 107: (528)	Body extremity	1.4 cm by 0.9 cm by 0.7 cm	The extremity is rounded, the foot is not marked. It is rather triangular in section and grey black in colour.	
DVI 108: (57051)	Body extremity	1.6 cm by 1.1 cm by 0.7 cm	This is a limb fragment, broken at both ends. It widens at the base. The sides are rather straight.	
DVI 109: (519)	Body extremity	1.2 by 0.7 cm by 0.5 cm	This body extremity is rather conical. It is broken at both ends.	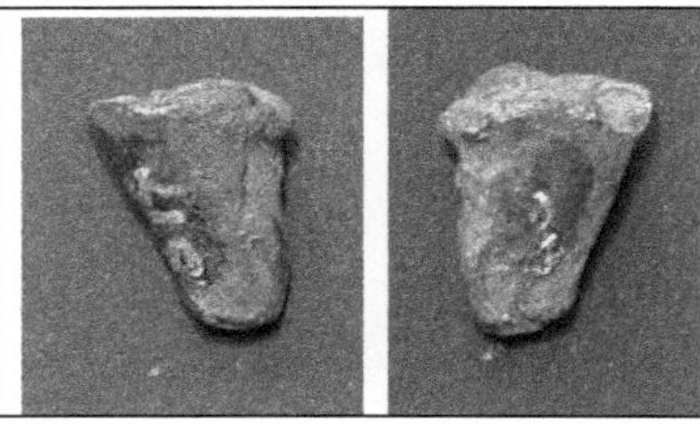

Dolni Vestonice I

DVI 110: (508)	Body extremity	1cm by 0.6 cm by 0.6 cm	Body extremity with a flattened foot and straight edges widening towards the top. It is cylindrical and a slight rim marks the foot.	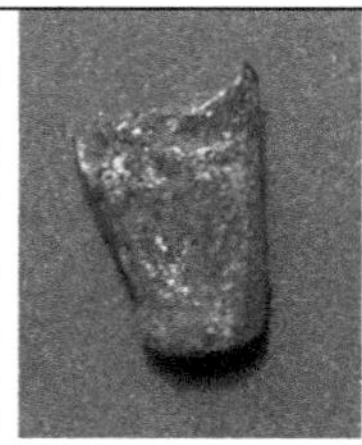 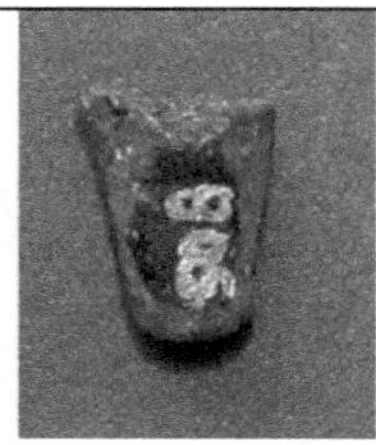
DVI 111: (54767)	Body extremity	1 cm by 1cm by 0.7cm	This piece is cylindrical. It is broken at both ends , with straight sides.	 
DVI 112: (502)	Body extremity	1 cm by 0.7 cm by 0.3 cm	This is a leg with a marked thigh and foot. It widens widely at top. The foot is flat underneath. It is dark grey in colour.	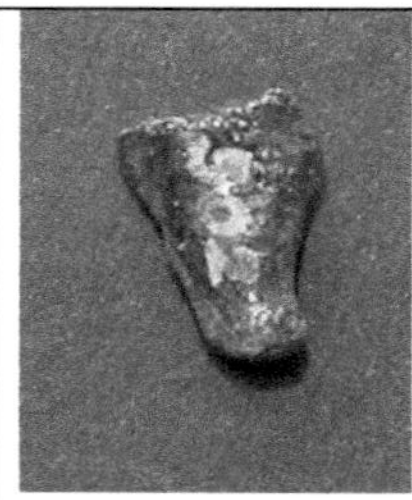 
DVI 113: (538)	Body extremity	0.8 cm by 0.6 cm by 0.5 cm	This is a limb fragment very similar to DVI 111. It is cylindrical, it widens and it is broken at both ends with straight sides.	 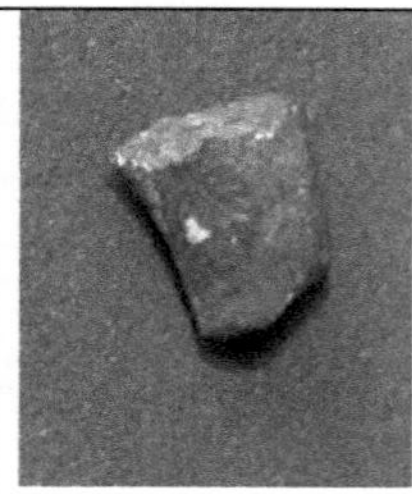
DVI 114: (50951)	Body extremity	1.8 cm by 1 cm by 1 cm	This body extremity widens towards one end and is thinner at the other, but both extremities are broken. It is black in colour and seems broken off a larger body.	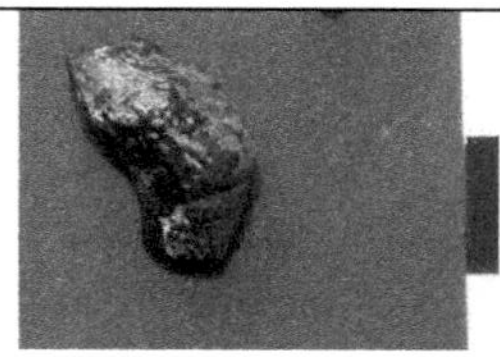 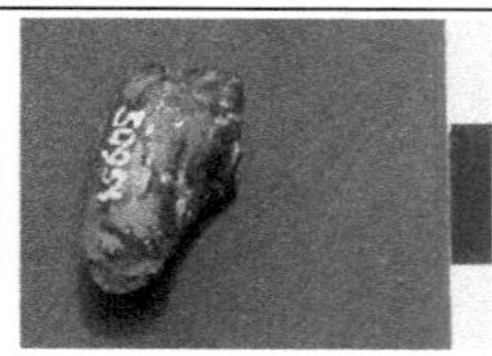

Dolni Vestonice I

DVI 115: (520)	Body extremity	1.3 cm by 0.8 cm by 0.6 cm	It is a body extremity shaped like a 'horn'. The end is very finely modelled, the base is larger, the sides are smoothed. It is black.	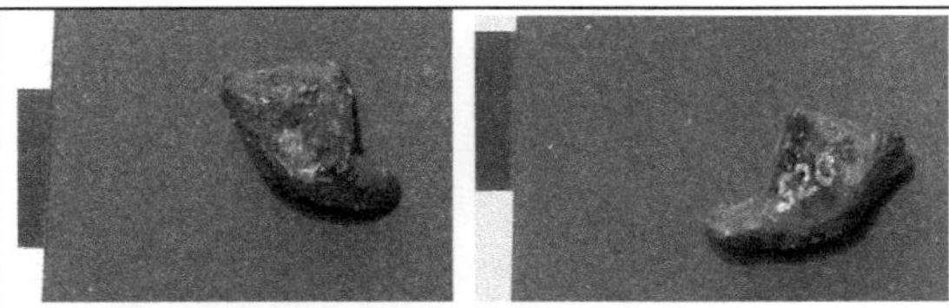
DVI 116: (527)	Body extremity	1.5 cm by 0.6 cm by 0.5 cm	Possible body extremity with one end flattened like a spatula. The other end is broken.	
DVI 117: (530)	Body extremity	2 cm by 0.5 cm by 0.5 cm	This piece is very pointy with the extremity broken. The foot is flattened underneath.	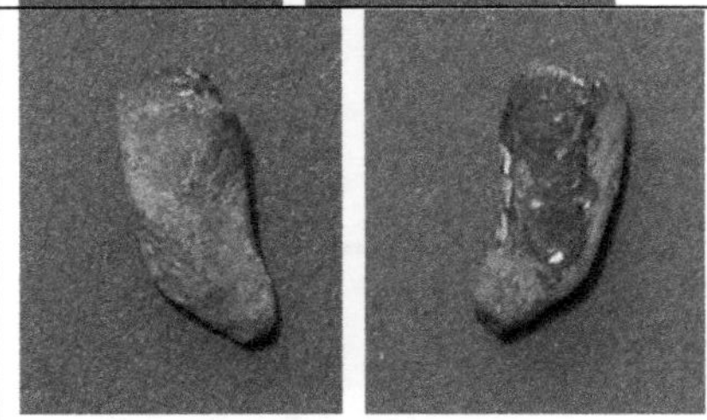
DVI 118: (615)	Body extremity	1.1 cm by 0.3 cm	This is a conical extremity, resembling a 'horn', curved and well rounded at one end. It is well smoothed and very small.	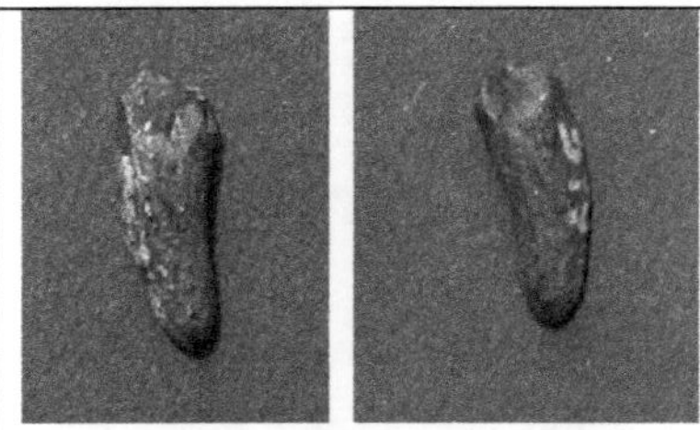
DVI 119: (529)	Body extremity	1.1cm by 0.5 cm by 0.4 cm	It resembles DVI 118 but is slightly larger. It is also horn-like.	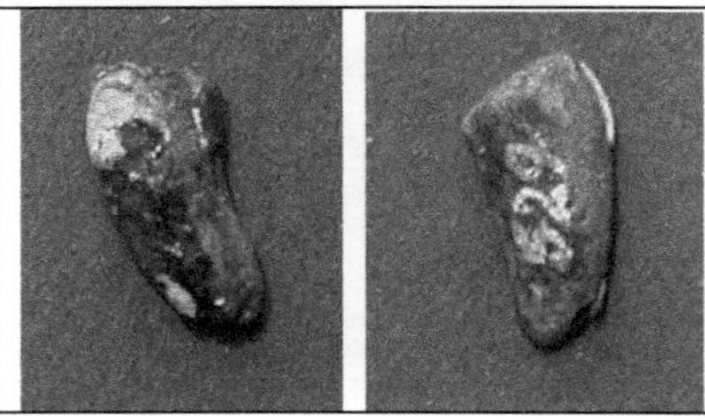

Dolni Vestonice I

DVI 120: (51751)	Body extremity	1.2 cm by 0.5 cm by 0.4 cm	This is a limb with a flattened foot forming an angle with the leg. The surface of the 'break' at the top is very smooth as if cut off.	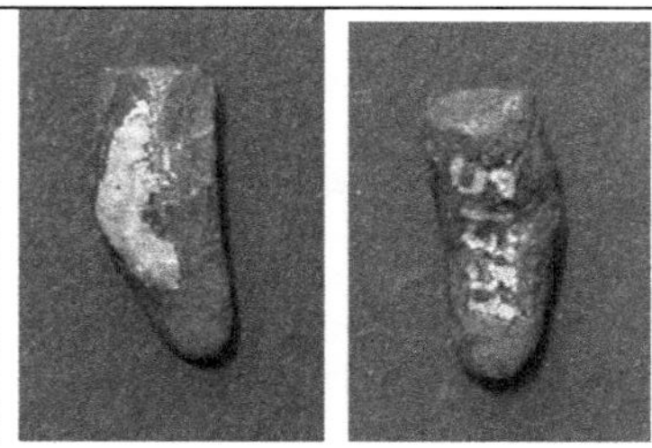
DVI 121: (53751)	Body extremity	1.8 cm by 1.1 cm by 0.5 cm	It is black. One broken extremity is rather circular; the other end is fully flattened and widens while keeping a rounded shape.	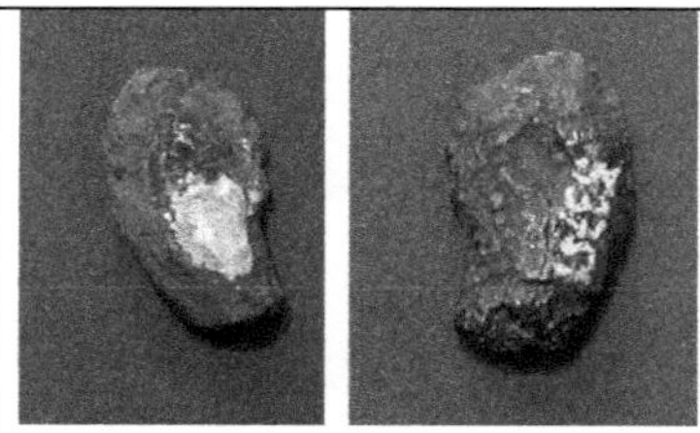
DVI 122: (544)	Body extremity	0.8 cm by 0.5 cm.	Piece is very pointed and flattened. It is ear-like, the base is broken. It is black in colour.	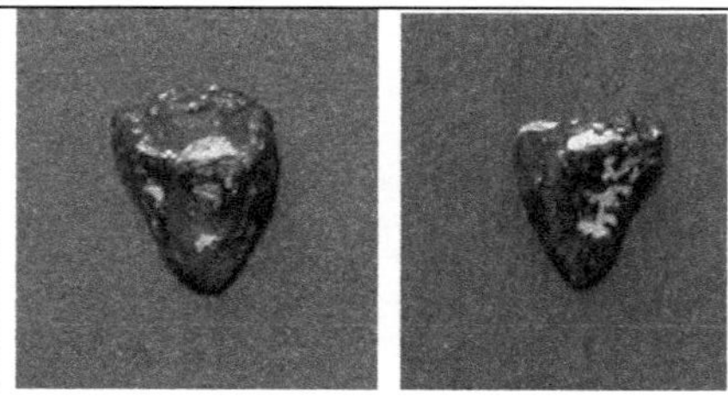
DVI 123: (53350)	Body extremity	1.8 cm by 0.6 cm by 0.6 cm	This body extremity is a leg with a flattened foot. It is conical. The foot makes an angle with the leg. It is broken on the side.	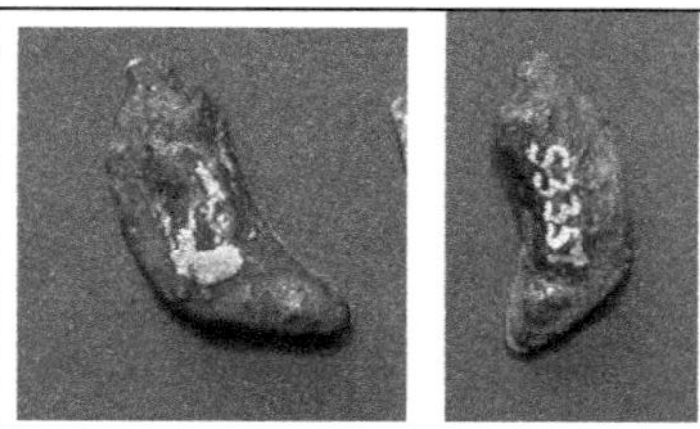
DVI 124: (555)	Body extremity	0.9 cm by 0.7 cm by 0.6 cm	This is a fragment of limb broken at both ends. It is cylindrical, one side is curved.	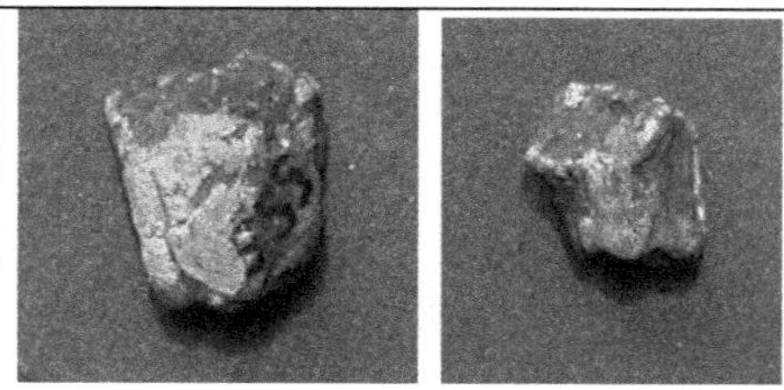

Dolni Vestonice I

DVI 125: (546)	Body extremity	0.8 cm by 0.5 cm	Extremity of a foot, finishing as a blunt point. It is flattened underneath for the sole.	
DVI 126: (532)	Body extremity?	1.7 cm by 1 cm by 0.6 cm	This piece is the side of a cylindrical fragment, broken at extremity and on the other side. The remaining surfaces are rather straight.	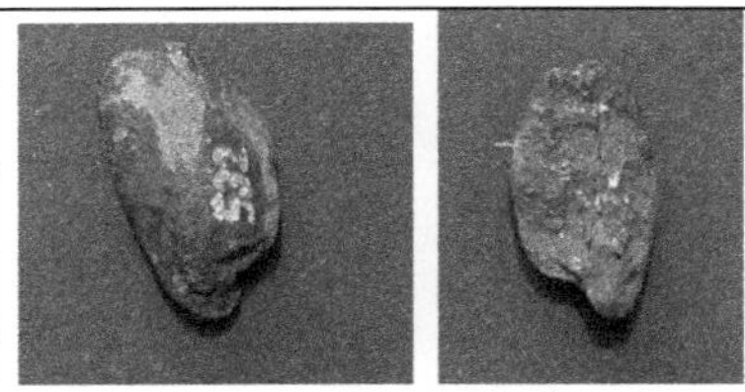
DVI 127: (523)	Body extremity	1.2 cm by 0.8 cm by 0.5 cm	This piece is a flattened cylinder, rounded and slightly curved at the foot. The section of the break at the top is rather neat.	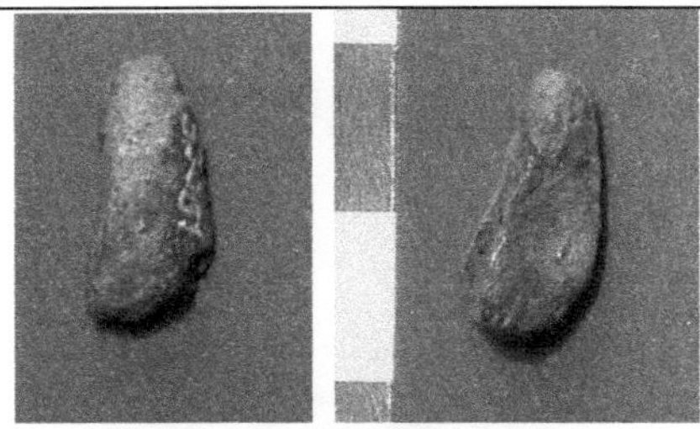
DVI 128: (53051)	Body extremity	1.6 by 0.7 cm by 0.6 cm	This piece has a triangular section. One side is scraped. It ends up as a rounded point.	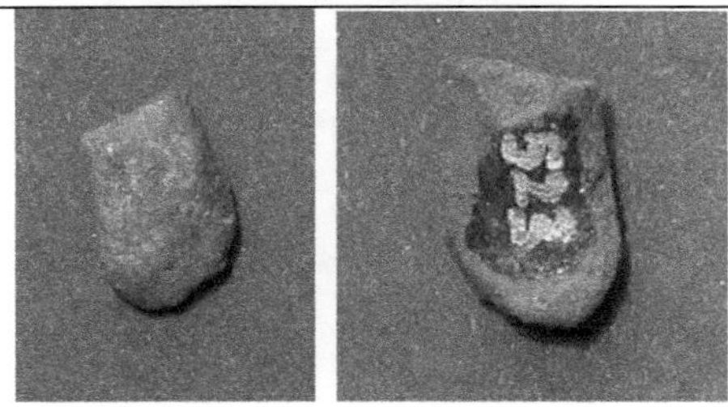
DVI 129: (515)	Body extremity	1.7 cm by 1cm by 0.8cm	This is a limb fragment broken at both ends and on one side. The remaining surfaces are straight. It widens towards the top.	

Dolni Vestonice I

DVI 130: (53451)	Body extremity	1.6 cm by 0.9 cm by 0.8 cm	On one side the surfaces are smoothed. The shape is elongated as for a limb, but the other side seems pinched between the fingers when still wet as it is deformed and thin. This piece is whole.	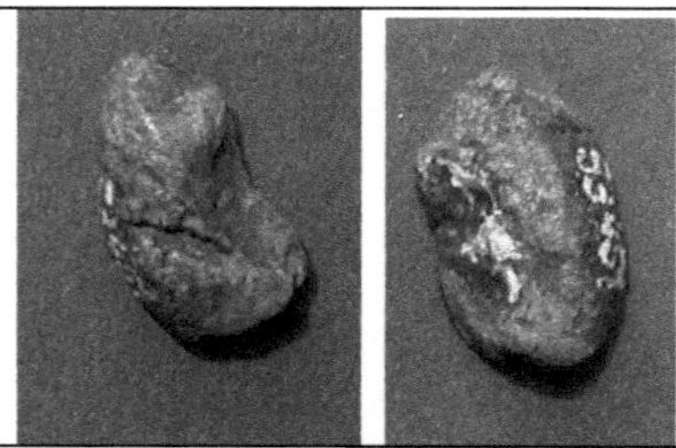
DVI 131: (52151)	Body extremity	1.9 cm by 0.8 cm by 0.6 cm	This piece is elongated. One end is smoothed and rounded; the other end is broken and deformed.	
DVI 132: (525)	Body extremity	1.7 cm by 0.8 cm	The extremity is rounded with straight edges. A slight angle underneath the foot could be the sole. It is deformed on one side and seems to have been pinched when wet.	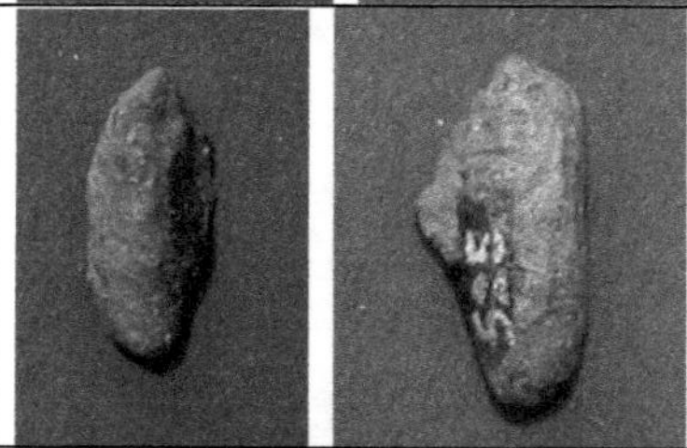
DVI 133: (57951)	Fashioned fragment	1.3 cm by 1 cm by 0.3 cm	This is an oval piece. One side is rounded like a small dome, the other surface is broken and irregular.	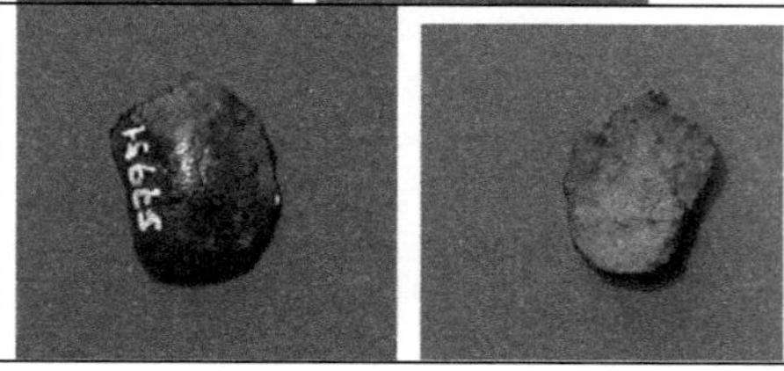

Dolni Vestonice II

Reference number	Graphic unit name	Size	Description	Illustration
DVII 1: (Obje 1289187)	Horse muzzle	Height: 1.4 cm; length: 2.6 cm; width: 1 cm	This piece represents the broken-off muzzle of a horse. It is well modelled to form the nose; the nostrils are indicated by two small protuberances on each side of the end of the muzzle; the chin is also well shaped. All the rest is broken off. The surface of the object is pierced by 3 deep circular holes. One of them is located where one of the nostrils would be, possibly to indicate it, but there is only one. The two other holes are placed on the cheeks, in symmetry, but they do not completely pierce the object. They have been made at an angle, which is possible to see because of the marks left on the edges of the holes. It is possible to see on the edge of the broken part that at least three more holes existed there. The object is golden yellow in colour. Some traces of scraping are visible under the chin. The unbroken surface is smoothed and well fashioned. Reference: Svoboda (ed.) 1991: 51	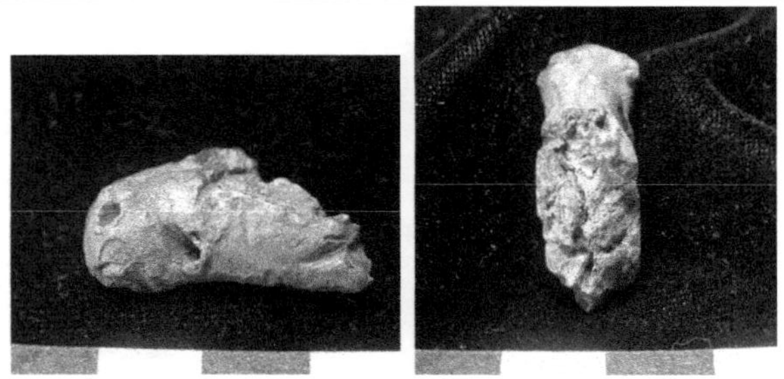
DVII 2: (DVII 8)	Fashioned fragment	Height: 1.4 cm; length: 3.7 cm; width: 2 cm	This object is very well smoothed. It is rounded on one side and broken on the other. It has been carefully fashioned but there is no detail allowing interpretation.	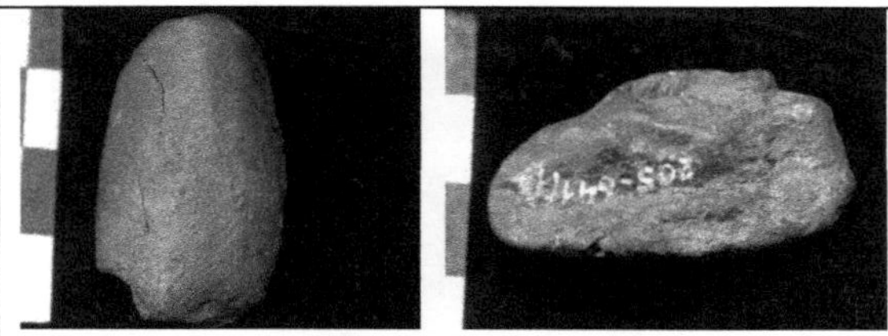
DVII 3: (DVII 9)	Animal head	Height: 1.5 cm; width: 1.3 cm	This object is a fragmentary animal head, with two small pointed ears, possibly a feline head. The entire front and the neck of the head are broken. The ears are the only detail indicated. It is golden yellow in colour, with well-smoothed surfaces.	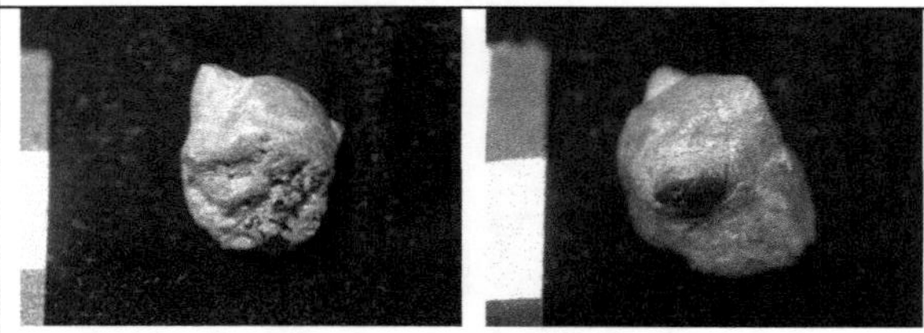
DVII 4: (DVII 50)	Possible feline head	Height: 1 cm; length: 2.8 cm; width: 1.4 cm	This piece is recognisable as a feline head because it has the usual conventions used in the representations of other feline heads in Dolni Vestonice: the eyes are represented by vertical slits; there are incisions to indicate the nose and the lips; the chin is well modelled. One ear is visible on the side of the head. The muzzle is unusual in the fact that it is very narrow and nearly pointed. Half of the head and the muzzle are missing on the other side. Various traces of scraping are visible on top of the head. It is yellow and grey in colour.	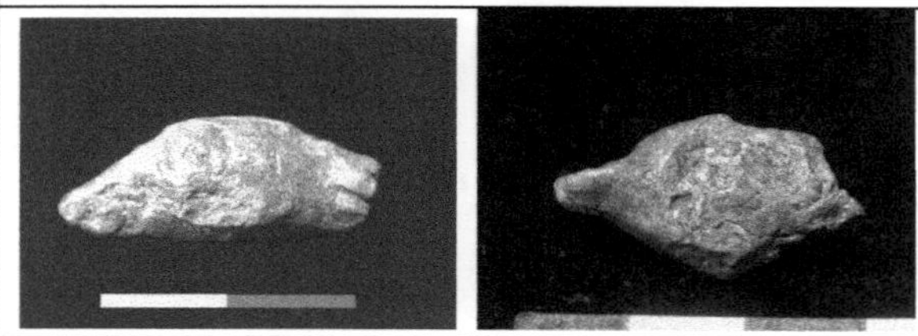

Dolni Vestonice II

DVII 5: (DVII 5)	Head of a female figurine	Height: 2 cm; diameter: 1.2 cm	This object is recognisable as the head of a female figurine because of its general shape that makes it an exact copy of head of the Black Venus. It has exactly the same four holes at the top. It is a rounded stump without any details visible. One side of the face is missing. It is also the same black colour as the Black Venus. The surfaces of the breaks at the neck and on one side seem very straight.	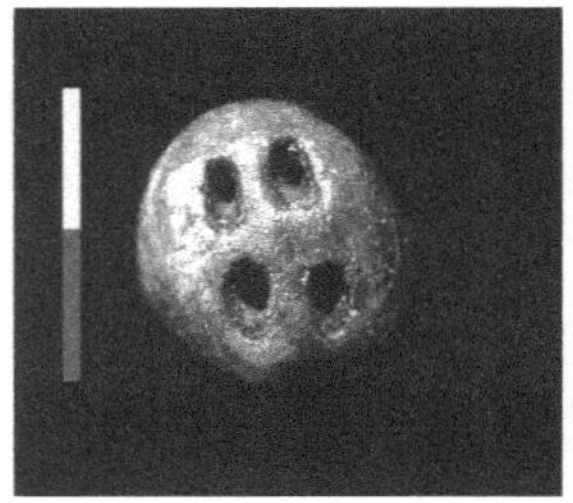 
DVII 6: (DVII 6)	Fragmentary female figurine	Height: 2 cm; width: 1.2 cm; maximal thickness: 0.8 cm	This is a very small fragmentary female figurine, recognisable by comparison with the other known figures from the site. It ends in a conical stump at one extremity, on which the legs are indicated by a deep incision at the front and at the back. On the front, which is lighter in colour, a regular pattern of impressed dots forms a belt around the area of the waist. The back is recognisable to the deep incision of the legs that continues on the back to mark the spine, just like for the Black Venus. This side is black. The top part is broken. Its general shape seen from the front has led to interpret also this piece as a schematic vulva. But the back is also modelled like on other female figurines.	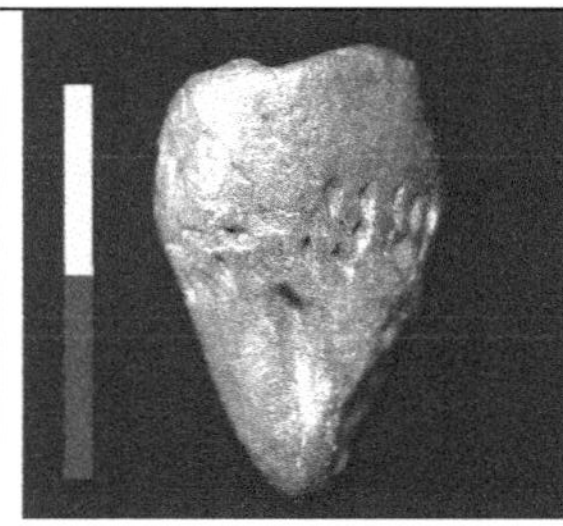 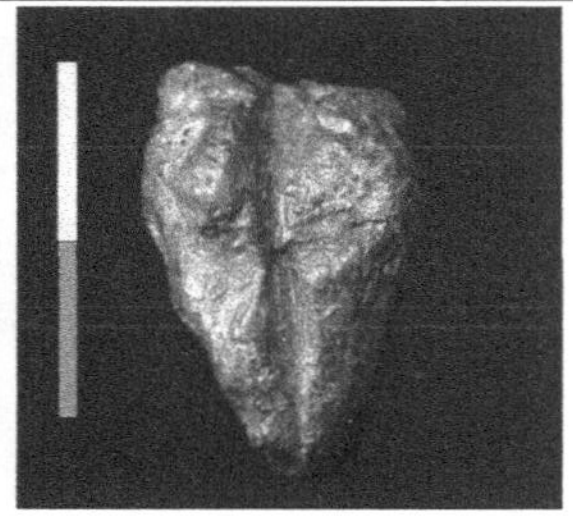
DVII 7: (DVII 7)	Large fashioned fragment	Height: 3 cm; length: 7 cm; width: 3.5 cm	This object is a large fashioned fragment with very irregular surfaces, some being clearly broken, some roughly fashioned. One protuberance, which is very smooth compared to the rest of the object, sticks out of the surfaces. Some incisions are also visible, but they are not organised. The material is different from everything seen up to now as it is really white in colour. It feels heavy and compact.	
6 Small rounded fragmen ts			They are all less than one centimetre in their larger dimension. They are smoothed and rounded as if they were worn out. They are grey to yellow in colour with some black traces. They seem very fragile as they have left some powder at the bottom of the box in which they are kept.	

Pavlov I

Reference number	Graphic unit name	Size	Description	Illustration
PI 1: (PI 393 or 103861)	Fragmentary female figurine	Height: 2.7 cm; width: 2.4 cm; thickness: 2.3 cm	This object represents a human head wearing a hat. It also wears a necklace at the base of the neck which is shaped by a series of small bumps, as if to indicate large beads nest to each other. It is possible to see from one remaining fragment below the necklace that it was attached to a body, but this part is now broken. The head is surrounded by the large rim of what appears to be a bell-shaped hat. It is also broken on one side of the head. No other features of the face are indicated. The object is yellow-brown in colour. The rim of the hat appears modelled by pinching with the fingers. The beads of the necklace are made by juxtaposing small triangular impressions with a tool on a rim of clay previously added. It looks like PI 2, but in much bigger.	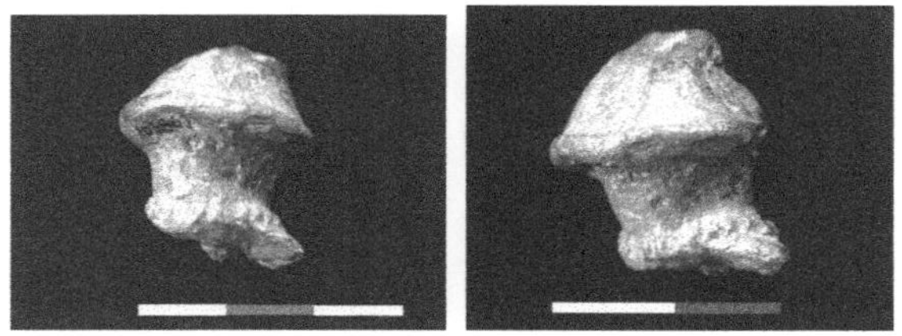
PI 2: (PI 394 or 557954)	Fragmentary female figurine	Height: 1.2 cm; maximal diameter: 1.2 cm	This object can be identified as a human head representation by comparison with the object PI 1 above as it has a similar general shape. It shows absolutely no detail, however. It is rounded on the top, with an edge reminding of the rim of a hat. It is broken at the neck. The material is grey in colour with darker stains.	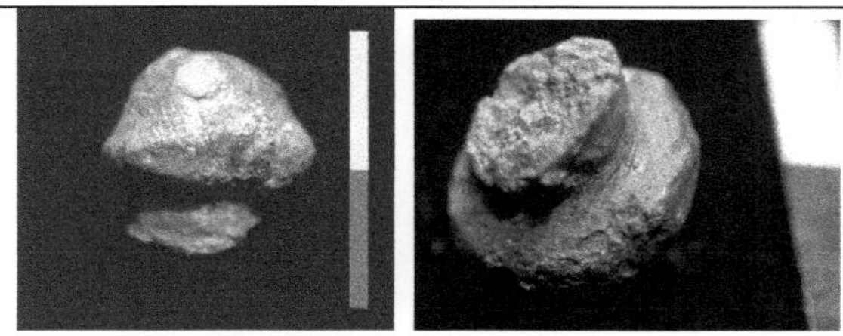
PI 3: (PI 395 or 552954)	Fragmentary female figurine	Height: 1.6 cm; width: 1.6 cm	This piece is a round head with the neck indicated, recognisable by comparison with other known female figurines. This object is entirely covered with small cupules on the upper rounded part of the head. They stop at what appears to be the neck. The head is broken from the lower part of the body, and also on one side. The shape of the impression is a sort of circle segment with one straight edge and a semi-circle on the other.	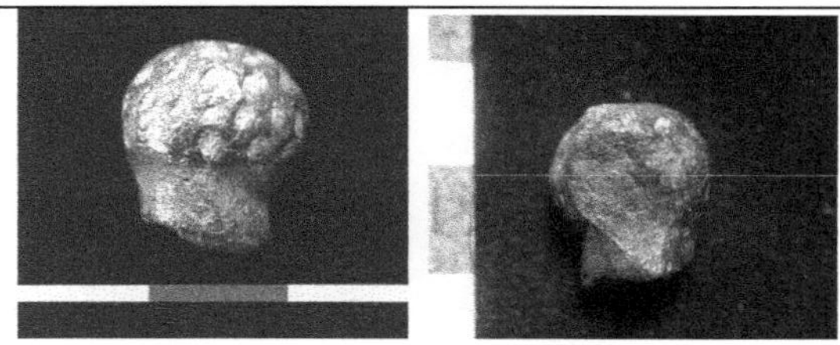
PI 4: (PI 396)	Small anthropo-morph	Height: 2.7 cm; width: 1.2 cm; thickness: 0.6 cm	This piece is recognisable as an anthropomorph by the indication of a neck and of a belt that runs around the waist of the figure. Both are formed by incisions. The head is also narrower than the rest of the body and it is rounded. There is a vertical incision in the middle of the face on one side. The entire representation is quite flat. The surfaces are smoothed and brown/black. No other detail is given. It has very short legs, stump-like; they seem finished and not broken: they are very schematic. The overall shape is also rather phallic, possibly as an intentional dual representation (Phallic/human)?	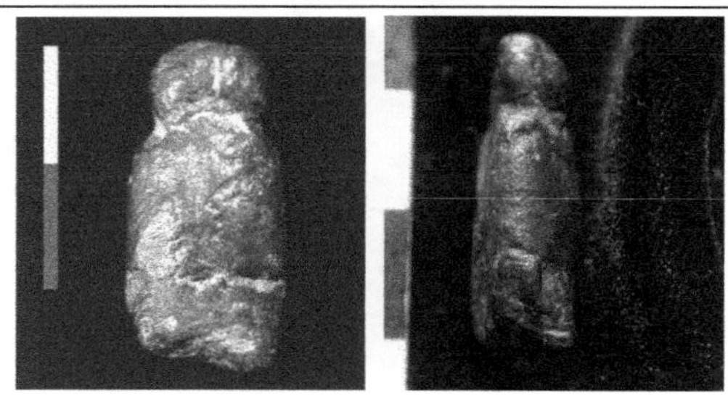

Pavlov I

PI 5: (PI 397)	Fragmentary female figurine	Height: 3.4 cm; width: 1.9 cm; thickness: 1.3 cm	This piece represents the middle part of a female representation, the upper part and the legs being broken. The front part is more damaged than the back, and one leg is visible up to the knee when the other is broken at the top. It is identifiable by comparison with the other female representations from the sites of Dolni Vestonice and Pavlov, notably by the presence of a well-detailed belt around the pelvis. The remaining intact surfaces are very finely and realistically fashioned. It seems that the figurine was represented in movement as it is possible to see that the knees were bent slightly. There is also an angle between the legs and the torso, visible in the line of the back, which seems to indicate it was leaning forwards slightly. The belt is made by a series of short parallel incisions that form a band in relief. It is also possible to see on the bottom of the back a hollow part that marks the spine, as in the Black Venus for example. The material is light brown. The broken surfaces show tearing traces, as if they had exploded in the fire. The waist is well-marked, quite slim.	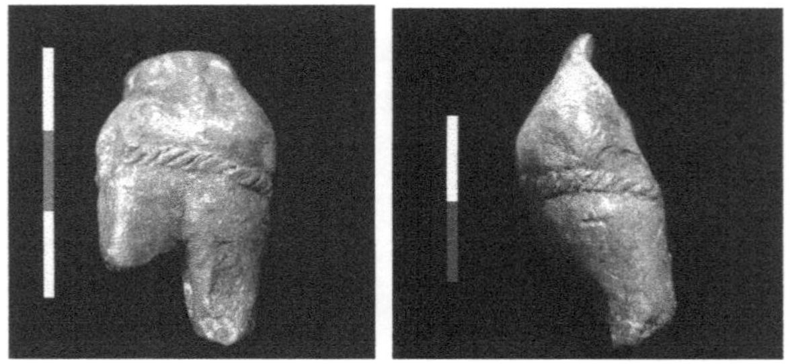
PI 6: (PI 198 or 586557)	Female figurine	Height: 2.3 cm; length: 2.6 cm; maximal thickness: 2 cm	This object is a small representation of a sitting female figure. It is fragmentary. Only one side of the body is well preserved, and the head is missing as well as the extremities of the legs. It is possible to see that the figure is represented with large buttocks and legs, and a smaller upper body, notably with a small thin arm, in a stylised manner that is however very realistic. Body ornaments are visible as an arm band on the remaining arm, and as decorations on the legs, just above the knees, forming two bands of incisions that appear to go around the limbs. The legs are defined at the front by incisions, and incisions also marked what appear to be the arms resting on the lap. This part is damaged however. The angle between the legs and the torso make it clear that the figurine is sitting. It was very carefully modelled with many details in spite of its very small size. The object has an irregular colour with black and golden light brown.	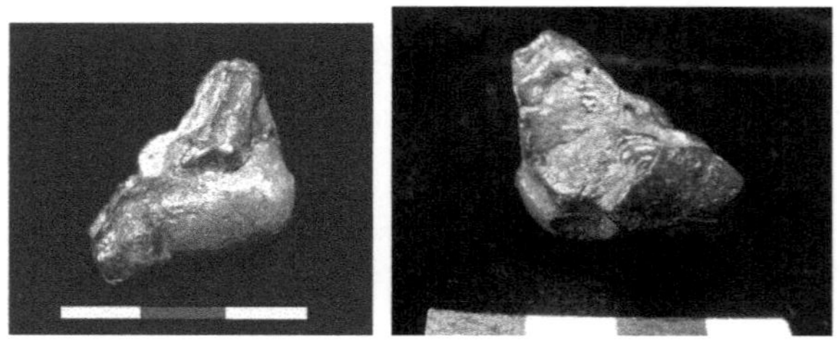
PI 7: (PI 399)	Fragmentary female figurine	Height: 3.5 cm; width: 1.7 cm; thickness: 2.4 cm	This piece forms a fashioned stump that can only be identified as a Venus because of its general shape and of the presence of a belt very similar to the belt of the objects PI 5. It is made in the same manner by a band of short parallel incisions forming a relief that runs all around the waist of the figure. The object is broken just above the waist and at the other end. The legs are not fashioned. It is rather yellow in colour. The surfaces are smooth on the non-broken areas. The belt is very high on the figure if the larger part is the bulk of the hips: at the sites normally low belts.	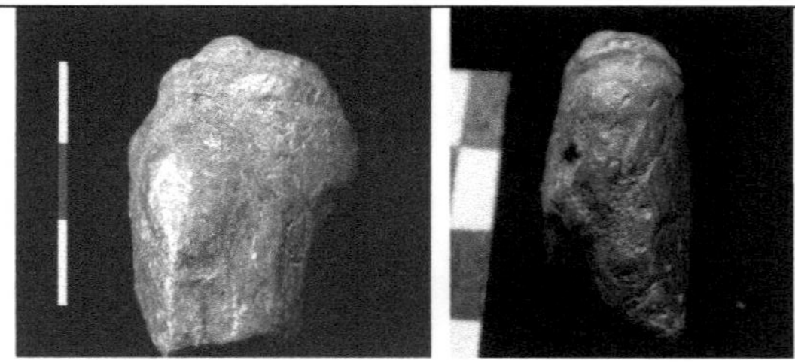

Pavlov I

PI 8: (PI 400)	Fragmentary female figurine	Height: 2.5 cm; width: 4 cm	This object is recognisable as a fragment of a pelvis because of the incisions that marked the separation at the top of the legs, but it is very fragmentary. All the surfaces appear damaged or irregular, with many inclusions. The figurine would have been quite large when entire in comparison to most of the pieces seen.	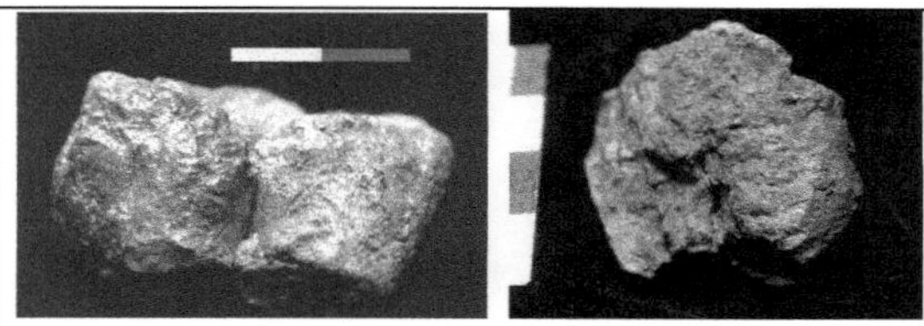
PI 9: (PI 401 or 650256)	Male figurine	Height: 2.7 cm; width: 1.5 cm; thickness: 1.2 cm	This object is a fragmentary human figure, interpreted as a male representation because of the small bump between the legs could represent the sex. The top part of the body is missing. The legs are actually whole with modelled extremity but they were originally fashioned as short stumps. It seems slightly bent forwards as the legs form an angle with the torso. The surface of the back is well smoothed.	
PI 10: (PI 450 or 615366)	Body extremity	Height: 1.2 cm by 0.7 cm	This object is a very small human foot with the heel and the ankle clearly indicated. There are some neat traces of scraping at the front of the leg and the foot. The sole of the foot is 0.5 cm long and the ankle 0.3 cm wide.	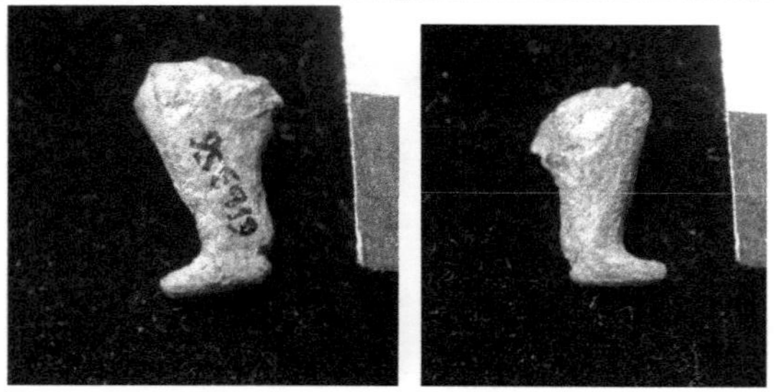
PI 11: (PI 402 or 485057)	Horse head	Height: 2.6 cm; length: 4.2 cm; width: 2 cm	This piece is the top part of the head of a horse, with the mane, ears, nostrils and eyes indicated. The two eyes are represented by simple holes made when the material was still wet: this can be seen because of the rim of material that formed around them. They are not symmetrical. The muzzle is very fine with small holes to indicate the nostrils. One ear is well marked in relief, the other is hardly visible. The mane forms a high relief on top of the head. It is shaped in such a way that it gives an impression of movement and life. It is yellow. The top of the mane is sculpted by areas represented by hollows and others in relief.	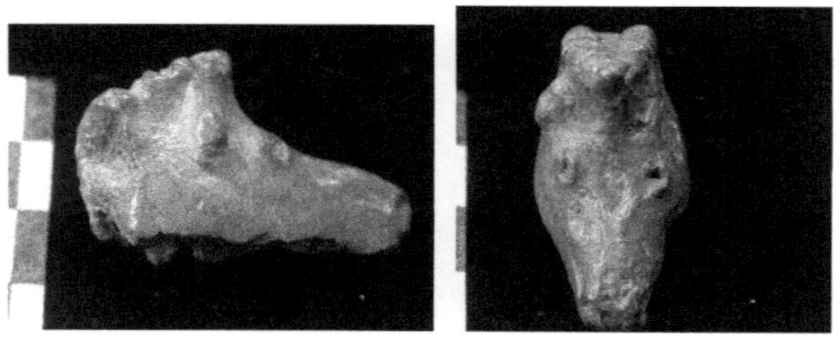

Pavlov I

PI 12: (PI 62660)	Fragmentary horse	Length: 3.8 cm; width: 2.8 cm; thickness: 1.5 cm	This object is recognisable as the top part of the hind side of a horse because of its general shape combined with the shape of the tail visible on it. It is very fragmentary. The intact surfaces are black in colour and well smoothed.	
PI 13: (PI 407)	Rhinoceros head	Maximal height: 3.3 cm; length: 3.1 cm; thickness: 1.2 cm	This object is a rhinoceros head recognisable by the general shape and by the bases of the two horns on the forehead and the hump on top of its head. It is very finely modelled, with clear traces left by the fashioning processes on it, such as a rim of material around the eye on one side and scraping traces forming small striations on the end of the muzzle. The eyes and the nostrils are indicated by holes. The mouth is made by a rather large incision at the bottom of which another thinner incision can be seen, probably due to the shape of the tool used. The material feels very dense and heavy. It is dark in colour. One side of the head is damaged.	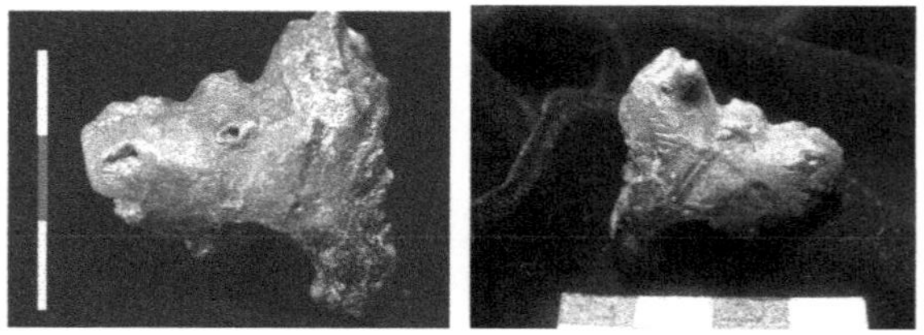
PI 14: (PI 418 or 53053)	Animal head	Height: 2.7 cm; length; 5.2 cm; width: 1.7 cm	The general shape of this object suggests a rhinoceros by comparison with the rhinoceros head PI 13 above, but it lacks all the details such as the base of the horns. It is certainly the head of a large herbivore however. The eyes are the only detail indicated. They are made by two impressed holes and there is no rim of material visible around them The back of the head is pinched between two fingers, narrowing for most of it to form the neck.	
PI 15: (PI 409 or 214556)	Head of a caprine	Height: 2.5 cm; length: 3.5 cm; maximal thickness: 2 cm	This object represents the head of a caprine animal, with the horns well indicated. The general shape is fashioned, but there are no details indicated. It is broken at the neck. One side of the head is better smoothed than the other. The horns are not exactly symmetrical.	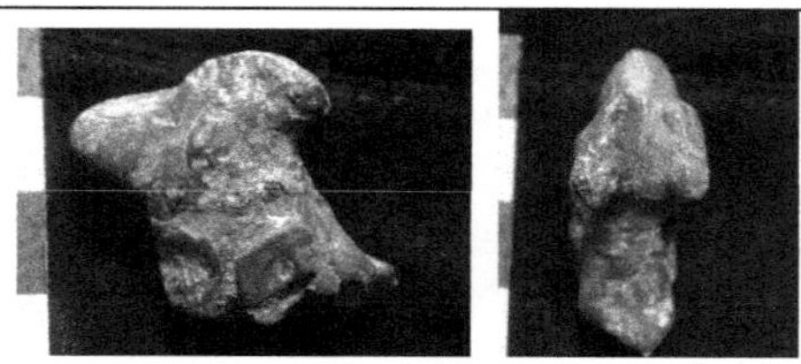
PI 16: (PI H3)	Horse head	Maximal height: 2.7 cm; length: 4.2 cm	This object is a roughly shaped animal head, broken at the neck and on top of the head. Two symmetrical holes indicate the eyes on each side of the head. The surfaces are very irregular. Traces of scraping are visible on the muzzle. Two very small holes mark the nostrils. The general shape correspond to that of a horse.	

Pavlov I

PI 17: (PI C2)	Feline head	Height: 1.5 cm; length: 3.5 cm; width: 2.4 cm	This piece represents the head of a feline. It follows the types already found in the feline representations in Dolni Vestonice with the same combination of incisions to shape the nose, the lips and the mouth. The chin is shaped in the same way, and the eyes are indicated by vertical slits. The ears are indicated in relief on the sides of the head, but on one side one ear seems to have moved out of place, as it is much lower than the other. The entire surface on this side is damaged, possibly indicating that it was deformed when the material was still wet, before firing. The neck is not broken, it is cone-like.	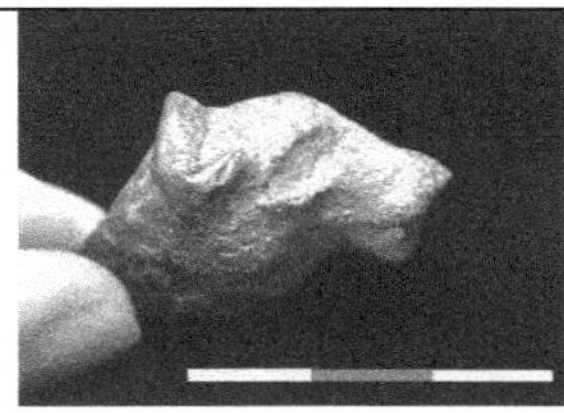 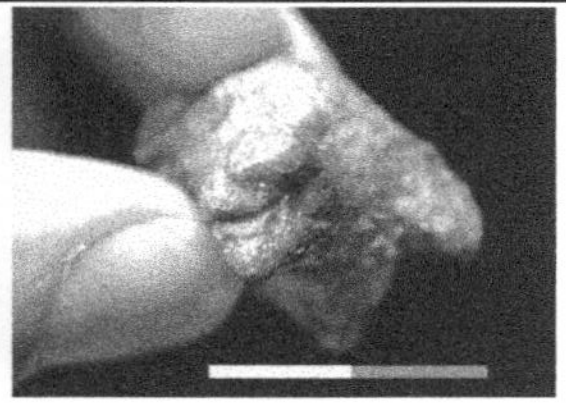
PI 18: (PI C3)	Reindeer head	Height: 1.5 cm; length: 2.5 cm; width: 1.2 cm	This figurine has an elongated muzzle. The mouth and the nose are marked by small slits. The eyes are in relief with small horizontal slits incised in them. The ears are completely on the side and form small protuberances. The surfaces are smooth. It is broken at the neck. The piece is finely modelled; the back of the head is broken. The head seems stretched towards the front.	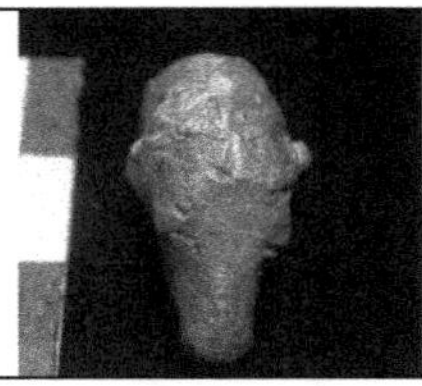 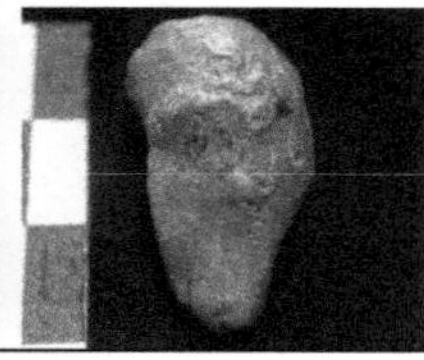
PI 19: (PI C1)	Feline head	Height: 0.8 cm; length: 2.1 cm; width; 1.3 cm	This object is a feline head. It is broken at the neck. It does not show the usual overall shape of feline's heads found in Dolni Vestonice and Pavlov but it is a feline without doubt. It rather resembles a feline head in ivory found in the Aurignacian of Vogelherd in Germany, with a square muzzle. It is very finely modelled. Only one ear visible on one side. The eyes were made by a triangular tool that impressed a deep hole without any rim of material visible around it. The volume of the cheeks is well modelled. The nose is in fine relief with two holes to mark the nostrils, then another relief for the lips, then two incisions to separate the mouth from the chin. The back of the head if well indicated as the neck is broken lower.	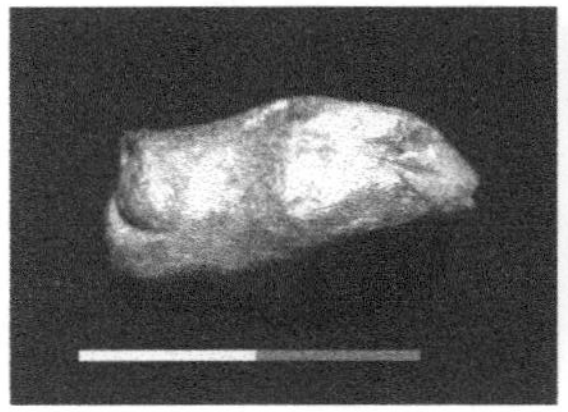 
PI 20: (PI 159563)	Feline muzzle	Length: 1.6 cm; thickness: 0.5 cm	This piece is a very small muzzle of a feline. There are two holes for the nostrils, but they are larger than for the figurine PI 19. The mouth is marked by an incision. The beginning of the modelled cheek can be seen. The rest is broken.	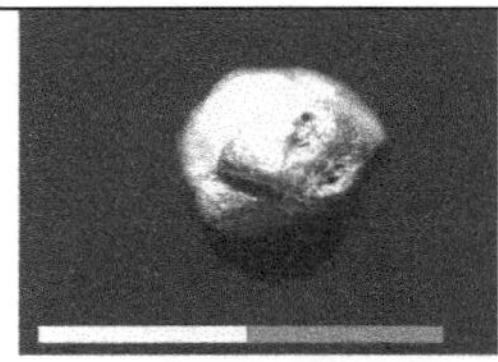 
PI 21: (PI 446062)	Feline muzzle	Height: 0.5 cm; length; 1.5 cm; width: 1 cm	This piece also represents the end of the muzzle of a feline. Two holes for the nose are still visible and two incisions to form the mouth. A small circle incised on the surface could indicate the eye. It is broken at an angle.	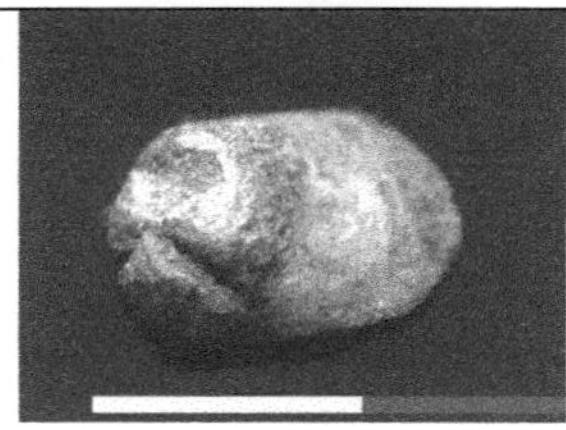

Pavlov I

PI 22: (PI 408 or M256)	Mammoth figurine	Height: 1.5 cm; length: 2 cm; Thickness : 0.8 cm	This object represents a small mammoth recognisable at the characteristic shape of the profile of the back and the head. It is roughly modelled, without detail. An appendix seems to indicate the trunk. The front legs are indicated by a stump. The rear side and legs form a large stump. The surfaces are very rough and irregular.	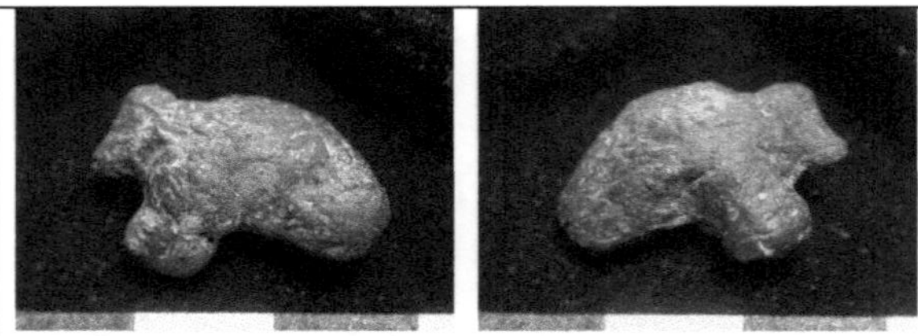
PI 23: (PI 325860)	Fragmentary mammoth	Height: 1.8 cm; length: 3.3 cm; width: 1.2 cm	This piece is mainly recognisable as a mammoth thanks to the typical profile of the back and the head. There is no detail. The surfaces are irregular and rough, although some traces of scraping are visible. The lower body does not seem to have been fashioned; the left side is smooth, the legs are very short, ball-like, the extremity of the trunk as well.	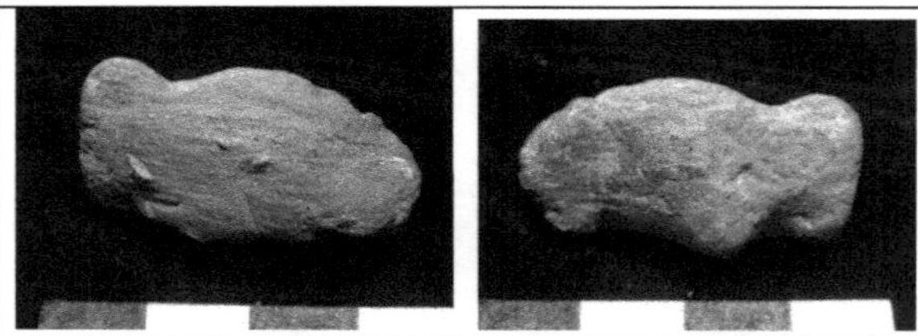
PI 24: (PI M3)	Mammoth figurine	Height: 4.2 cm; length: 5 cm; width: 2.5 cm	This object represents a mammoth recognisable at the typical profile of the back and head and at the general bulk of the body. It is nearly whole, with two full legs, which are conical in shape and about 1.5 cm long. The extremities of the trunk and of the other legs are missing. Again, there are no details indicated. The surfaces are roughly scraped and are quite irregular. It is yellow-brown in colour with some red traces around the legs.	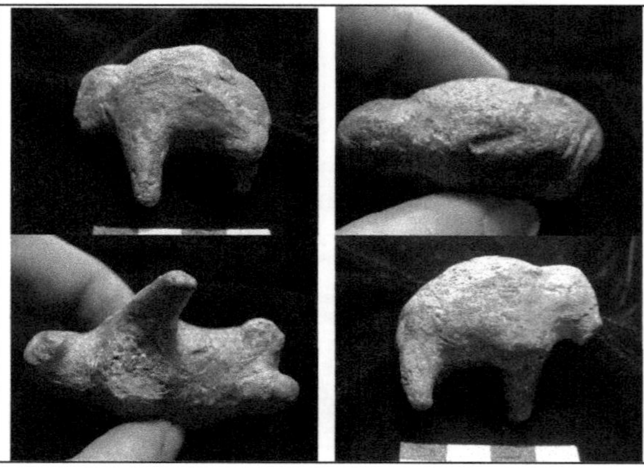
PI 25: (PI 417)	Animal body	Height: 3 cm; length: 4 cm	This object is the fragmentary body of a bulky animal, possibly a mammoth or a bear, without enough details to allow deciding. The head and the limbs are missing. The tail is indicated. It is quite smooth. Two deep holes can be seen on one of the sides. They seem to be impressions made with a triangular small tool. Irregular surfaces.	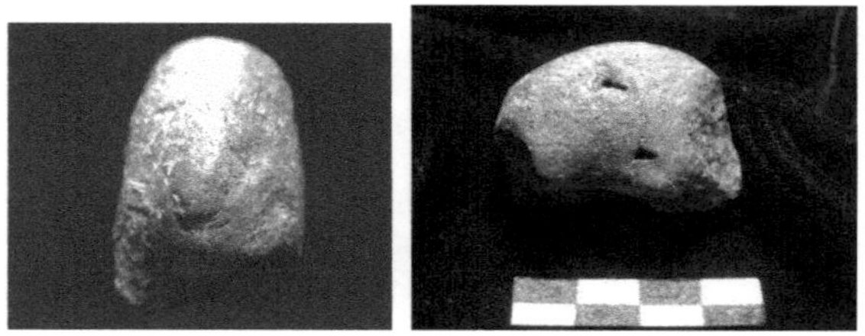

Pavlov I

PI 26: (PI 411? or M154)	Large mammoth	Maximal height: 5.5 cm; length: 8 cm; maximal width 3 cm	This piece is a mammoth figurine. It is large in comparison to the average size of the figurines of the site. It has been reconstituted from broken fragments, but many parts are still missing. It has a nearly full leg, the other are broken. It is still possible to see that it was finely modelled, with details. The small tail is visible for example. The belly underneath is carefully shaped. There are red traces, possibly of red ochre, on the flanks of the animal and on the belly. On the side of the forehead, there is one eye represented by a protuberance. There are light traces of scraping on the intact remaining surfaces, which are very smooth. It is brown-black.	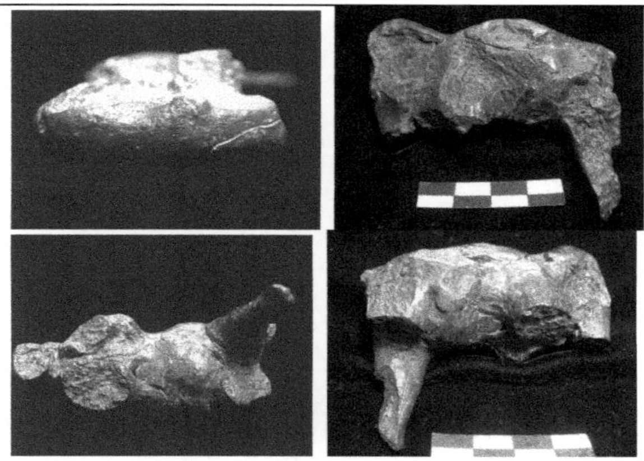
PI 27: (PI 145060 or 145962?)	Animal body	Height: 1.5 cm; length: 3 cm; width: 2 cm	This object is very fragmentary: only the rear side of the animal survives, the legs and the front of the body are broken. It represents a bulky animal with a small rounded tail. It is brown-red in colour.	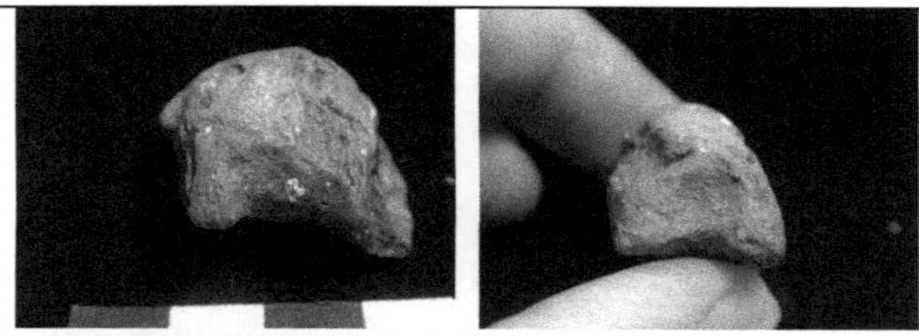
PI 28: (PI 453 or 10396)	Body extremity	Length: 1.7 cm; width: 0.7 to 0.5 cm; thickness: 0.7 cm	This body extremity represents the end of the trunk of a mammoth. It is represented twisted, as if in movement. It is very finely modelled and realistic, with good proportions.	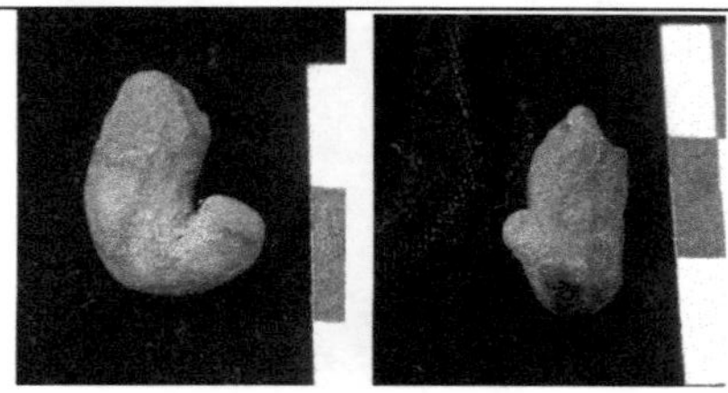
PI 29: (PI 406 or 363956)	Animal head	Length: 2 cm	This object is very roughly shaped as an animal head with the eyes indicated by holes. The surfaces are very irregular. The nose is well marked. The general shape of the chin, which is well modelled and ball-like, evokes a feline head.	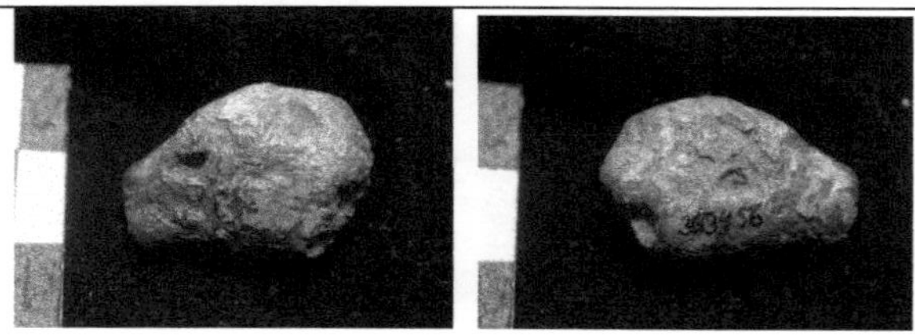

Pavlov I

PI 30: (PI 410 or 513756)	Animal body	Height: 2.3 cm; length: 4.2 cm	This object forms the rear side of a flattened animal. The legs and the front of the body are broken. The small tail in indicated by a protuberance. Svoboda thinks it could be the body of the head PI 29.	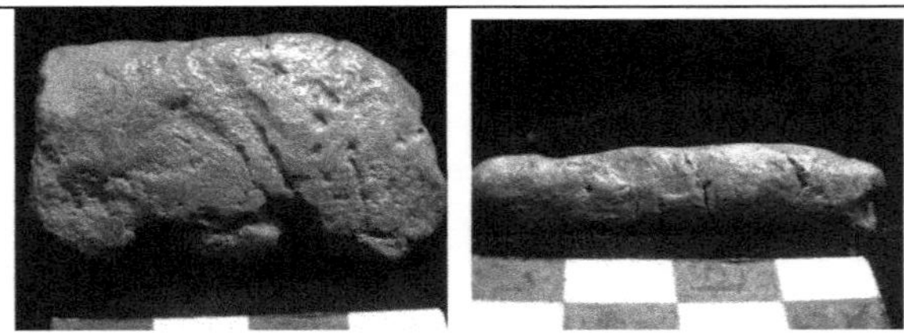
PI 31: (No reference)	Flattened animal body	Length: 5 cm	This object is the rear of an animal body, distorted and flattened on one side. The rear legs are visible. All the front part is broken.	
PI 32: (PI 412 or 369757)	Body extremity	Length: 4.5 cm; width: 1.7 cm at top; 0.5 cm at the foot	This object is the very well fashioned leg of an herbivore. It is very fine and carefully modelled, with indication of the tip of the foot and of the knee for example. Few finger prints are visible on it. The surface is very smooth. It is yellow on one side, black on the other.	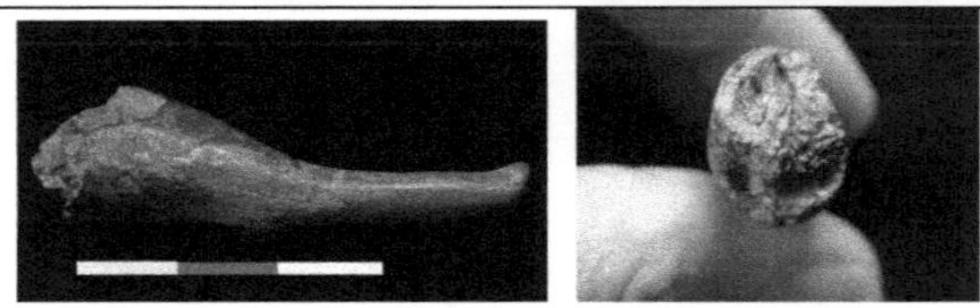
PI 33: (PI 757)	Body extremity	Length: 3.8 cm; width: 1.8 cm; Thickness : 1.6 cm	This body extremity follows the mammoth leg type as it is cylindrical in shape with straight sides and a marked circular flattened foot. The surfaces are smooth. It is broken on all its side.	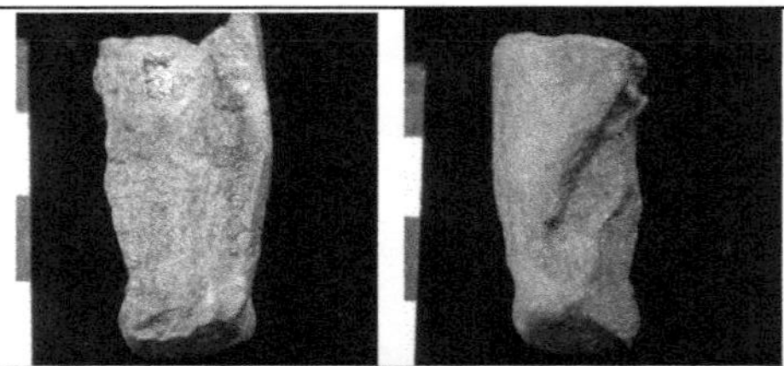
PI 34: (PI 414 or 589857)	Animal body	Length: 3.2 cm long; height: 1.4 cm; thickness: 0.7 cm	This piece represents the body of an animal. One side of it is missing. The other side is modelled with the belly, a relief for the top of the legs and a rounded rear side. The entire spine is marked with short parallel incisions from the back to the head. The head is small and shapeless. One ear can be seen formed by a small protuberance in relief. The surface is well smoothed, yellow in colour.	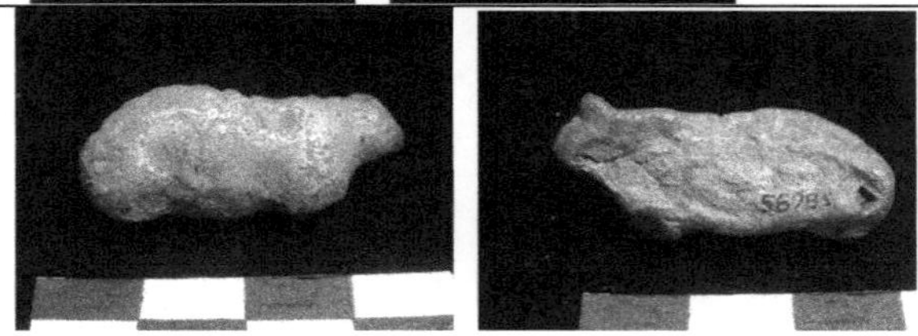

Pavlov I

PI 35: (PI 348 or 586757)	Body extremity	Length: 3.8 cm; Thickness : 1 cm by 0.9	This object forms an elongated limb, which is well smoothed on one side on nearly all the length. It bears marks of scraping whose colour and texture indicate a possible reworking after the firing process. The limb is broken at both ends. The articulation of the knee is marked.	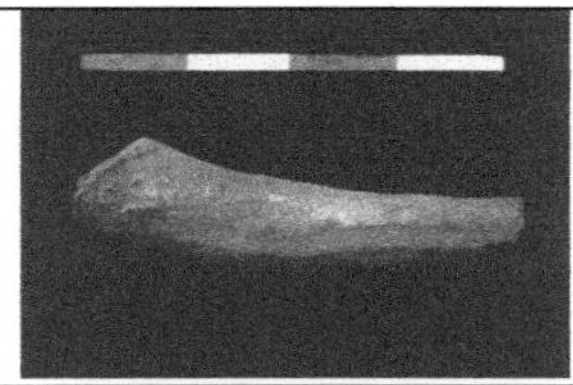
PI 36 (PI 416 or 586657)	Body extremity	Length: 4.1 cm ; thickness: 2.5 cm by 1.6 cm	This body extremity is possibly an animal leg. It is rather conical in shape, with a very wide extremity in comparison to the other. It is brown-yellow in colour. The remaining part of body attached to the leg is massive. The foot is whole, circular. This piece differs from others as it is attached to a (fragmentary) body: it is not just an isolated leg.	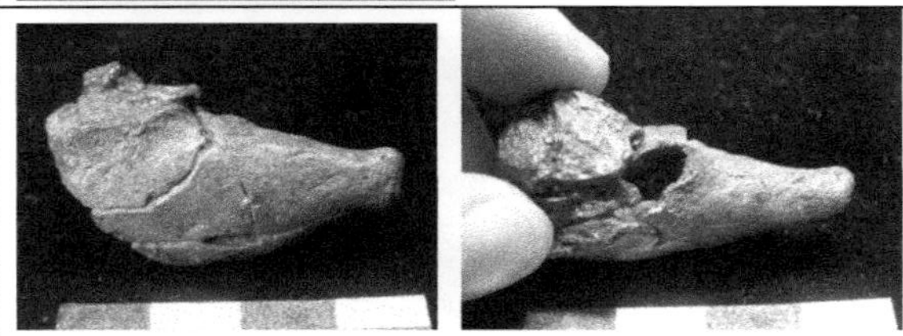
PI 37: (PI 438 or 419761)	Body extremity	Height: 0.7 by 0.4 by 0.3 cm	This object forms a very small foot similar to larger ones also found at the site. It is cylindrical at the top.	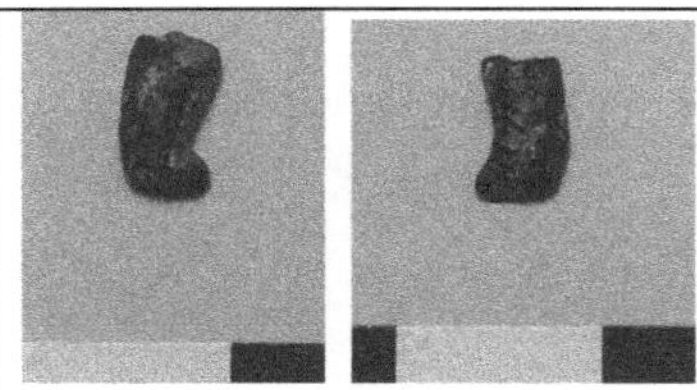
PI 38: (PI 404 or 5956)	Body extremity	Height: 2.1 cm. Diameter of the foot 0.7 cm	This body extremity is a leg of the mammoth type. It is slightly conical in shape, with very straight sides and a circular flattened foot.	
PI 39: (PI 403 or 1602956)	Body extremity	Length: 2.5 cm; wide 1.5 cm	Body extremity of the mammoth type. It has a slightly conical form that widens again at one end to form the foot with a circular and flattened surface underneath.	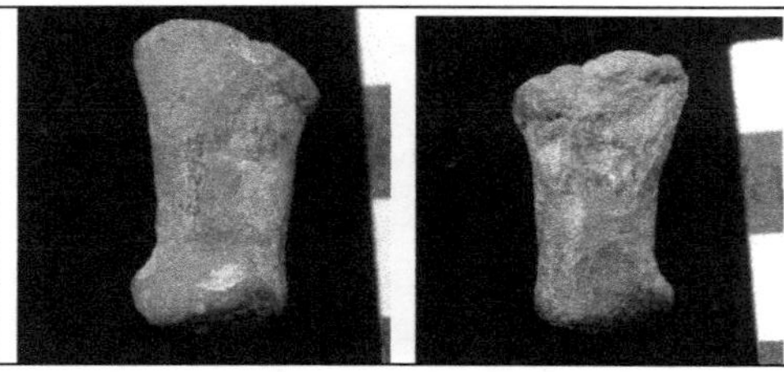

Pavlov I

PI 40: (PI 761)	Broken animal body	Height: 3.6 cm; length: 2.6 cm; width: 2.8 cm	This object is very roughly shaped with scraping traces visible. The very beginning of a rear leg and a tail are marked. It is much damaged with very irregular surfaces.	 
PI 41: (PI 405)	Incised flattened fragment	Height: 2.6 cm; length: 3.2 cm; width: 1.2 cm	This object has been fashioned with smoothed surfaces but does not represent anything recognisable. It bears deep incisions on one side. It is black on one side, yellow on the other.	 
PI 42: (PI 416 or 119052)	Fragment with parallel incisions	Length: 2.6 cm; width: 2.6 cm; thickness: 1 cm	It is a flat fragment that bear five parallel incisions regularly spaced every 4 mm.	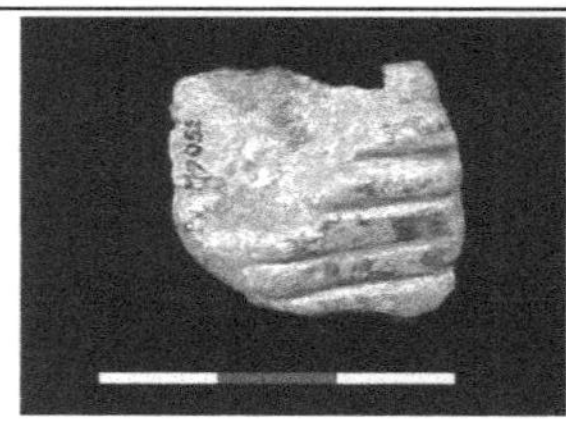 
PI 43: (PI 368263)	Fragment with holes	Length: 2.5 cm	This piece is a shapeless fragment bearing traces of modelling, with five holes made on it. Two of them are clearly made from impressions with a tool.	
PI 44: (PI 456)	Body extremity	1.5 cm high	This object is a leg with straight edges belonging to the mammoth type. The surfaces are well smoothed.	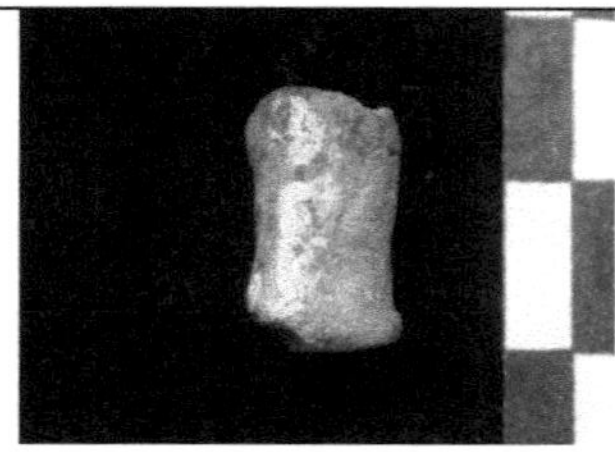

Pavlov I

PI 45: (No reference)	Possible animal head	Length: 2.8 cm	This object resembles the muzzle of an animal. It is very roughly shaped with irregular surfaces. The nostrils are indicated by light marks. One ear is represented.	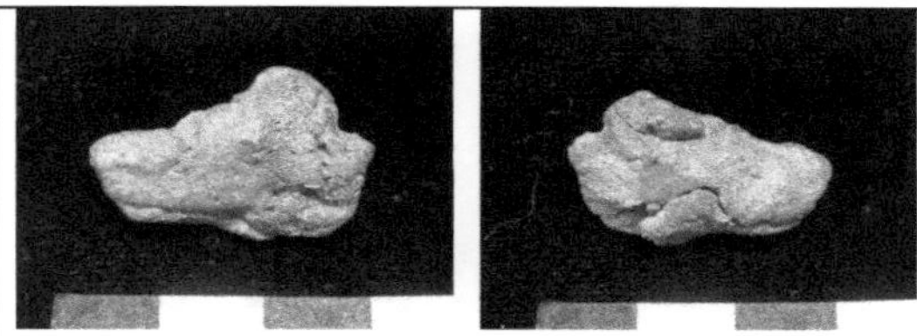
PI 46: (No reference)	Pierced red fragment.	Length: 2.8 cm by 2.5 by 0.8 cm	This object has clearly been fashioned and flattened on each side. There seems to have been an attempt to drill a hole. The piece broke alongside this hole.	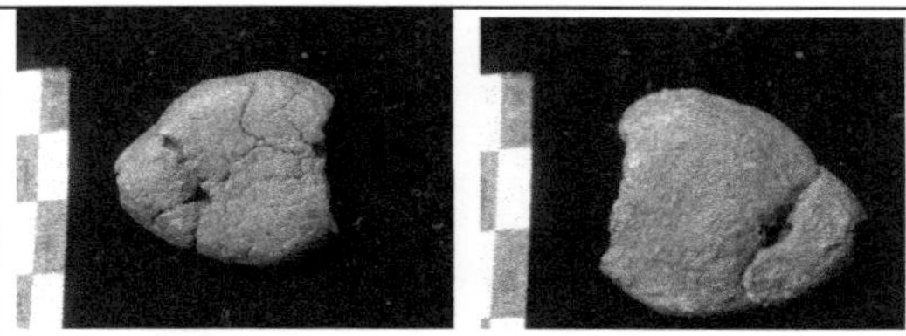
PI 47: (343956)	Possible animal head	2.5 cm by 2.2 cm by 1.5 cm	This object represents a head, with the head above the neck as in humans. The face, however, is caricatural and has nothing human. It is completely broken on one side. The eye is indicated by a hole. One ear is visible and the cheek is well shaped in relief.	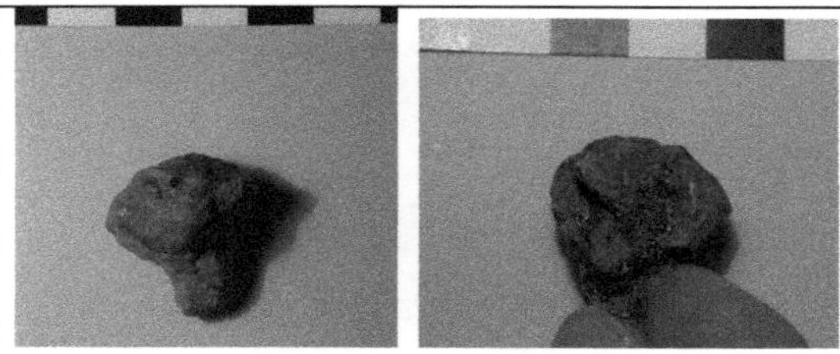
PI 48:	Shapeless fragment	4.2 cm by 4.3 cm	This is a broken fragment, glued back together. There are traces of a well-smoothed surface but it is much damaged.	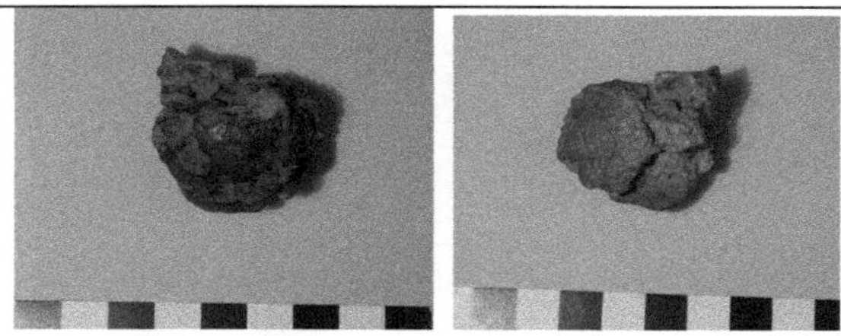
PI 49: (761)	Shapeless fragment	3.8 by 2.7 cm	Shapeless fragment of fired clay with some worked surfaces	

Pavlov I

PI 50: (558954)	Animal body	2 cm by 1.2 cm by 1 cm	The piece is flattened and twisted. It appears to be an animal body with the back legs, the line of the belly and the beginning of the front limbs. The rest is then broken.	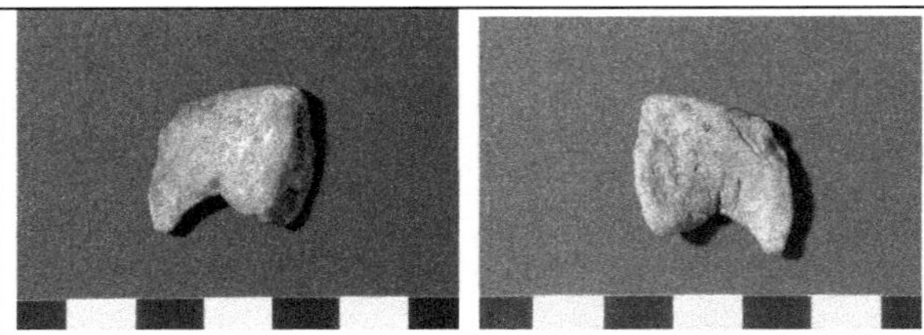
PI 51: (567650)	Semi-circular fragment	2.2 cm by 1.4 cm by 0.4 cm	This fragment is rounded into a semi circle. It is rather black and broken. The surfaces are damaged and irregular.	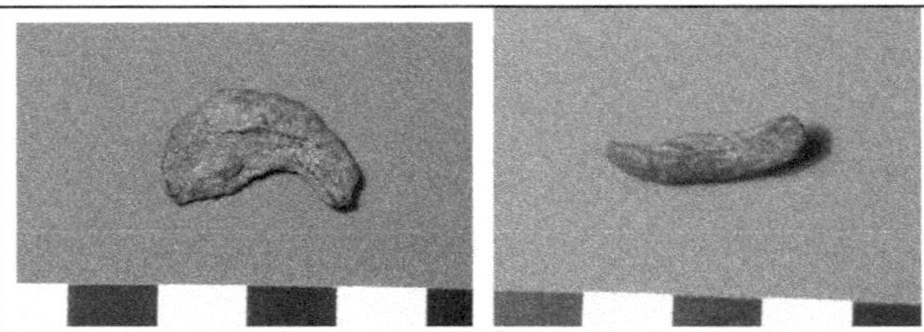
PI 52: (588557)	Animal body	3.9 cm by 1.8 cm by 1.3 cm	This object is broken and glued back. It is possibly the fragment of the rear side of an animal. The remaining intact surfaces are well smoothed. The piece is slightly curved. The extremity is nearly whole.	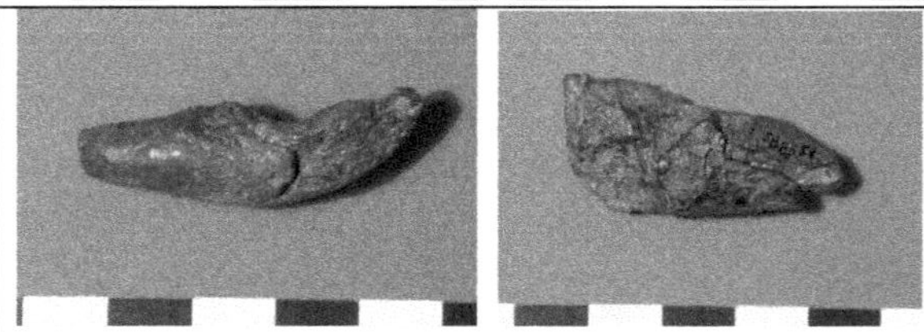
PI 53: (553754)	Possible animal body	2.1 cm by 1.4 cm by 1.4 cm	This is possibly a fragmentary body. The surfaces are well smoothed. The beginning of the legs is visible.	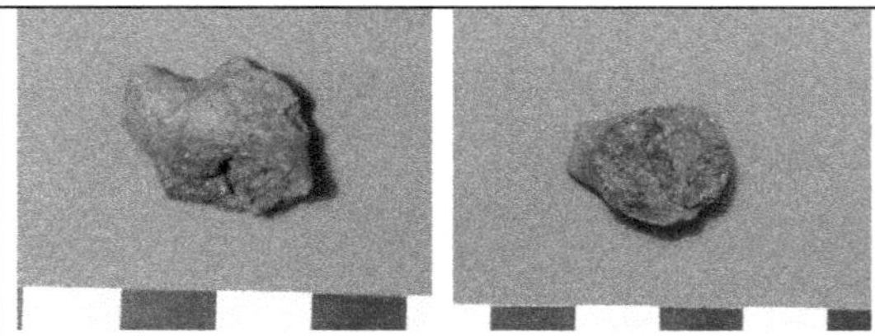
PI 54: (588858)	Flattened fragment	3.8 cm by 2.4 cm by 1.1cm	It is black and flat with traces of fashioning. One extremity is rounded.	

Pavlov I

PI 55: (4856)	Fragmentary animal body	3.5 cm by 2.2 cm by 0.9 cm	This piece shows a flattened body with small cracks on the surface. One edge is all smoothed and rounded. One surface is well smoothed and flat. The rest is broken.	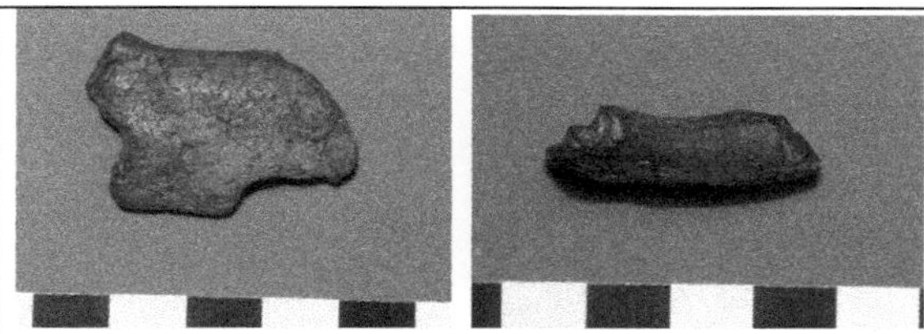
PI 56: (558854)	Shapeless fragment	4.7 cm by 4 cm by 2.5 cm	Shapeless fragment, very light in colour. It is unusual, possibly with some fashioning traces.	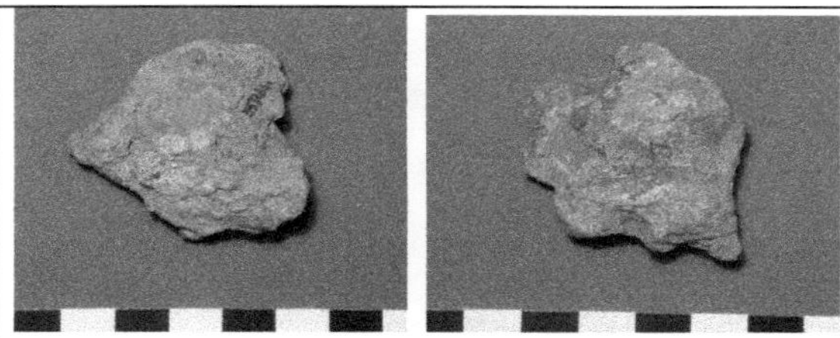
PI 57: (558654)	Shapeless fragment	5.7 cm by 3.1 cm by 2.3 cm	This is a large shapeless black fragment, much broken. Some surfaces appear smoothed on one side. It has been restored.	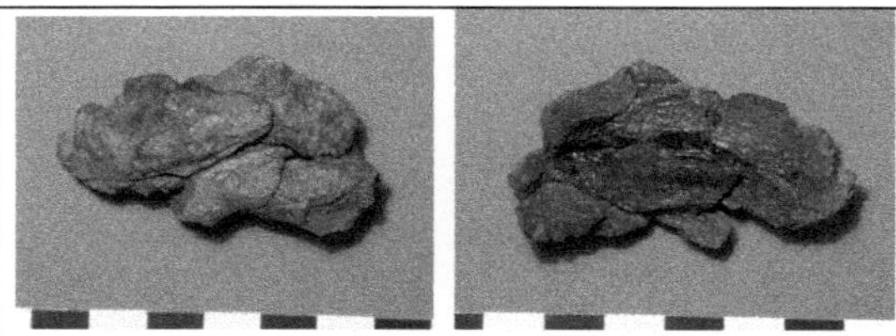
PI 58: (60760)	Fashioned fragment	3 cm by 2.3cm by 1.8 cm	This piece is yellow ochre in colour. The surfaces are very smooth and shiny. It is worked all around, but the shape is not recognisable. There are two parallel protuberances next to each other on one side.	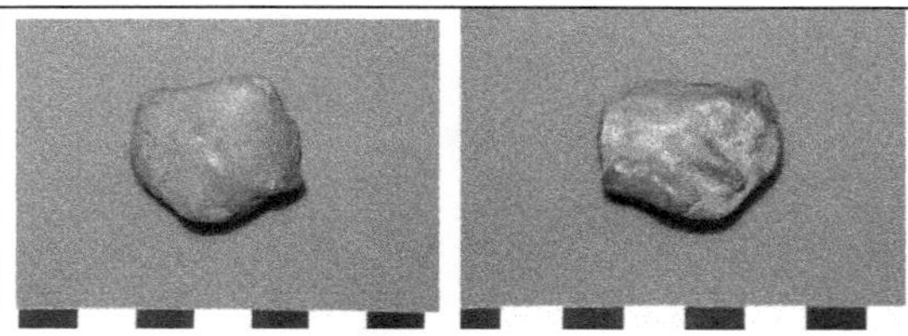
PI 59: (60560)	Fragmentary animal body	2.5 cm by 1.5 cm by 1cm	This is a fragmentary animal body with one limb still attached to it. It is black in colour.	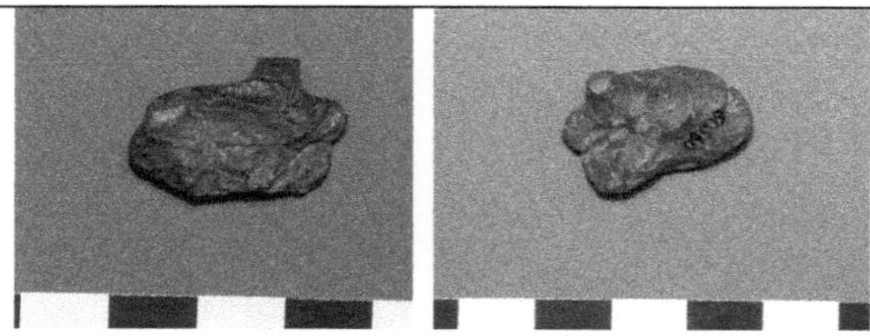

Pavlov I

PI 60: (558554)	Fragment with parallel incisions	2.3 cm by 2.1 cm by 1.2 cm	One side is smoothed and curved with 6 parallel lines. Two lines are whole and are 1.5 cm long, the others are broken. The distance between the lines is 0.2 to 0.3 cm. There is a pointed edge between the two sides. The other side is possibly scraped.	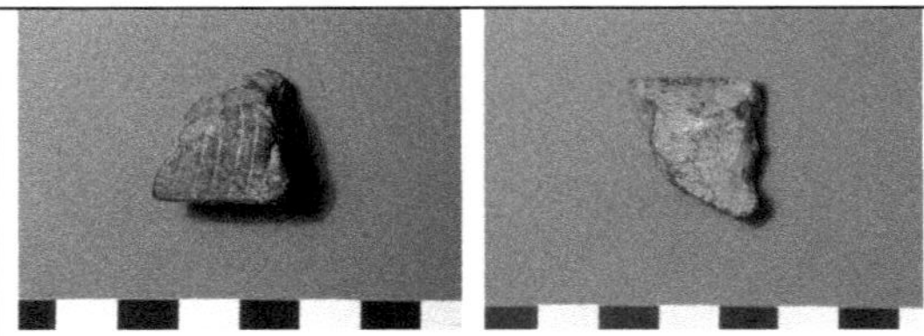
PI 61: (723656)	Body extremity	2.1 cm by 1.2 cm by 1.1 cm	This piece is greyish and elongated. One end is worked and smooth, the other is broken. It is slightly curved, with a ball-like shape at the extremity.	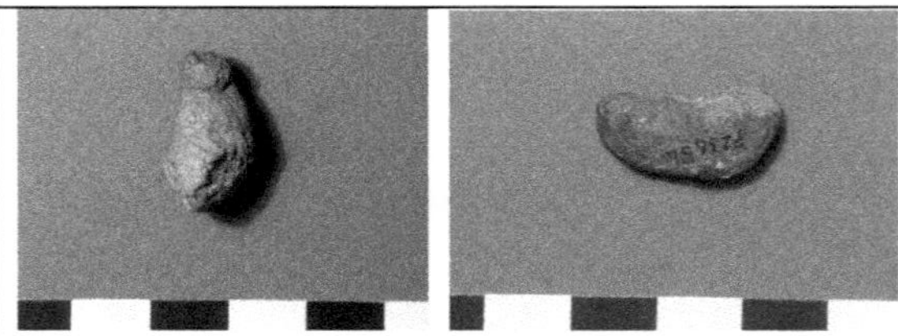
PI 62: (588657)	Body extremity	2.5 cm diameter at base; 2.2 cm long	This is a conical object. It is massive, pointy and seems broken off a larger object at the base. It is damaged on the side of the cone.	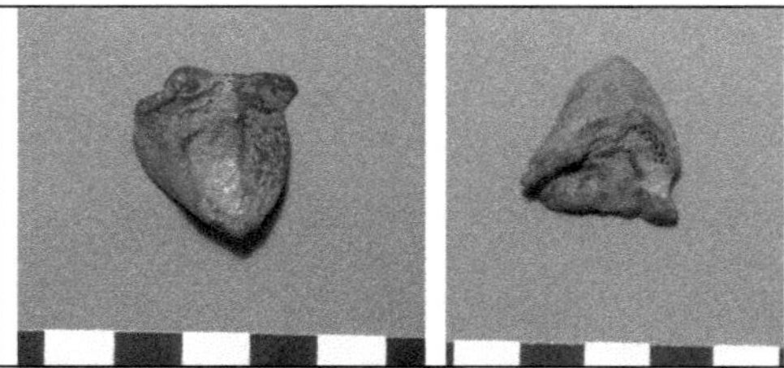
PI 63: (567960)	Body extremity	2.3 cm by 1.1 cm by 0.8 cm	This is a thin pointed object, broken only at one end. The point is very fine: it resembles a horn. Two incisions are visible on it.	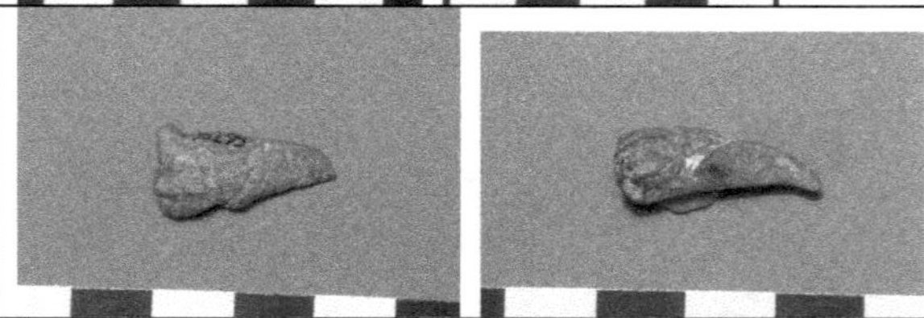
PI 64: (723556)	Shapeless fragment	4.5 cm by 2.7 cm by 1.7 cm	This is a large greyish shapeless piece. Some surfaces appear worked.	

Pavlov I

PI 65: (705957)	Fashioned fragment	3.8 cm by 2.2 cm by 1.6 cm max thickness	This is a large light brown piece, much damaged. Some worked surfaces are visible.	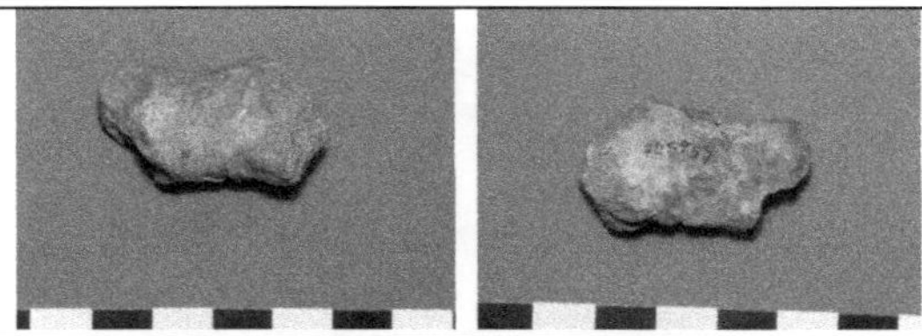
PI 66: (503456)	Possible fragmentary animal body	3.1 cm by 2.5 cm by 1.1 cm	This is a large light brown piece, flattened and smoothed, with the beginning of a limb visible.	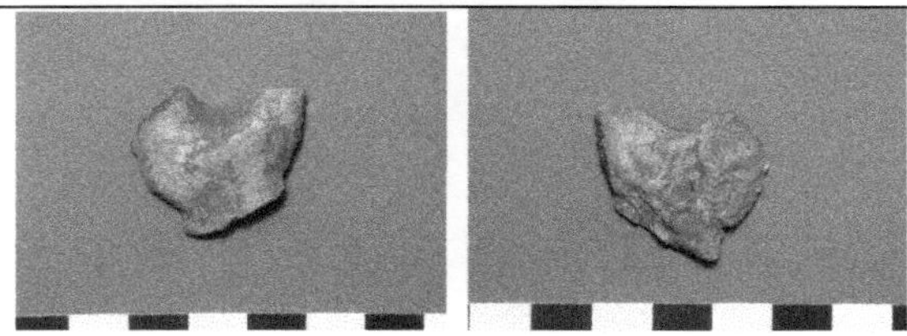
PI 67: (60160)	Fashioned fragment	6.3 cm by 3.6 cm by 2.5 cm	Large black shapeless fragment with much damaged surfaces. It is slightly rounded. Some worked surfaces are visible. It is large in size.	
PI 68: (563360)	Body extremity	3.7 cm by 1.3 cm at top and 0.9 cm at other end	This is a leg with a foot marked as a ball on one side. The surfaces are well smoothed; the extremity of the foot is broken, the top of the leg is quite wide, but is broken as well as a part of the side of the leg. It is light brown in colour with black traces.	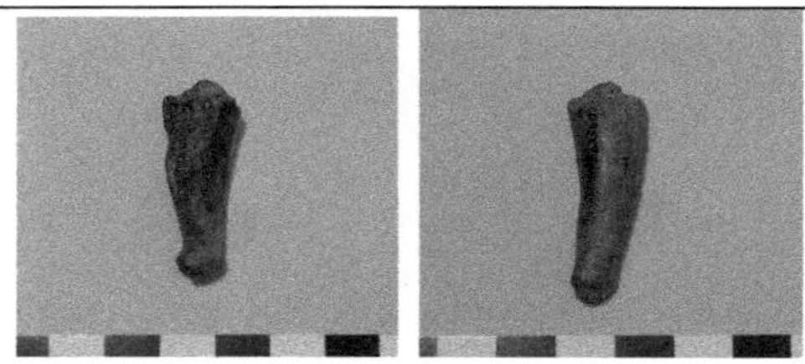
PI 69: (567460)	Body extremity	3.2 by 1.3 and 1.7 cm by 1 cm and 0.6 cm	This is a limb fragment. The overall shape is conical. It is broken at both end and on the side. The remaining sides are straight. It widens towards one end.	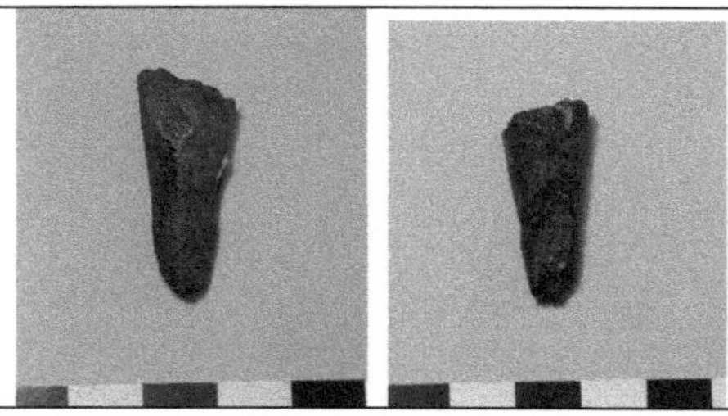

Pavlov I

PI 70: (221560)	Body extremity	2.6 cm by 1.2 cm by 0.8 cm and 0.7 cm	Leg with a well fashioned foot. An angle indicates the beginning of the thigh. Underneath the foot is flattened, slightly in hollow. It is thinned and finishes as a point. A 2 mm-wide hole is pierced above the foot at the ankle. It does not perforate.	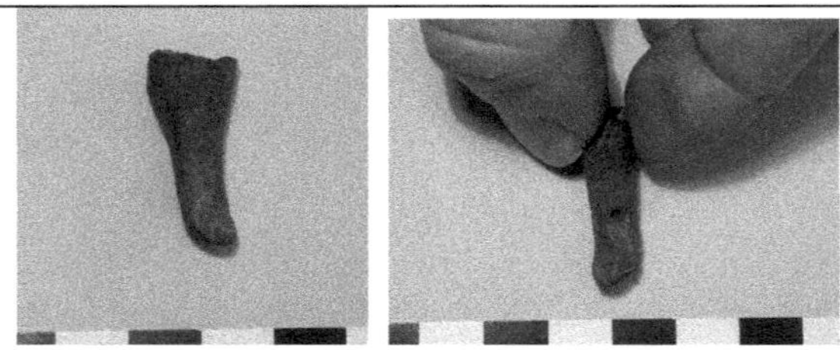
PI 71 (563060)	Body extremity	2.4 cm by 0.9 cm by 0.6	This piece resembles PI 70. It is a leg with a foot. There is a slight change in angle for the foot which is well rounded and widens towards the top.	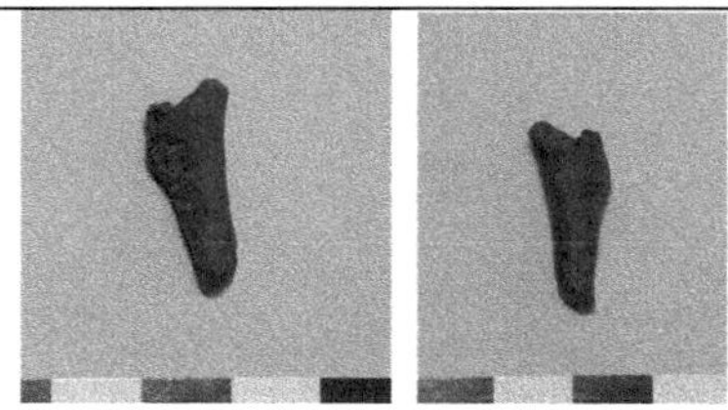
PI 72: (1954)	Possible animal body	4.3 cm by 3.2 cm by 1.1 cm	This is a large flattened piece with many traces of smoothing on it. Some were clearly made on a dry material. The beginning of a leg is possibly visible on one side, but it is much deformed. Some fingerprints are also visible on it.	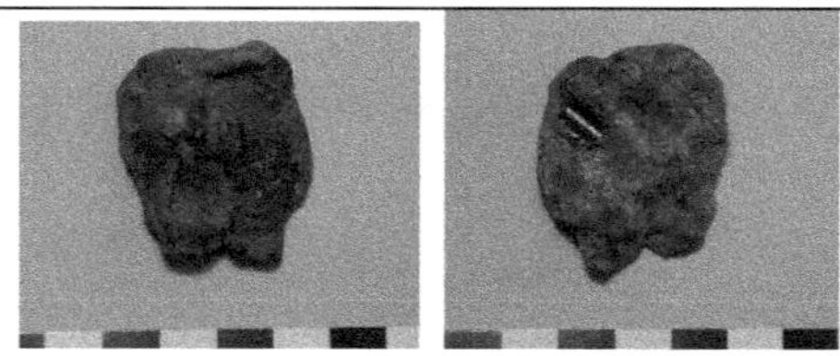
PI 73 (56..56)	Body extremity	1.6 cm by 0.5 cm by 0.2cm	This is a very thin black leg with a well marked foot forming an angle with the leg. The top part seems pinched on wet clay, it is not broken. The part where a body should have been attached is flattened.	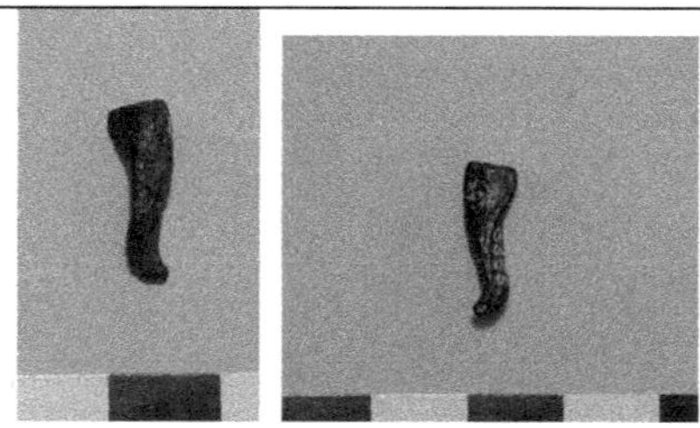
PI 74: (421856)	Body extremity	2.6 cm by 1.3 cm by 0.6 cm	This is a conical curved piece. It is light brown, short and quite massive at the top.	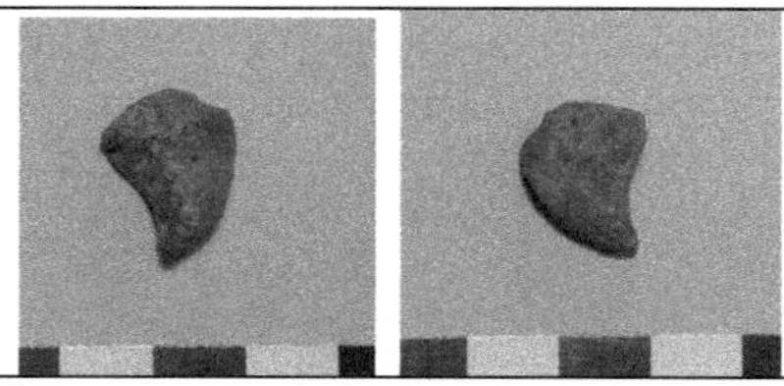

Pavlov I

PI 75: (101750)	Body extremity	1.7 cm by 1.5 cm by 0.7 cm	A very small ball-like protuberance sticks out of a smoothed and well rounded surface, broken on each side and at the back. Previous interpretations of this piece as a fragmentary breast, the small protuberance being seen as a nipple. This interpretation is far from obvious when looking at the piece.	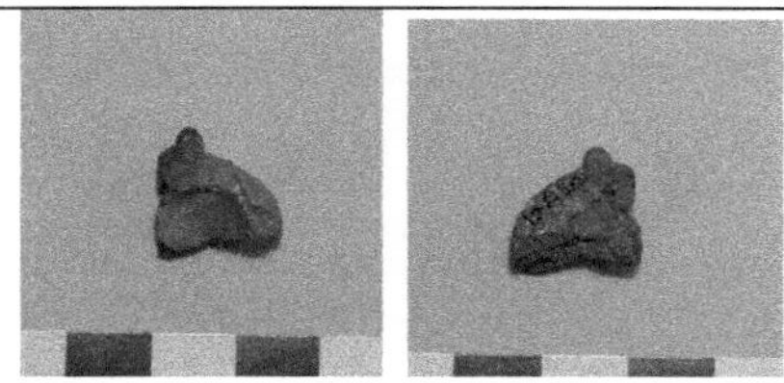
PI 76: (324360)	Body extremity	1.6 cm by 0.9 cm by 0.7 cm	This piece is the end of a limb, black, with a rounded extremity. It is quite massive and broken at one end. A series of small parallel lines are visible on the length of the piece: it is possibly an impression.	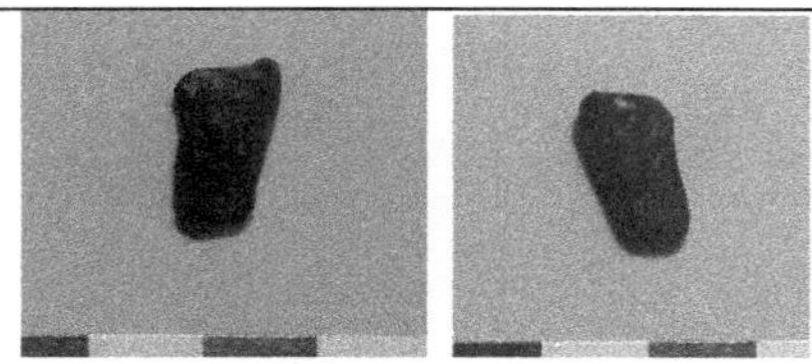
PI 77: (324860)	Body extremity	1.5 cm by 1cm by 0.6 cm	This piece is dark grey. This is the extremity of a short leg. It is rounded and widens at the top, broken on one side and slightly flattened.	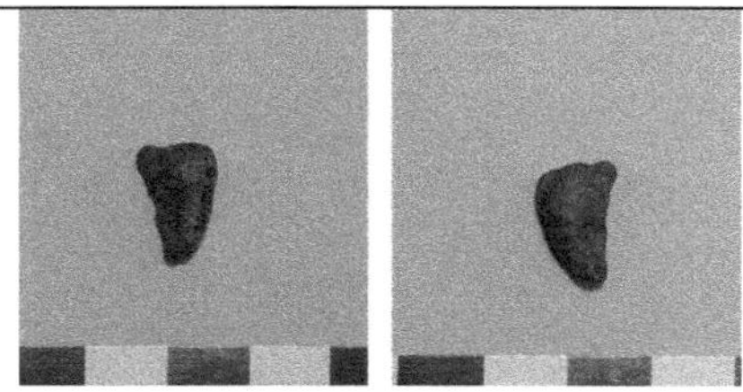
PI 78 (588257)	Body extremity	1.2 cm by 0.4 cm by 0.3 cm	This is a tiny extremity, very fine, bent at the foot and rounded at the end. It is slightly flattened and the break at the top is very neat.	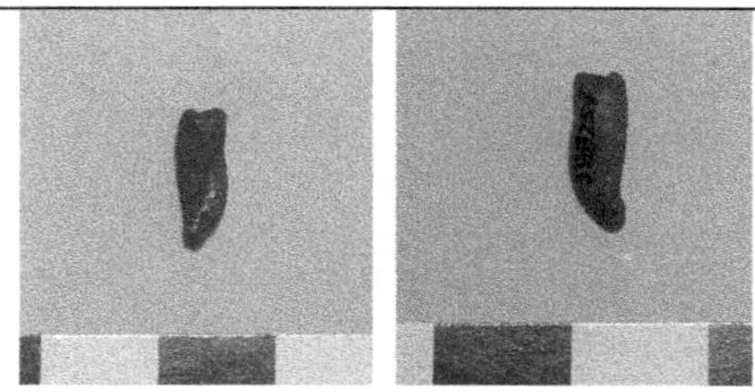
PI 79: (587956)	Body extremity	1.1 cm by 0.4 cm by 0.3 cm	This piece is a small cylinder with straight sides. It is slightly bevelled at the end of the foot.	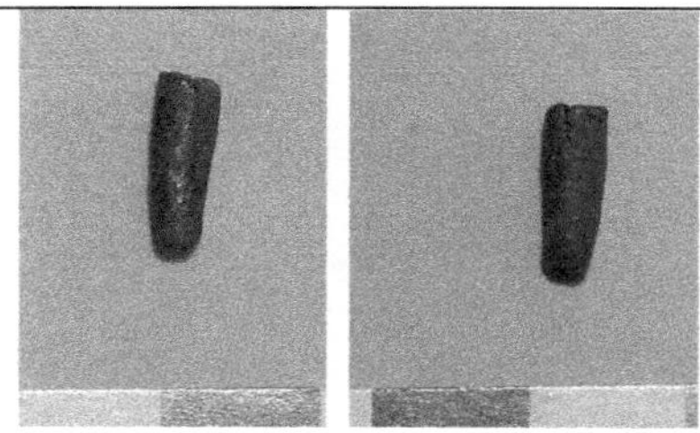

Pavlov I

PI 80: (5878)	Body extremity	1.4 cm by 0.5 cm by 0.5 cm	This is a cylindrical piece, curved towards the end and slightly flattened. It widens at the other. The thin extremity is bevelled. The foot is not indicated. The top is broken. It is brown in colour.	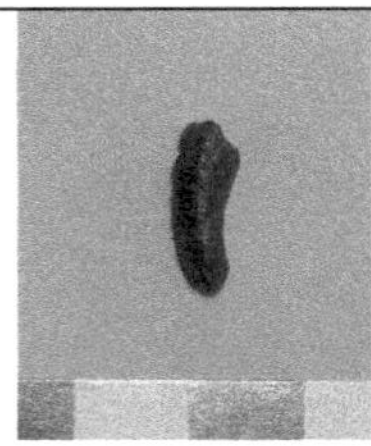 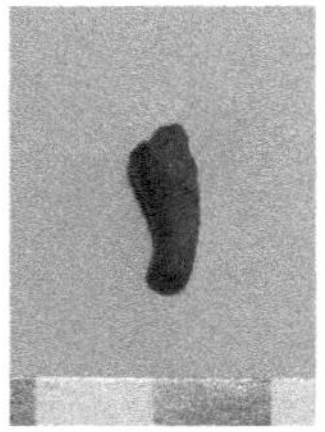
PI 81 (51964)	Body extremity	2.2 cm by 1 cm by 0.8 cm	This is a thick leg with the foot slightly flattened with a hollow underneath. It has a flattened cylinder shape, with an angle at the front of the foot. It is greyish.	 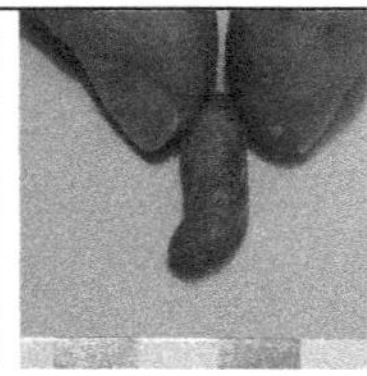
PI 82: (721056)	Body extremity	2.1 cm by 1.4 cm by 1 cm	This is a leg extremity, flattened underneath the foot. It widens well at the top. The foot is small and curved.	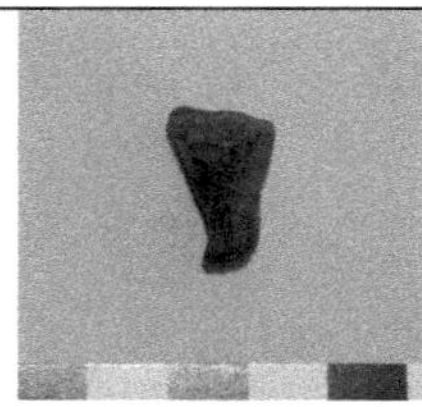 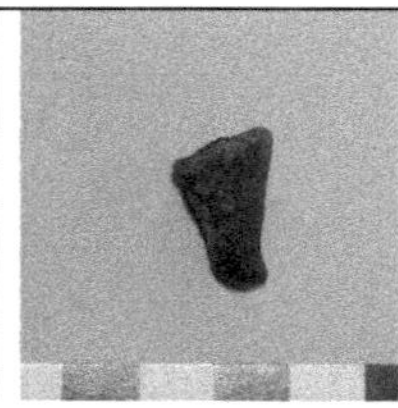
PI 83: (317361)	Body extremity	2.6 cm by 2 cm by 1.2 cm	This is a leg segment, much broken. Some straight smooth surfaces remain. They are not parallel but converging.	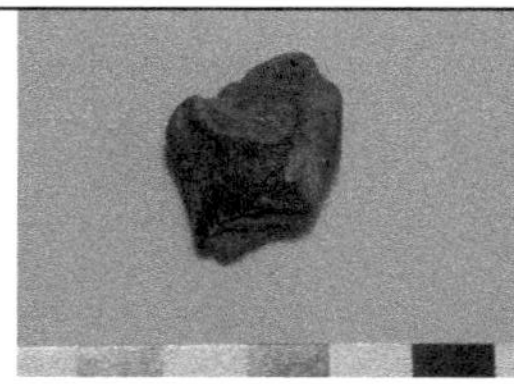 
PI 84: (733356)	Body extremity	1 cm by 1.2 cm	This is a whole leg and the beginning of a body. The end is bevelled. The beginning of the body shows a completely flat and smoothed surface where it should have been attached to a wider body, as if it was actually cut off it with a tool.	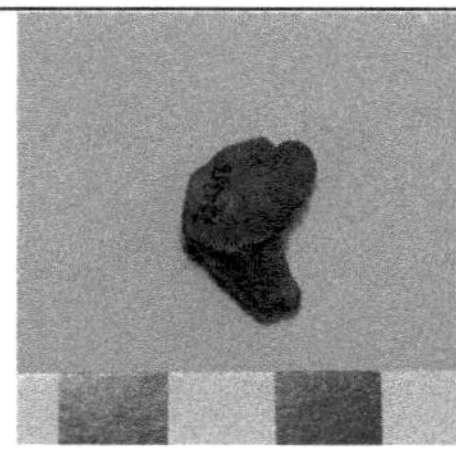 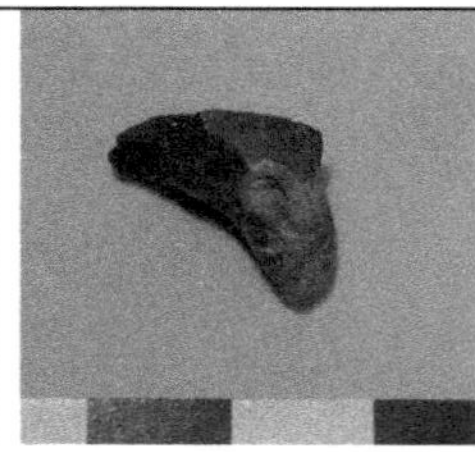

Pavlov I

PI 85: (……60)	Body extremity	2.2 cm by 1 cm by 0.6 cm	This is a leg with foot. It widens at the foot. It is broken on the side on both edges and the remaining sides are parallel and straight. The foot underneath seems flattened.	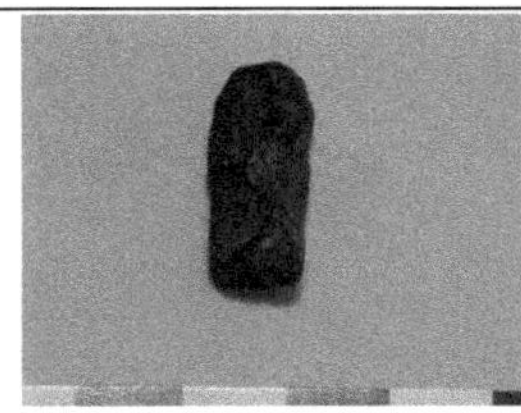 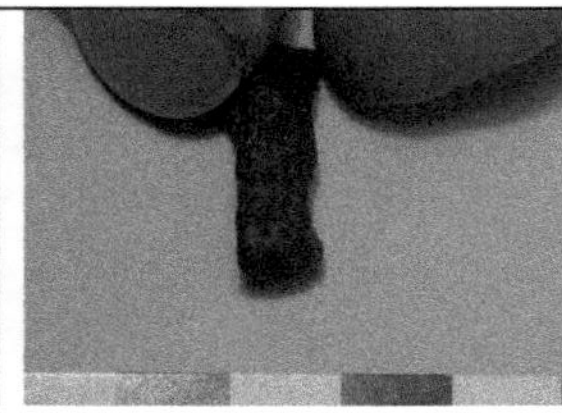
PI 86: (104261?)	Body extremity	2.5 cm by 1.3 cm by 1.2 cm	This is a segment of a cylindrical limb, slightly conical, with straight edges. It is broken at both ends.	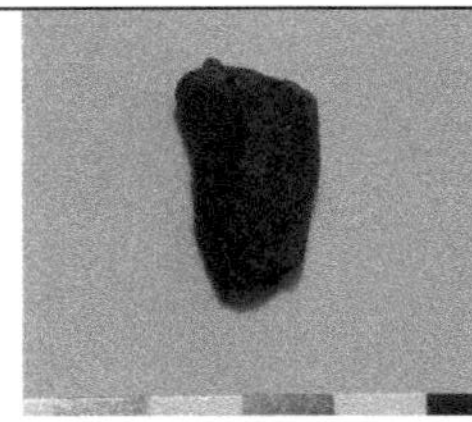 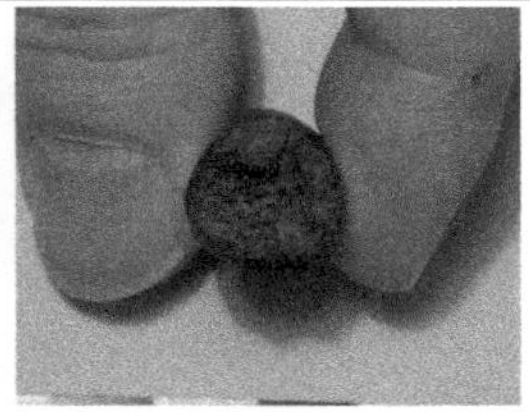
PI 87:	Body extremity?	0.8 cm by 1 cm by 0.7 cm	The base is very large, roughly circular. The rest of the piece is conical with a very thin end, slightly flattened at the extremity. Traces of nails are visible on it.	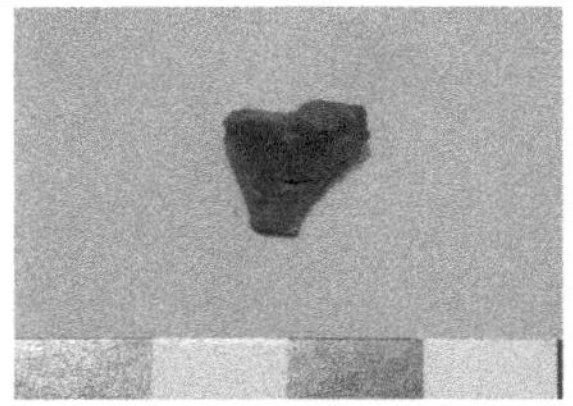 
PI 88 : (221460)	Body extremity	1.5 cm by 0.8 cm by 0.7 cm	This is a segment of limb. The section at one end is oval. The other end is broken and rather shapeless.	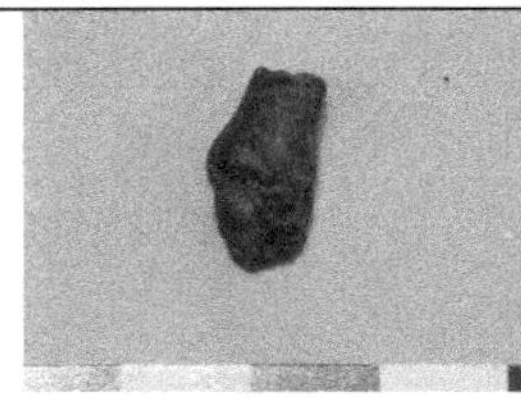 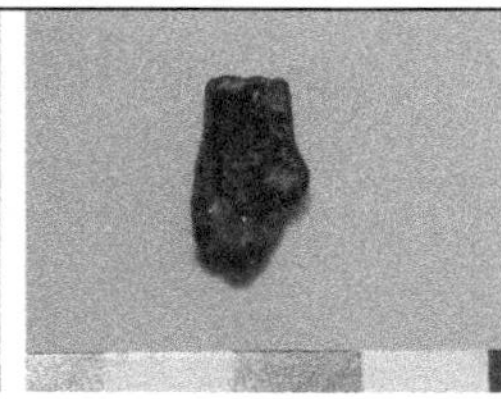
PI 89: (588057)	Body extremity	1.7 cm by 0.5 cm by 0.5 cm	This is a very thin leg. The angle of the knee is marked. The extremity is rounded. The foot is not marked. The break is quite large at the top.	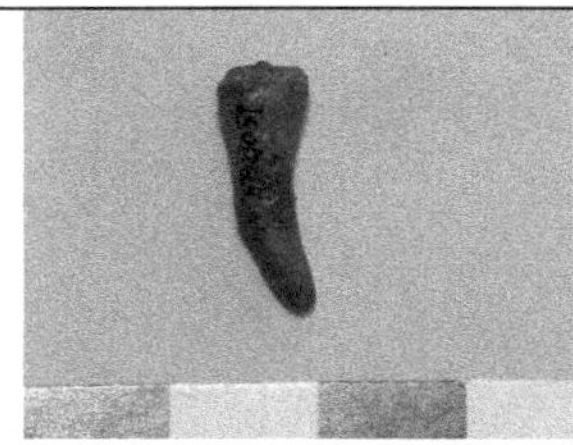 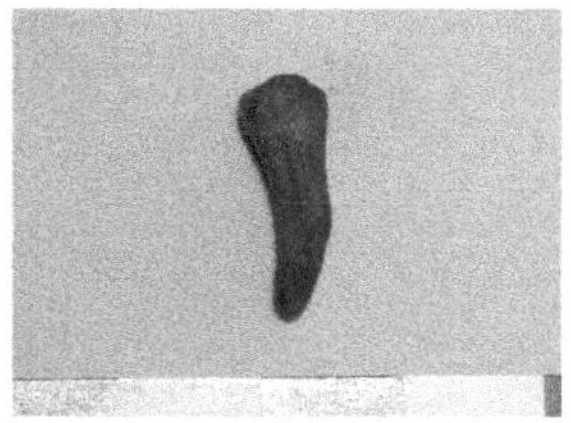

Pavlov I

PI 90 (160663)	Body extremity	2.3 cm by 0.7 cm by 0.4 cm	Fragment of a cylindrical limb broken at both ends and on the length of it. Clear traces of scraping mark each surface, making angles between these surfaces.	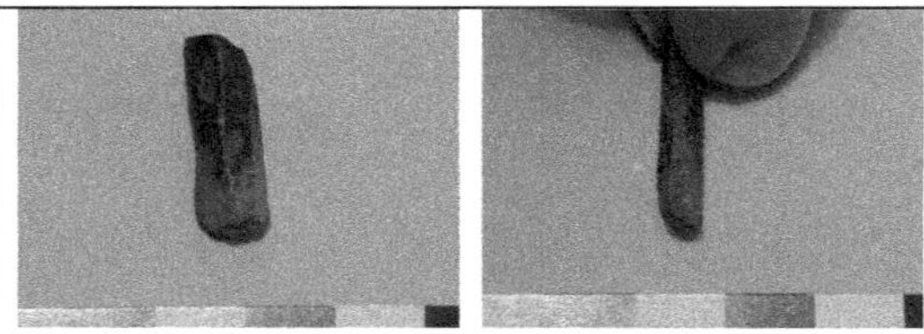
PI 91: (723455)	Body extremity	2.2 cm by 1 cm by 0.8 cm	Cylindrical segment, flattened and deformed at one end. Grey in colour.	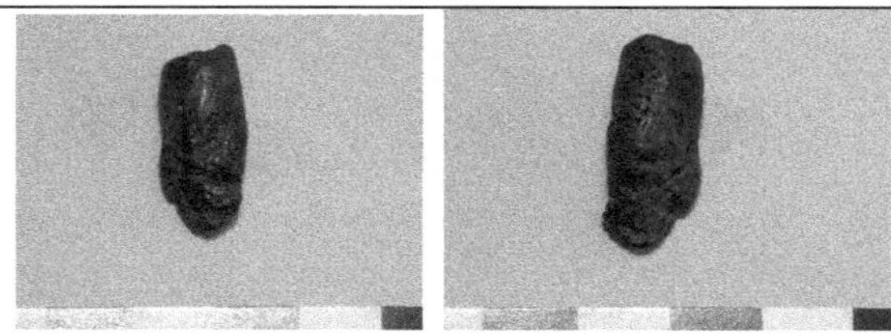
PI 92: (159863)	Body extremity	1.8 cm by 0.7 cm by 0.7 cm	This is a leg with a rounded foot. A flat area marks the foot that forms a cylinder. It widens slightly for the top. There is a marked angle for foot.	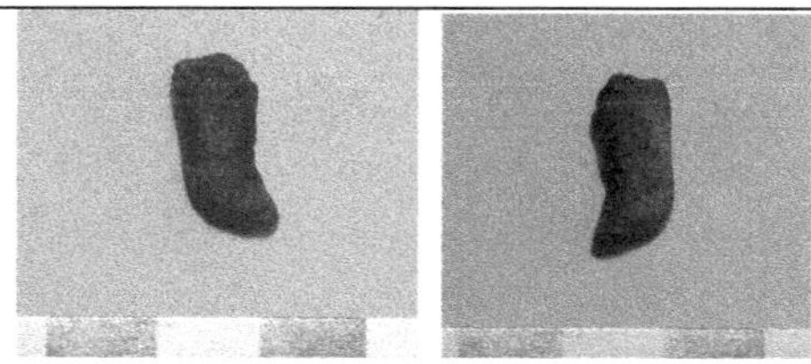
PI 93: (701956)	Body extremity	1.6 cm by 1 cm and 0.7 cm	This is a leg with foot of the mammoth type. It forms a slightly flattened cylinder. The foot is marked as a ball. It is flat and circular underneath.	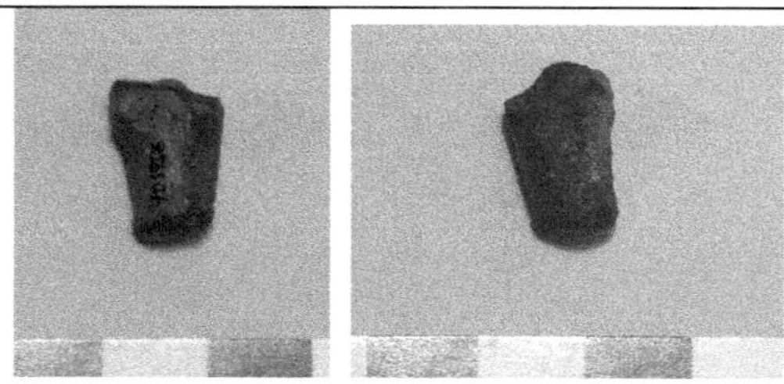
PI 94: (368364)	Body extremity	2.5 cm by 1.3 cm by 1 cm	This piece is a segment of a cylinder, slightly flattened and rounded at one end. It is black with parallel edges.	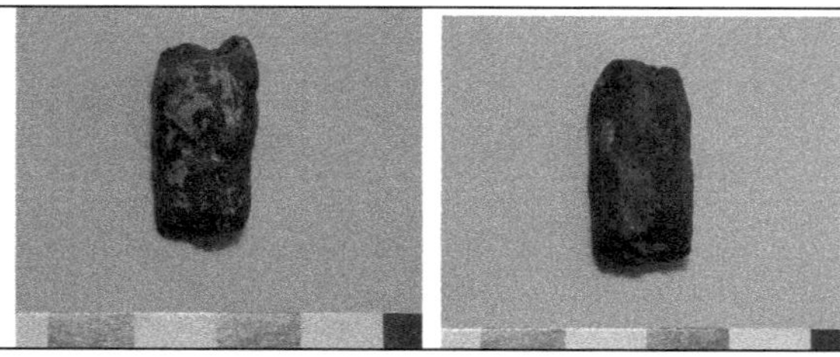

Pavlov I

PI 95: (159963)	Body extremity	1 cm by 0.6 cm by 0.5 cm	This is the end of a leg of the mammoth type. The foot is marked by a slight swelling; the sides of the cylinder are straight and parallel; the foot underneath is flat and circular.	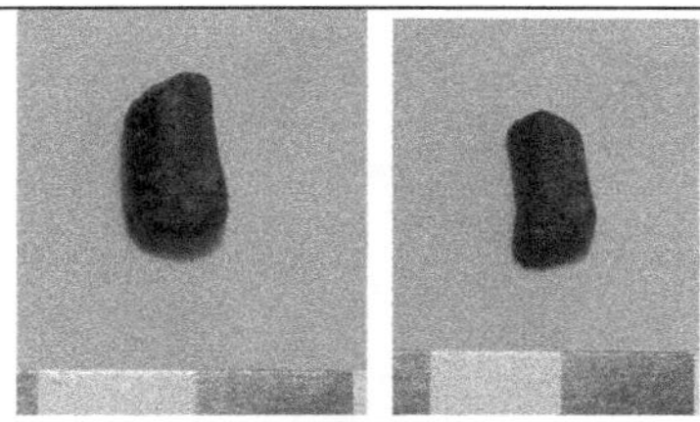
PI 96: (221060)	Body extremity	1.5 cm by 1 cm by 1 cm	This is a quite massive flattened piece. It is curved at the foot and circular at top.	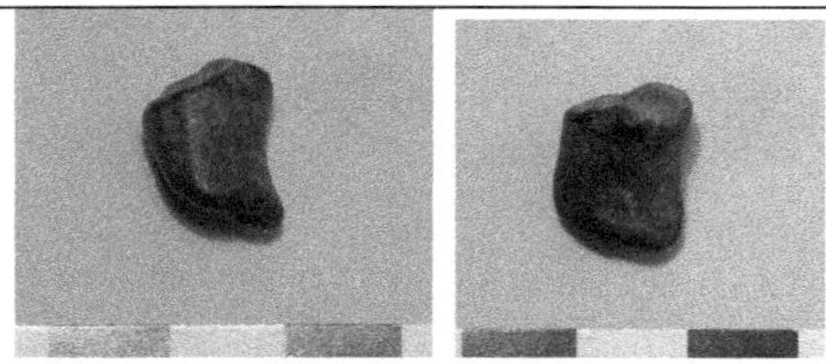
PI 97: (427261)	Body extremity	2.2 cm by 1.4 cm by 1 cm	This is a cylindrical segment that widens at the top. Both ends are broken, as well as one side.	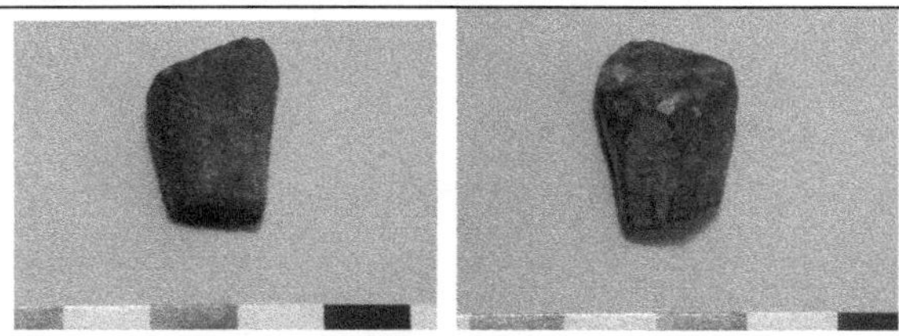
PI 98: (721756)	Body extremity	2.3 cm by 1 cm by 0.9 cm	This is a leg with a bevelled foot forming an angle with the leg. It is broken at the heel and at the calf. The top of the leg is quite wide. There is a break on one side. It is black.	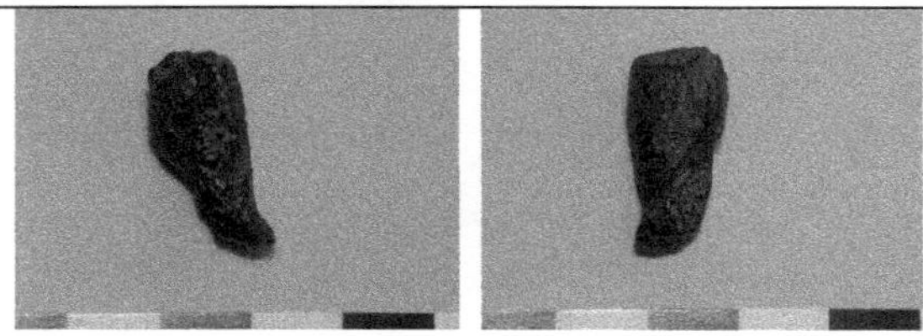
PI 99:	Body extremity	2 cm by 1.2 cm by 0.9 cm	This piece is a fragment of a cylindrical limb that widens at one end.	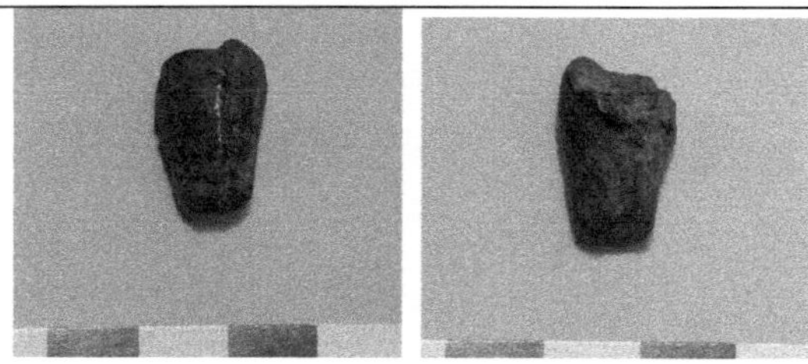

Pavlov I

PI 100: (427161)	Body extremity	2.7 cm by 1.2 cm by 0.7 cm	This is a cylinder broken on the length (half remains) and at both ends. Marked traces of scraping or fashioning with a tool are visible on it. The break at top is rather flat.	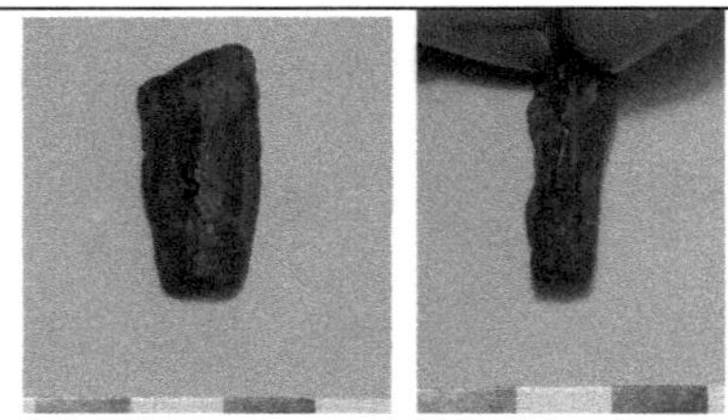
PI 101: (650356)	Body extremity	1.2 cm by 0.6 cm by 0.6 cm	This is a whole leg, very wide at one end and flattened. The section is circular, slightly flattened. It is very small.	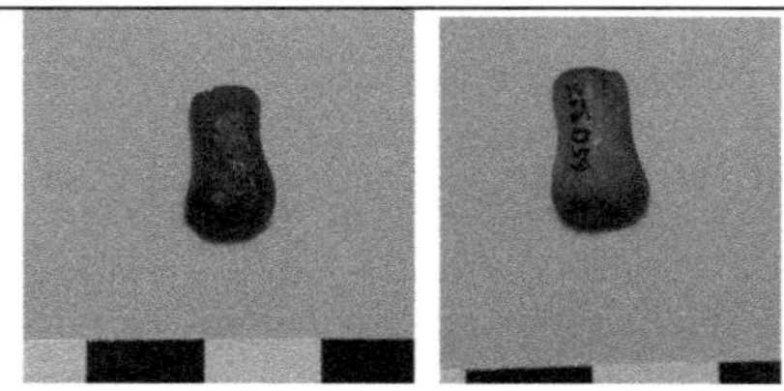
PI 102: (723156)	Body extremity	1 cm by 1cm by 0.7 cm	This is a worked extremity, slightly rounded with small relief, then broken.	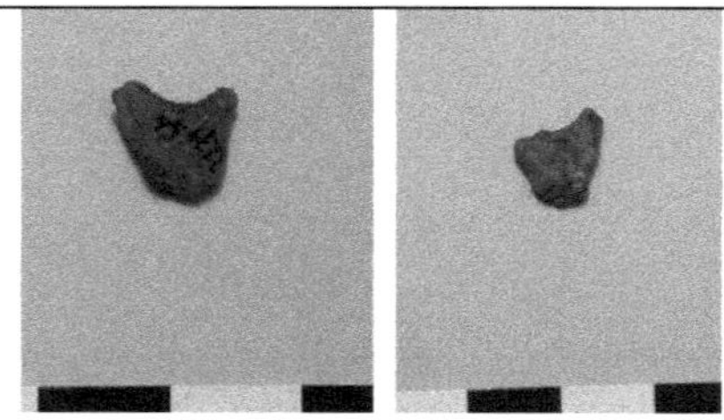
PI 103: (62956)	Body extremity	1.4 cm by 0.8 cm by 0.8 cm	This is a rounded extremity, pointy, that forms an angle with the main part of the piece. The section at the top is triangular.	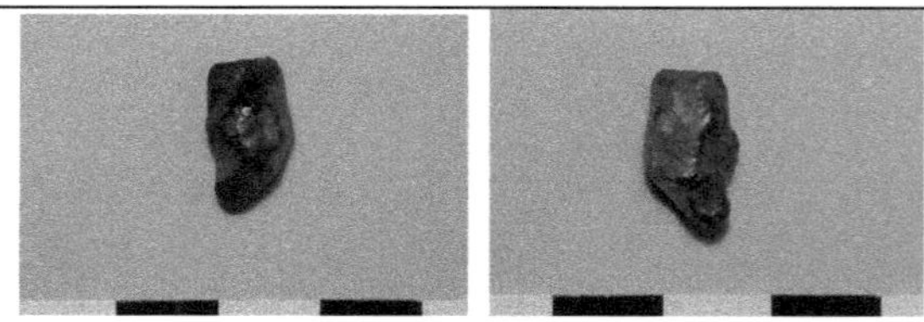
PI 104: (63460)	Body extremity	1.4 cm by 0.8 cm by 0.8 cm	This is a leg of the mammoth type, cylindrical section. The foot is flattened underneath and rounded. It widens at the foot, with straight edges.	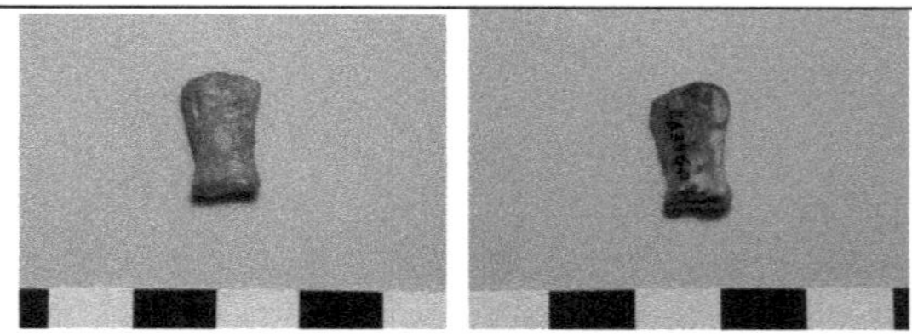

Pavlov I

PI 105: (…8?)	Body extremity	1.2 cm by 0.7 cm by 0.6 cm	This is a slightly curved fragment, short, black, broken.	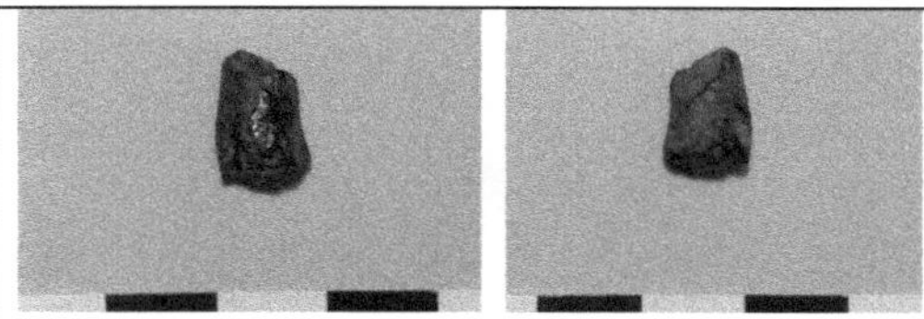
PI 106: (210952)	Body extremity	2.3 cm by 0.9 cm by 0.7 cm	This is a cylindrical piece, slightly flattened, rounded at one end, with no foot marked.	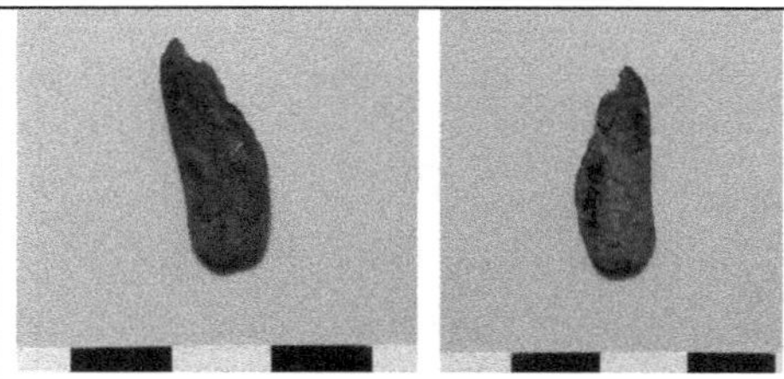
PI 107: (722356)	Body extremity	2 cm by 0.8 cm by 0.8 cm	This is a cylinder, flattened at one end, slightly curved, with straight edges.	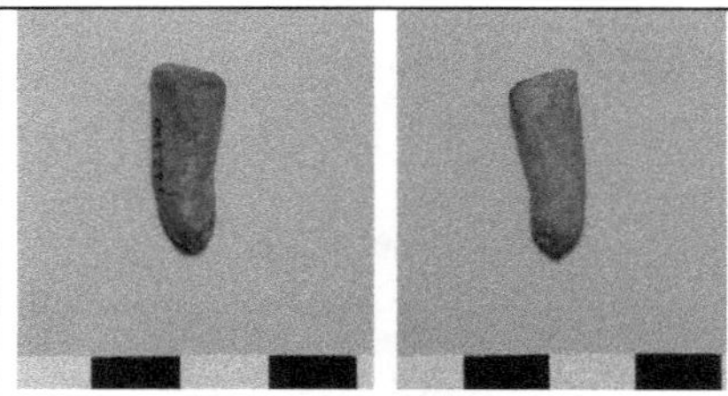
PI 108: (428161)	Body extremity	1.9 cm by 0.8 cm by 0.5 cm	This is the segment of a limb, flattened, black; the two ends are damaged with the beginning of a bevelled foot.	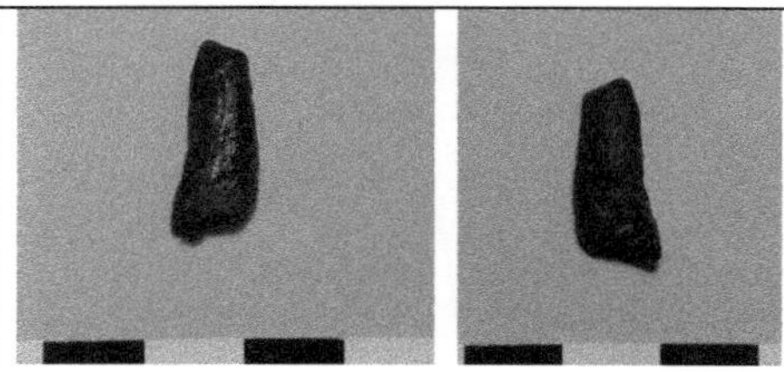
PI 109: (316961)	Body extremity	1.5 cm by 1cm by 0.6 cm	This piece is a large foot fragment, broken on the side. The ball forming the foot is partly remaining on one side.	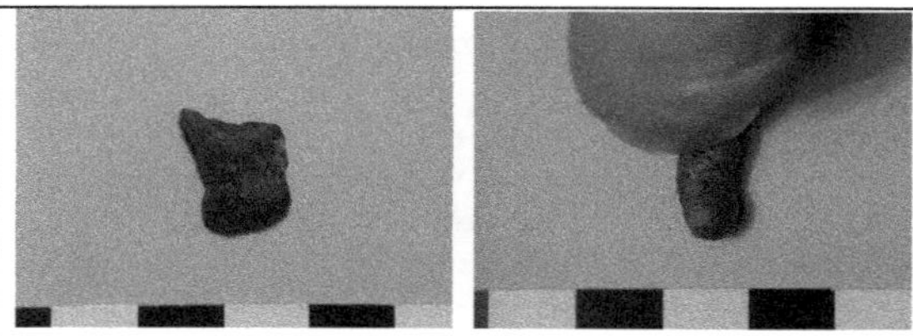

Pavlov I

PI 110: (577456)	Body extremity	1.5 cm by 1 cm by 0.7 cm	The cylindrical flat extremity seems to have been cut off. It widens at the top.	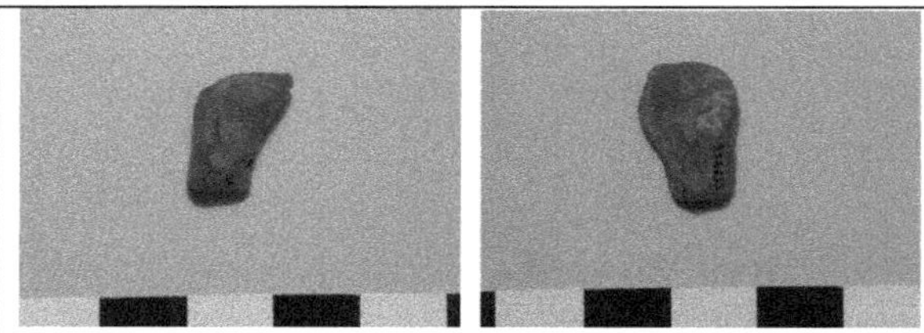
PI 111 : (574360)	Body extremity	1.5 cm by 0.7 cm by 0.8 cm	This extremity has a marked and elongated foot; the heel is marked, the end is rounded. It looks like a boot.	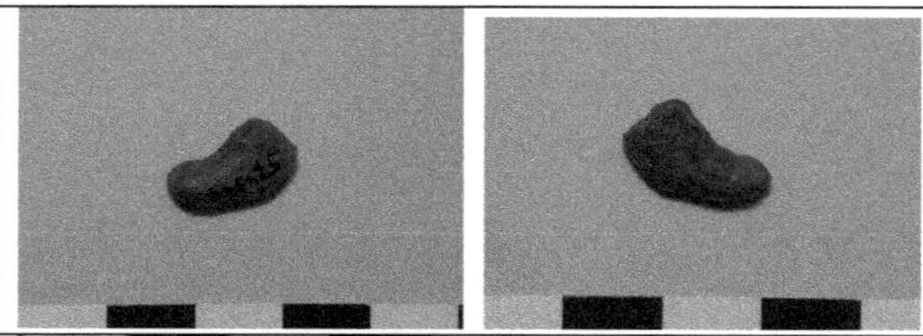
PI 112: (587757)	Body extremity	1.6 cm by 1 cm by 0.6 cm	This piece is a rounded extremity. It is clearly cut off in bevel shape at the top where it should have been connected to a body. The piece is not broken. The surface of the object is decorated by a series of small impressions, 2 to 3 mm long, possibly to indicate the fur. This is very unusual among the ceramic pieces.	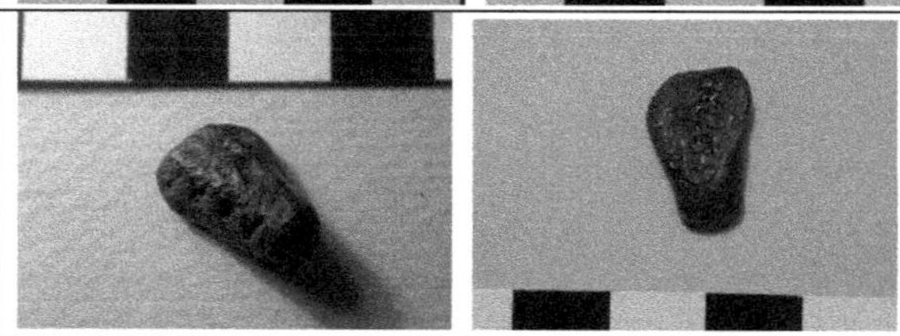
PI 113: (723256)	Body extremity	1.9 cm by 1.1 cm by 1cm	This is a possible limb fragment, curved, much damaged at both ends and at on the length.	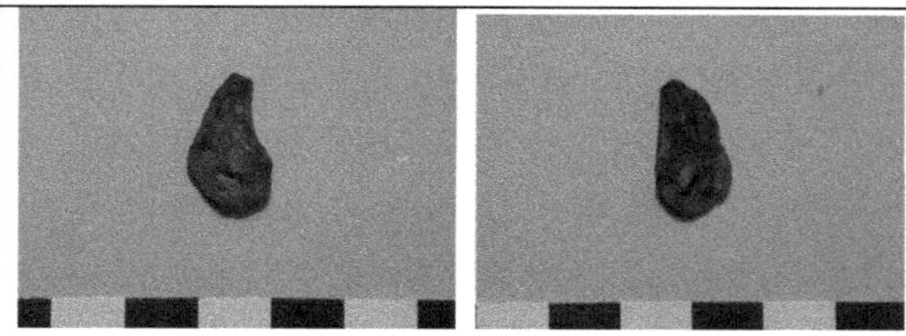
PI 114: (588357)	Fragmentary animal body	1.4 cm by 1 cm by 0.6 cm	This is a small conical thin extremity attached to a wider piece, possibly the rear part of an animal.	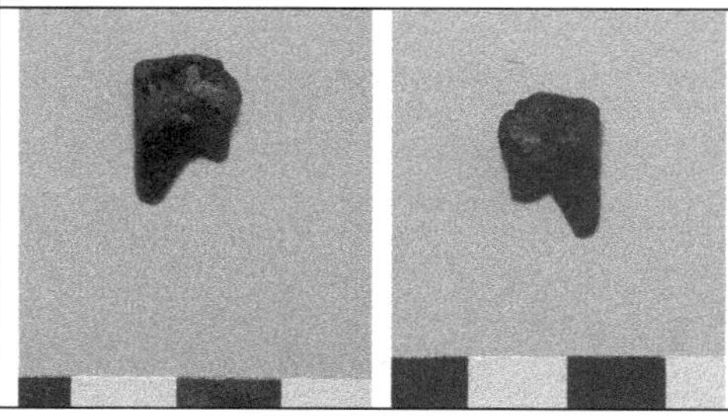

Pavlov I

PI 115: (215856)	fragmentary anthropo-morph	2.2 cm by 1 cm by 0.6 cm	This object represents a fragmentary anthropomorph; only the torso with the beginning of the legs is recognisable. One leg is a thin pointy cone and seems whole; only the beginning of the second leg is visible.	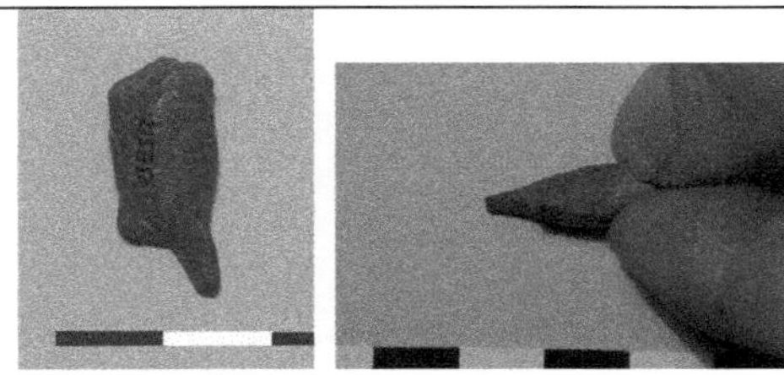
PI 116: (64856)	Body extremity	2.4 cm by 1.2 cm by 1 cm	This piece is a segment of a circular object broken on the side and at both ends; it has been restored. The print of a small stick is visible towards the middle.	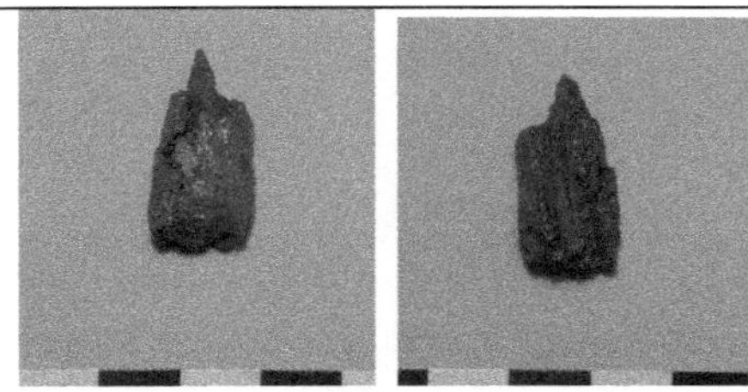
PI 117: (722656)	Body extremity	2 cm by 1.3 cm by 0.8 cm	This object is much broken. It is possibly a fragment from a cylindrical piece.	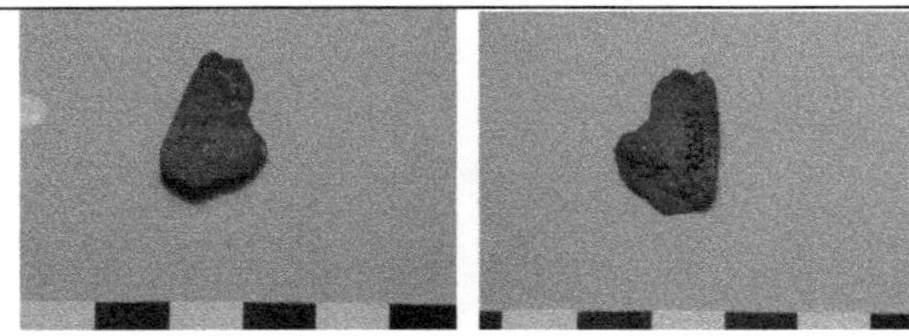
PI 118: (353656)	Body extremity	2.2 cm by 1 cm by 0.7 cm	This massive leg widens at the top and as well for the foot. Traces of scraping are visible. It is flattened underneath.	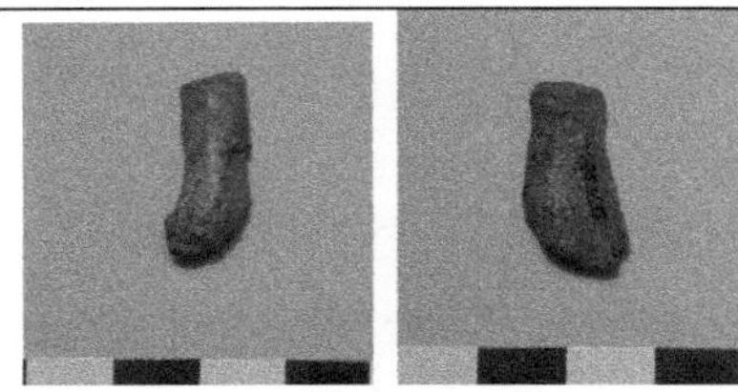
PI 119: (581456)	Body extremity	1.4 cm by 0.6 cm by 0.6 cm	This piece is the rounded extremity of a limb. The foot is not marked and one side is slightly flattened. It widens towards the top, then it is broken.	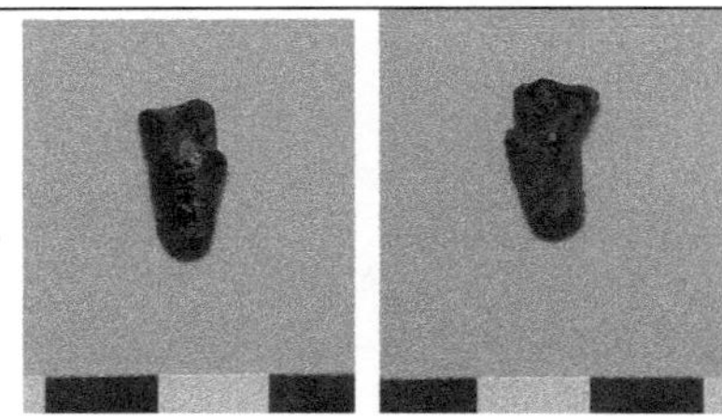

Pavlov I

PI 120: (722956)	Body extremity	1.7 cm by 0.7 cm by 0.6 cm	This is a bevelled foot with the leg and the beginning of the thigh. It is cylindrical. It widens at the top then is broken.	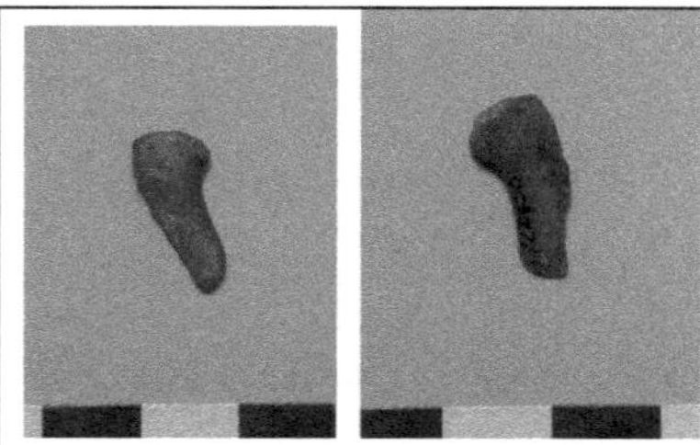
PI 121: (6934?)	Body extremity	1.5 cm by 0.8 cm by 0.7 cm	This is a leg with a bevelled foot, circular in section. Underneath the foot is flattened then broken. It widens at the top.	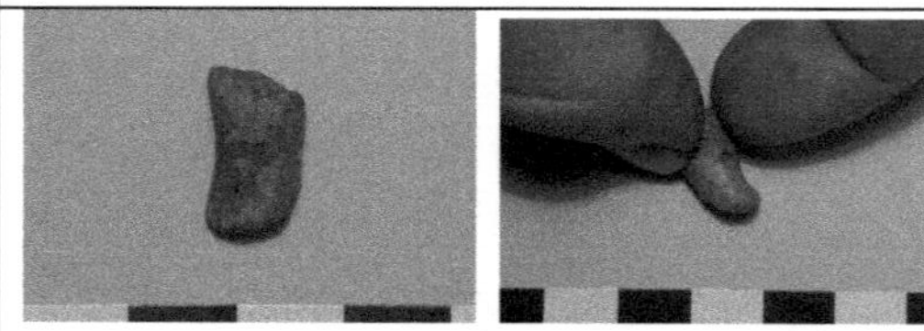
PI 122: (583656)	Body extremity	0.9 cm by 0.6 cm and 0.3 cm	This is a leg with a bevelled foot. The small flattened cylinder widens at the top. It is curved at one extremity. The breakage surface is neat and flat.	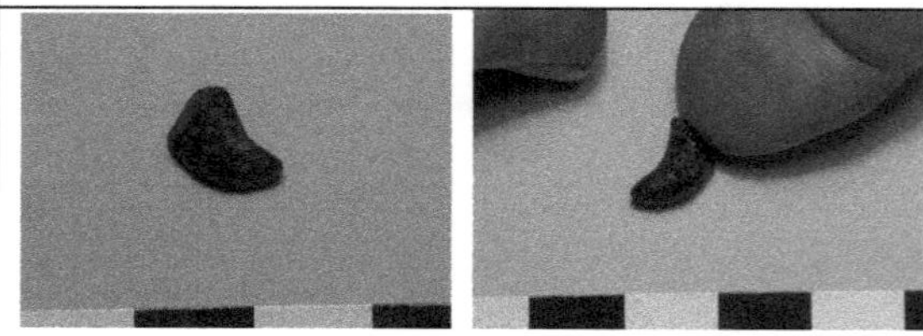
PI 123: (588157)	Body extremity	1.8 cm by 0.6 cm by 0.5 cm	This cylindrical limb widens at top. The extremity is very thin and slightly curved; the foot is not marked; the top does not seem broken: it is smooth with scraping traces.	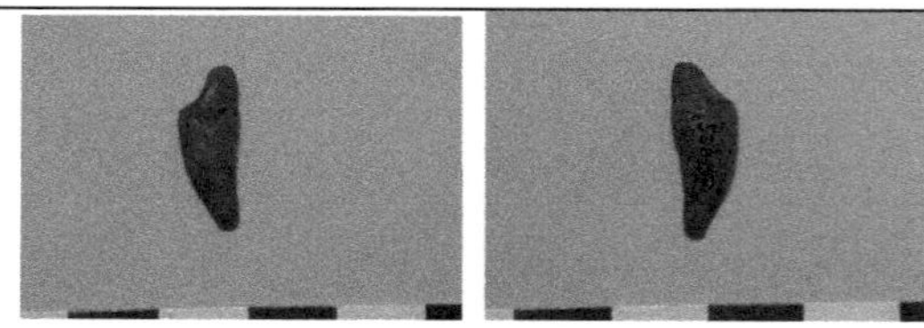
PI 124: (514256)	Body extremity	1.5 cm by 0.7 cm by 0.3 cm	This piece is a leg with foot, shaped as a flattened cylinder with a rounded curved foot.	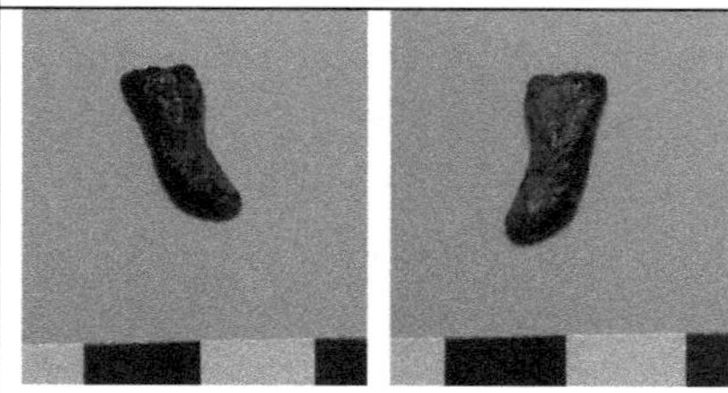

Pavlov I

PI 125: (722856)	Fragmentary animal body	1.7 cm by 1 cm by 0.6 cm	This object represents the hind side of an animal with a very thin back leg which appears whole. The body is flattened. The front part is broken.	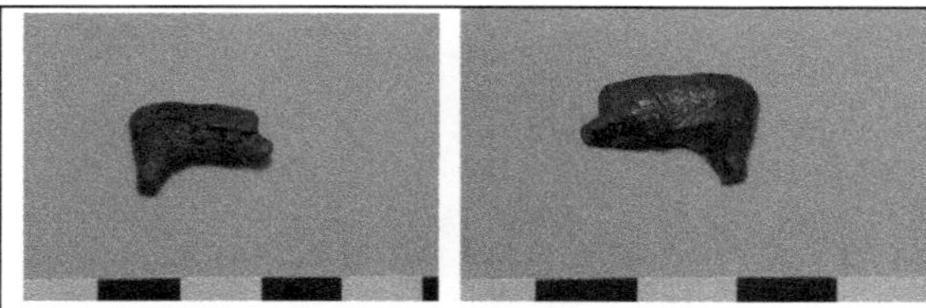
PI 126: (722756)	Body extremity	0.7 cm by 0.6 cm by 0.5 cm	This straight cylinder widens slightly toward one end. The foot is hardly marked. It is very small.	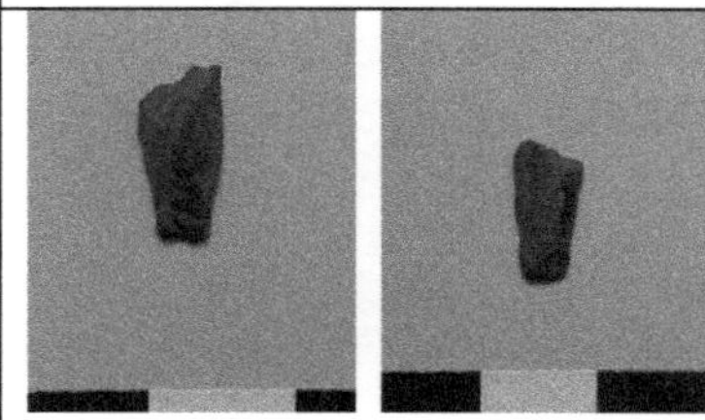
PI 127: (574160)	Body extremity	1.4 cm by 0.7 cm by 0.7 cm	This body extremity shows a bevelled foot, flattened for the sole, with a rounded extremity. It is rather massive with a marked heel and resembles a boot.	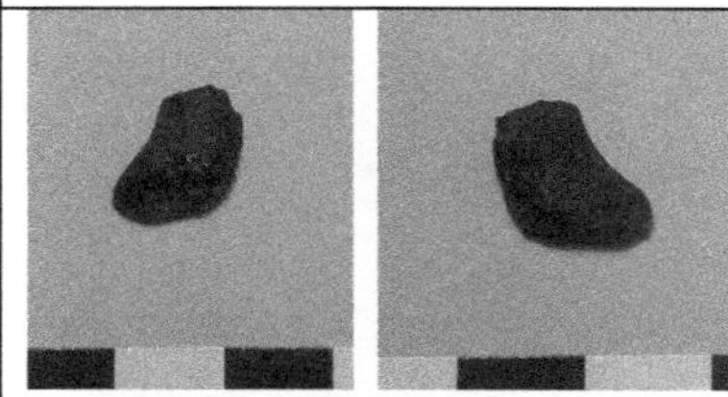
PI 128: (587457)	Body extremity	1.4 cm by 1 cm by 0.7 cm	This object is very similar to PÏ 127. The foot is a large rounded bevel and it is boot-like.	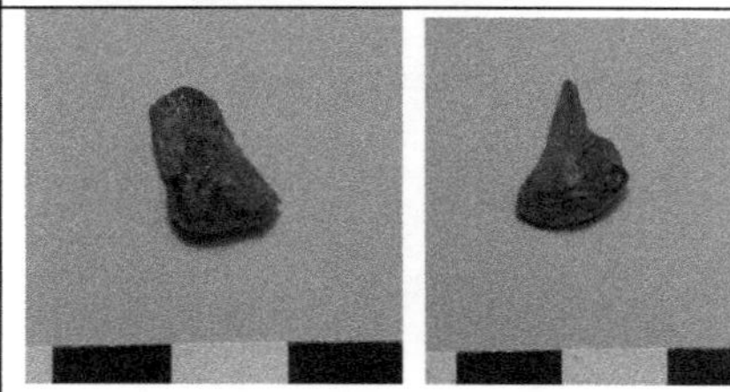
PI 129: (722556)	Body extremity	1.6 cm by 1 cm by 0.7 cm	This is a limb with the beginning of the thigh cut off in a bevelled and flattened section. The end of the foot is flat.	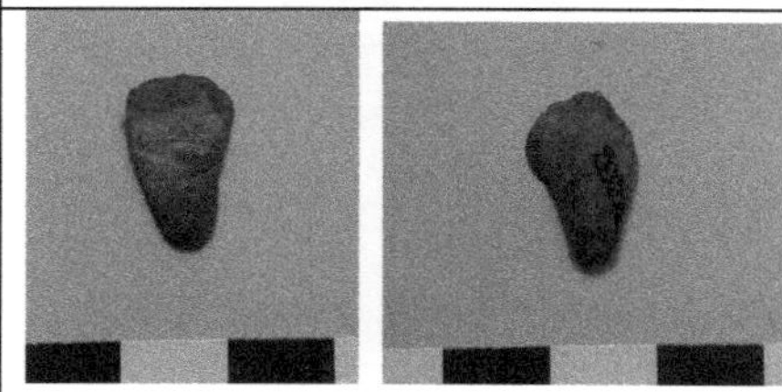

Pavlov I

PI 130: (427561)	Body extremity	1.1 cm by 0.7 cm by 0.6 cm	This piece is a fragment of a cylindrical flattened limb. It starts to widens for the foot then it is broken off underneath. It reminds of a boot.	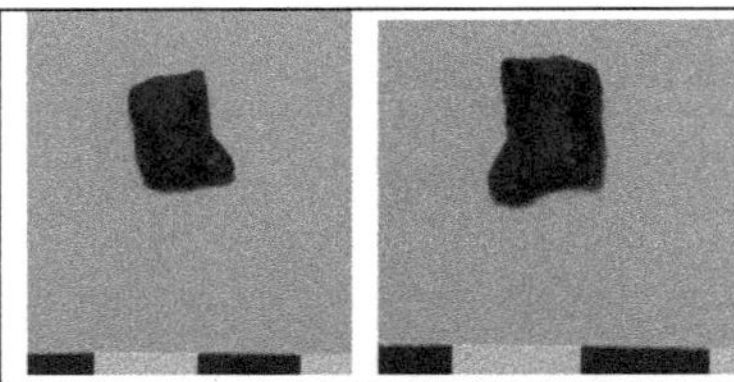
PI 131: (16239?)	Body extremity	1.3 cm by 1 cm by 1 cm	This cylindrical extremity is of the mammoth type. It widens at the top with a circular flattened foot.	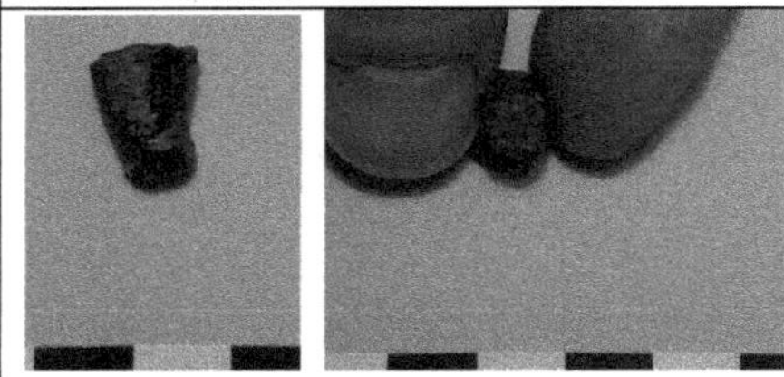
PI 132: (396356)	Body extremity	1.2 cm by 1 cm by 0.8 cm	This piece is a cylindrical extremity of the mammoth type. It widens at the top with a circular flattened foot.	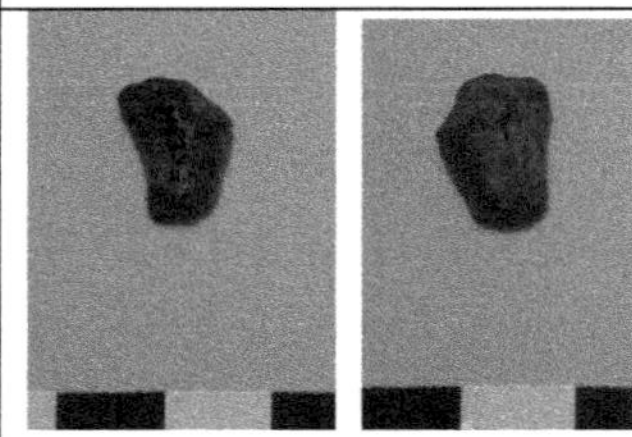
PI 133: (661960)	Body extremity	1.8 cm by 1.1 cm by 1 cm	This object is a fragment of cylindrical limb. It widens slightly at one end as if to mark a foot, but it is damaged with rough surfaces.	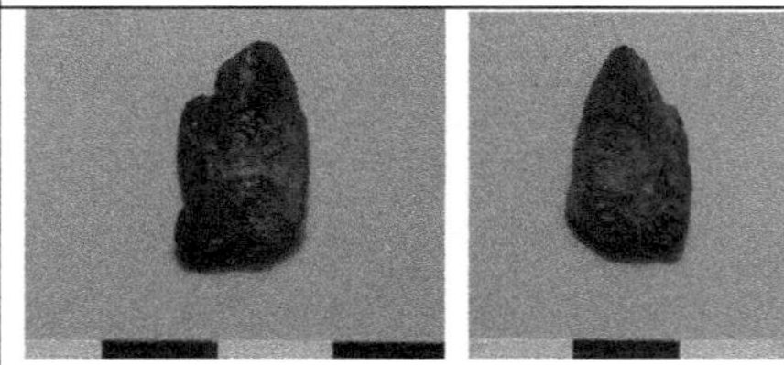
PI 134: (586857)	Body extremity	1.9 cm by 0.8 cm by 0.6 cm	Body extremity with a well-marked foot, flattened with a slight bevel. The piece is slightly flattened. It is circular underneath the foot which widens at the front.	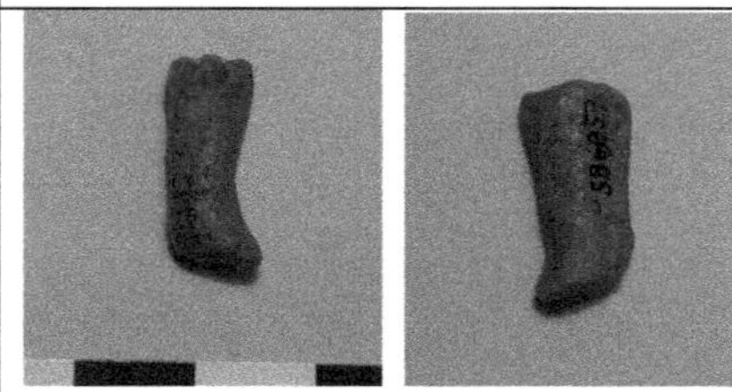

Pavlov I

PI 135: (319261)	Body extremity	1.7 cm by 1 cm by 0.9 cm	This piece is a wide foot with a rim, oval and flat underneath. It is broken at one end.	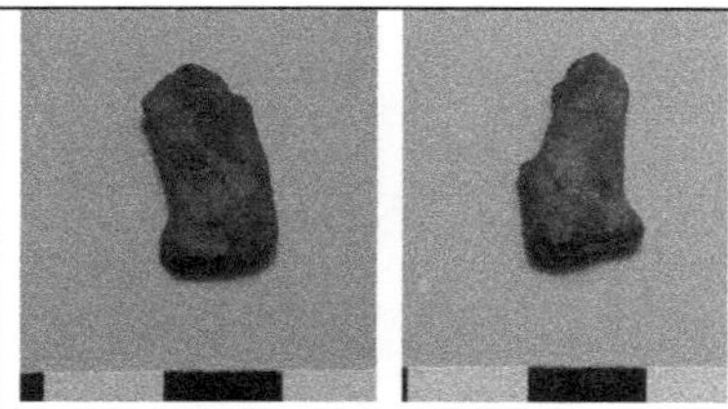
PI 136: (722456)	Body extremity	1.7 cm by 0.9 cm and 0.7 cm	This cylinder has straight edges and shows scraping traces. The foot is rounded. The surfaces are damaged.	
PI 137: (519958)	Body extremity	2 cm by 1.2 cm by 1.2 cm	This body extremity is a pointed cone with a rounded end, possibly a leg with the beginning of the body. It widens a lot at the top.	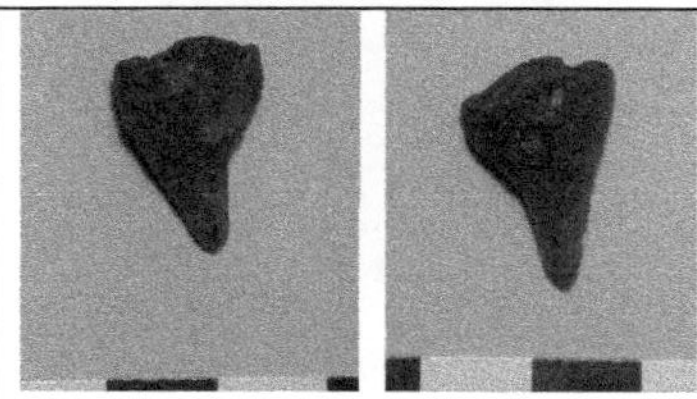
PI 138: (642456)	Body extremity	1.6 cm by 1.3 cm by 0.8 cm	This piece is very similar to PI 137. One extremity forms a thin pointed cone, slightly broken; the other extremity is much wider, possibly being part of a body to which it was attached before the break.	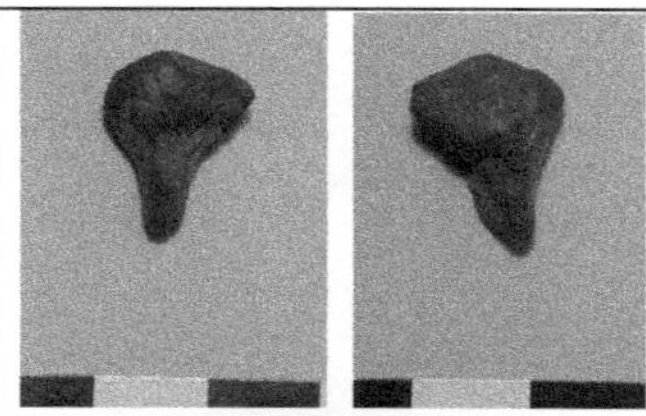
PI 139: (722256)	Body extremity	1.2 cm by 0.7 cm by 0.6 cm	This body extremity if a leg with a well marked rounded foot. It is cylindrical with straight edges. The foot is rather flat underneath, with a slight bevel to join the back of the leg.	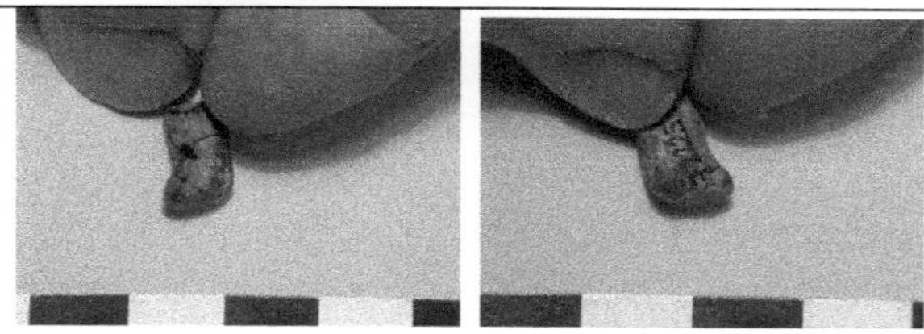

Pavlov I

PI 140:	Body extremity	2.2 cm by 0.7 cm by 0.6 cm	This leg has a bevelled foot at the end. It is black in colour and broken on some edges.	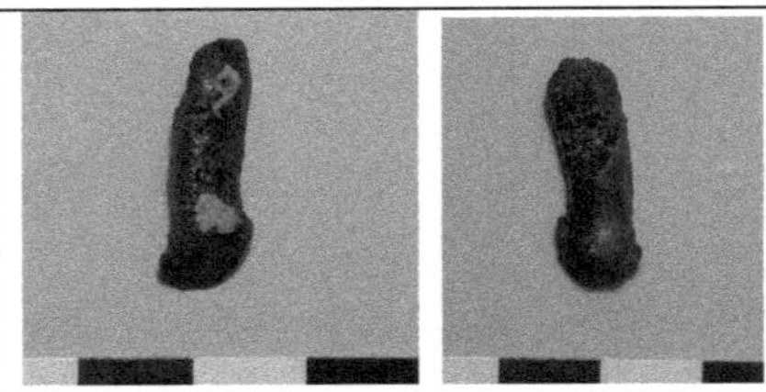
PI 141: (722156)	Body extremity	1.4 cm by 0.7 cm by 0.6 cm	This rounded extremity has straight edges and is slightly conical.	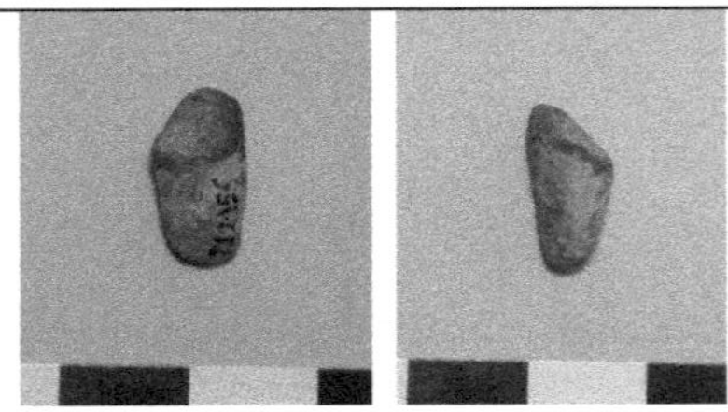
PI 142: (371156)	Body extremity	1 cm by 0.8 cm by 0.6cm	This piece is a rounded extremity. It was attached to a larger base then broke off.	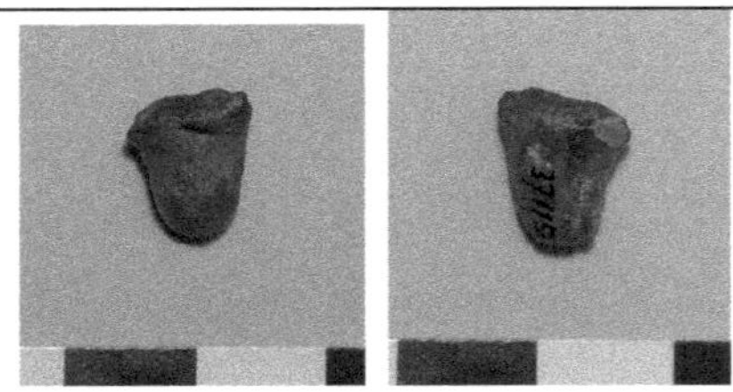
PI 143: (324460)	Body extremity	2.1 by 0.7 cm by 0.5 cm	This piece is a very fine and long extremity. It is curved. It is damaged on one side and there seem to be a foot at the end. It is black in colour.	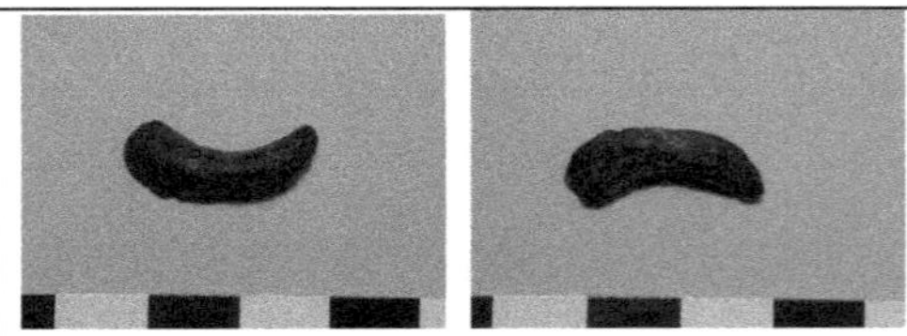
PI 144: (347556)	Fragmentary animal	1.8 cm by 1 cm by 0.5 cm	This object appears to be a flattened animal body with a stylised back leg and the front part broken.	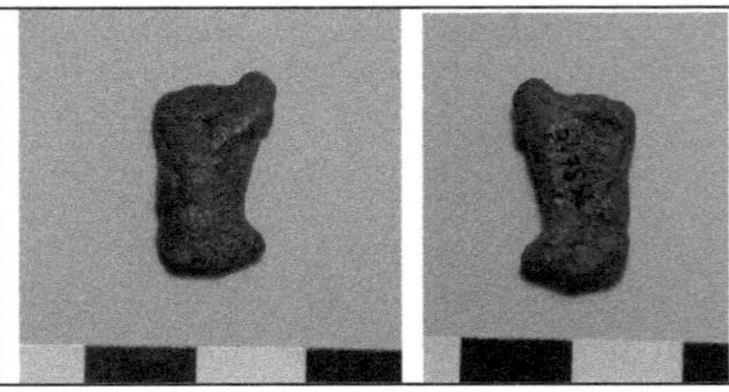

Pavlov I

PI 145: (159860)	Body extremity	1.6 cm by 1.5 cm by 1 cm	This body extremity resembles a boot. It is slightly flattened, quite massive and continued by a well-marked thin rim. It is flat underneath. Scraping traces are visible on one side.	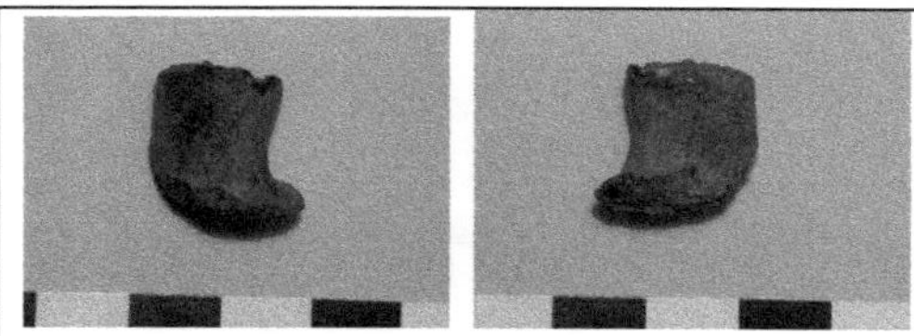
PI 146: (721650)	Body extremity	1.7 cm by 1 cm by 0.6 cm	This leg is fashioned with a foot at one end. The general shape is that of a flattened cone that widens towards the top. The foot is well-marked, bevel-like. The piece is black.	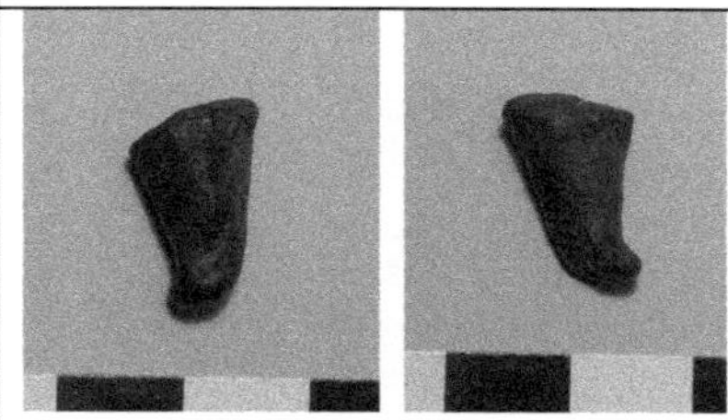
PI 147: (587557)	Body extremity	1.3 cm by 0.7 cm by 0.6 cm	This piece is the extremity of a leg. The main part it flattened leg then it becomes circular and larger.	
PI 148: (721956)	Body extremity	1.9 cm by 1.1 cm by 0.5 cm	This body extremity has a rounded bevelled foot that is flat underneath. Part of the body it was attached to is still visible.	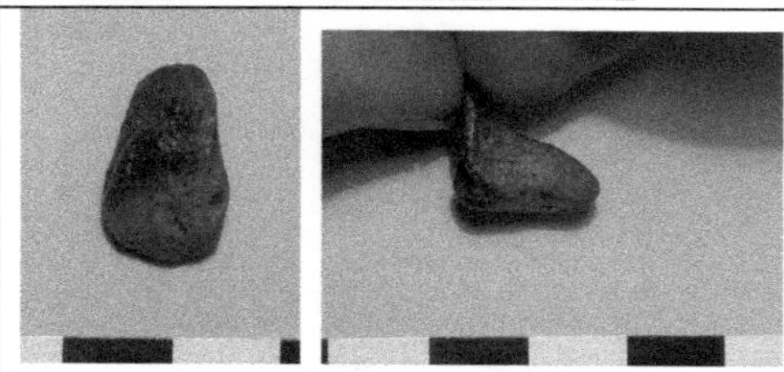
PI 149:	Body extremity	1.8 cm by 1.2 cm by 1 cm	This body extremity represents a clearly fashioned boot. The foot sole is flattened and even slightly arched and is perpendicular with the axis of the leg. There is a large hole pierced at the ankle when the material was still wet material.	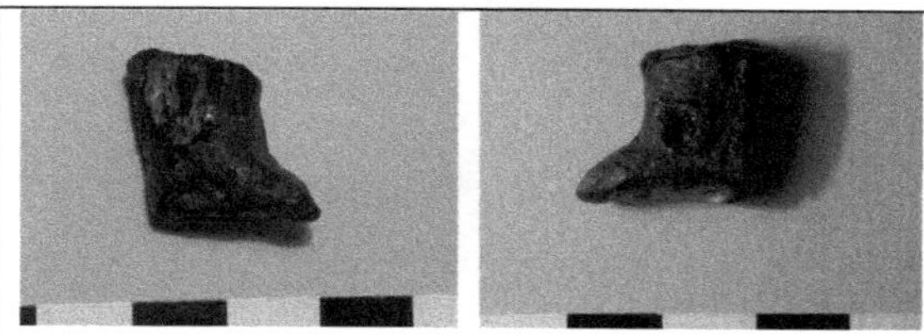

Pavlov I

PI 150:	Body extremity	2.3 cm by 1.6 cm by 1 cm	This body extremity is fragmentary. It widens toward one end to mark the beginning of the thigh. It was restored.	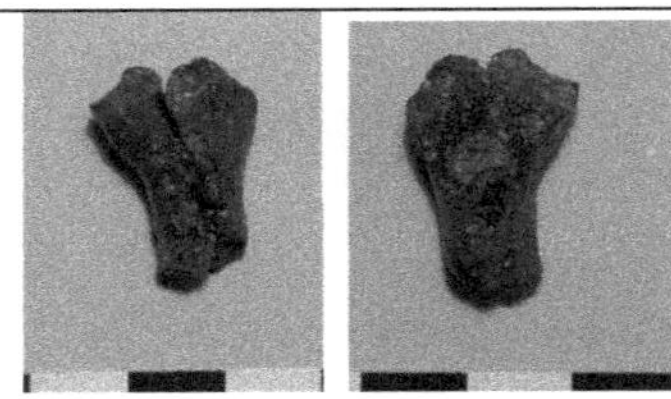
PI 151: (669756)	Body extremity	1.7 cm by 1 cm by 0.6 cm	This leg has a marked circular foot. It widens towards one end. It is slightly hollow underneath the foot.	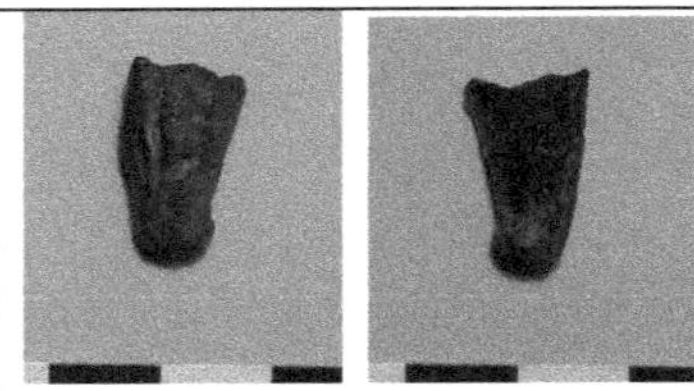
PI 152: (721656)	Body extremity	2 cm by 1cm by 0.8 cm	This piece is a limb fragment, curved, with possibly the knee indicated. It is black.	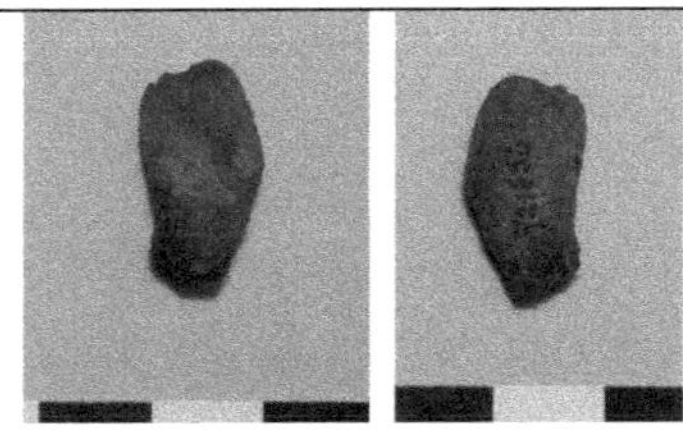
PI 153: (722056)	Fragmentary animal?	1.3 cm by 1 cm by 0.6 cm	This object is the fragment of a flattened cylinder with two protuberances visible on each side of one of the extremities. They could indicate one leg and/or tail of an animal.	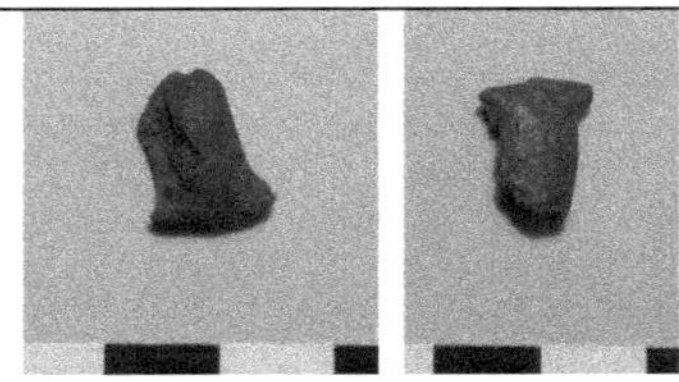
PI 154: (22?58)	Body extremity	1.7 cm by 0.7 cm by 0.7 cm	This piece is a leg with a foot. It is cylindrical and widens at top. The foot is thin and well-marked. The edges are straight. It is black.	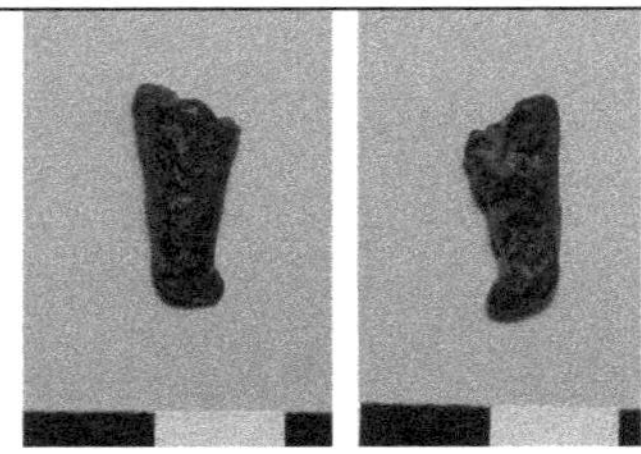

Pavlov I

PI 155: (721556)	Body extremity	2.6 cm by 1.5 cm by 1.5 cm	This is a large extremity, flattened with possibly a foot indicated. The edges are straight.	
PI 156: (587357)	Body extremity	1.8 cm by 1 cm by 1 cm	This extremity is a leg with a foot. It is flattened and has straight edges. The foot is little marked.	
PI 157: (255456)	Possible body extremity	2 cm by 1.3 cm by 0.6 cm	This object is an elongated piece. A worked surface is visible on one side.	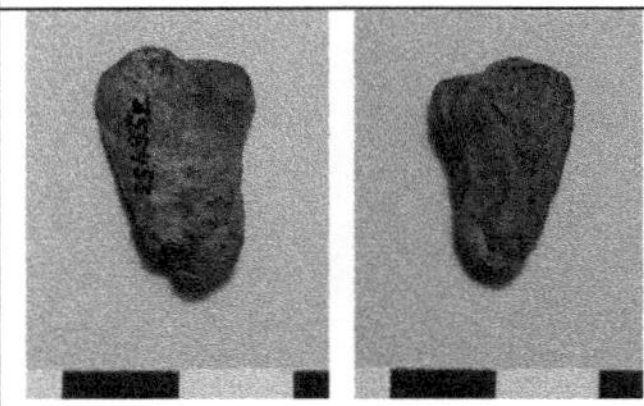
PI 158: (721456)	Body extremity	2 cm by 1.1 cm by 0.7 cm	Body extremity of the mammoth type. It is slightly flattened. The foot is circular and flat underneath, with a marked rim. The leg widens at the top.	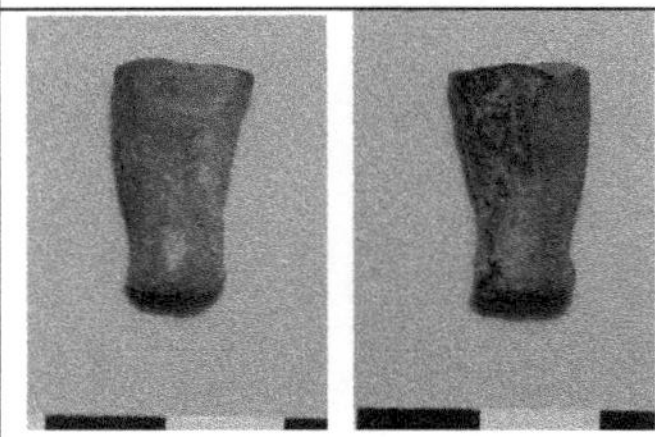
PI 159: (108?56)	Body extremity	2.4 cm by 1.2 cm by 0.9 cm	Body extremity with the beginning of a foot. It is much broken. The piece is cylindrical at the top. It was restored. The sole of the foot is missing but it is possible to see it was bevelled.	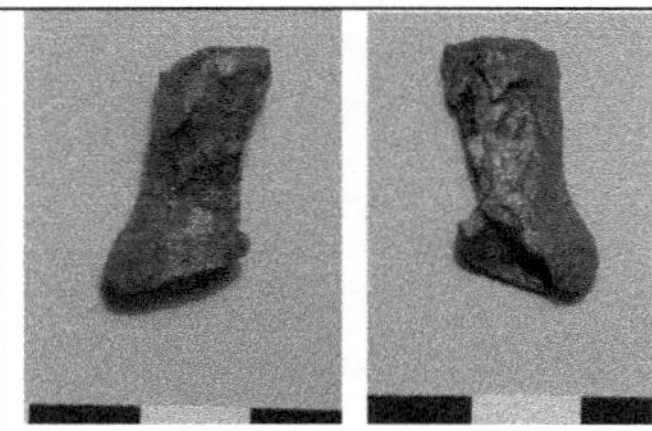

Pavlov I

PI 160: (721356)	Body extremity	3.5 cm by 2.3 and 1.6 by 1.7 and 1.8	wide extremity; widens at top, foot flattened underneath but the sides are not marked; on the side it looks like the beginning of a hole, 6 mm in diameter but the rest is broken.	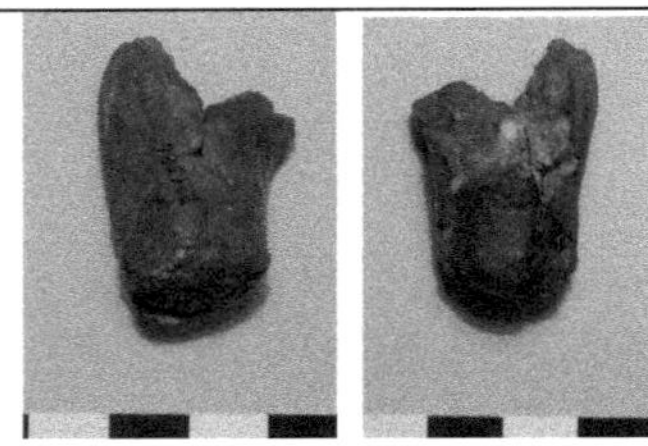
PI 161: (33160)	Possible animal body	2.6 cm by 1.2 cm by 0.8 cm	This is a cylindrical piece with the beginning of what could be two limbs at the rear of an animal body. It is much broken.	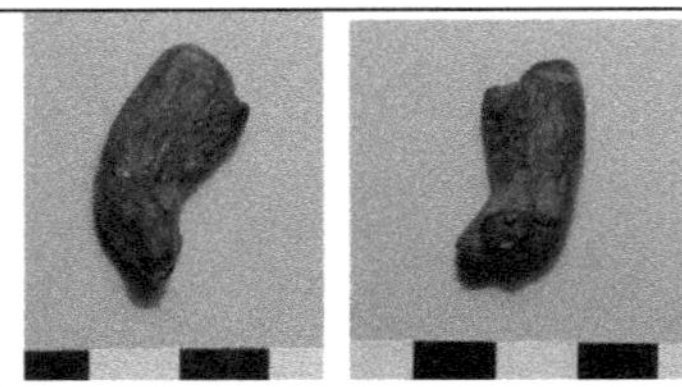
PI 162 (346062)	Body extremity	3.4 cm by 2.2 cm by 1.5 cm	This leg is still attached to the beginning of a body. It is straight and rounded at the end.	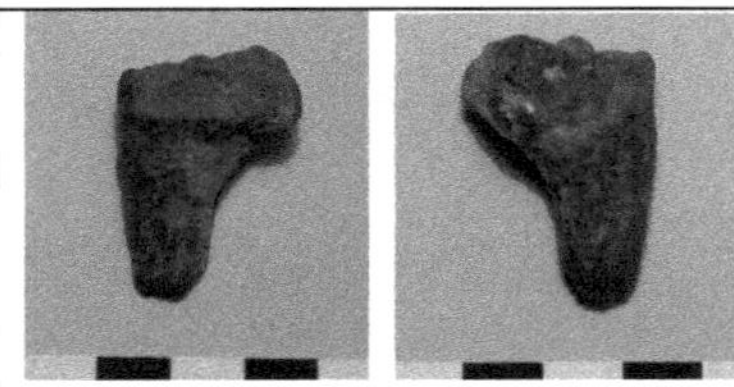
PI 163: (587657)	Body extremity	1.4 cm by 1.1 cm by 0.8 cm	This body extremity is long and thin. The surface is deformed. It is flattened at the top and broken at both ends.	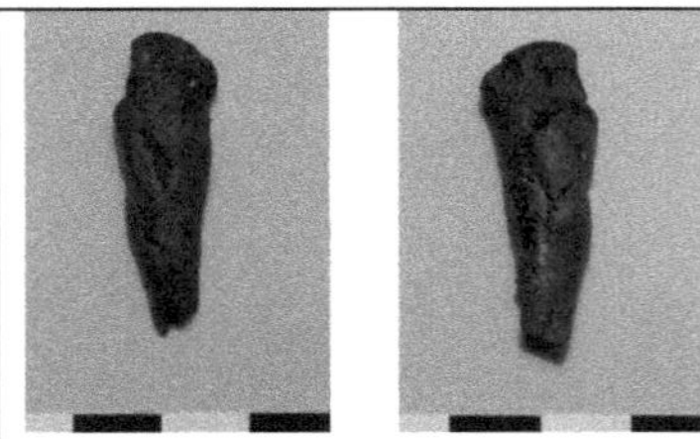
PI 164: (587157)	Body extremity	2.6 by 2 cm by 1.1 cm	This piece is similar to PI 162. It is a leg still attached to part of a body. The foot is flattened underneath.	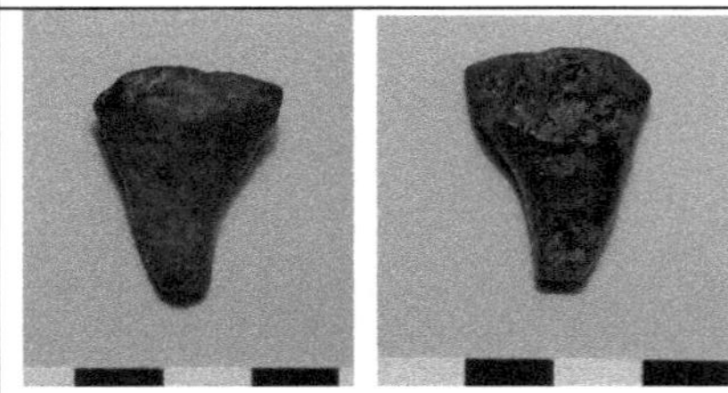

Pavlov I

PI 165: (587057)	Body extremity	1.7 cm by 1.7 cm by 0.7 cm	This body extremity is flattened. It widens at the top then it is broken.	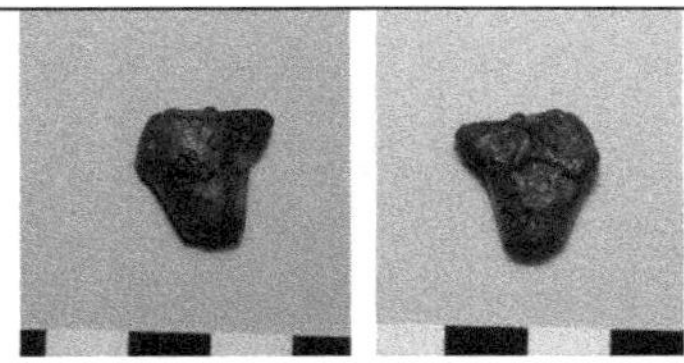
PI 166: (558454)	Body extremity	1 cm by 1 cm by 0.8 cm	This piece is the end of a large foot of the mammoth type. It is circular and flattened.	
PI 167: (198356)	Body extremity	3.2 cm by 1 cm by 0.8 cm	This piece is a leg with a foot. It is very straight and widens at the top. It is broken on the side. The foot forms a strong angle with the leg. It is long and well-marked.	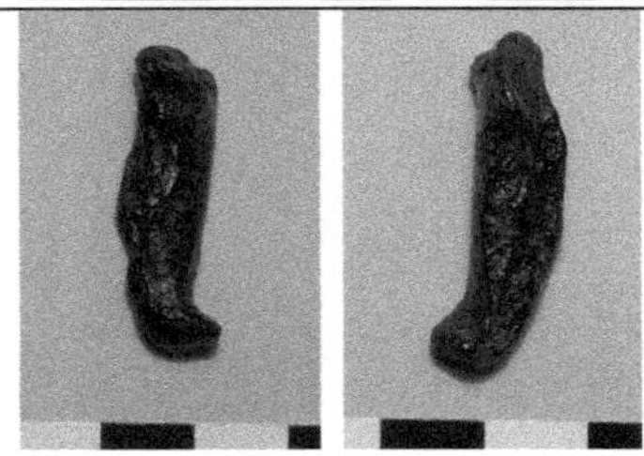
PI 168: (609657)	Body extremity	1.8 cm by 1 cm by 0.5 cm	Fragment of a limb with a flat end. It widens at the top. It is reddish in colour.	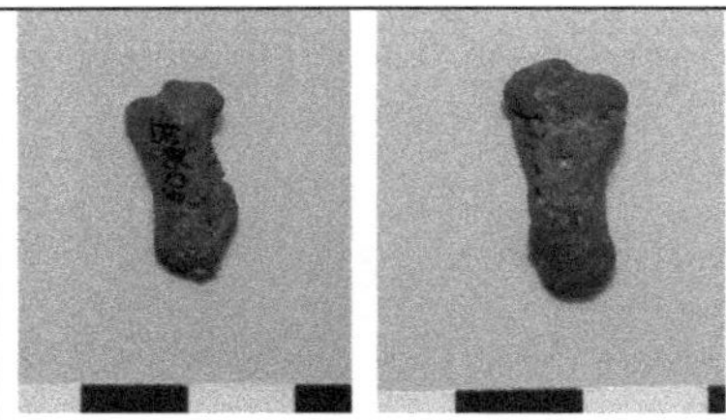
PI 169: (376963)	Possible animal head	1.9 cm by 1.1cm by 0.9 cm	This is a possible animal head with a pointed muzzle and two protuberances for the ear. The neck is marked and smooth on the entire surface. The object is not broken.	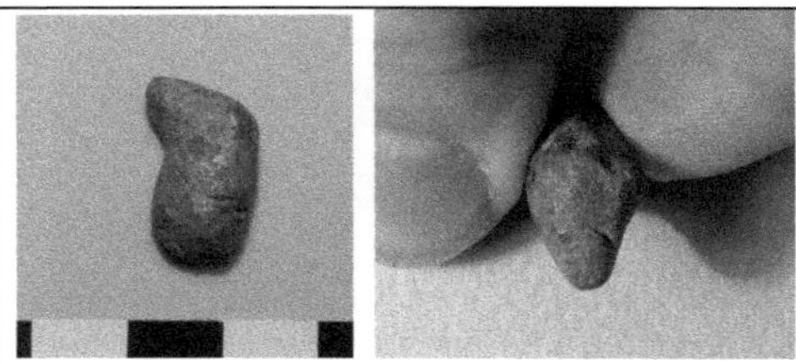

Pavlov I

PI 170: (687257)	Body extremity	1.6 cm by 1.1 cm by 1 cm	This piece is a boot-like extremity with a thin elongated foot. Striations are visible on the side: they seem to result from some form of impression.	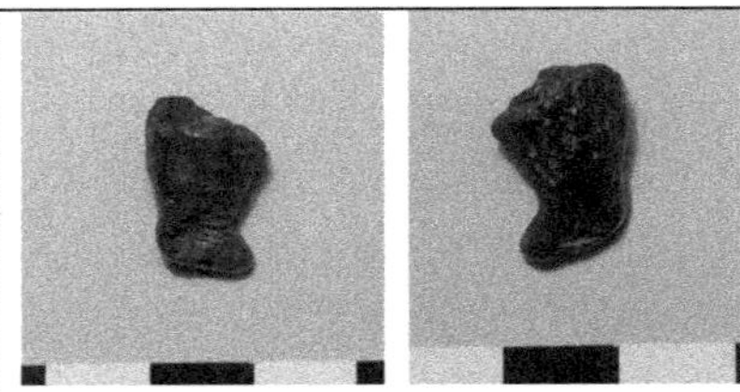
PI 171: (723056)	Body extremity	0.8 cm by 0.9 cm by 0.5 cm	Boot-like extremity that is flat underneath, pointed at the front, large with an oval section at the top.	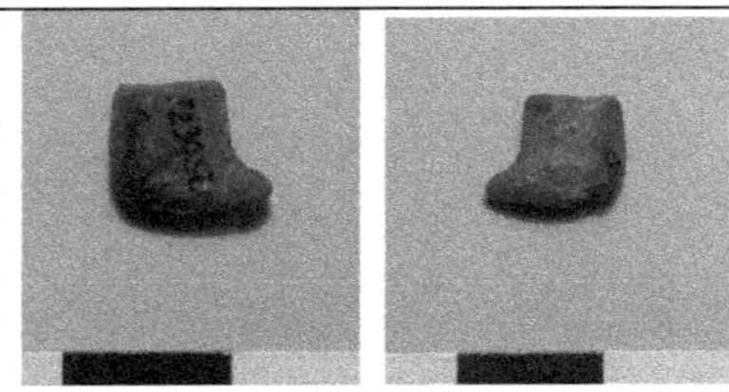
PI 172: (353256)	Body extremity	1 cm by 0.7 cm by 0.7 cm	This body extremity is rounded but fragmentary.	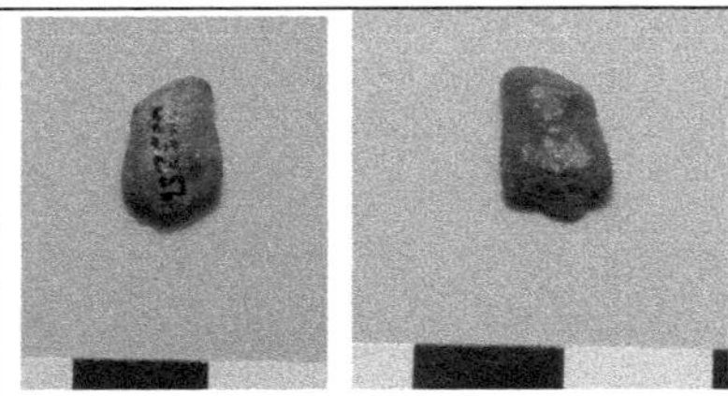
PI 173: (721256)	Body extremity	1.5 cm by 1.2 cm by 0.8 cm	This large extremity shows a bevelled foot with a rim.	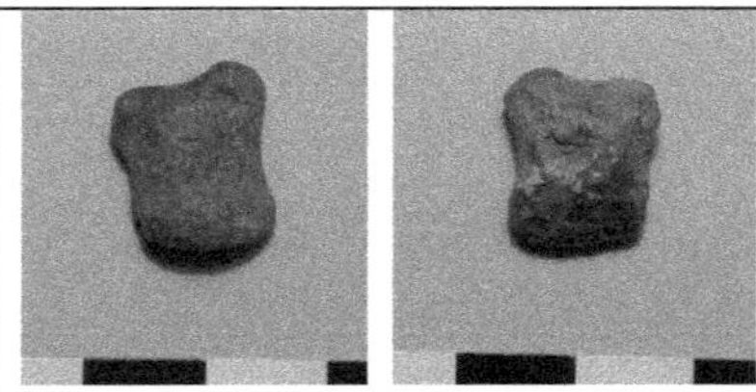
PI 174: (623757)	Body extremity	1.3 cm by 1 cm by 0.8 cm	This object is a segment of a cylindrical leg with straight edges. It widens slightly at one end but it is broken at each extremity.	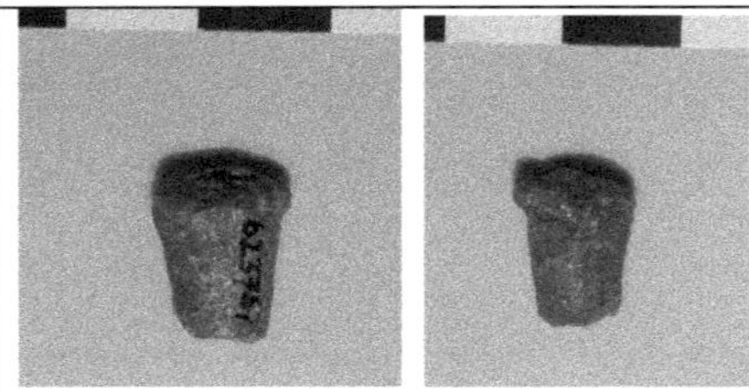

Pavlov I

PI 175: (558354)	Body extremity	1.8 cm by 0.8 cm by 0.8 cm	Curved body extremity with a thin rounded end. It is black.	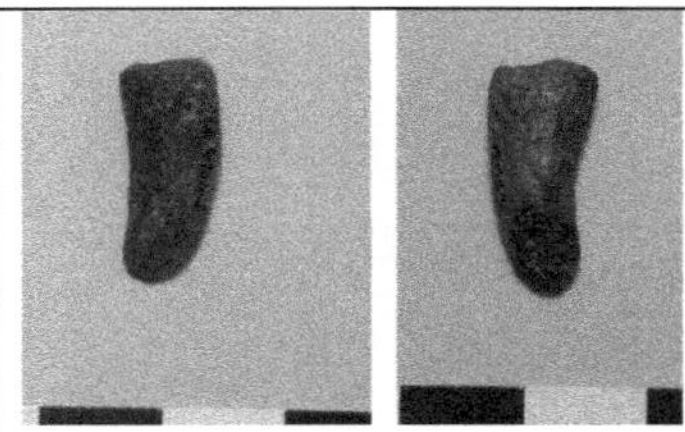
PI 176: (586957)	Body extremity	2 cm by 1.3 cm by 1 cm	Bevelled body extremity with a flattened sole. It is thinned at the ankle and resembles a boot.	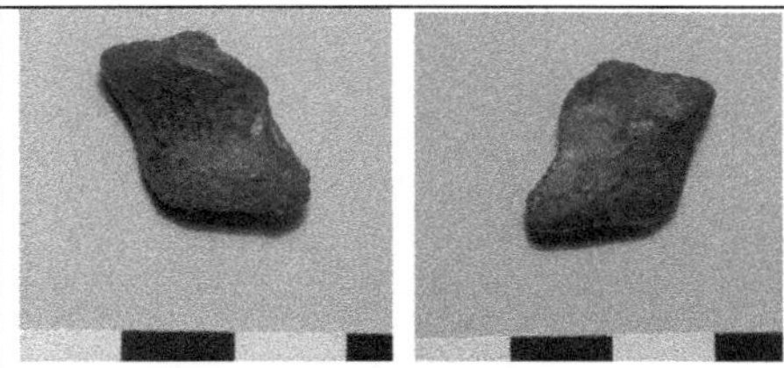
PI 177:	Body extremity	2.5 cm by 1 cm by 0.8 cm	This body extremity has a flattened cylindrical form. It is broken at both ends.	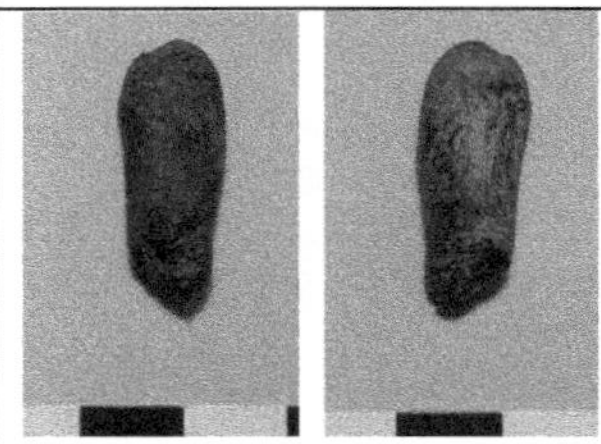
PI 178: (215456)	Body extremity	1.9 cm by 1.5 cm by 1 cm	This piece is a fragment of a large foot of the mammoth type, flat and circular underneath the foot. It is fragmentary.	

TABLE OF CONTENT

www.ingramcontent.com/pod-product-compliance
Lightning Source LLC
LaVergne TN
LVHW072121210926
848482LV00028B/725